Soybean Breeding

Felipe Lopes da Silva • Aluízio Borém
Tuneo Sediyama • Willian Hytalo Ludke
Editors

Soybean Breeding

Springer

Editors
Felipe Lopes da Silva
Federal University of Viçosa
Viçosa, Minas Gerais
Brazil

Aluízio Borém
Federal University of Viçosa
Viçosa, Minas Gerais
Brazil

Tuneo Sediyama
Federal University of Viçosa
Viçosa, Minas Gerais
Brazil

Willian Hytalo Ludke
Federal University of Viçosa
Viçosa, Minas Gerais
Brazil

ISBN 978-3-319-57432-5 ISBN 978-3-319-57433-2 (eBook)
DOI 10.1007/978-3-319-57433-2

Library of Congress Control Number: 2017943161

Printed on acid-free paper

This Springer imprint is published by Springer Nature
The registered company is Springer International Publishing AG
The registered company address is: Gewerbestrasse 11, 6330 Cham, Switzerland

Foreword

Soybean is the fourth most widely grown crop in the world. Originally from China it is a major crop in the USA, Brazil, Argentina, China, and many other countries. The improvements undergone by soybeans are not only a self-success, but it also triggers new technologies to other crops and regions around the world. As a source of oil, biodiesel, etc., soybeans are currently grown from low to high latitudes.

This book addresses the most recent and best technologies applied to soybean breeding. Some of the authors such as Dr. Vernetti and Dr. Sedyiama were pioneers in shaping the current generation of soybean breeders in countries like Brazil. This is a great opportunity to celebrate. Dr. Francisco de Jesus Vernetti, a personal friend and former major advisor, introduced and encouraged me to work under Dr. Edgar E. Hartwig, the icon in soybean breeding. Dr. Hartwig's contribution to soybean breeding in the USA and Brazil cannot be overemphasized. Dr. Hartwig was the most prominent soybean breeder in history. He was the first scientist to adapt this crop to areas closer to the equator, understanding soybean flowering in short days. This trait, today called "long juvenile stage," is well studied and its genetic control well understood (a few recessive genes), which allowed its use to develop cultivars for new agricultural frontiers in countries like Brazil and elsewhere.

Brazil is the second largest soybean producer and this is due to the breeding done with this crop in the last 60 years. Some aspects distinguish the work done in Brazil, since the crop had to be adapted to marginal areas with diverse cropping systems. Brazilian breeders had the additional challenge to develop cultivars resistant to the highly aggressive and prevalent pests typical of tropical regions, besides dealing with the soil acidity and seed quality, among other defies.

Another interesting area is breeding for nutritional quality, especially for oil and protein, besides reduction of antinutritional factors.

The large genetic progress observed in the last decades was mainly due to the mechanization of the breeding operations in planting, evaluating, and harvesting field trails, besides the use of computer power increasing the ease and speed to organize and analyze data and spreadsheets.

The GM cultivars made it easier to control weeds and are very promising in pest control. Additionally, the genetic studies and detection of molecular markers will also contribute to soybean breeding in the near future.

As always, breeders are optimistic and I personally believe that traditional breeding, helped by new tools, will continue to be a great asset to soybean future.

Romeu Kiihl

Synopsis

This book was written by experts, in a very detailed way, and aims to provide support to soybean breeders, seed farmers, and students interested in the techniques and methods for improvement of this species. Here we describe the most modern and recent technologies applied to the development of new soybean cultivars. It is, therefore, an essential work to all those interested in the improvement of this species.

Contents

Contributors

Lilian Cristina Andrade de Araújo Teixeira, D.Sc. State University of São Paulo (UNESP), São Paulo, Brazil

Carlos Alberto Arrabal Arias, M.S., Ph.D. Embrapa Soja, Londrina, Parana, Brazil

Leonardo de Azevedo Peixoto, M.S., D.S. The Federal University of Viçosa, Viçosa, Minas Gerais, Brazil

Rosângela Maria Barbosa Department of Agronomy, Federal University of Viçosa, Viçosa, Brazil

André Ricardo Gomes Bezerra, M.S. Federal University of Viçosa, Viçosa, Minas Gerais, Brazil

Leonardo Lopes Bhering, M.S., D.S. The Federal University of Viçosa, Viçosa, Minas Gerais, Brazil

Aluízio Borém, M.S., Ph.D. Federal University of Viçosa, Viçosa, Minas Gerais, Brazil

Adriano Teodoro Bruzi, M.S., D.S. Federal University of Lavras, Lavras, MG, Brazil

Thays Vieira Bueno, M.S. Federal University of Viçosa, Viçosa, Minas Gerais, Brazil

Pedro Crescêncio Souza Carneiro, M.S., D.S. Federal University of Viçosa, Viçosa, Minas Gerais, Brazil

Rita de Cássia Teixeira Oliveira, M.S. Campo Experimental Bacuri, Viçosa, Minas Gerais, Brazil

Sueli Ciabotti, D.Sc. Federal Institute of Education, Science and Technology of Triângulo Mineiro, Uberaba, Brazil

Cosme Damião Cruz, M.S., D.S. The Federal University of Viçosa, Viçosa, Minas Gerais, Brazil

Stênio Andrey Guedes Dantas Department of Agronomy, Federal University of Viçosa, Viçosa, Brazil

Murilo Viotto Del Conte, M.S. Federal University of Viçosa, Viçosa, Minas Gerais, Brazil

Marcelo Fernandes de Oliveira, M.S., Ph.D. Embrapa Soja, Londrina, Parana, Brazil

Paulo Afonso Ferreira, M.S., D.S. Federal University of Viçosa—Campus Universitário do Araguáia, Barra do Garças, Brazil

Silvana da Costa Ferreira, M.S., D.S. Federal University of Viçosa, Viçosa, Minas Gerais, Brazil

Amilton Ferreira da Silva, M.S., D.S. Federal University of São João Del-Rei (UFSJ), Campus de Sete Lagoas, Sete Lagoas, Minas Gerais, Brazil

Ana Cristina Pinto Juhász, D.Sc. Agricultural Research Company of Minas Gerais (EPAMIG), Uberaba, Brazil

Laércio Junio da Silva, M.S., D.S. Federal University of Viçosa, Viçosa, Minas Gerais, Brazil

Rita de Kássia Siqueira Teixeira, M.S. Federal University of Lavras, Lavras, MG, Brazil

Felipe Lopes da Silva, M.S., D.S. Federal University of Viçosa, Viçosa, Minas Gerais, Brazil

Alisson Santos Lopes da Silva Federal University of Viçosa, Viçosa, Minas Gerais, Brazil

André Luiz Lourenção, M.S., D.S. Phytosanitary Center, Agronomic Institute of Campinas, Campinas, São Paulo, Brazil

Willian Hytalo Ludke, M.S. Federal University of Viçosa, Viçosa, Minas Gerais, Brazil

Éder Matsuo, M.S., D.S. Federal University of Viçosa - Campus Rio Paranaíba, Rio Paranaíba, Brazil

Antonio Orlando Di Mauro, M.Sc., D.Sc. Universidade Estadual Paulista, Jaboticabal, São Paulo, Brazil

Pedro Patric Pinho Morais, M.S., D.S. Monsanto Brazil, São Paulo, SP, Brazil

Fabiana Mota da Silva, M.Sc., D.Sc. Universidade Estadual Paulista, Jaboticabal, São Paulo, Brazil

Heloisa Rocha do Nascimento Department of Agronomy, Federal University of Viçosa, Viçosa, Brazil

Felipe Maniero Nazato, M.S. Department of Genetics, College of Agriculture "Luiz de Queiroz", University of São Paulo, Piracicaba, São Paulo, Brazil

José Aírton Rodrigues Nunes, M.S., D.S. Federal University of Lavras, Lavras, Minas Gerais, Brazil

Guilherme de Sousa Paula Department of Agronomy, Federal University of Viçosa, Viçosa, Brazil

José Baldin Pinheiro, M.S., D.S. Department of Genetics, College of Agriculture "Luiz de Queiroz", University of São Paulo—ESALQ, Piracicaba, São Paulo, Brazil

Magno Antonio Patto Ramalho, M.S., D.S. Federal University of Lavras, Lavras, Minas Gerais, Brazil

Marcos Deon Vilela de Resende, M.S., D.S. Brazilian Agricultural Research Corporation/Federal University of Viçosa, Viçosa, Minas Gerais, Brazil

Haroldo Silva Rodrigues, M.S., D.S. The Federal University of Viçosa, Viçosa, Minas Gerais, Brazil

Renato Domiciano Silva Rosado, M.S., D.S. The Federal University of Viçosa, Viçosa, Minas Gerais, Brazil

Hamilton Carvalho dos Santos Júnior Federal University of Viçosa, Viçosa, Minas Gerais, Brazil

Ivan Schuster, M.S., D.S. Dow AgroSciences, Cravinhos, São Paulo, Brazil

Tuneo Sediyama, M.S., Ph.D. Federal University of Viçosa, Viçosa, Minas Gerais, Brazil

Francisco Charles dos Santos Silva, M.S. Federal University of Viçosa, Viçosa, Minas Gerais, Brazil

Reginaldo Castro de Souza Júnior Federal University of Viçosa, Viçosa, Minas Gerais, Brazil

Francisco Terasawa FT Sementes, Ponta Grossa, Parana, Brazil

José Maurício Terasawa, M.S., Ph.D. FT Sementes, Ponta Grossa, Parana, Brazil

Milene Möller Terasawa, M.S., D.S. FT Sementes, Ponta Grossa, Parana, Brazil

Sandra Helena Unêda-Trevisoli, M.Sc., D.Sc., Ph.D. Universidade Estadual Paulista, Jaboticabal, São Paulo, Brazil

Natal Antonio Vello, M.S., D.S. Department of Genetics, College of Agriculture "Luiz de Queiroz", University of São Paulo, Piracicaba, São Paulo, Brazil

José Djair Vendramim, M.S., D.S. Department of Entomology and Acarology, College of Agriculture "Luiz de Queiroz", University of São Paulo, Piracicaba, São Paulo, Brazil

Francisco de Jesus Vernetti, M.S. Embrapa Clima Temperado, Pelotas, Rio Grande do Sul, Brazil

Francisco de Jesus Vernetti Junior, M.S., D.S. Embrapa Clima Temperado, Pelotas, Rio Grande do Sul, Brazil

Leonardo Volpato Department of Agronomy, Federal University of Viçosa, Viçosa, Brazil

Chapter 1
Economic Importance and Evolution of Breeding

Francisco Charles dos Santos Silva, Tuneo Sediyama, Rita de Cássia Teixeira Oliveira, Aluízio Borém, Felipe Lopes da Silva, André Ricardo Gomes Bezerra, and Amilton Ferreira da Silva

Abstract Originating in northeast China, soybean (*Glycine max* (L.) Merrill) in the last decades has become the main oilseed in the world due mainly to being a great source of vegetable oil and protein. This crop currently occupies an area more than 118 million hectares worldwide. Due to the projections for the increase of the world population and the high purchasing power of this, the demand for the soybean grain will not decrease in the next years. In Brazil, soybean found adequate conditions for its development in the South. From this region, since the 1970s, soybean has been expanding in the Brazilian savannah, called Cerrado, a process known as the "tropicalization of soybean". The genetic breeding of the species was one of the main factors responsible for the success of its expansion in Brazil. The evolution of soybean genetic breeding in Brazil can be divided into two phases. The first extends the introduction of the crop until the mid-1980s, marked by the participation of the public sector in the development of varieties, seed supply and technology transfer. This period was mainly marked by the adaptation of the soybean to the conditions of the Cerrado and regions of low latitudes. In the 1990s, a new phase for the soybean breeding in Brazil was started, which was marked by the participation of large groups of international capital. This period has been seeking, in addition to the

---

F.C. dos Santos Silva, M.S. (✉) • T. Sediyama, M.S., Ph.D. • A. Borém, M.S., Ph.D.
F.L. da Silva, M.S., D.S.
Federal University of Viçosa, Viçosa, Minas Gerais 36570-900, Brazil
e-mail: fcasantossilva-ma@hotmail.com; tuneo@ufv.br; borem@ufv.br; felipe.silva@ufv.br

R. de Cássia Teixeira Oliveira, M.S.
Campo Experimental Bacuri, Viçosa, Minas Gerais, Brazil
e-mail: cebacuri@uol.com.br

A.R.G. Bezerra, M.S.
Fundação MS Para Pesquisa e Difusão de Tecnologias Agropecuárias, Maracaju, Brazil
e-mail: bezerra.agro@yahoo.com.br

A.F. da Silva, M.S., D.S.
Federal University of São João Del-Rei (UFSJ), Sete Lagoas Campus, Sete Lagoas, Minas Gerais, Brazil
e-mail: amiltonferreira@ufsj.edu.br

F. Lopes da Silva et al. (eds.), *Soybean Breeding*,
DOI 10.1007/978-3-319-57433-2_1

development of adapted and productive varieties, new materials of greater market value from the development of transgenic events.

Keywords *Glycine max* (L.) Merrill • Tropicalization of soybean • Soybean breeding • Soybean yield

Introduction

Soybean (*Glycine max* (L.) Merrill) is one of the most important legumes in the world, especially for Brazil, which is the second largest producer in the world, due to an entire productive and technological chain developed around this species.

The crop originated in the northeastern region of China, but, in Brazil, the first soybean commercial crops were conducted in the southern region. From this region, since the 1970s, soybean crops have been expanding over the Brazilian Savanah, currently occupying areas that span the country from South to North.

The success and expansion of soybeans in Brazil are due to several factors, including the breeding of the species, with the development of more productive cultivars, adapted and resistant to various diseases.

Economic Importance

The economic importance of soybeans, worldwide, is due to the fact that the crop is an excellent source of vegetable oil and protein, with contents around 20 and 40%, respectively, whose production system has already been established.

According to the USDA (2015), world soybean production in the 2014/2015 crop season was 317.25 million tons, covering an area of 118.135 million hectares. In Brazil, in the 1960s, soybeans were only relevant in the southern region; however, after 55 years, production has reached 95 million tons in 31.57 million hectares, which places the Brazil as the second largest producer and the world's largest exporter of this crop.

In Brazil, the great expansion of soybean from the South Region to the Cerrado Region occurred in the 1970s. According to Bonato and Bonato (1987) and the MAPA (2015a, b) reports, from the 1969/1970 crop season to 2015, the crop production in the main producing states of Brazil has undergone significant increases. In Rio Grande do Sul, production has increased from 976,807 to 6,117,000 tons, from 90,086 to 3,957,000 tons in São Paulo, from 52,998 to 318,900 tons in Santa Catarina, from 368,006 to 14,766,000 tons in Paraná, in Minas Gerais it has gone from 1806 to 8159 tons, from 25 to 281,800 tons in Bahia, from 9817 to 807,300 tons in Goiás, in Mato Grosso, which until 1977 formed a single state with Mato Grosso do Sul, production increased from 8159 to 1,778,200 tons, whereas, in Mato Grosso do Sul production reached 8159 tons. In Maranhão, the first production record dates from 1977/1978 crop season, with a production of 55 tons, and reached 2,057,700 tons in the 2014/2015 crop season.

In recent years, the soybean production chain has had a large share of the Brazilian trade balance. Revenue from exports of soybeans and their by-products, such as meal and oil, in the 2014, exceeded US$ 31.4 billion, making soybeans the main export item in Brazil (MAPA 2015a, b).

Its economic importance is reflected in some social aspects of Brazil, that is, for each job generated by soybeans, the number of workers rises to 12.66, when taking into account the entire grain production chain, without accounting for the trade and rendering of services. The social impact of soybean production can also be measured by salary income, that is, for each real paid as salary in the field, the amount rises to R$ 33.23, when adding up the entire production chain (MTE 2010).

One of the factors to which the success of soybeans can be attributed is the increase in the world population, which grows at a rate of 70 million inhabitants per year; the increase in income per capita, which is associated with increases in the consumption of animal protein, which, in most part, is produced from soybean meal. Moreover, alternative products can be produced from soybeans, such as raw material for paints, lubricants, plastics, etc.

Due to the quality of the soybean protein and its relatively low production cost, this oilseed has become important as a source of protein supplementation, especially in developing countries. Soybean is the main source of protein meal in the world. In total, more than 188 million tons are produced, equivalent to 67% of the total supplied worldwide. Among the sources of vegetable oils available, soybean oil accounts for 56% of the total supply, which corresponds to 281 million tons (Bezerra et al. 2015).

In Brazil, soybean contributes to the incorporation of biodiesel into the energy mix. In 2013 it became mandatory, by law, to increase from 2 to 5% the amount of biodiesel in diesel oil. To date, soybean oil has been the main raw material for the production of biodiesel, followed by animal fat and cotton oil. In 2015, of the 2.6 million cubic meters of biodiesel produced, 78% derived from soybean oil (ABIOVE 2015).

In the coming years, the world's population and purchasing power will continue to grow in emerging economies and, especially, in Asian countries, which concentrates the highest consumption potential. Studies forecast that, by 2050, the earth's population will reach nine billion, which will demand 333.674 million tons of food (Alexandratos and Bruinsma 2012). Due to the great importance of soybeans in animal feed, in addition to its many industrial applications, the demand for its grain will not decrease in the coming years. In this context, the future scenario for Brazilian soybean is extremely promising.

Projections from MAPA (2015a, b) estimate that in 2024/2025 Brazilian soybean production will reach 126.2 million tons, which represents an increase of 32.8% compared to 2014/2015 crop season. Exports will be close to 66.5 million tons, an increase of nearly 42.1%. The production of soybean meal is expected to increase by 26.2% and that of oil by 21.1%. Exports of meal will increase by 17.4% between 2014/2015 and 2024/2025, while oil exports are expected to fall by 6.5%. Domestic consumption is expected to be the main factor to boost soybean oil production in the coming years, which is expected to grow at the end of the decade to close to 8.5 million tons, a 30.6% increase in relation to 2014/2015 crop season.

Such level of perspective of soybean growth in Brazil is due to the increase in global demand and the fact that few countries are able to supply it. The USA and Argentina, for instance, will not be able to increase the area planted like Brazil. Moreover, the need to produce other crops and livestock activity prevents them from expanding the soybean area (Bezerra et al. 2015).

However, in Brazil, the soybean area is expected to increase by 9.7 million hectares over the next 10 years and reach, in 2025, 41.2 million hectares. It is the crop whose area is expected to expand the most in the next decade. Such expansion must occur mainly in areas of great production potential, such as those of *Cerrado*, in the region currently called MATOPIBA, for it comprises lands located in the states of Maranhão, Tocantins, Piauí, and Bahia. The area planted with grains in this region is expected to expand 18.7% over the next 10 years. This is equivalent to reaching the area of 8.7 million hectares, which, at its top limit, can reach 11.4 million hectares. Grain production in the states comprising this region is expected to increase from 19.4 million tons in 2014/2015 to 22.5 million in 2024/2025. At its top limit, production at the end of the period can reach 27.9 million tons of grain (MAPA 2015a, b).

Breeding Evolution in Brazil

Soybean breeding, developed by several public and private institutions, always seeking new and more productive varieties that are adapted to the Brazilian conditions, was and has been the one greatly responsible for the occupation and expansion of the soybeans in Brazil. Originated and domesticated from Northeast China, also known as Manchuria, soybeans arrived in Brazil in 1882, in the state of Bahia. This was the beginning of the process known as "tropicalization of soybeans in Brazil."

The following will describe the evolution of soybean breeding programs in each Brazilian state since its arrival in the country.

Rio Grande do Sul

The first reference of soybean introduction in Rio Grande do Sul is Minssen's (1901), which reports the planting of this legume before 1901, in the town of Dom Pedrito, by the agronomist A. Welhauser (Magalhães 1981).

As for soybean research, studies started in the state in the 1930s, in the former Colony Phytotechnical Experimental Station, in Veranópolis. This work gave rise to the first soybean cultivar created in Rio Grande do Sul, called "Pioneira," which was launched in 1960 (Feres and Gomes 1981).

From 1946, research was extended to other experimental stations and also to the University of Rio Grande do Sul, which intensified its research program and established partnerships with the Agronomic Institute of Campinas (IAC) and the State Secretariat of Agriculture, with the introduction of collections from the Federal University of Viçosa (Sediyama et al. 2005).

In the 1950s, these institutions expanded their research by introducing a significant number of accessions from the USA, Japan, and various parts of Brazil (Feres and Gomes 1981). The purpose of the work at that time was to select soybean cultivars with high yield, plant height and insertion of the first pods suitable for mechanization of the crop, resistant to lodging and to the natural dehiscence of the pods, resistant to diseases and with good seed quality (Sediyama et al. 2005).

The result of this study was the gradual substitution of outdated cultivars such as Amarela Comum, Dorman, and Dortchsoy 67A by cultivars Serrana, Majos, Santa Rosa, Industrial, and Jubileu. These cultivars have contributed greatly to increase the average yield of Rio Grande do Sul crops (Miyasaka and Medina 1981).

In 1948, the Ministry of Agriculture began its research on soybean breeding in Rio Grande do Sul, in the city of Pelotas, at the *Instituto Agronômico do Sul* [Southern Agronomic Institute] (IAS), later called the *Instituto de Pesquisas Agropecuárias do Sul* [Southern Agricultural Research Institute] (IPEAS) (Sediyama et al. 2005). From 1953 to 1959, the program was limited to the introduction and evaluation of cultivars, and it was only after 1963 that the first hybridizations were performed (Feres and Gomes 1981).

Genetic material from the USA has always been the one to best adapt to the ecological conditions of southern Brazil. Thus, while hybridization and selection work was underway, farmers from Rio Grande do Sul used the best cultivars introduced from the USA, such as Hill, Hood, Majos, Bragg, Davis, Jew 45, Hampton, Hardee, and Bienville. The purposes of this program only began to be reached after 1968 when the cultivar Campos Gerais was launched (Feres and Gomes 1981).

Rio Grande do Sul is also marked as the state where the first transgenic soybean crops were grown on Brazilian soils. That fact took place in the 1998/1999 crop with the cultivation of RR soybeans, tolerant to Roundup herbicide, both developed by Monsanto.

Nonetheless, until that date, the commercial production of transgenic soybeans in Brazil was not allowed. Accordingly, due to the illegality of RR soybean farming, many crops in the state were incinerated and banned for 180 days (Lima 2005). In the 2000/2001 season, new clandestine RR soybean crops were found in the state. The first bags of RR soybeans to arrive in the state were smuggled from Argentina and became known to producers as "Maradona soybeans."

The definitive release of the farming of GMOs in Brazil occurred only in March 2005, with the approval of the Biosafety Law.

Santa Catarina

The first statistical data on soybeans in the state of Santa Catarina refer to 1951/1952 crop season, which was introduced by farmers who came from Rio Grande do Sul and started their activities in the West and in the Vale do Rio do Peixe (Doldatelli 1981).

The breeding work began in 1966/1967 and was coordinated by the *Instituto de Pesquisa Agropecuária do Sul* [Institute for Agricultural Research in the South] (IPEAS). The results allowed for the recommendation, in 1970, of cultivars with high yield and good agronomic traits such as Davis, Bragg, Hampton, Bienville, CNS-4, Hardee, Lee, and Santa Rosa, and this latter was for a long time the most cultivated in the state. After the establishment of the Santa Catarina company *Empresa Catarinense de Pesquisa Agropecuária S. A.* (EMPASC), the cultivars Paraná, Sulina, BR-1, and BR-3 were indicated (Orrego 1981).

Paraná

In the state of Paraná, soybean farming began in 1954, when about 2000 sacks were imported from São Paulo to be sown and soybean plants used as green manure in coffee plantations (Unfried and Braga 2011). However, until 1964 the cultivars used came mainly from Rio Grande do Sul and São Paulo (Kaster et al. 1981).

After 1965, due to the ease commercialization of soybeans, soybean research was implemented by the Ministry of Agriculture and the IRI Research Institute, and studies evaluated lines and cultivars provided by the IPEAS and the Federal University of Viçosa (UFV) (Unfried and Braga 2011).

From this program derived the cultivars Viçoja and Mineira, selected by the Federal University of Viçosa, as well as the launch of Campos Gerais, Paraná, Florida, and Sant'Ana. Moreover, the program allowed the recommendation of the American varieties Bragg, Davis, Hardee, Hill, and Hood and of Brazilian varieties Santa Rosa and Industrial (Sediyama et al. 2005).

In 1976, through an intergovernmental agreement, the *Centro Nacional de Pesquisa da Soja* [National Center for Soybean Research] (CNPSo) took over the coordination and development of soybean research in the state. The program developed by CNPSo gained national dimension and began to develop cultivars for the north, northeast, and central regions of Brazil (Unfried and Braga 2011). The program also contributed to the development of numerous cultivars such as Doko and later Doko RC, with significant cultivated areas in the *Cerrado* (Sediyama et al. 2015).

Among the soybean breeding programs that were developed in the state of Paraná, *FT-Pesquisa e Sementes*, founded by Francisco Terazawa, in the late 1960s, in Ponta Grossa, should be highlighted. It was the first Brazilian private company dedicated to soybean research and breeding. Cristalina was one of the first cultivars that, in 1993/1994 crop season, was responsible for 52% of soybean production in Brazil (Sediyama et al. 2015).

One believes that the cultivar FT-Cristalina originates from a natural crossing between UFV-1 and an unknown parent. However, Hiromoto and Vello (1986) reported that the most probable genealogy of FT-Cristalina is Davis × UFV-1, due to the similarity between the parents.

According to Santini (2002), throughout the 1980s, its soybean cultivars began to predominate from the South to the Midwest of Brazil. In 1996, the soybean segment of *FT-Pesquisa e Sementes* was acquired by Monsanto, thus creating Monsoy.

Another important institution in soybean breeding in the state of Paraná is the *Cooperativa Central de Pesquisa Agrícola* [Central Cooperative for Agricultural Research] (COODETEC), created in 1995, which originated from the former research department in the 1970s, by the *Organização das Cooperativas do Estado do Paraná* [Organization of Cooperatives of the State of Paraná]—OCEPAR. COODETEC develops cultivars and licenses the protected materials to other partners that multiply the seeds. Due to the technological partnership agreement with Monsanto, COODETEC was allowed to launch six cultivars with the glyphosate-tolerant RR gene, including cultivar CD 219RR, one of the first GM cultivars, resistant to glyphosate, cultivated in Central Brazil (Cordeiro et al. 2007).

In 2014, all sectors of COODETEC were acquired by the American company Dow AgroSciences, a subsidiary of the Dow Chemical Company.

São Paulo

The first reports of soybeans in São Paulo date back to 1892, with experiments carried out by the Agronomic Institute of Campinas (IAC). However, its planting began to spread in the state from 1908, with the arrival of the first Japanese immigrants in the state, who used the legume as food (Sediyama et al. 2005).

Soybean breeding in São Paulo started in 1921, when Henrique Lobbe, director of the Experimental Station of São Simão-SP, introduced the varieties Hermann, Arksaben, Ebony, Biloxi, Guelph, Hato, Changai, Haberlandt, Hamilton, Easycook, Chiquita, and Hoosier. Subject to field conditions, the plants were selected among these materials, which resulted in three new cultivars, Jomichel, Julieta, and Joalo, very early, with development cycles close to 90 days, with very low plant structure. By means of crossings, the researcher also obtained another cultivar that received the name of Artrofi, which, although productive, also featured limited plant size (Miranda et al. 1981).

In 1936, Neme Abdo Neme, with the introduction of a new collection of materials from the USA and the Japanese colony based in Brazil, launched the cultivar Abura, for industry use, and Otootan, with black seeds used for fodder (Miranda et al. 1982).

With the introduction of materials from the Southern USA in 1951, José Gomes da Silva identified several promising materials for the state, including Acadian, Pelican, Improved Pelican, Yelnando, Volstate, N 46-2652, and D 49-772 (Miranda et al. 1982).

As early as 1952, a new breeding program was developed by Shiro Miyasaka under the guidance of Leonard F. Williams. This program gave rise to cultivars IAC-1, Santa Rosa, and Industrial. Later, by means of the hybridization between Yelnando and Aliança Preta, they obtained the cultivar IAC-2, which was one of the most cultivated in Central Brazil in the open regions of the *Cerrados* in the 1970s (Miranda et al. 1981).

In 1957, Shiro Miyasaka identified two cultivars less sensitive to the short days, Santa Maria-1 and Aliança Preta, which suggested the possibility of searching for cultivars that best adapted to the latitudes closer to the equator. This hypothesis inspired Romeu Afonso de Sousa Kiihl, a former IRI researcher, at IAC, in 1966, to undertake a postgraduate course at the University of the Mississippi under the guidance of researcher Edgar Hartwig. They decided to focus on a strategy line that would enable the expansion of soybeans in tropical regions, successfully implemented in the following decades (Unfried and Braga 2011).

The *Instituto Agronômico de Campinas* [Agronomic Institute of Campinas] (IAC), in 1967, received from Dr. Kirk L. Athow, hybrids he created at the Federal University of Viçosa, through crossings between Pelicano and IAC-2, with the Hardee variety. Romeu Kiihl worked with these populations and selected the line IAC 70-559 (IAC-4) (Miranda et al. 1982). Later, still in the IAC, Romeu Kiihl performed other hybridizations and backcrossings between American cultivars and PI 240.664 and launched cultivars with long juvenile period for floral induction, called IAC-6 and IAC-7 (Miranda et al. 1981).

According to Sediyama et al. (2005), other breeding programs were under development in the state of São Paulo, such as that of *Cooperativa de Produtores de Cana-de-Açúcar e Álcool do Estado de São Paulo* [São Paulo Sugarcane and Alcohol Producers Cooperative] (Coopersucar), the main purpose of which was to develop early varieties to allow the cultivation of two sugarcane plantations in the off-season, and that of the Higher School of Agriculture Luiz de Queiroz (ESALQ), which has emphasized, among other purposes, the development of cultivars suitable for human consumption.

Minas Gerais

The first references of soybeans in the state of Minas Gerais date back to 1930s, when it was studied at the old Experimental Substation of Lavras (Hunnicutt 1930).

More detailed studies on the behavior of cultivars were initiated in 1956/1957 crop season by Brandão (1961), at the Federal University of Viçosa (UFV). However, the beginning of soybean breeding in the state occurred in 1963, at

UFV, by means of the UFV/Purdue University (USA) Project, with Henry Shands as coordinator. At that time, a great number of varieties and lines of the South of the USA were introduced; however, most of the introductions did not show good adaptation, due to the sensitivity of the species to the length of the day (Sediyama 1981).

In 1965 the first hybridizations were performed by Kirk L. Athow, of Purdue University, with the collaboration of Marvin L. Swearingin and Elton R. da Silva and Tuneo Sediyama. This work gave rise to the first two cultivars developed in Minas Gerais, named Viçoja (Viçosa soybean) and Mineira, which derived from lines provided by Kuell Hinson of the Experimental Station of Gainesville, Florida (USA), and which had the variety Improved Pelican as common ancestor. These two cultivars occupied significant areas of the Triângulo Mineiro, São Paulo, Paraná, and Mato Grosso do Sul (Sediyama et al. 1983).

In 1968, the breeding program had the collaboration of Marvin L. Swearingin and, in the following years, also of Tuneo Sediyama, Carlo S. Sediyama, and Múcio S. Reis (Sediyama 1981).

In 1969, in an experimental field of Viçoja variety, Tuneo Sediyama selected a plant that had superior traits to those of that variety, mainly because it features a longer cycle, higher plant height, and higher height of the first pod. All other traits were similar to those of Viçoja. Due to the lack of knowledge regarding its pedigree and the absence of segregation in later generations, which indicated that it did not result from a natural crossing, and as there was no other line or variety with similar traits in that locality, Viçoja was believed to be a mutation. Thus, it was first called "Viçoja mutante," and later, in 1973, it was as a variety under the name of UFV-1 (Sediyama 1981).

UFV-1 was extensively cultivated in the 1970s to 1980s, mainly in the Cerrado of Minas Gerais, Goiás, Mato Grosso do Sul, and Mato Grosso and, to a lesser extent, in the states of São Paulo and Paraná (Sediyama et al. 2015).

Up to 2015, during this program, 53 adapted cultivars were developed, mainly to the edaphoclimatic conditions of the southeast and Midwest regions of Brazil. Since 2013, the program has been led by Felipe Lopes da Silva, with the collaboration of Tuneo Sediyama and Carlos S. Sediyama.

Another ongoing program at the Federal University of Viçosa is that of molecular genetics in soybean breeding for the agroindustry (Moreira et al. 1995). Through this program, in collaboration with the researchers of the Department of Phytotechnology of the UFV, the first cultivars with traits more suitable for human consumption, with the removal of lipoxygenases (Lox 1, Lox 2 and Lox 3), were developed. The cultivars were named UFVTN 101, UFVTN 102, UFVTN 103, and UFVTN 104.

In 1988, the *Fundação Triângulo de Pesquisa e Desenvolvimento* was established. Among the several cultivars developed by the institution, the one that stood out the most was MG/BR 46 Conquista, which was launched in 1995, and became one of the most cultivated in Central Brazil (Sediyama et al. 2015).

Cooperativa Agropecuária do Alto Paranaíba Ltda-Coopadap, founded in 1994, in São Gotardo, MG, is another important institution in soybean breeding in the state (Sediyama et al. 2005). Among the several cultivars developed by this program, which are mainly intended for the oil and meal industry, the CAC-1 stands out.

In 1997, under the leadership of Osvaldo Toshiyuki Hamawaki, the Federal University of Uberlândia (UFU) started a breeding program aimed at developing new conventional cultivars that were productive, resistant to biotic and abiotic stresses, and featured high phenotypic adaptability and stability. The first cultivars launched were named UFUS Impacta and UFUS Milionária (UFU 2010).

Goiás

The introduction of soybeans in the state of Goiás occurred in 1950, by Znamenskiy (1965), who conducted experiments with the crop for 15 years.

In 1969, the Secretariat of Agriculture began the tests for evaluation of cultivars in partnership with institutions of other states. Among the materials evaluated, varieties Santa Rosa, Viçoja, Mineira, Hardee, and IAC-2 (Santos and Costa 1981) stood out. However, it was only in the 1973/1974 crop season that the breeding program in the state started by means of hybridization, when the cultivar Jupiter was crossed with the variety IAC-2, originating cultivars with the prefix GO (Costa et al. 1981).

The contribution of the *Centro de Pesquisa Agropecuária do Cerrado* [Center for Agricultural Research] (SPA), with the participation of the CNPSo, as well as of the *Empresa Goiânia de Pesquisa Agropecuária* [Goiânia Agricultural Research Company] (EMGOPA), universities, and the private initiative, should be highlighted in the program conducted in Goiás. Cooperative System of Agricultural Research (SCPA). Among the factors that considerably accelerate the development of new cultivars in the state, the possibility of producing two crops per year should be pointed out, as the region does not present frost problems (Spehar et al. 1993).

This program gave rise to the varieties Doko, Numbaira, BR-9 (Savana), and EMGOPA. Moreover, several varieties were developed for the region near the equator, which effectively consolidated soybean adaptation in the states of Maranhão and Piauí. Among these cultivars, we highlight BRS Tracajá, BRS Sambaíba, BRS Carnaúba, BRS Barreiras, and BRS Babaçu (Sediyama et al. 2015).

It should be noted that the first cultivar selected in Goiás was FT-Cristalina, which resulted from a selection made by Francisco Terasawa, in a field of the cultivar UFV-1, in Londrina, PR, at Serrana Farm, during the winter of 1972. Initially, the derived line was identified as M-4 and tested in the municipality of Unaí, MG. In 1974, the line was evaluated for 2 years at Vereda Farm, in Cristalina, GO. After this period, M-4 started to be called Cristalina, in homage to the municipality in which its evaluations were conducted (Terazawa and Silva 1981).

Mato Grosso

In the state of Mato Grosso, the challenges for the implementation of the soybean crop were greater, due to the territorial extension of the state, the high initial crop implementation costs, and the inexistence of production infrastructure and market logistics for transportation of the crop (Unfried and Braga 2011).

The first commercial plantation initiatives in the state date back to the early 1970s, with cultivation, mainly, of cultivars IAC-2, UFV-1, and Santa Rosa, in areas ranging from 5 to 100 ha (Paro and Santos 1981). In association with soybean producers in 1981, EMATER, CPAC, and UFV developed the first cultivar especially for Mato Grosso, which was called UFV-Araguaia and derived from the crossing between Hardee and IAC-2 (Sediyama et al. 1981).

In 1993, the Foundation for Support to Agricultural Research of Mato Grosso, *Fundação MT*, formed by 23 farmers was established and had researcher Dario Minoru Hiromoto as president. The first cultivars developed include MT/BRS Xingu, BRSMT Uirapuru, BRSMT Pintado, FMT Tabarana, and FMT Tucunaré (Sediyama et al. 2015).

In 2001, *Fundação MT* and *Tropical Melhoramento Genético* (TMG), established by researcher Romeu Afonso de Sousa Kiihl, started a partnership dedicated to the development of soybean cultivars. The headquarter of TMG is in Cambé, PR. Annually, about 45,000 progenies are evaluated, and the first cultivars were launched in 2005. In 9 years of research, 24 soybean cultivars were developed. TMG was the first company to introduce Inox technology, which confers resistance to Asian rust (*Phakopsora pachyrhizi* Syd. e P. Syd), with cultivars TMG 801 and TMG 803 (Fundação MT 2011).

Maranhão, Piauí, Tocantins, and Bahia (MAPITOBA—MATOPIBA)

The acronym MAPITOBA or MATOPIBA, formed by the initials of the states of Maranhão, Piauí, Tocantins, and Bahia, refers to the new frontier of occupation of soybeans in the Brazilian Cerrado.

In this region, soybean research began in 1968 in the region of Irecê, BA (Sediyama et al., 2005). However, in the following decades, little was done for the development of cultivars adapted to this area. Most of the work was limited to the introduction and selection of cultivars and lines developed in research centers in regions where soybean farming was already consolidated.

However, since the mid-1990s, with the increase in agricultural activity in the region, several companies started to indicate cultivars for the region such as Monsanto, TMG, Brasmax, Pioneer, Soy Tech Seeds, and Donmario. The performance of *Fundação de Apoio à Pesquisa do Corredor de Exportação Norte* [Foundation for Support to the Research of the Northern Export Corridor]

(FAPCEN), which, in partnership with Embrapa, has developed specific cultivars for this region, should also be highlighted.

Multinationals in Soybean Breeding

Until the 1980s, national public and private companies were still responsible, in a hegemonic way, for the creation and transfer of technology and for the seed supply to the entire domestic soybean market. However, in the 1990s, the globalization process significantly affected the agricultural sector. Moreover, in the same decade, with neoliberal political guidelines, the current government chose to remove public power from the productive sphere of the economy, which gave rise to the depression of public research institutions. As a result new players emerged, identified as large international capital corporations (Unfried and Braga 2011).

In addition to the aforementioned facts, the enactment of the Plant Variety Law (Law No. 9456/1997), which allows the financial return to companies for the creation of new varieties, and Law 10.814/2003, which approved the commercial release of GMOs, has made the market for cultivar development more conducive to the setting up of multinationals.

In 1996, Monsanto began its soybean breeding activities in Brazil, first with conventional cultivars, and, as of 2005, it modified its genetic base to GM cultivars. Later, other companies such as Pioneer, Syngenta, Basf, Bayer, and Dow also implemented their soybean breeding programs (Sediyama et al. 2015).

According to Unfried and Braga (2011), all such institutional structure triggered a process of acquisitions and mergers of companies in the field of seeds, fertilizers, and other inputs. The main acquisitions of companies associated with the soybean market, which occurred in Brazil or that have had a negative impact on the domestic market, are depicted in Table 1.1.

Many of these companies, due to the market power of soybeans in Brazil, now the second largest producer in the world, and the possibility of aggregating revenues from royalties, have been making investments in the biotechnology sector for the development of GM traits. According to Sediyama et al. (2015), in 2011 alone, about 85% of all Brazilian production came from soybean with the trait RR, glyphosate-tolerance technology developed by Monsanto.

Table 1.2 lists some of the technologies already approved for use in soybeans in Brazil and some that are still in the test phase.

Final Remarks

The current production and productivity level of soybeans in Brazil, as well as the crop occupation in regions previously unsuitable for growing the legume, is due to several factors such as the development of production and management technology

Table 1.1 Main acquisitions of companies associated with the soybean market that impacted the Brazilian market up to the year 2015

Year	Company acquiring	Origin	Acquisition	Origin
1996	Monsanto	USA	FT—Pesquisa e Sementes (creating Monsoy)	Brazil
1998	Dow	USA	Mycogen	USA
1998	Monsanto	USA	Braskalb Agropecuária Brasileira Ltda., Dekalb's representative	Brazil
1998	Monsanto	USA	Divisão internacional de sementes da Cargill na América Latina	Brazil
1999	Du Pont	USA	Pioneer Sementes do Brasil	Brazil
1999	Agrevo	Germany	Sementes Ribeiral	Brazil
1999	Du Pont	USA	Dois Marcos	Brazil
2000	Syngenta	Switzerland	Novartis + Astra Zeneca	Switzerland
2000	Aventis	France	Agrevo	Germany
2002	Bayer CropScience	Germany	Aventis	France +
2002	Nidera	Netherlands	Aventis (Agrevo + Rhone Poulenc)	Germany
2005	Nidera	Netherlands	Bayer CropScience (corn and soybean division)	France +
2014	Dow AgroSciences	USA	COODETEC	Germany

Source: Adapted by the authors

Table 1.2 GM technologies developed for soybeans by the main companies operating in the Brazilian market

Technology	Action	Developer company	Year of approval in Brazil
Roundup Read® (RR1)	Tolerance to glyphosate herbicide	Monsanto	1998
Cultivance®	Tolerance to the herbicides of imidazolinone group	Embrapa and BASF	2009
Liberty Link™	Tolerance to glufosinate-ammonium herbicide	Bayer CropScience	2010
Intacta RR2 PRO®	Resistance to caterpillars and tolerance to glyphosate herbicide	Monsanto	2010
Enlist E3®	Tolerant to herbicides glyphosate, glufosinate-ammonium and 2,4-D	Dow AgroSciences MS Technologies	2015
Dicamba	Tolerant to dicamba herbicide	Monsanto	–
HPPD	Tolerant to herbicides that inhibit 4-hydroxyphenylpyruvate dioxygenase (HPPD) enzyme	Syngenta e Bayer CropScience	–

Source: Adapted from Fundação MT (2014) and CBI (2015)

and also breeding, with the development of varieties that are more productive, adapted to the farming conditions and resistant to the main diseases.

The evolution of soybean breeding in Brazil occurred in two periods. The first occurred from the introduction of soybeans in the country until the 1980s. In this stage, the public sector was responsible for the development of varieties, seed supply, and technology transfer. In the period, important research centers were created, mainly in the states of Rio Grande do Sul, São Paulo, and Minas Gerais, where it was possible to adapt soybeans to the conditions of the Cerrado and regions of low latitudes, as well as the development of varieties resistant to various diseases.

After the 1990s, with the globalization and enactment of the Plant Protection Law and due to the position of the then current government, a new era for the improvement of soybeans in Brazil began, marked by the participation of large corporations with international capital such as Monsanto, Pioneer, Syngenta, Basf, Bayer, Dow, and others that, in addition to the development of adapted and productive varieties, have sought genotypes of greater market value from biotechnology, with the introduction of GM traits, for instance.

References

Alexandratos, N., Bruinsma, J.: World agriculture towards 2030/2050: the 2012 revision. ESA Working Paper No. 12-03. FAO, Rome (2012)

Associação Brasileira das Indústrias de Óleos Vegetais (ABIOVE): Estatística Mensal do Complexo Soja—November, 2015. http://www.abiove.org.br/site/index.php?page=estatistica&area=NC0yLTE (2015). Accessed 30 Sept 2015

Bezerra, A.R., Sediyama, T., Borém, A., Soares, M.M.: Importância econômica. In: Sediyama, T., Silva, F., Borém, A. (eds.) Soja: do plantio à colheita, pp. 9–26. UFV, Viçosa (2015)

Bonato, E.R., Bonato, A.L.V.: A soja no Brasil: história e estatística, p. 32. EMBRAPA-CNPSo, Londrina (1987)

Brandão, S.S.: Contribuição ao estudo de variedades de soja. Experientia. **1**(2), 119–199 (1961)

Conselho de Informação Sobre Biotecnologia (CBI): Eventos aprovados. http://cib.org.br/biotecnologia/regulamentacao/ctnbio/eventos-aprovados/ (2015). Accessed 4 Nov 2015

Cordeiro, A., Perez, J., Guazzelli, M.J.: Impactos potenciais da tecnologia terminator na produção agrícola: depoimentos de agricultores brasileiros, p. 33. Pesquisa contratada ao Centro Ecológico pelo Grupo ETC, Florianópolis (2007)

Costa, A.V., Monteiro, P.M.F.O., Rolim, R.B.: Melhoramento de cultivares no Brasil—No Estado de Goiás. In: Miyasaka, S., Medina, J.C. (eds.) A soja no Brasil, 1st edn, pp. 334–340. Instituto de Tecnologia de Alimentos, Campinas (1981)

Doldatelli, D.: Introdução e evolução da soja no Brasil—No Estado de Santa Catarina. In: Miyasaka, S., Medina, J.C. (eds.) A soja no Brasil, 1st edn, pp. 20–22. Instituto de Tecnologia de Alimentos, Campinas (1981)

Feres, J., Gomes, J.E.S.: Melhoramento de cultivares no Brasil—No estado do Rio Grande do Sul. In: Miyasaka, S., Medina, J.C. (eds.) A soja no Brasil, 1st edn, pp. 279–282. Instituto de Tecnologia de Alimentos, Campinas (1981)

Fundação de Apoio à Pesquisa Agropecuária do Mato Grosso—Fundação MT: Boletim de Pesquisa de Soja, p. 492. Rondonópolis (2011)

Fundação de Apoio à Pesquisa Agropecuária do Mato Grosso—Fundação MT: Boletim de Pesquisa de Soja, p. 346. Rondonópolis (2014)
Hiromoto, D.M., Vello, N.A.: The genetic base of Brazilian soybean (*Glycine max* (L.) Merrill) cultivars. Rev. Bras. Genet. **9**(2), 295–306 (1986)
Hunnicutt, B.H.: Feijão soja. Bol. Agric. Zootech. Vet. **3**(9), (1930)
Kaster, M., Queiroz, E.F., Terazawa, F.: Introdução e evolução da soja no Brasil—No Estado do Paraná. In: Miyasaka, S., Medina, J.C. (eds.) A soja no Brasil, 1st edn, pp. 22–25. Instituto de Tecnologia de Alimentos, Campinas (1981)
Lima, C.P.: As percepções dos agricultores que cultivam a soja transgênicas no município de Não-Me-Toque—RS, Brasil: Um estudo de caso mediante metodologia "Q", 100f. Dissertation (Master in Rural Extension), Universidade Federal de Santa Maria, Santa Maria (2005)
Magalhães, C.M.: Introdução e evolução da soja no Brasil—No Estado do Rio Grande do Sul. In: Miyasaka, S., Medina, J.C. (eds.) A soja no Brasil, 1st edn, pp. 18–20. Instituto de Tecnologia de Alimentos, Campinas (1981)
Ministério da Agricultura, Pecuária e Abastecimento (MAPA): Estatísticas e dados básicos de economia agrícola agosto de 2014. http://www.agricultura.gov.br/arq_editor/file/vegetal/Estatistica/Estat%C3%ADticas%20e%20Dados%20B%C3%A1sicos%20de%20Economia%20Agr%C3%ADcola/Pasta%20Marco%20-%202015.pdf (2015a). Accessed 30 Oct 2015
Ministério da Agricultura, Pecuária e Abastecimento (MAPA): Projeções do Agronegócio: Brasil 2014/2015 a 2014/25 projeções de longo prazo, p. 132. AGE/MAPA, Brasília (2015b)
Ministério do Trabalho e Emprego (MTE): Brasileiras e Brasileiros no Campo. Informações úteis. Brasília. http://www.mte.gov.br/ (2010). Accessed 29 Oct 2015
Minssen, S.: A soja. Rev. Agric. **5**(1), (1901)
Miranda, M.A.C., Miyasaka, S., Mascarenhas, H.A.A., Rossetto, C.J.: Melhoramento de cultivares no Brasil—No Estado de São Paulo. In: Miyasaka, S., Medina, J.C. (eds.) A soja no Brasil, 1st edn, pp. 311–324. Instituto de Tecnologia de Alimentos, Campinas (1981)
Miranda, M.A.C., Miyasaka, S., Mascarenhas, H.A.A.: Melhoramento da soja no Estado de São Paulo. In: Miyasaka, S., Medina, J.C. (eds.) A soja no Brasil Central, 2nd edn, pp. 75–112. Fundação Cargill, Campinas (1982)
Miyasaka, S., Medina, J.C.: A soja no Brasil, 1st edn, p. 1062. Instituto de Tecnologia de Alimentos, Campinas (1981)
Moreira, M.A., Barros, E.G., Sediyama, C.A., Sediyama, T.: Genética molecular no melhoramento de soja para agroindústria. 1° Encontro sobre Pós-Graduação em Genética e Melhoramento da UFV. Programa e Resumos, pp. 27–28. Imprensa Universitária—UFV, Viçosa (1995)
Orrego, F.O.Y.: Melhoramento de cultivares no Brasil—No estado de Santa Catarina. In: Miyasaka, S., Medina, J.C. (eds.) A soja no Brasil, 1st edn, pp. 294–300. Instituto de Tecnologia de Alimentos, Campinas (1981)
Paro, H., Santos, E.N.: Introdução e evolução da soja no Brasil—No Estado do Mato Grosso. In: Miyasaka, S., Medina, J.C. (eds.) A soja no Brasil, 1st edn, pp. 255–265. Instituto de Tecnologia de Alimentos, Campinas (1981)
Santini, G.A.: Reestruturação da indústria de sementes no Brasil: o novo ambiente concorrencial dos segmentos de milho híbrido e soja. Dissertação de mestrado, Departamento de Engenharia de Produção (DEP), Universidade Federal de São Carlos (UFSCar), São Carlos (2002)
Santos, G., Costa, A.V.: Introdução e evolução da soja no Brasil—No Estado De Goiás. In: Miyasaka, S., Medina, J.C. (eds.) A soja no Brasil, 1st edn, pp. 41–46. Instituto de Tecnologia de Alimentos, Campinas (1981)
Sediyama, T.: Melhoramento de cultivares no Brasil—No Estado de Minas Gerais. In: Miyasaka, S., Medina, J.C. (eds.) A soja no Brasil, 1st edn, pp. 334–340. Instituto de Tecnologia de Alimentos, Campinas (1981)
Sediyama, T., Sediyama, C.S., Marins, O., Pereira, M.G., Reis, M.S., Athow, K.L., Spehar, C.A., Costa, A.V.: UFV-Araguaia cultivar de soja para o cerrado do Brasil Central, p. 5. UFV, Viçosa (1981)

Sediyama, T., Sediyama, C.S., Pereira, M.G., Borém, A., Gomes, J.L.L., Reis, M.S., Dutra, J.H., Béring, M.S.: Vinte anos de melhoramento na Universidade Federal de Viçosa, p. 4. UFV, Viçosa (1983)
Sediyama, T., Teixeira, R.C., Reis, M.S.: Melhoramento da Soja. In: Borém, A. (ed.) Melhoramento de espécies cultivadas, 2nd edn, pp. 553–604. UFV, Viçosa (2005)
Sediyama, T., Matsuo, E., Oliveira, R.C.T., Glasenapp, J.S.: Origem e evolução. In: Sediyama, T. (ed.) Melhoramento genético da soja, pp. 11–18. Mecenas, Londrina (2015)
Spehar, R., Monteiro, P.M.F.O., Zuffo, N.L.: Melhoramento genético da soja na região Centro—Oeste. In: Arantes, N.E., Souza, P.I.M. (eds.) Cultura da soja nos cerrados, pp. 229–253. Associação Brasileira para Pesquisa da Potassa e Fosfato, Piracicaba (1993)
Terazawa, F., Silva, P.M.F.: O cultivar Cristalina nos Cerrados. In: Miyasaka, S., Medina, J.C. (eds.) A soja no Brasil, 1st edn, pp. 334–340. Campinas, Instituto de Tecnologia de Alimentos (1981)
Unfried, J.R., Braga, N.R.: Os desafios da soja tropical no Brasil. In: Boletim de Pesquisa de Soja, pp. 32–43. Fundação MT, Rondonópolis (2011)
United State Department of Agriculture (USDA): Economic Research Service. http://www.ers.usda.gov/ (2015). Accessed 1 Nov 2015
Universidade Federal de Uberlândia (UFU): Programa de Melhoramento de Soja UFU, p. 4. Uberlândia (2010)
Znamenskiy, V.: As plantas forrageiras e a agropecuária de Goiás. Cerne. 71 (1965)

Chapter 2
FT Sementes and the Expansion of Soybeans in Brazil

Francisco Terasawa, José Maurício Terasawa, and Milene Möller Terasawa

Abstract In Paraná State, research to develop new soybean cultivars began in 1964, at the Experimental Station of Ponta Grossa. Initially, the study included a small number of cultivars introduced from the United States. Later, other American soybean lines were introduced through collaboration agreements with universities located further South and center of the United States, such as the universities of Florida, Mississippi, Georgia, Tennessee, Arkansas and North Carolina, which provided materials with good adaptation to the South Region of Brazil. In 1972, after years working for the Brazilian Federal Government, coordinating research and development of soybean cultivars in South of Brazil, Francisco Terasawa founded FT Pesquisa e Sementes in Londrina, in North of Paraná State. The frost in the coffee region of Northern Paraná, together with better adapted and more productive cultivars, such as Bragg, Davis, IAS-1, IAS-4, IAS-4, IAS-5, Paraná, Mineira and Visoja, accelerated and facilitated the dissemination of soybean crop in Paraná. The cultivar FT CRISTALINA was one of the first releases of FT Pesquisa e Sementes, and, over time, it started to be called by farmers as the "Queen of the Cerrado." This cultivar was one of the main pillars to make soybean production economically viable in the Brazilian Cerrado between the late 1970s and early 1990s. In addition, FT-ABYARA contributed significantly to soybean farming in the South Region and part of the Southwest region of Brazil, due to its good yield, lodging resistance and wide adaptation. At the time, it was considered the best grain-transforming line, that is, it featured the highest correlation of dry matter weight vs. grain weight. The international success of FT Pesquisa e Sementes has been proven by the wide diffusion of its cultivars in several Latin America countries, such as Mexico, Guatemala, Nicaragua, Paraguay, Argentina, Bolivia, Colombia and Venezuela and also in Africa, including countries like Mozambique, Angola and Ivory Coast. In 1995/1996 Monsanto acquired FT Pesquisa e Sementes's soybean germplasm bank that was the first soybean germplasm trade transaction in Brazil. In 1997, the Plant Protection Law was approved in Brazil,

F. Terasawa • J.M. Terasawa, M.S., Ph.D. (✉) • M.M. Terasawa, M.S., D.S.
FT Sementes, Ponta Grossa, Parana, Brazil
e-mail: francisco.terasawa@ftsementes.com.br; mauricio.terasawa@ftsementes.com.br; milene.moller@ftsementes.com.br

F. Lopes da Silva et al. (eds.), *Soybean Breeding*,
DOI 10.1007/978-3-319-57433-2_2

which attracted more national and international seed companies to get involved in soybean breeding in Brazil, which generated numerous high-yielding soybean cultivars. Since its foundation, FT SEMENTES launched several non-GMO, glyphosate-resistant and *Bt*-tolerant soybean cultivars. In 2017, FT SEMENTES reached 45 years of history and will continue to work hard to compete in the globalized market of soybean cultivar development, always committed to meeting the farmers' real needs.

Keywords FT Sementes • High-yield soybeans • Cerrado • FT CRISTALINA • FT-Abyara • Soybean breeding

In Paraná, research to develop new soybean cultivars began in 1964, at the Experimental Station of Ponta Grossa. At that time, the station belonged to the *Instituto de Pesquisa e Experimentação Agropecuária do Sul* [Institute of Agricultural Research and Experimentation of the South (IPEAS)], an agency associated to the Ministry of Agriculture and headquartered in the city of Pelotas, RS. During this period, soybean research was coordinated by researcher Francisco Jesus Vernetti, MSc.

Initially, the study included a small number of cultivars introduced from the United States, such as Hill, Hood, Ogden, Arksoy, Roanoke, etc., all of which considered early-cycle cultivars. These cultivars yielded at most 1,800 kg ha^{-1}. Later, other American lines were introduced through technical agreements with universities located further south and center of the United States, such as the universities of the States of Florida, Mississippi, Georgia, Tennessee, Arkansas, and North Carolina, which allowed a good adaptation of the crop in the Southern Region of Brazil.

One of these lines introduced, called N-45-2994, received from North Carolina, presented good performance in the southern region of Paraná, and it was developed the cultivar named Campos Gerais (Terasawa and Vernetti 1973). Its adaptation greatly contributed to the development of soybeans in the Mid-South region of Paraná and, especially, in the municipality of Guarapuava, at the *Cooperativa Agrícola Agrária* [Agrária Agricultural Cooperative], a highly technified colony of German immigrants. Cultivar Campos Gerais was planted for more than 30 years with excellent yield performance.

In 1968, IPEAS was dismembered in the state of Paraná and part of the state of São Paulo, now known as *Instituto de Pesquisa e Experimentação Agropecuária Meridional* (IPEAME). Two agricultural stations were transformed into Experimental Stations and received the name of *Estação Experimental de Londrina* and *Estação Experimental de Maringá*. At that time, Paraná was the state with the largest coffee production in Brazil. Its crops covered the hottest region of the state and occupied specially high-fertility soils.

In 1969, agronomist Francisco Terasawa was transferred from Ponta Grossa to Londrina, with the responsibility of coordinating experimental works on soybeans, wheat, and cotton from the two Experimental Stations, as well as acquiring an area for the implementation of the great North Paraná station. In the meantime, the government of the state of Paraná also showed interest in the creation of its own agricultural research institute in the northern region of the State of Paraná. The state government was faster than the federal government, and in 1972 it inaugurated the *Instituto Agronômico do Paraná* [Agronomic Institute of Paraná (IAPAR)], in Londrina, first CEO of which, agronomist Raul Juliatto, had previously been president of IPEAME.

When the task for which he was appointed ended, in 1972, Terasawa resigned from the Federal Public Service and started a new endeavor, to which he has dedicated till today, that is, the private breeding of soybean cultivars. At that time, there was no copyright protection law in enforced for breeders and no intellectual property guarantee. Therefore, there was no guarantee of economic return to breeders. For this reason, when Terasawa sought advice from his colleagues, they unanimously advised against his pioneering initiative to create the first private soybean research company in Brazil. Such was his desire to produce a more productive soybean cultivar and better adapted to Brazilian conditions that, in 1972, Francisco Terasawa founded FT Pesquisa e Sementes in Londrina, in the north of Paraná. At that time, there were four main company goals (Terasawa 1982)

- To contribute in the best way possible to Brazil through agricultural research.
- To offer safety to farmers by recommending new cultivars, that is, that featured greater productivity and stability in all environments.
- To search for new successions of crops and economically viable techniques.
- To assist in the technical training of new researchers-breeders, making them aware of their future practical-scientific leaderships.

While Francisco continued his activities as a farmer, growing soybeans, wheat, and coffee, he also provided services such as plowing, harrowing, and harvesting. In order to pay for his research activity, he carried out the first experimental works of FT Pesquisa e Sementes at Serrana Farm in Londrina. Also included in his plans was an associative work model with cooperatives and seed producers, in which these entities would make available agricultural technicians and agronomists, with exclusive dedication, to assist in the agricultural experimentation and for the development of soybean cultivars of FT Pesquisa e Sementes. In those days, such initiative was something really unprecedented. His greatest difficulty was to demonstrate and also prove that soon these producers and cooperatives would be the first to receive the basic seed of the new soybean cultivars for multiplication.

It should be noted that, then, FT Pesquisa e Sementes belonged exclusively to Francisco Terasawa. In the beginning, the experimentation work was entirely manual: opening of furrows, fertilization, sowing, seed coverage, harvesting, and threshing. Although laborious, the manual harvesting of the plants was very valuable, for it allowed to observe the plant root system. This method contributed greatly to the selection and obtaining of plants with the best root system, cycle after cycle of cultivation. Once harvested, plants were placed in used coffee bags, hand threshed, and then sieved, just as one would do with coffee beans.

At that time, 3 years of testing were required for soybean cultivars to be officially recommended, one of which was a preliminary test and two for tests or final evaluations in the official experimental network, conducted in ten different environments, and required a good performance of the lines, in accordance with preestablished rules. Once these requirements had been met, farmers were allowed to use them, and official rural credit was granted to those who cultivated them. Thus, the breeding entity would produce the basic seed, the sale of which happened only once.

Confident in the potential of FT research and development, even before cultivars had been officially recommended, three entities signed a technical agreement with FT Pesquisa e Sementes: Cooperativa Agropecuária Mista do Oeste Ltda. (COOPAGRO), Sementes Dois Marcos, and Granjas Unidas.

The 1972 frost, in the coffee region of Northern Paraná, together with better adapted and more productive cultivars, such as Bragg, Davis, IAS-1, IAS-4, IAS-4, IAS-5, Paraná, Mineira, and Viçosa, accelerated and facilitated the dissemination of soybean farming in Paraná. After the frost of 1972, Francisco Terasawa received seeds of a line called Mutasoja, from Mr. Thomas Owens, of Itaporã, State of Mato Grosso do Sul, MS. At the beginning of August 1972, he sowed this line at Serrana Farm in Londrina. In this advance, he selected 42 plants with distinct traits of the Mutasoja line, which presented higher height, longer cycle, and pubescence and different flower colors. This soybean planting in Londrina during the winter was groundbreaking in Brazil, with its cycle reduced due to the photoperiod. If that first winter crop in 1972 had been completed one month later, all the work would have been lost due to the severe frost that occurred in July 1972, and, thus, the history of soybean in the Cerrado would have been different. Francisco later seeded the 42 selected plants, now during the normal summer planting season, and each plant was individually named M-1, M-2, M-3, and so on, to M-42, the letter "M" being one reference to the original line. This Mutasoja line was later recommended as a cultivar and was named UFV-1. From that time, the purpose with those lines was to evaluate them in the Brazilian *Cerrado* environment, in order to determine the behavior and the individual traits of each one of the lines, in order to identify better adapted materials and with traits of interest for the central region of the country.

As a pioneer and trailblazer, in November 1973, Terasawa arrived at Unaí, in the state of Minas Gerais, bordering the city of Cristalina, in the state of Goiás, and, in a pasture area, he prepared the place for experimentation with the 42 lines and used the cultivar IAC-2 as a comparative standard. In the transversal direction of the sowing, agronomist Terasawa evaluated different formulas of fertilizers as well as micronutrients. The soil analysis showed that those soils were extremely acidic and chemically limited, not very suitable for cultivation, similar to the soils of Campos Gerais in Paraná, and even poorer in organic matter.

Terasawa was assisted by Sementes Trisol, of Mr. Luiz Souza Lima, a farmer and seed producer in the region of Leópolis, Paraná. This work, in partnership with Dr. Pedro Moreira da Silva Filho, was quite promising, especially for line M-4, whose productivity was higher than that of control IAC-2. In November of 1974, already in Goiás, at Vereda Farm, acquired by Mr. Souza Lima, the tests were repeated, and, again, the highlight was M-4. After 1975, M-4 started to be called FT CRISTALINA, in tribute to the municipality of Goiás (Terasawa 1982).

The soybean cultivar FT CRISTALINA was one of the first launches of FT Pesquisa e Sementes, and, over time, it started to be called by farmers as the "Queen of the Cerrado." The pompous character of the denomination has its raison d'être, ratified by the facts. This cultivar was one of the main pillars to make soybean production economically viable in the Brazilian Cerrado between the late 1970s and early 1990s. By extension, it was a valuable component for the feasibility of agriculture in the Midwest. In the agricultural year 1978/1979, for example, the cultivar yielded, in experiments, 4780 kg ha^{-1}, at Vereda Farm, in Cristalina, according to data of the Embrapa Cerrados Agricultural Research Center, associated to the Ministry of Agriculture and Supply (Gilioli et al. 1988).

Cultivar FT CRISTALINA is described as the most used in the region of Cerrados of Central Brazil; it occupies about 95% of the area currently cultivated with soybean in the geo-economic region of the federal district. It is widespread in São Paulo, Minas Gerais, Mato Grosso do Sul, Goiás, and Rio Grande do Sul. This cultivar has few limitations, and due to its excellent qualities, it still tends to expand in Brazil (Terasawa 1983). As previously mentioned, FT CRISTALINA derives from a segregating plant found in the winter multiplication of UFV-1. The cultivar has determined growth habit, purple flower, gray pubescence, bright light yellow seed integument and light brown hilum. It is resistant to the frogeye spot (*Cercospora sojina*), to bacterial pustule (*Xanthomonas campestres* pv. *glycines*), and to wildfire (*Pseudomonas syringae* pv. *tabaci*) and moderately resistant to bacterial blight (*Pseudomonas syringae* pv. *glycineae*) and mildew (*Peronospora manshurica*).

In the agricultural year 1977/1978, for the first time, FT Pesquisa e Sementes soybean lines were evaluated in official trials in the state of Paraná. After presenting excellent results in 3 years of testing in the experimental network, in 1980, the first line identified as PR8104 was raised to the category of variety and named FT-1. Thus, FT Pesquisa e Sementes initiated its contributions to the Southern Region of Brazil with the cultivar FT-1, which in addition to presenting 5.6% higher yield than that of the cultivar Paraná, had as its main trait the tolerance to the root-knot nematode (*Meloidogyne javanica*) (Terasawa 1982). The cultivar is resistant to the frogeye spot (*Cercospora sojina*), to bacterial pustule (*Xanthomonas campestres* pv. *glycines*), and to wildfire (*Pseudomonas syringae* pv. *tabaci*) and features moderate resistance to bacterial blight (*Pseudomonas Syringae pv. Glycineae*) and moderate resistance to mildew (*Peronospora manshurica*) (Terasawa 1983, 1984).

Then, a new cultivar, FT-2, was recommended, and its tolerance to drought was remarkable when compared to other lines (Terasawa 1982). Such extremely desirable trait is the result of its deep and aggressive root system, even in soils with moderate levels of aluminum. FT-2 featured another relevant trait for the time, since its closing point was completed with extreme brevity, that is, its leaf development is fast, which reduced the distance between lines, making it difficult to develop invasive plants.

That trait was very important, as the conventional cropping system adopted at that time did not have many alternatives for chemical control of weeds. This cultivar maintained a very good performance in all trials conducted, and its average yield in Paraná trials was 12% higher than that of Bossier cultivar and 15% higher than that of BR-1. Cultivar FT-2 is the result of a selection of cultivar IAS-5, a widely used cultivar in the Southern Brazil. IAS-5 is the result of Hill's artificial crossing with D52–810, conducted at the University of North Carolina. Based on the cultivar stability analysis method of Eberahart and Russel (1966), according to Dr. José Francisco Ferraz, cultivar FT-2 remains stable and with high yields in the most representative environments with this crop. The cultivar features resistance to the frogeye spot (*Cercospora sojina*), bacterial pustule (*Xanthomonas campestres pv. Glycines*), and wildfire (*Pseudomonas syringae* pv. *tabaci*) and tolerance to bacterial blight (*Pseudomonas syringae* pv. *glycineae*), to brown spot (*Septoria glycines* Hemmi), to mildew (*Peronospora manshurica*), to the common mosaic virus, and to the purple seed stain (*Cercospora kikuchii*) (Terasawa 1982, 1983).

After 1974, FT Pesquisa e Sementes has initiated the artificial hybridizations program. The difficulty in obtaining parental seeds in Brazilian governmental institutions, coupled with the contacts that Terasawa had with researchers in several foreign universities, caused the majority of the seeds of the parents used in the crossings to be derived from other countries, especially from the United States.

FT Pesquisa e Sementes breeding program could have contributed much more to the development of soybean farming in several states of Brazil, but, unfortunately, there was delay in the recommendation of the cultivars in some locations, due to the strong governmental obstruction and aversion to private research at that time.

The state research institutions that regulated, conducted, and standardized the evaluations of the new cultivars did not allow the inclusion of new soybean lines from FT Pesquisa e Sementes in official tests. As an example of such difficulty, in the state of São Paulo, the company had to wait for 5 years for a vacancy for the line test in the official trials of that state.

In the 1980s and early 1990s, the company had already developed 102 cultivars, including FT CRISTALINA, FT-1, FT-2, FT-3, FT-4, FT-5, FT-6, FT-7, FT-8, FT-9, FT-10, FT-11, FT-12, FT-13, FT-14, FT-15, FT-16, FT-17, FT-18, FT-19, FT-20, FT-21, FT-100, FT-101, FT-102, FT-103, FT-104, FT-105, FT-106, FT-107, FT-108, FT-109, FT-110, FT-111, FT-2000, FT-2001, FT-2002, FT-2003, FT-2004, FT-2005, FT-2006, FT-2007, FT-2008, FT-2009, FT-2010, FT ABYARA, FT MANACÁ, FT IRAMAIA, FT SARAY, FT SIRIEMA, FT JATOBÁ, FT COMETA, FT ESTRELA, FT GUAÍRA, FT BAHIA, FT CANARANA, FT MARACAJU, FT CANAVIEIRA, etc. (Terasawa 1982, 1983, 1984, 1985, 1986, 1987, 1988, 1989, 1990).

As previously pointed out, it was through partnerships with cooperatives and seed companies that they built an infrastructure that allowed the regional targeting of breeding work as well as the possibility of training and maintenance of a specialized team in the area. Two examples of regionalization of the scientific work were the creation of the soybean cultivar FT GUAÍRA, to meet the specific demands of Paraná and São Paulo, and the development of the variety FT SARAY, specially created for Rio Grande do Sul. Featuring an early cycle, with higher yield, cultivar FT GUAÍRA was perfectly integrated to the rotation system of soybean and corn crops (Terasawa 1988), whereas the FT SARAY variety was recommended in the 1993/1994 crop as a high-yield option with high potential for resistance to common diseases in the state of Rio Grande do Sul. As an example of the importance and proportion of that work, in those days there was an expressive number of entities partnering with the FT Pesquisa e Sementes breeding program. At that time, in 1988, the company's staff included 25 agronomists and five agricultural technicians allocated by 35 partner entities (Terasawa 1988). FT Pesquisa e Sementes has expanded the scope of its research work, both nationally and internationally.

Always innovating in research, one of the cultivars indicated for farming was FT COMETA, in 1988, the result of diligent plant breeding work, aiming at early maturity group soybean for Brazilian conditions (Terasawa 1988). The work to obtain this cultivar was carried out in Londrina, PR, in 1978, as a result of the hybridization between genotypes Williams and FT420. In official evaluations, coordinated by EMBRAPA National Center for Research on Soybean, this line was identified as FT81–1866. FT COMETA was the first indeterminate growth habit early maturity group cultivar in Brazil. FT COMETA cultivar features white flower, brown pubescence, bright light yellow seed and black hilum. The cultivar is tolerant to the root-knot nematode (*Meloidogyne javanica*), resistant to the frogeye spot (*Cercospora sojina*), bacterial pustule (*Xanthomonas campestres* pv. *glycines*), and to wildfire (*Pseudomonas syringae* pv. *tabaci*), and features moderate resistance to bacterial blight (*Pseudomonas syringae* pv. *glycineae*), to mildew (*Peronospora manshurica*), to septoriosis (*Septoria glycines*), to the purple seed stain (*Cercospora kikuchii*), and to the common mosaic virus. Its yield, at the time, was always equal to or superior to that of cultivar Paraná.

Another cultivar, FT ABYARA (Abyara is a Tupi-Guarani word that means something exceptional, unusual) contributed significantly to soybean farming in the South Region and part of the southwest of Brazil, due to its yield potential and lodging tolerance (Terasawa 1988). At the time, it was considered the best grain-transforming line, that is, it featured the highest correlation of dry matter weight vs. grain weight. This cultivar was a market leader, and the first variety to exceed 5000 kg ha^{-1}. Such top yields were achieved in the north, west, and southwest regions of Paraná, in high-fertility soils. The cultivar FT ABYARA is the result of the artificial hybridization between genotypes União and Sant'Ana. It is a cultivar of the medium-maturity group, and its cycle is similar to that of cultivars Bossier and FT-2, around 126 days in the state of Paraná. Its phenotypic features are purple flower, brown pubescence, medium intensity bright light yellow seed integument, and brown hilum. Its growth habit is determined, with good resistance to pods' dehiscence and physiological quality seeds. As mentioned above, one of the most relevant traits of this cultivar is its high lodging tolerance. In relation to diseases, it is resistant to the frogeye leaf spot (*Cercospora sojina*), bacterial pustule (*Xanthomonas campestres* pv. *glycines*), and wildfire (*Pseudomonas syringae* pv. *tabaci*) and moderately resistant to bacterial blight (*Pseudomonas syringae* pv. *glycineae*), to mildew (*Peronospora manshurica*), to septoriosis (*Septoria glycines*), to the purple spot (*Cercospora kikuchii*), and to the common mosaic virus.

The international success of FT Pesquisa e Sementes has been proven by the wide diffusion of its cultivars in many Latin American countries, such as Mexico, Guatemala, Nicaragua, Paraguay, Argentina, Bolivia, Colombia and Venezuela and also in Africa, including countries like Mozambique, Angola and Ivory Coast.

An example of the success of the company in Brazil is the fact that FT varieties were spread over all Brazilian states that produce this legume. In the 1993/1994 crop season, data released by CONAB (2016) evidenced an average participation of 52% of the cultivars developed by FT Pesquisa e Sementes in 11 Brazilian states, whose percentages can be checked in Fig. 2.1.

The cultivars that stood out the most in the south region and in the state of São Paulo were FT10-PRINCESA, FT5-FORMOSA, FT ABYARA, FT COMETA, FT SARAY, and FT JATOBÁ. In the other regions of the country, the varieties FT SERIEMA, FT ESTRELA, FT104, FT106, as well as the super-varieties FT CRISTALINA were the ones to have contributed decisively to the consolidation of soybean farming at the Brazilian Cerrado (Terasawa 1988).

In 1994, in Brazil, 850 companies produced and marketed soybean seeds. Out of this total, less than 0.5% developed new varieties. The large majority, therefore, only invested in the reproduction of soybean cultivars without recognizing the copyrights of the breeders of cultivars. Of this total of companies working with soybean seeds, only 35 had some type of agreement with FT Pesquisa e Sementes. Specific legislation on this matter was required to protect the technology generated by the few breeding companies at the time. Many attempts took place, all of them,

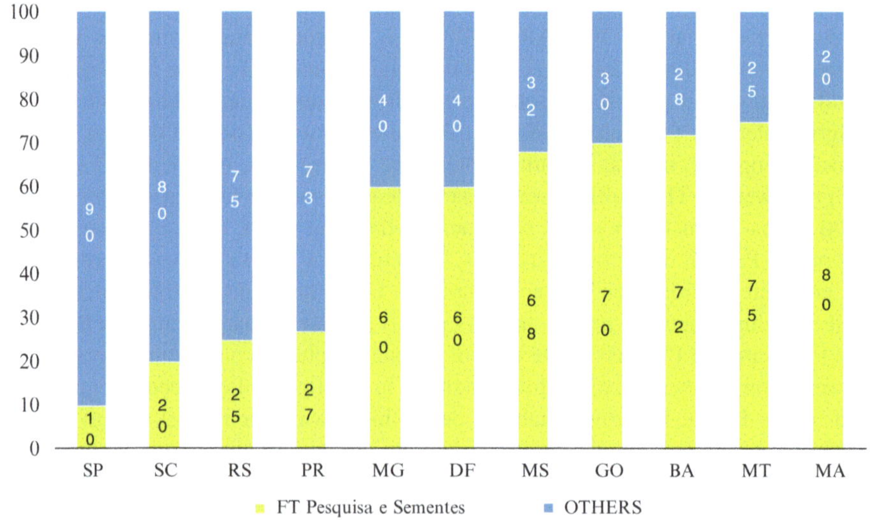

Fig. 2.1 Market share of the cultivars developed by FT Pesquisa e Sementes in the 1993/1994 harvest in 11 units of the federation. Fonte: Adapted from CONAB (2016)

however, unsuccessful. At the time, researchers and breeders, for the most part, were either federal or state civil servants and understood that research would be something exclusive to the government. Also, the seed producers, in their entirety, did not agree to pay a small fee to breeders because they believed such additional cost would render their activities unfeasible and would not provide an adequate financial return. At the time, seed producers were not convinced of the benefits that a more productive and resistant new crop could generate for both the seed market and producers.

FT innovated once again. For the first time, a soybean breeding bank was traded. The multinational Monsanto acquired FT Pesquisa e Sementes's soybean germplasm bank, in 1995/1996. According to Londres and Almeida (2009), Monsanto made a high investment in Brazil, by buying important companies not only in the soybean segment but also in the segment of corn and cotton seeds. As a result, in 1995 Monsanto became the largest sowing company in the world, and, currently, it is a leader in the production of GM seeds. In 1997, the Plant Protection Law (Brazil 1997) was approved, which allowed the entry of more national and international research companies in this market, which generated and continue to generate numerous high-yield soybean cultivars.

In 2006, FT resumed its soybean breeding activities with a new structure and team under the name FTS SEMENTES, under the command of Terasawa's son, agronomist José Maurício Terasawa. Since then, several cultivars have been launched with RR technology, released for farming in Brazil in 2003/2004 crop season. Among them the highlights are FTS GRACIOSA RR, with multiple resistance to Soybean Cyst Nematode (SCN) and to Root-Knot Nematode (RKN), FTS TRIUNFO RR with resistance to 11 cyst nematode races and low reproduction factor to *Pratylenchus sp.*, FTS MASTER RR with resistance to 11 cyst nematode races, and FTS PARAGOMINAS RR and FTS CAMPO MOURÃO RR, both with tolerance to high levels of aluminum in depth and with good tolerance to high temperatures and water stress (Grupo Cultivar 2016).

Soybean is the agricultural crop that has shown the greatest growth in the last decades. Currently, soybean farming corresponds to 49% of grain area planted in Brazil (MAPA 2016). Such significant increase is associated with the technological advances that the crop has presented. Always following the latest trends and technologies available to farmers, FTS SEMENTES has invested in the integration of several traits originating from biotechnology, such as Monsanto's Intacta Technology, and incorporated into its portfolio cultivars of various maturity groups, adapted to all macro-regions of Brazil.

According to Monsanto (2016), Intacta RR2 PRO provides protection against caterpillars due to the presence of *Bt* protein (*Cry1Ac*), which is highly effective against the velvetbean caterpillar (*Anticarsia gemmatalis*), the soybeans looper (*Chrysodeixis includens* and *Rachiplusia nu*), the soybean budborer (*Crocidosema aporema*), the tobacco budworm (*Heliothis virescens*), lesser cornstalk borer (*Elasmopalpus lignosellus*), and corn earworm (*H. zea* and *H. armigera*). Moreover, tolerance to glyphosate is provided by Roundup Ready (RR) 2 technology, which makes it easier for farmers to manage weeds. In the 2015/2016 crop season,

Table 2.1 Number of cultivars of FTS SEMENTES registered with the RNC (National Register of Cultivars) between 2005 and 2016

Year	Conventional	RR[a]	IPRO[b]	Total
2005	3	–	–	3
2006	1	–	–	1
2007	4	12	–	16
2008	2	–	–	2
2009	–	3	–	3
2010	–	5	–	5
2011	–	3	–	3
2012	–	11	–	11
2013	–	–	–	0
2014	–	–	–	0
2015	–	4	1	5
2016	–	–	5	5
Total	10	38	6	54

Source: Data adapted from RNC (2016)

[a]Roundup Ready (Trait GTS-40-3-2): Genetically modified soybean tolerant to glyphosate herbicide

[b]Intacta RR2PRO: MON87701 × MON89788: Genetically modified soybean, insect resistant and tolerant to glyphosate herbicide

FT SEMENTES launched its first Intacta cultivar called FTR 4160 IPRO. Since 2005 till presently, FT SEMENTES has featured in its portfolio several cultivars registered with National Register of Cultivars (RNC), with ten conventional cultivars, 38 RR cultivars, and six cultivars featuring Intacta Technology (Table 2.1) (FT Sementes 2016). In 2017, FT SEMENTES will complete 45 years of history and will continue to work hard to compete in the globalized market of soybean cultivar development, always committed to meeting the farmers' real needs.

References

Brazil: Lei n° 9.456, de 25 de abril de 1997. Institui a Lei de Proteção de Cultivares. http://www.planalto.gov.br/ccivil_03/leis/L9456.htm (1997). Accessed 10 May 2016

Companhia Nacional de Abastecimento (CONAB): http://www.conab.gov.br (2016). Accessed 10 May 2016

Eberhart, S.A., Russell, W. A.: Stability Parameters for Comparing Varieties. Crop Science 6:36–40 (1966)

FT Sementes: http://www.ftsementes.com.br (2016). Accessed 13 May 2016

Gilioli, J.L., Terasawa, F., Trecenti, R., Artiaga, O.P., Ferreira, B.M.: Descrição e uso de cultivares de soja para a região dos cerrados, p. 27. FT Pesquisa e Sementes, Brasília (1988)

Grupo Cultivar: http://www.grupocultivar.com.br (2016). Accessed 5 May 2016

Londres, F., Almeida, P.: Impacto do controle corporativo no setor de sementes sobre agricultores familiares e sistemas alternativos de distribuição: Estudo de caso do Brasil. Agricultura Familiar e Agroecologia (ASPTA), Rio de Janeiro (2009)

Ministry of Agriculture, Livestock and Supply—MAPA: http://www.agricultura.gov.br (2016). Accessed 13 May 2016

Monsanto: Tecnologia. http://www.monsanto.com (2016). Accessed 13 May 2016

National Variety Register-RNC: http://www.agricultura.gov.br/guia-de-servicos/registro-nacional-de-cultivares-rnc (2016). Accessed 13 May 2016
Terasawa, F.: FT Pesquisa e Sementes—Relatório Técnico. Ponta Grossa (1982)
Terasawa, F.: FT Pesquisa e Sementes—Relatório Técnico. Ponta Grossa (1983)
Terasawa, F.: FT Pesquisa e Sementes—Relatório Técnico. Ponta Grossa (1984)
Terasawa, F.: FT Pesquisa e Sementes—Relatório Técnico. Ponta Grossa (1985)
Terasawa, F.: FT Pesquisa e Sementes—Relatório Técnico. Ponta Grossa (1986)
Terasawa, F.: FT Pesquisa e Sementes—Relatório Técnico. Ponta Grossa (1987)
Terasawa, F.: FT Pesquisa e Sementes—Relatório Técnico. Ponta Grossa (1988)
Terasawa, F.: FT Pesquisa e Sementes—Relatório Técnico. Ponta Grossa (1989)
Terasawa, F.: FT Pesquisa e Sementes—Relatório Técnico. Ponta Grossa (1990)
Terasawa, F., Vernetti, F. de J.: Cultivar Campos Gerais. vol. 88, p. 6. Ministério da Agricultura, Departamento Nacional de Pesquisa Agropecuária, Instituto de Pesquisas Agropecuárias do Sul. Boletim Técnico, Pelotas (1973)

Chapter 3
Center for Diversity and Genetic Resources

Marcelo Fernandes de Oliveira and Carlos Alberto Arrabal Arias

Abstract Although evidence suggests that cultivated soybeans (*Glycine max* (L.) Merr.) have been domesticated in northern China, current theories suggest that soybeans could have been domesticated in the same way in southern China in the medium or low "Yellow River valley" of central China, northeastern China, or simultaneously in multiple centers. For soybeans, centers of genetic diversity are considered important sources of genetic variability. Plant genetic resources include cultivated plants and wild species of proven value or even potential. These resources have great importance to humankind because they make possible the solutions of numerous agriculture problems in the present and future, which can be found in plant genetic resources. In the case of soybean, an annual species in the subgenus *Soja* and 22 perennial species within the subgenus *Glycine* have been reported as related wild species. Therefore, genetic resources are considered for purposes of genetic improvement of the crop. Much of the genetic variability of this crop has been maintained and conserved in Germplasm Banks (BG) in several eastern and western countries. In Brazil, the formation of the Soybean Germplasm Active Bank (BAG/Soybean) started in September 1975, at Embrapa Soybean, in order to gather pure and characterized stocks of genotypes for the use of breeders and geneticists.

Keywords Wild species • Genetic variability • Plant introduction

Introduction

Soybean is an autogamous plant that belongs to the family Leguminosae, subfamily Papilionoideae, Phaseoleae tribe, and *Glycine* genus. The latter is subdivided into two subgenres: the subgenus *Glycine* and the subgenus *Soja*. Within the subgenus *Soja*, there are two recognized species, *Glycine max* and *Glycine soja*, with $2n = 40$ chromosomes. The cultivated soybean belongs to the species *G. max*.

M.F. de Oliveira, M.S., Ph.D. (✉) • C.A.A. Arias, M.S., Ph.D.
Embrapa Soja, Londrina, Parana, Brazil
e-mail: marcelofernandes.oliveira@embrapa.br; carlos.arias@embrapa.br

F. Lopes da Silva et al. (eds.), *Soybean Breeding*,
DOI 10.1007/978-3-319-57433-2_3

This chapter is intended to describe the center of diversity and genetic resources of the genus *Glycine*.

Origin of Soybeans

In addition to *G. max* and *G. soja*, an intermediate form known as *G. gracilis* Skvortz has been described, first proposed as a new species by Skvortzow (1927). However, *G. gracilis* is not recognized as a new species by the "International Legume Database and Information Service" (ILDIS) and by the US Department of Agriculture—Germplasm Resources Information Network (USDA-GRIN) (Hymowitz 2004). Many authors consider *G. gracilis* a variant of *G. max*. According to the study of Chen and Nelson (2004), *G. gracilis* accessions can be distinguished from *G. max* and *G. soja* based on phenotypic markers and DNA, but do not necessarily support the designation of another species, and *G. gracilis* is more related to *G. max* accessions than to *G. soja*.

As to the origin of the genus *Glycine* (with $2n = 20$), and based on a broad review on works involving taxonomic, cytological, cytogenetic, and molecular systematic aspects, Hymowitz (2004) hypothesized that the probable ancestor for the genus *Glycine* $2n = 20$ would have occurred in Southeast Asia (Fig. 3.1). However, such a parent either would be extinct or had not yet been collected and identified in Cambodia, Laos, or Vietnam. This ancestral species is believed to have undergone a process of autopolyploidization or allopolyploidization ($2n = 2x = 40$) before or after the dissemination from the ancestral region into the Australian continent or to China.

Perennial wild species that have adapted to the ecological niches of the Australian continent have not been domesticated. The route of migration from the ancestral region to China from a common ancestor is believed to have occurred initially through a wild perennial species ($2n = 4x = 40$, unknown or extinct), followed by an annual wild species ($2n = 4x = 40$, *G. soja*) until finally giving rise to the annual cultivated species ($2n = 4x = 40$, *G. max*, cultivated). Morphological data, cytogenetics, analysis of mitochondrial DNA fragments, ribosomal RNA, chloroplastic DNA, and ITS sequences of the ribosomal nuclear DNA region have considered *Glycine soja* as the ancestor of *G. max* (Chen and Nelson 2004).

All currently described species of the genus *Glycine* exhibit diploid-like meiosis and are generally endogamous, producing seeds by cleistogamy. Allopolyploidization (interspecific hybridization followed by chromosome duplication) has probably played an important role in the speciation of the genus *Glycine*, in which *Glycine* species with 40 chromosomes and *G. tabacina* and *G. tomentella* with 80 chromosomes are tetraploid and octoploid, respectively. The expression of four ribosomal DNA loci in *G. curvata* and *G. cyrtoloba* (Singh et al. 2001) reinforces this hypothesis of the origin of the allotetraploid.

As for the origin of the cultivated species [*Glycine max* (L.) Merr.], linguistic, geographical, and historical evidences suggest that soybean [*Glycine max* (L.)

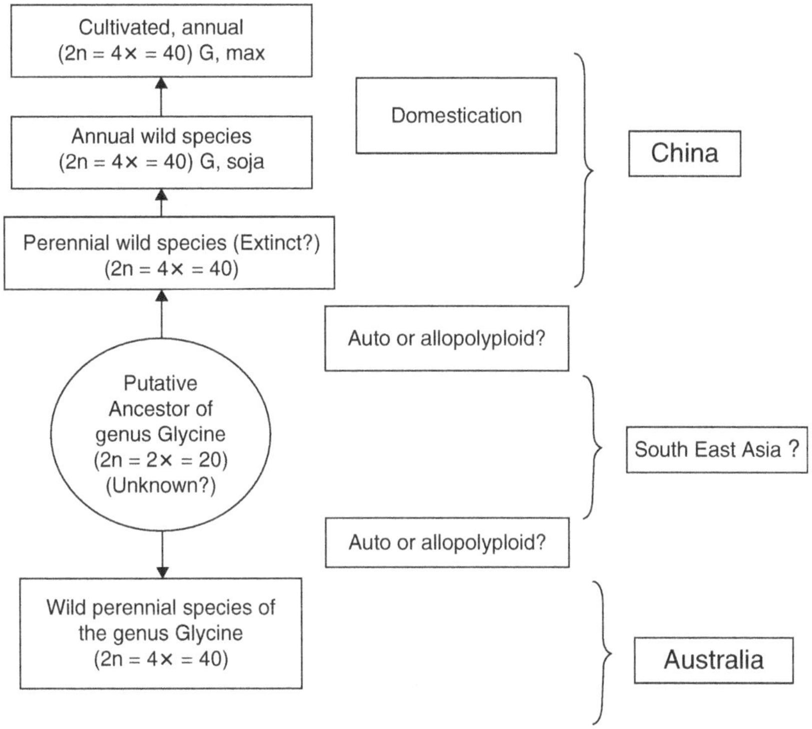

Fig. 3.1 The origin of genus *Glycine*. Source: Hymovitz (2004)

Merr.] was domesticated during the Zhou dynasty, in the Eastern China around the eleventh century BC. Around the first century, soybeans probably expanded through Central and Southern China and also through Korea. The movement of soybean germplasm within this region, primary diversity center, is associated with the development and consolidation of the territories and the degeneration of the Chinese dynasty. From the first century until the mid-fifteenth century AD, soybeans were introduced in Japan, Indonesia, the Philippines, Vietnam, Thailand, Malaysia, Myanmar, Nepal, and Northern India, where many landraces (name given to cultivars prior to the date of scientific improvement) were developed. These regions comprise the secondary diversity center.

Despite evidences that soybeans were domesticated in Northern China, current theories show that this legume could have been domesticated in the same way in Southern China, in the middle or lower "Yellow River Valley" of Central China (Zhou et al. 1999), Northeast China, or simultaneously in multiple centers (Lu 1977). However, there is no consensus on this issue, since wild soybeans commonly grow in Eastern China in the latitude range 24–53 N, Japan, Korea, and the Far East of Russia. Theoretically, soybean domestication could have occurred in any region of China. However, many researchers accept the hypothesis that soybean was domesticated in the "Yellow River Valley" or "Yangtze River" valleys of Central or Southern China.

Two recent studies illustrate the arguments for these areas as centers of origin. In the first one, DNA markers were used to compare landraces and accessions of wild soybean in order to infer patterns of domestication using germplasm from the South, Central, and Northwest Regions of China. The premise of the study was that *G. max* should be more related to the gene pool of *G. soja*, from which it was domesticated. Gai et al. (1999) evaluated chloroplasts and mitochondria of 200 *G. max* accessions and 200 of *G. soja* using molecular markers such as restriction fragment length polymorphism (RFLP). The authors showed that *Glycine max* accessions from all regions of China were more related to those of *Glycine soja* from the southern region of the "Yangtze River" valley than to accessions of *G. soja* from other regions. These data support the idea that *G. max* was domesticated from *G. soja* from the southern region and then disseminated into other regions. It was also possible to infer that the beginning of the expansion of domesticated soybean to the Central and Northeast Region of China may have favored the significant soybean variability regarding the total cycle within these geographic regions, which has allowed the adaptation of the crop to the spring, summer, and fall planting system. The fact is that the diversity among the DNA markers of the three regions of China was much broader than among the different maturity groups within a region.

In a second study, Zhou et al. (1999) used the Vavilov concept that the greatest genetic diversity for a species should be that of the domestication center. They evaluated 15 morphological and biological traits of 22,695 accessions of *G. max* from China and concluded that the soybean diversity center resides within a corridor between Southwest and Northeast China, which includes the provinces of Sichuan, Shaanxi, Shanxi, Hebei, and Shandong. This corridor connects two agriculture cradles in China, the "Yellow River Valley" and "Yangtze River Valley." These two ancient centers of agriculture have a long history of soybean and millet farming, and the agricultural exchange between these two areas was intense (Gai and Guo 2001).

Genetic Variability

In the case of soybeans, genetic diversity centers are considered important sources of genetic variability. Accessions were grouped into four major groups: the first formed by China (primary diversity center), the second formed by Korea and Japan (secondary diversity center), the third formed by the other countries of South Asia (Asia, India, Indonesia, and Vietnam) and Russia (secondary diversity center), and the fourth formed by all non-Asian countries where soybean cultivation is recent, when compared to the millennium data of the other groups (Africa, Americas, Europe). Such grouping was based on several studies carried out by several authors, who compare the geographical origin of soybean accessions, as well as the genetic distance between the accessions, which suggests the importance of this criterion (Griffin and Palmer 1995; Li and Nelson 2001). The consistency of the results provides strong evidence that accessions from Japan and South Korea are genetically similar and distinct from China's accessions. These studies also indicate that the soybean gene pool from Japan and South Korea was probably derived from few introductions from China. Based on these results, one observes that Japan and South Korea are secondary sources of soybean germplasm, but distinct from China's gene pool. They also show that germplasm accessions from various regions of Asia were categorized into four groups: Korea and Japan, China and Eastern Russia, Southeast Asia, and Central South Asia.

According to the same authors, within China the diversity of the groups formed generally reflects the geographical origin of accessions. In the study of Li and Nelson (2001), the degree of genetic variation revealed by RAPDs within China was more extensive than within Japan and South Korea. As a result, accessions from China were divided into four subgroups, according to the provincial origin. The first subgroup was composed of accessions from the Northeast region of China; the second, by the accessions of the Huanghe-Huaihe-Haihe region; the third, by the accessions of the South region of China; and the fourth group by accessions whose province data are unknown.

Soybean variability for physiological, morphological, and agronomic features is considered quite broad. This allows the use thereof in breeding and selection programs of cultivars with high adaptability and stability to the most diverse edaphoclimatic conditions and production systems. Chinese farmers were responsible for this genetic variability, since, well before history was written, they used this genetic diversity so that the soybean crop could reach high levels, by increasing yield and resistance to diseases and pests and by adapting the crop to extreme climates. Historical data reveal the importance of farmers for the development of phenotypically diverse varieties during the domestication process. These varieties that preceded scientific breeding were defined as landraces, which are irreplaceable genetic material, inasmuch as they represent the genetic diversity gathered by farmers over 3000 years, which included the conversion of wild soybeans to modern crop. By the beginning of the twentieth century, Chinese farmers had already grown between 20,000 and 45,000 different soybean genotypes. Many of these landraces have been collected by germplasm curators and breeders and are preserved in extensive germplasm collections, and make up the largest source of genetic diversity within soybean germplasm collections. The best breeders in the world use such diversity as a basis for soybean breeding (Singh and Hymowitz 1999; Carter et al. 2004).

Traditionally, the evaluation of genetic diversity in soybean has been based on differences in morphological and agronomic traits or genealogy information, which has provided important information for germplasm management and evaluation. Phenotypic diversity in soybean is extensive and is under the genetic control of qualitative and quantitative traits. Qualitative traits provide tags for the identification of accessions and classification of differences. Most economic characters and plant traits have quantitative inheritance.

Brazil is among the countries with the greatest diversity of species in the world, and yet it is extremely dependent on germplasm from other countries, as most cultivated species have their centers of origin in other continents. With grounds on this assumption, the area of genetic resources stands out for its importance in the search for new species not native to Brazil, but cultivated in the world, in the conservation of these introductions and multiplications and in the search for genetic variability of introduced and native species.

Genetic Resources

Plant genetic resources include plants cultivated and wild species of proven value or even potential. These resources are of great importance to humankind, for it is possible that solutions to numerous agriculture problems, both current and future, may be found in plant genetic resources. Therefore, conservation and rational use of genetic resources can contribute to the eradication of hunger and poverty.

Despite the great diversity, the number of plants used by man is minimal when compared to that of plant species in nature. Throughout its history, man has known about 300,000 edible species, but only 300 have been used as food. Currently, only 15 species contribute with 90% of human nutrition, in that, eight of them are responsible for 75% of the vegetal contribution to human energy (Nass 2001). The dependence on a limited number of species shows the risk of food safety and points out the importance of the management, conservation, and use of genetic resources.

The maintenance of these resources is done by means of the establishment of environmental protection areas and through the collection and maintenance of these materials, which are now known as germplasm. Germplasm is defined by Allard (1971) as the sum total of the materials of each species. Thus, these may be in the form of plants, anthers, pollen, seeds and tissues (meristem, callus), cells, or even simple structure.

Each germplasm unit should represent a single copy of the genetic material and the living organism of current or potential interest. Consequently, germplasm is the element of genetic resources that represents the inter- and intraspecific genetic variability, to conserve and use it in research in general, especially in breeding programs. Thus, genetic resources comprise the diversity of the genetic material contained in primitive, obsolete, traditional, or modern varieties, in wild relatives of the target species (wild species or primitive lines) that can be used now or in the future for food, agriculture, and other purposes.

Germplasm collections were proposed to preserve the genetic diversity of cultivated species, due to the adoption of modern cultivars as a replacement to primitive ones.

Frankel and Brown (1984) described three phases that have characterized genetic resource activities over the past 60 years: the first emphasized biogeography, taxonomy, and evolution; the second, the collection and conservation of germplasm; and the third, the characterization and use of germplasm. In the first two phases, the strategy was to collect and preserve as much of the allelic diversity as possible. Higher-value alleles have been found to be rare and often restricted to specific geographic areas. This has resulted in farms to ensure perfect collection with the maximum diversity of germplasm. Accordingly, for many species, germplasm collections have become large and diffuse. The size of the collection and the limit of financial resources available have reduced the efficiency of germplasm evaluation, which has discouraged the enrichment and use thereof.

Genetic resources can be conserved by maintaining the species in their *habitats* (in situ) or outside them (ex situ). The conservation method to be adopted depends on the needs, the possibility of conservation, and the target species. The purpose of ex situ conservation is to maintain accessions without alterations in their genetic constitutions. Thus, the use of germplasm banks ensures conservation, making accessions genetically stable and at the reach of users.

Gene banks around the world maintain collections of plant genetic resources for long-term conservation and to facilitate accession for plant breeders, researchers, and other users. In recent decades there has been remarkable progress in gathering and conserving these resources (van Hintun et al. 2000). However, in order to efficiently use the genetic potential maintained in the germplasm collections, detailed knowledge of the collection, including characterization and evaluation, is required. This contributes to the prevention of possible genetic losses, such as those that may occur during multiplications of the accessions collected, and allows for the establishment of sites or collection areas that contain greater variability, thus helping in the planning of new collections.

In the case of soybeans, an annual species within the subgenus *Soja* and 22 perennial species within the subgenus *Glycine* have been reported as related wild species (Table 3.1). Out of the 22 species, seven were described before 1981, and 15 additional ones after this period. These species are, therefore, considered as genetic resources for breeding purposes (Hymowitz 2004). Much of the genetic variability of this crop has been maintained and conserved in germplasm banks (GB) in several eastern and western countries. According to data collected and updated by the International Plant Genetic Resources Institute (IPGRI), over 170,000 accessions of *Glycine max* are maintained by more than 160 institutions in approximately 70 countries. China has the largest soybean germplasm collection in the world with nearly 26,000 accessions of *G. max* and 6200 accessions of *G. soja*, located at the Institute of Crop Germplasm Resources of the Chinese Academy of Agricultural Science in Beijing (Carter et al. 2004). The US Department of Agriculture (USDA) germplasm collection is the second largest, with 18,570 *G. max* accessions, 1116 *G. soja* accessions, and 919 accessions of perennial species of *Glycine*. All perennial species are native to Australia, which has the largest collection in the world, with over 2100 accessions. Currently, more than 3500 accessions of the 22 perennial species of *Glycine* have been maintained in nine collections worldwide. Most of the accessions of *Glycine max* and *Glycine soja* available in the world were collected in China and Japan.

Table 3.1 List of species in the genus *Glycine* Willd., three-letter code, 2*n*, genome symbol, and distribution

	Code	2*n*	Genome[a]	Distribution
Subgenus *Glycine* (perennial)				
G. albicans Tind. & Craven	ALB	40	II	Australia
G. *aphyonota* B. Pfeil	APH	40	?	Australia
G. arenaria Tind.	ARE	40	HH	Australia
G. *argyrea* Tind.	ARG	40	A2A2	Australia
G. *canaescens* F. J. Herm.	CAN	40	AA	Australia
G. clandestina Wendl.	CLA	40	A1A1	Australia
G. curvata Tind	CUR	40	C1C1	Australia
G. *cyrtoloba* Tind	CYR	40	CC	Australia
G. *dolichocarpa* Tateishi and Ohashi	DOL	80	?	Taiwan
G. falcata Benth	FAL	40	FF	Australia
G. *hirticaulis* Tind. & Craven	HIR	40	H1H1	Australia
		80	?	Australia
G. *lactovirens* Tind & Craven	LAC	40	I1I1	Australia
G. latifolia (Benth.) Newell & Hymowitz	LAT	40	B1B1	Australia
G *latrobeana* (Meissn.) Benth	LTR	40	A3A3	Australia
G. microphylla (Benth.) Tind.	MIC	40	BB	Australia
G. *peratosa* B. Pfeil & Tind.	PER	40	?	Australia
G. *pindanica* Tind. & Craven	PIN	40	H2H2	Australia
G. *pullenii* B. Pfeil, Tind. & Craven	PUL	40	?	Australia
G. rubiginosa Tind. & B. Pfeil	RUB	40	?	Australia
G. *stenophita* B. Pfeil & Tind.	STE	40	B3B3	Australia
G. *tabaciana* (Labill.) Benth	TAB	40	B2B2	Australia
		80	Complex[b]	Australia, west, central, and southern Pacific Islands
G. tomentella Hayata	TOM	38	EE	Australia
	TOM	40	DD	Australia, Papua New Guinea
		78	Complex[c]	Australia, Papua New Guinea
		80	Complex[d]	Australia, Papua New Guinea
Subgenre *Soybeans* (Moench F. J. Herm.(annual)				
G. soja Sieb & Zucc.	SOJ	40	GG	China, Russia, Taiwan, Japan, Korea (wild soybeans)
G. max (L.) Merr.	MAX	40	GG	Cultivated (soybean)

Source: Hymowitz (2004)
[a]Genomically similar species carry the same symbol (letter)
[b]Allopolyploids (genomes A and B) and segmental allopolyploids
[c]Allopolyploids (D and E, A and E, or other unknown combination)
[d]Allopolyploids (genomes A and D or other unknown combination)

Prior to 1949, no effort was made to preserve soybean germplasm in the United States, and many domestic introductions and varieties were wasted. The Soybean Germplasm Collection was established in 1949 to collect and maintain the entire soybean variability in the world, with an emphasis on East Asian landraces, where soybeans originated. When the collection was established in 1949, all available lines in the USDA and Canada were pooled and, thus, the collection started. Each accession entered into the USDA system is identified with the prefix PI and a number from the Plant Introduction Office. Most of the collection refers to the introduction of plants (16,981 accessions). Accessions of the US Soybean Germplasm Bank, after being characterized, evaluated, and multiplied, are made available to the scientific community and can be requested through the website [http://www.ars-grin.gov/npgs/orders.html]. All accesses of *Glycine max* are kept as pure lines. Each access in the collection is descended from a single seed of the original seed lot, and multiple accesses are preserved from samples of heterogeneous introductions. The only genetic variation that can exist within an access is a result of the heterogeneity of the original seed. As *Glycine max* is autogamous, accession within the collection can be considered homogeneous and homozygous. To maintain such integrity, extensive numbers of descriptors have been used to allow most contaminants to be easily detected and removed.

In addition to these germplasms, other institutions working in soybean breeding maintain smaller, but genetically important, collections to develop their own programs. Some of the collections in South America are also large, but they have been established more recently. Records indicate that accessions that belong to these collections derive, virtually exclusively, from other collections. European collections are smaller but may be genetically important because they contain landraces or their derivatives that were introduced from Asia over 100 years ago.

Accessions that precede scientific breeding are probably the greatest resource of genetic diversity in *G. max* collections and are good measure of the effective size of the collection. Not all *G. max* germplasm currently preserved in Eastern Asian countries precedes scientific breeding, but it is estimated that approximately 40,000 of the 93,000 soybean accessions in Asia may fall into this category. Accessions falling within this description originated in Eastern Asia, in the region that comprises India in the West, Japan in the East, Indonesia in the South, and Russia in the North. In the center of this region is China, the most important source of accessions.

In Brazil, the Soybean Active Germplasm Bank (BAG/Soja) was established in September 1975, at Embrapa Soja, with the purpose of gathering pure and characterized genotype stocks for the use of breeders and geneticists, in order to incorporate all the germplasm in the country and for acquisition of new accessions of agronomic importance. Initially, the bank received genotypes from UEPAE/Pelotas, FECOTRIGO, and IPAGRO, totaling 1200 new accessions, which were multiplied and characterized in the 1975/1976 agricultural season. In 1976, genotypes from the IAPAR collection and from the collection of UEPAE, in Ponta Grossa, were incorporated into the BAG/Soja. In 1977, cultivars from Delta Branch Experimentation Station, Stoneville, Mississippi, were introduced in Brazil. In 1979, 118 genotypes were collected from the UEPAE of Pelotas, which constitute the collection of genetic types. In 1980, at least 28 genotypes were introduced from

Table 3.2 Species and number of soybean accessions maintained in Embrapa Soja Germplasm Bank, Londrina, Paraná

Subcollection of *Glycine max*	21,404 accessions
Subcollection of *Glycine soja*	1,116 accessions
Subcollection of perennial species	919 accessions
Subcollection of domestic cultivars	
Old cultivars	208 accessions
Modern cultivars	509 accessions
Private cultivars	60 accessions
Subcollection of genetic stocks	
Contour lines	640 accessions
Genetic type	195 accessions
Improved lines	181 accessions

the US Regional Soybean Laboratory, Urbana, Illinois. Eight other genotypes were introduced in Brazil in 1981, derived from IITA Nigeria. In 2007, the process of importing the soybean collection from the US Germplasm Bank began, with 20,614 accessions. The vast majority of these accessions were introduced mainly from China, Japan, and other countries where diversification of species had occurred. These accessions are identified by the acronym PI, initials for plant introduction. In addition to the PI accessions, the BAG/Soja germplasm collection also has public domain lines and cultivars developed in Brazil and the North American collection of genetic types (identified by the "T") of all genes already studied and described.

Currently, Embrapa has two germplasm collections, totaling more than 25,200 soybean accessions, consisted mainly of plant introductions (PIs), national improved lines, and cultivars adapted to tropical and subtropical areas (Table 3.2). An active collection—Active Germplasm Bank—is maintained for short and medium periods at Embrapa Soja (Londrina, PR), where morphological and agronomic characterization, evaluation, and multiplication are also carried out, as well as the activities of regeneration and increase of stocks and the accessions are stored in cold rooms to maintain a temperature of 5 °C and 25% RH. The second, the soybean base collection (CBS), is maintained at Embrapa Recursos Genéticos e Biotecnologia—CENARGEN (Brasília, DF) under long-term storage conditions. The introduction, exchange, and quarantine of postentry are carried out exclusively by CENARGEN, in the case of material coming from outside or from regions not allowed in the country. The base collection is stored in aluminized packages, at a temperature of −18 °C.

Expansion to the Western World: Soybeans in Brazil

Soybean was introduced in the western world as of the eighteenth century when, in 1739, soybean seeds from the Paris Botanical Garden were experimentally planted in Europe. In the American continent, its introduction occurred in 1765 by Samuel Bowen in Savannah, Georgia, and the earliest account of its behavior dates back to

1804. Despite being known and intensely exploited in Eastern diet, for over 5000 years, the West ignored its cultivation until the second decade of the twentieth century, when the United States began its commercial exploitation—first as forage and then as grain. In 1940, at the peak of soybean cultivation as forage, about two million hectares were cultivated in that country for that purpose. Beginning in 1941, the planted grain area surpassed that cultivated for forage, which declined fast until it disappeared in the mid-1960s, while the area planted to grain production grew exponentially, not only in the United States but also in Brazil and Argentina, especially after the 1970s (Bonato and Bonato 1987).

Bonato and Bonato (1987) presented a complete review of the introduction and of the first experiments with soybeans in Brazil. The first experimental soybean planting reference in Brazil dates back to 1882, when cultivars were introduced and tested by Gustavo D'Utra in Bahia. The germplasm was brought from the United States and was not adapted to the low latitude conditions of that state (latitude 12°S), so, the crop was unsuccessful in the region. Ten years later the first studies were carried out in São Paulo and, in 1901, in Rio Grande do Sul (RS).

Similar to what happened in the United States, during the 1920s and 1940s, the first soybean cultivars introduced in Brazil were studied mostly intended for the evaluation of their performance as forage than as grain-producing plants for the meal and vegetable oil industries. However, commercial farming began much later. The first official statistics began to be reported in 1941 in RS, and the first soybean processing industry was set up, in 1945, in the State of São Paulo. The real drive to large-scale production in Brazil occurred in the mid-1950s, with the official decision to provide tax incentives for wheat production, which also benefited soybeans due to the perfect combination of the two crops, considering a technical viewpoint such as, for instance, the economic one.

In the 1960s, the crop initially expanded in latitudes between 30° S and 20° S, mainly in the States of Rio Grande do Sul, Santa Catarina, Paraná, and São Paulo. At the same time, about 200 North American lines were introduced and began to be tested in Ponta Grossa, Londrina, and Maringá, cities located in the State of Paraná. From this introduction originated the recommendations of cultivars Viçoja and Mineira, and the launch of Campos Gerais, Paraná, Florida and Sant'Ana. Moreover, it allowed the recommendation of American cultivars Bragg, Davis, Hardee, Hill, and Hood and of the Brazilian cultivars Santa Rosa and Industrial (Kaster et al. 1981). Given the identification of the cultivars with better adaptation, breeders began to combine the traits of these cultivars through hybridization.

Despite the significant production growth during the 1960s, it was in the following decade that soybean production grew further and was consolidated as the main crop of Brazilian agribusiness, having gone from 1.5 million tons (1970) to more than 15 million tons (1979). Such growth was due not only to the increase in cultivated area (1.3–8.8 million hectares) but also to the significant increase in yield (1140–1730 kg/ha). More than 80% of Brazilian soybean production in the late 1970s was still concentrated in the three southern states, but the central region of Brazil already showed signs of becoming a major producer. Until 1970, commercial soybean crops worldwide were restricted to temperate and subtropical climates whose latitudes were close to or above 30°. Its expansion toward the low latitudes of Brazil was limited in those days, due to the response of the crop to the photoperiod, since it is classified among short-day plants, as it will delay flowering until the day length is shorter than a critical photoperiod, which is specific for each cultivar. In the 1970s, plant size, dependent on the varietal response to the photoperiod, became the main limiting factor in the expansion of soybean farming in Brazilian tropics (Bonetti 1981).

Only after the 1970s was it possible to obtain genotypes featuring the late-blooming trait in lower latitudes, as a result of hybridizations between cultivars adapted to high-latitude regions and sources of genes for late flowering under short-day conditions. These genes known as genes for long juvenile period were present in genotype PI 240664 from the Philippines. With the breaking of the too early flowering barrier, Brazilian breeding programs developed germplasm adapted to the tropical conditions and enabled soybean cultivation in any region of the national territory (Kiihl et al. 1985).

In 1970, less than 2% of the national soybean production was harvested in regions with latitudes below 20°S. In 1980, such percentage rose to 20%, in 1990 it already exceeded 40%, and in 2004 it reached 64%. Such transformation upheld and consolidated the State of Mato Grosso as the national leader in soybean production.

Final Considerations

The inter- and intraspecific genetic variability of the genus *Glycine* conserved in germplasm banks in Brazil and in the world is essential to be used in research in general, especially in breeding programs.

References

Allard, R.W.: Princípios do melhoramento genético de plantas, p. 331. USAID, Rio de Janeiro (1971)

Bonato, E.R., Bonato, A.L.V.: A soja no Brasil: história e estatística, p. 61. EMBRAPA-CNPSo, Londrina (1987). (EMBRAPA-CNPSo. Documentos, 21)

Bonetti, L.P.: Distribuição da soja no mundo. In: Miyasaka, S., Medina, J.C. (eds.) A Soja no Brasil, pp. 1–6. ITAL, Campinas (1981)

Carter, T.E., Nelson, R.L., Sneller, C.H., Cui, Z.: Genetic diversity in soybean. In: Boerma, H.R., Specht, J.E. (eds.) Soybeans: Improvement, Production and Uses, 3rd edn, pp. 303–416. ASA, Madison, (2004). (Agronomy Monograph 16)

Chen, Y., Nelson, R.L.: Genetic variation and relationships among cultivated, wild, and semiwild soybean. Crop Sci. **44**, 316–325 (2004)

Frankel, O.H., Brown, A.H.D.: Plant genetic resources today: a critical appraisal. In: Holden, J.H. W., Williams, J.T. (eds.) Crop Genetic Resources: Conservation and Evaluation, pp. 249–257. G. Allen, London (1984)

Gai, J., Guo, W.: History of Maodou production in China. In: Proceedings of Second International Vegetable Soybean Conference, pp. 41–47. Washington State University, Tacoma (2001)
Gai, J., Xu, D., Gao, Z., Zhao, T., Shimamoto, Y., Abe, J., Fukushi, H., Kitajima, S.: Genetic diversity of annual species of soybeans and their evolutionary relationship in China. In: Proceedings of the World Soybean Research Conference VI, p. 515. University of Illinois/ Soybean Research & Development Council, Chicago (1999)
Griffin, J.D., Palmer, R.G.: Variability of thirteen isozyme loci in the USDA soybean germplasm collections. Crop Sci. **35**, 897–904 (1995)
van Hintum, T.J.L., Brown, A.H.D., Spillane, C., Hodgkin, T.: Core Collection of Plant Genetic Resources, p. 48. IPGRI, Rome (2000). (IPGRI. Technical Bulletin, 3)
Hymowitz, T.: Speciation and cytogenetics. In: Boerma, H.R., Specht, J.E. (eds.) Soybeans: Improvement, Production and Uses, 3rd edn, pp. 97–136. ASA, Madison (2004). (Agronomy Monograph 16)
Kaster, M., Queiroz, E.F., Teresawa, F.: Introdução da soja nas principais regiões agrícolas. In: Miyasaka, S., Medina, J.C. (eds.) A soja no Brasil, pp. 22–24. ITAL, Campinas (1981)
Kiihl, R.A.S., Bays, I.A., Almeida, L.A.: Soybean breeding for the Brazilian tropics. In: Proceedings of the Symposium [On] Soybean in Tropical and Subtropical Cropping Systems, 1983, pp. 141–143. AVRDC, Shanhua (1985)
Li, Z., Nelson, R.L.: Genetic diversity among soybean accessions form three countries measured by RAPDs. Crop Sci. **41**, 1337–1347 (2001)
Lu, S.L.: The origin of cultivated soybean (*G. max*). In: Wang, J.L. (ed.) Soybean. Shanxi People's Press, Shanxi (1977)
Nass, L.L.: Capítulo 2: Utilização de recursos genéticos vegetais no melhoramento. In: Nass, L.L., Valois, A.C.C., de Melo, I.S., Valadares-Inglis, M.C. (eds.) Recursos Genéticos e Melhoramento-Plantas, pp. 29–56. Fundação MT, Rondonópolis (2001)
Singh, R.J., Hymowitz, T.: Soybean genetic resources and crop improvement. Genome. **42**, 605–616 (1999)
Singh, R.J., Kim, H.H., Hymowitz, T.: Distribution of rDNA loci in the genus *Glycine* Willd. Theor. Appl. Genet. **103**, 212–218 (2001)
Skvortzow, B.V.: The soybean-wild and cultivated in Eastern Asia. Proc. Manchurian Res. Soc. Pub. Ser. A Nat. History Sec. **22**, 1–8 (1927)
Zhou, X., Peng, Y., Wang, G., Chang, R.: Studies on the center of genetics diversity and origin of cultivated soybean. In: Proceedings of the Sixth World Soybean Research Conference, p. 510. University of Illinois/Soybean Research & Development Council, Chicago (1999)

Chapter 4
Vegetative and Reproductive Morphology

Éder Matsuo, Silvana da Costa Ferreira, Aluízio Borém, and Tuneo Sediyama

Abstract Knowledge of the vegetative and reproductive morphology of soybean is important for cultivar development. The breeder should be able to distinguish each trait and understand the variation in the soybean germplasm before pursuing the goals in cultivar development. To efficiently select lines in a breeding program requires familiarity with the traits and the range of available variation. In this chapter the root system, stem, leaves, flower, fruit, and seeds of the soybean plant with breeding purpose in mind are addressed.

Keywords *Glycine max* • Phenotypic traits • Flowering • Root system • Leaves

Introduction

The production of soybean crops (*Glycine max* (L.) Merr.) is estimated to be 100.9 million tons of beans for the 2015/2016 harvest in Brazil, which will keep the country in second place in the ranking of the largest producers in the world of this oilseed (CONAB 2015). The United States Department of Agriculture has stated that there is an increase in North American production of soybean, which is estimated to be 107.1 million tons (CONAB 2015). However, regarding the productivity (kg/ha) of soybean cultivation, Brazil has been leader in the productivity ranking for some years, having obtained 2960 kg/ha in the last harvest, as opposed to 2660 kg/ha and 2280 kg/ha, respectively, that were obtained by the United States of America and Argentina (Bezerra et al. 2015).

The considerable increase in production and productivity of beans is the result of the seriousness with which soybean improvement programs are conducted, of rural businessmen and of the experts in the various areas relating to technology and crop production (Matsuo et al. 2015c). In this context, knowledge of the vegetative and

É. Matsuo, M.S., D.S. (✉)
Federal University of Viçosa - Campus Rio Paranaíba, Rio Paranaíba, Brazil
e-mail: edermatsuo@ufv.br

S. da Costa Ferreira, M.S., D.S. • A. Borém, M.S., Ph.D. • T. Sediyama, M.S., Ph.D.
Federal University of Viçosa, Viçosa, Minas Gerais 36570-900, Brazil
e-mail: silvanacferreira@ufv.br; borem@ufv.br; tuneo@ufv.br

F. Lopes da Silva et al. (eds.), *Soybean Breeding*,
DOI 10.1007/978-3-319-57433-2_4

reproductive morphology of soybean plants is extremely important for achieving success with the crop, especially with regard to obtaining high productivity and profit for the farmers. What is more, knowing the morphology of the soybean plant is extremely important for carrying out the tests for distinguishableness, homogeneity, and stability (DHE) of soybean crops. This is one of the documents that is needed so that new cultivar may be granted protection (Santos and Pacheco 2011). The instructions for carrying out the distinguishableness, homogeneity, and stability trials on soybean crops are available in MAPA (2013a). Details regarding the DHE test and Registry for Protecting Cultivars can be found in Chap. 23.

Bearing this in mind, in this chapter the root and stem systems, leaves, flowers, fruit, and seeds of the soybean plant will be described with relation to their morphology, with the focus on breeding. For those who wish to increase their knowledge regarding the vegetative and reproductive morphology of soybean plants, the works of Müller (1981), Lersten and Carlson (2004), Nogueira et al. (2009) and Matsuo et al. (2015d) are suggested.

Vegetative Morphology

Root System

This is made up of the main root and secondary roots (Sediyama et al. 1985). The lateral roots originate in an endogenous manner, out of tissues in the central cylinder, following the typical pattern of eudicots. In the axillary root, the branches appear a few centimeters away from the root apex, and in the lateral ones and the upper ones, the branches develop nearer the apex. As a rule, the new root is smaller in diameter than the one it originated from (Müller 1981; Lersten and Carlson 2004).

The development of the root system starts at germination and lasts until the plant reaches physiological maturity (Müller 1981). The growth and development of the roots is divided into phases 1, 2, and 3, and they take place simultaneously during the vegetative growth of the plant (Mitchell and Russel 1971). Phase 1: The rootlet is transformed into the axillary root, which can reach to depths of 45–60 cm at an approximate rate of 2.5–5.0 cm per day (This phase takes place during vegetative growth); Phase 2: The axillary root continues to grow and is able to reach a depth of 75 cm, and lateral roots are formed along the first 25 cm of the axillary root. At the end of this phase, 4–6 secondary roots normally stand out from the others because of their diameter and greater development. These spread out horizontally or slightly slanted and can reach up to 75 cm away from the axillary root. Later, their geotropic behavior suddenly changes, and they plunge the ends of the roots vertically into the ground and these develop vigorously with positive geotropism (This phase takes place during heightened development of the aboveground part of the plant when it is beginning to bloom and the pods are forming); Phase 3 is characterized by the slowing down of the growth of the axillary root, and there is greater development and penetration of the main secondary roots. In soil that has good permeability and drainage, the axillary root can reach depths of 150 cm, and the secondary ones reach up to 180 cm (This phase takes place when the pods are forming, the beans are filling out, and the plant is maturing) (Mitchell and Russel 1971).

The depths to which the root system can reach is more important, in order to make the most of the water that is available in the soil, than its horizontal development, as the plant is capable of using large quantities of water from the subsoil (Müller 1981). Besides compaction, the temperature, humidity, aeration, and availability of nutrients can inhibit root growth (Camargo and Alleoni 1997), and the genetic material (the cultivar) can also influence root development (Müller 1981; Sediyama et al. 1985). When analyzing the root system in terms of soil compaction, in the direct planting system, it was identified by Cardoso et al. (2006) that compacted soils lessen the ability of the soybean root system to explore and that compacted layers in the soil profile, when there is no water restriction, do not affect the productivity of the beans or dry material to accumulate.

In the production system, it is interesting to have available cultivars that feature root systems that develop well under stressful conditions (biotic or abiotic) and/or that demonstrate efficiency when using resources. This is because it is desirable for the producer who cultivates genetic material to show high profitability, even under conditions that are limiting for the crop. Details regarding breeding focused on stress tolerance or efficiency in the use of resources can be found in Fritche-Neto et al. (2011).

It is important to point out the biological fixing of nitrogen that occurs on the nodules formed on the roots of the plants (Finoto et al. 2009), with emphasis on the nodule development temperature and the fixing of nitrogen (Barros and Sediyama 2009). The roots can be the targets of several pathogens, such as: cyst nematodes (*Heterodera glycines* Ichinohe), root-knot nematode (*Meloidogyne incognita* (Kofoid and White) Chitwood and *M. javanica* (Treub) Chitwood), root lesions (*Pratylenchus brachyurus* (Godfrey) Filipjev and Schuurmans Stekhoven) and reniform (*Rotylenchulus reniformis* Linford and Oliveira) (Matsuo et al. 2015a), charcoal rot (*Macrophomina phaseolina* (Tassi) Goid.), phytophthora root rot (*Phytophthora sojae* Kaufm. and Gerd.), red root rot or sudden death syndrome (*Fusarium tucumaniae*), and collapse and wilting from sclerotic (*Sclerotium rolfsii*) (Matsuo et al. 2015b).

Stem System

The stem is of the herbaceous, straight, pubescent, branched type (Sediyama et al. 1985). It starts to develop from the axis of the embryo after the seed has germinated (Müller 1981). The stem starts growing during the development of the epicotyl, which results in a gradual separation from the cotyledon node and from the feathery part together with the beginnings of the primary leaves. After the epicotyl, the internodes are formed in the direction of the apex of the plant, and, on each node, a leaf develops, and, in their axis, there is a lateral bud that can become first-order branches, as well as inflorescence (Müller 1981). In most cultivars, the stem has orthotropic growth. However, when the plants are cultivated in low light, and especially in long photoperiods, they become voluminous and have a slim stem that can grow more than three meters in length; they need to be guided to grow vertically (Nogueira et al. 2009).

The slant of the lateral branches—the branching—is what determines how the growth pattern is classified (Sediyama et al. 2015). For this trait, the plants can be classified as straight, semi-straight, or prostrate (MAPA 2013a). When the slant of the lateral stem is less than 30° with relation to the main stem, the plant is classified as straight to semi-straight; when it is 30°–60°, the plant is called semi-straight to horizontal (prostrate); and when it is more than 60°, it is called horizontal (prostrate) (Sediyama et al. 2015). In the breeding programs, it is preferable to develop cultivars that have straight or semi-straight growing patterns, as light and herbicides penetrate the canopy of the plants better, and this achieves improved efficiency and less bean loss during mechanized harvesting (Sediyama et al. 2015).

Soybean plants can also be characterized with reference to the type of growth, which defines the end limit of the stem's development (Müller 1981), and it can be determined, semi-determined, or indeterminate (MAPA 2013a). The phenotype variations are set out in Sediyama et al. (2005) in the following manner: The cultivars of the determined growth type are characterized by the presence of axillary and terminal raceme inflorescence, and, soon after the first flowers have appeared, the growth of the plant practically ceases (in some cases it can still grow approximately 10% of its final height); cultivars that present semi-determined growth type also show axillary and terminal raceme inflorescence, but after florescence has started, the plant can still grow about 30% of its final height; and in the cultivars of the indeterminate growth type, the plant only shows axillary inflorescence, without any terminal inflorescence. After florescence has started, it can double in height. Sediyama et al. (2005) also describe that with the determined growth cultivars and the semi-determined ones, the first flowers appear in the upper mid third and the pods tend to mature from the top downward. This fact has not been observed in cultivars with indeterminate growth, which start to mature in the lower third of the plant. Cultivars with indeterminate growth are mainly grown in the more southern regions of Brazil or, when a very early cultivar variety is intended, in the central region of Brazil. Brazilian cultivars are predominantly classified as determined and indeterminate, but rarely as semi-determined, because the indeterminate ones are easily recognizable, but it is difficult to distinguish between the determined and semi-determined ones, given the similarity between the plants that have this type of growth pattern (Nogueira et al. 2015).

The presence or lack of autocyne pigmentation on the hypocotyl (Fig. 4.1) and the intensity thereof (weak, medium, or strong) are essential factors for describing varieties of soybean (MAPA 2013a). Lack of pigmentation is identified phenotypically by the color green, while the presence of pigmentation is identified by purple or bronze coloration. Cultivars that have bronze colored hypocotyls should have white flowers and brown or light brown stem pubescence (MAPA 2013a).

Fig. 4.1 Autocyne pigmentation of the hypocotyl of soybean plants. (**a**) Lack; (**b**) Presence. Photo: Éder Matsuo

Autocyne pigmentation of the hypocotyl plays an important role in soybean artificial hydration (Matsuo et al. 2015c).

Scientific studies have reported the genetic variability between cultivars for various phenotypical features, among which are the length of the hypocotyls (Costa et al. 1999; Nogueira et al. 2008). Following these studies, others like those done by Matsuo et al. (2012a, b), Silva (2013) and Oliveira (2014) analyzed, among several traits of the soybean plant, the length of the hypocotyl (distance between the soil and the cotyledonal node) and of the epicotyl (distance between the cotyledonal node and the insertion node of unifoliolated leaves).

Leaves

During their development, soybean plants can produce four different types of leaves: cotyledon, unifoliate, trifoliate, and the sessile leaves (Müller 1981; Lersten and Carlson 2004).

The cotyledons, when totally stretched out, have an elliptical oval shape, contribute with reserves toward the development of the seedling, and remain on the plant until their reserves have been used up, at which point they become yellow, wither, and fall off (Müller 1981; Lersten and Carlson 2004).

The unifoliate leaves occur in the opposite way, precisely at the first node above the cotyledon node (Müller 1981; Lersten and Carlson 2004). They have a wide or narrow base; are auriculate, truncate, or lanceolate in shape (Fig. 4.2); and have an obtuse to acuminate apex (Brandão 1961 *Apud* Nogueira et al. 2009; Müller 1981). They usually have a pair of stipules at the base of each leaf, a pulvinus at the insertion point of the leaf, a petiole that is 1–2 cm long (Müller 1981; Lersten and Carlson 2004; Nogueira et al. 2009), and a pulvinus at the base of the petiole (Müller 1981).

Fig. 4.2 Base of unifoliate leaf (**a**) Auriculate; (**b**) Truncate; and (**c**) Lanceolate. Photo: Éder Matsuo

The shape of the base of the unifoliate leaves (cuneate, truncate, and auriculate) is well defined, but there are certain variations among the same cultivar (Dorchester 1945). Brandão (1961) *apud* Nogueira et al. (2009) distinguished unifoliate leaves from 25 cultivars and reported that the shape of the base of the unifoliate leaf is a trait that shows variability among the cultivars within the same cultivar and even in the same plant. Nogueira (2007) and Silva (2013) found results similar to those of Brandão (1961) *apud* Nogueira et al. (2009). Gonçalves et al. (2015) evaluated a series of soybean cultivars with regard to the shape of the base of the unifoliate leaf, aiming to select homozygous plants. The authors managed to have—after a cycle of self-fertilization of plants that had been previously selected for double auriculate, double truncate, and double cuneate—the opportunity to select plants from the BG 4177, BRSGO 7560, and TMG 1176 RR cultivars, respectively, such as double-based cuneate, auriculate, and truncate of unifoliate leaves, in other words, without any variation within the same cultivar. However, further studies must be carried out, the aim of which is to add to the knowledge on the subject.

The trifoliate leaves develop above (subsequent nodes) the unifoliate leaves, show alternate phyllotaxy, and are composed of three leaflets (one at the end and two lateral ones), and they can be observed on the main stem and on the branches (Müller 1981; Sediyama et al. 1985; Lersten and Carlson 2004). The lateral leaflets can be classified thus as narrow lanceolate, lanceolate, triangular, pointed-oval, or rounded-oval, and there are three sizes—small, medium, and large (MAPA 2013a).

There is a positive correlation between cultivars with lanceolate leaves and a larger number of seeds per pod, and in recent years there has been a tendency to develop cultivars with lanceolate leaves and other phenotypical traits, the aim of which is to improve efficiency when dealing with the various pathogens, and therefore lower the number of applications of agrochemicals (Sediyama et al. 2015). What is more, they report that, despite the correlation stated above, the most important factors for bean productivity are the number of pods per plant and the total weight of seeds per plant.

The sessile leaves are found at the base of the branches and do not have any pulvinus or petiole (Carlson 1973).

Reproductive Morphology

Flower

The development of the flower starts with physiological induction; it is induced only when the plant is exposed to conditions in which the number of hours of light (photoperiod) is lower than or equal to a maximum critical photoperiod, since the plant is able to perceive the variation in the length of the day; this means that it has passed the juvenile period (Barros and Sediyama 2009).

Soybean flowers develop in axillary or terminal, from 2 to 35 in number per inflorescence, and they are white or purple in color (Fig. 4.3); inside them is the apocarp, formed of a single pistil that features a short curved hairless stylus ending in a headlike stigma that is bifid, globose, plumose, and covered in papillae, a superior ovary that is sessile, oval, and pubescent and contains one to five campilidrome ovules and the androecium, formed of ten diadem stamens. Protecting the reproductive structures, there is a perianth made up of the corolla (with five petals) and the cup (tubular and composed of five sepals or unequal lobes, partially connected and persistent), and this is considered to be a complete flower (Müller 1981; Sediyama et al. 1981, 2005; Nogueira et al. 2009).

The studies involving flower morphology are important both for the production system and for breeding. In the production system, they are useful for identifying atypical plants in seed-producing fields, and in breeding they are useful for developing segregant populations by way of artificial hybridization and when registering cultivars at the Ministry of Agriculture, Livestock, and Supply.

In the context of artificial hybridization, specifically in the hybridization process, the features of the plant and flower are extremely important, such as the colors of the hypocotyl (green or purple), of the flower (white or purple), and of the pubescence (gray or brown) and how tolerant they are to glyphosate, as this enables hybrid plants to be identified (Matsuo et al. 2015c). Furthermore, it is necessary to be aware of how sensitive the plant is to photoperiod, its juvenility (long or short),

Fig. 4.3 Color of soybean flower. (**a**) *White*; and (**b**) *Purple*. Photo: Éder Matsuo

and the temperature conditions in the place where the hybrid block is installed (Barros and Sediyama 2009).

Important references on artificial hybridization are Fehr (1978), Sediyama et al. (1981), Lersten and Carlson (2004), Sediyama et al. (2005), Borém et al. (2009) and Matsuo et al. (2015c).

Fruit

The fruit of the soybean is a type of legume, commonly known as a pod (Fig. 4.4). Their development starts at the time of fertilization, and the pod is the result of a completely developed ovary (Müller 1981).

One pod can contain from one to five seeds. However, most of the cultivars have pods with two or three seeds in them (Sediyama et al. 1985). In some genetic material, it is possible to see pods with four to seven seeds in each. The number of seeds per pod can be increased through breeding techniques, mainly with relation to the choice of parent plants, and by conducting segregant populations (e.g., recurrent selection) (Matsuo et al. 2015c). Furthermore, excellent knowledge of cultivation and the capacity to identify superior plants can contribute toward increasing the

Fig. 4.4 Soybean pods. (**a**) Pod with three or four seeds developing; (**b**) Pod containing four dry seeds. Photo: Éder Matsuo

number of seeds per pod and, mainly, toward developing improved cultivars with respect to productivity in breeding programs.

The fruit, leaves, stem, stipules, petiole, flowers, and any other organs (except the cotyledons), in almost all the varieties, are covered with trycomas that are called pubescence (Singh et al. 1971). The color and density of the pubescence on soybean plants are also considered to be descriptors and must be classified in terms of color, gray, light brown, or mid brown (the colors should be observed by holding the stem against a white background in a well-lit place but not in direct sunlight and using a magnifying glass), and the density in terms of low, medium, or high (the density should be observed in the third mid-node from the apex of the plants) (MAPA 2013a).

Seed

The mature embryo is made up of an axis that contains two cotyledons, the epicotyl and the hypocotyl, which refer to the axil below the cotyledons when root formation occurs, and this makes a distinction between hypocotyl and root (Plambin 1963). The mature embryo, its nutritional reserves (endosperm), and its outer casing (integument) make up the seed. In soybeans there is a scar on the outside of the seed called a thread, which is the result of degeneration of the funiculus (Müller 1981).

The thread can be classified into different colors (gray, yellow, greenish yellow, green, light brown, mid brown, dark brown, or black) and the intensity of the shine on the seed into low, medium, or high (MAPA 2013a). Furthermore, the seed can be classified into four shapes: spherical, flattened spherical, elongated, and flattened (Randall and Peiying 1989, *apud* MAPA 2013a).

The outer casing (integument) of a soybean seed is permeable. That is why, a short time after being put into the soil, under favorable environmental conditions in terms of temperature, aeration, and water supply, the process of germination starts (Müller 1981). There are several factors that affect germination, and they can be divided into intrinsic ones (genotype, vitality and longevity, viability and maturity, physiological potential, and quality of health) and extrinsic ones (water, temperature, light, oxygen, soil condition, and planting depth) (Nogueira and Sediyama 2013).

Stages of Development

Standardization of the terminology used to describe the stages of soybean development is essential for reaching a better understanding between the professionals and the farmers who work with the crop (Nogueira et al. 2013). Furthermore, the possibility of studies being carried out in different places being compared and the ease of orientating and decision-making when handling all justify the use of scales for plant development (Nogueira et al. 2013). The stages of soybean development are comprised of two main phases—the vegetative and reproductive ones. The duration of each phase is genetically controlled and influenced by environmental conditions, and the stages of development are classified based on observing the leaves, flowers, and how the pods and seeds that are found on the nodes on the main stem of the plant develop (Nogueira et al. 2013). Thus, a correct understanding and application of knowledge about the vegetative and reproductive morphology of soybean plants is fundamental for using the Stages of Development for Cultivation Table correctly. There are currently two stages of development tables for soybean cultivation: the one by Fehr and Caviness (1977) and the other one by MAPA (2013b).

Final Considerations

Understanding vegetative and reproductive morphology contributes greatly toward soybean breeding, mainly when correctly describing the features of the plants.

Ability to recognize the variations is essential for selecting the plants that have desirable phenotype traits when aiming to improve productivity and integrated handling and to preserve the environment.

References

Barros, H.B., Sediyama, T.: Luz, umidade e temperatura. In: Sediyama, T. (ed.) Tecnologias de produção e usos da soja, pp. 17–27. Mecenas, Londrina (2009)

Bezerra, A.R.G., Sediyama, T., Borém, A., Soares, M.M.: Importância econômica. In: Sediyama, T., Silva, F., Borém, A. (eds.) Soja: do plantio à colheita, pp. 9–26. UFV, Viçosa (2015)

Borém, A., Almeida, L.A., Kiihl, R.A.S.: Hibridação em soja. In: Borém, A. (ed.) Hibridação artificial de plantas, 2nd edn, pp. 514–536. UFV, Viçosa (2009)

Camargo, O.A., Alleoni, L.R.F.: Compactação do solo e o desenvolvimento das plantas, p. 132. ESALQ, Piracicaba (1997)

Cardoso, E.G., Zotarelli, L., Piccinin, J.L., Torres, E., Saraiva, O.F., Guimarães, M.F.: Sistema radicular da soja em função da compactação do solo no sistema de plantio direto. Pesq. Agrop. Brasileira. **41**(3), 493–501 (2006)

Carlson, J.B.: Morphology. In: Caldwell, B.E. (ed.) Soybeans: Improvement, Production, and Uses, pp. 17–95. American Society of Agronomy Inc, Madison (1973)

Companhia Nacional de Abastecimento (CONAB): http://www.conab.gov.br/OlalaCMS/uploads/arquivos/15_10_16_10_52_19_safras_outu_2015.pdf (2015). Accessed 22 Oct 2015

Costa, J.A., Pires, J.L.F., Thomas, A.L., Alberton, M.: Comprimento e índice de expansão radial do hipocótilo de cultivares de soja. Ciência Rural. **29**, 609–612 (1999)

Fehr, W.R., Caviness, C.E.: Stages of Soybean Development, p. 12. Iowa State University of Science and Technology, Iowa (1977). (Special Report 80)

Fehr, W.R.: Breeding. In: Norman, A.G. (ed.) Soybean Physiology, Agronomy, and Utilization, pp. 119–155. Academic, New York (1978)

Finoto, E.L., Sediyama, T., Barros, H.B.: Fixação biológica de nitrogênio e inoculação com *Bradyrhizobium*. In: Sediyama, T. (ed.) Tecnologias de produção e usos da soja, pp. 59–69. Mecenas, Londrina (2009)

Fritche-Neto, R., Dovale, J.C., Cavatte, P.C.: Melhoramento para tolerância a estresse ou para eficiência no uso de recursos? In: Fritche-Neto, R., Borém, A. (eds.) Melhoramento de plantas para condições de estresse abióticos, pp. 29–38. UFV, Viçosa (2011)

Gonçalves, V.A.R., Matsuo, É., Gonçalvez, F.A.R., Alves, G.F., Nogueira, J.P.G., Ferreira, S.C.: Seleção de plantas homozigotas para forma da base de folhas unifolioladas de soja. In: Simpósio de Integração Acadêmica—Campus Rio Paranaíba. Anais, Rio Paranaíba (2015)

Lersten, N.R., Carlson, J.B.: Vegetative morphology. In: Boerma, H.R., Specht, J.E. (eds.) Soybeans: Improvement, Production and Uses, pp. 15–57. American Society of Agronomy, Inc., Crop Science Society of America, Inc., Soil Science Society of America, Inc., Madison (2004)

Matsuo, É., Sediyama, T., Cruz, C.D., Oliveira, R.T.C.: Análise da repetibilidade em alguns descritores morfológicos para soja. Ciência Rural. **42**, 189–196 (2012a)

Matsuo, É., Sediyama, T., Cruz, C.D., Oliveira, R.C.T., Cadore, L.R.: Estimates of the genetic parameters, optimum sample size and conversion of quantitative data in multiple categories for soybean genotypes. Acta Sci. Agron. **34**, 265–273 (2012b)

Matsuo, É., Sediyama, T., Dias, W.P., Glasenapp, J.S.: Resistência aos nematoides. In: Sediyama, T. (ed.) Melhoramento genético da soja, pp. 211–238. Mecenas, Londrina (2015a)

Matsuo, É., Sediyama, T., Glasenapp, J.S.: Resistência aos fungos. In: Sediyama, T. (ed.) Melhoramento genético da soja, pp. 149–180. Mecenas, Londrina (2015b)

Matsuo, É., Sediyama, T., Ferreira, S.C., Glasenapp, J.S.: Hibridação. In: Sediyama, T. (ed.) Melhoramento genético da soja, pp. 57–72. Mecenas, Londrina (2015c)

Matsuo, É., Ferreira, S.C., Sediyama, T.: Botânica e fenologia. In: Sediyama, T., Silva, F., Borém, A. (eds.) Soja: do plantio à colheita, pp. 27–53. UFV, Viçosa (2015d)

Ministério da Agricultura, Pecuária e Abastecimento (MAPA): http://www.agricultura.gov.br/arq_editor/file/vegetal/RegistroAutorizacoes/Formularios%20Prote%C3%A7%C3%A3o%20Cultivares/SOJA_FORMULARIO_NOVO_29OUT2009_P.doc (2013a) Accessed 23 Nov 2013

Ministério da Agricultura, Pecuária e Abastecimento (MAPA): http://www.agricultura.gov.br/arq_editor/file/vegetal/RegistroAutorizacoes/Formularios%20Prote%C3%A7%C3%A3o%20Cultivares/SOJA_ANEXO_I_EST%25C1DIOS_P.doc (2013b) Accessed 30 July 2013

Mitchell, R.L., Russel, W.J.: Root development and rooting patterns of soybeans (*Glycine max* (L.) Merrill) evaluated under field conditions. Agron. J. **63**, 313–316 (1971)

Müller, L.: Taxonomia e morfologia. In: Miyasaka, S., Medina, J.C. (eds.) A soja no Brasil, pp. 65–104. Instituto de Tecnologia de Alimentos, Campinas (1981)

Nogueira, A.P.O.: Análise discriminante na caracterização de novos descritores em soja [*Glycine max* (L.) Merrill]. p. 96. Dissertação de M.S., Mestrado em Genética e Melhoramento. Universidade Federal de Viçosa, Viçosa (2007)

Nogueira, A.P.O., Sediyama, T.: Desenvolvimento, morfologia e germinação da semente. In: Sediyama, T. (ed.) Tecnologia de produção de sementes de soja, pp. 45–72. Mecenas, Londrina (2013)

Nogueira, A.P.O., Sediyama, T., Cruz, C.D., Reis, M.S., Pereira, D.G., Jangarelli, M.: Novas características para diferenciação de cultivares de soja pela análise discriminante. Ciência Rural. **38**, 2427–2433 (2008)

Nogueira, A.P.O., Sediyama, T., Barros, H.B., Teixeira, R.C.: Morfologia, crescimento e desenvolvimento. In: Sediyama, T. (ed.) Tecnologias de produção e usos da soja, pp. 7–16. Mecenas, Londrina (2009)

Nogueira, A.P.O., Sediyama, T., Oliveira, R.C.T., Destro, D.: Estádios de desenvolvimento. In: Sediyama, T. (ed.) Tecnologia de produção de sementes de soja, pp. 15–44. Mecenas, Londrina (2013)

Nogueira, A.P.O., Sediyama, T., Oliveira, R.C.T., Glasenapp, J.S.: Caracteres qualitativos. In: Sediyama, T. (ed.) Melhoramento genético da soja, pp. 33–56. Mecenas, Londrina (2015)

Oliveira, S.C.: Repetibilidade de descritores adicionais em diferentes estádios de desenvolvimento da soja. p. 32. Monografia de graduação em Agronomia, Universidade Federal de Viçosa—*Campus* Rio Paranaíba, Rio Paranaíba (2014)

Pamplin, R.A.: The anatomical development of the ovule and seed in the soybean. Diss. Abst. **63**, 5128 (1963)

Santos, F.S., Pacheco, L.G.A.: Testes de DHE. In: Ministério da Agricultura, Pecuária e Abastecimento. Proteção de Cultivares no Brasil, pp. 161–167. MAPA/ACS, Brasília (2011)

Sediyama, T., Almeida, L.A., Miyasaka, S., Kiihl, R.A.S.: Genética e melhoramento. In: Miyasaka, S., Medina, J.C. (eds.) A soja no Brasil, pp. 209–226. Instituto de Tecnologia de Alimentos, Campinas (1981)

Sediyama, T., Pereira, M.G., Sediyama, C.S., Gomes, J.L.L.: Cultura da soja—Parte I, p. 96. Universitária da Universidade Federal de Viçosa, Viçosa (1985)

Sediyama, T., Teixeira, R.C., Reis, M.S.: Melhoramento da soja. In: Borém, A. (ed.) Melhoramento de espécies cultivadas, 2nd edn, pp. 551–603. UFV, Viçosa (2005)

Sediyama, T., Matsuo, É., Oliveira, R.C.T., Glasenapp, J.S.: Caracteres agronômicos de cultivares. In: Sediyama, T. (ed.) Melhoramento genético da soja, pp. 73–82. Mecenas, Londrina (2015)

Silva, F.C.S.: Influência do tamanho de sementes e de características agronômicas em descritores adicionais de soja. p. 46. Dissertação de M.S., Mestrado em Fitotecnia, Universidade Federal de Viçosa, Viçosa (2013)

Singh, B.B., Hardly, H.H., Bernard, R.L.: Morphology of pubescence in soybeans and its relationship to plant vigor. Crop Sci. **11**, 13–16 (1971)

Chapter 5
Qualitative Traits in Breeding

Francisco de Jesus Vernetti and Francisco de Jesus Vernetti Junior

Abstract Inheritance of qualitative characters has been found out since the beginning of the twentieth century. It plays an outstanding role for a soybean breeder, as a tool to confirm the effectiveness of the crosses made between/among selected germplasm cultivars/lines. Morphological characters of the plants, as root, stem, pubescence, leaf, and petiole, carefully watched may corroborate the result. As well, reproductive traits like flower, legumes, seed, and its hilum attentively observed bring similar outcomes. The feature of the watching are presence/absence, form, shape, type, position, length, number, and pigments. Refined technology allowed to evaluate several chemical substances that are present in the seed: proteases, flavonoids, proteins and enzymes, peroxidases, ureases, phytohemagglutinins, lecithin, lipoxygenase, and phytic acid. They vary in level, presence/absence, migration, mobility, and activity, mainly. Also, special cultivar development may be obtained with their end traits, as fatty acids palmitic, stearic, linolenic, linoleic, and oleic. A comprehensive list of genes has been published along the years by many authors in papers, reviews, and at least eight books. Inheritance of reaction nutrients, insects, diseases, and herbicides should have been presented in Chaps. 16, 17, and 18. Fertility and sterility is mentioned elsewhere in this book.

Keywords Soybean breeding • Morphological and reproductive traits • Chemical compounds • Gene lists • Reaction to nutrients • Herbicides • Diseases and insect reactions

Introduction

The qualitative traits of soybeans, like those of other species, are codified by one or a few genes whose phenotypic expression is marked and little influenced by the environment. They are distributed along the adult plant and roots to the maturity of the grains. The great majority plays an important role in the cultivar development process, not only as markers or indicators of homozygosis or heterozygosis of

F. de Jesus Vernetti, M.S. • F. de Jesus Vernetti Junior, M.S., D.S. (✉)
Embrapa Clima Temperado, Pelotas, Rio Grande do Sul, Brazil
e-mail: francisco.vernetti@embrapa.br

F. Lopes da Silva et al. (eds.), *Soybean Breeding*,
DOI 10.1007/978-3-319-57433-2_5

individual plants and rows of plants during selection but also due to the intrinsic relevance of each trait to the desired phenotype as the ultimate goal of planning.

Important morphological traits for soybean researchers include flowering and maturity, stem and petiole growth, plant height and dwarfness, leaf shape, and type of pubescence, among others. Significant physiological traits such as absorption of nutrient, flavonols, and isoflavones, chlorophyll deficiency, pigmentation, sterility, nodulation, and nitrogen fixation are also included. In the same way, the protein and oil content, fatty acids, phytate, and phosphatide concentration, for example, stand out in the composition of the seed.

Thus, traits linked to biotic and abiotic stresses are very important, such as resistance to diseases, pest insects, pathological viruses, nematodes, soil drench and drought, and tolerance to Al, Cl, Mn, and Fe toxicity.

Plant breeding is the art and science of improving plant heredity in relation to its economic utility (Fehr 1987); art is the practical part of the definition, such as that held by prehistoric farmers.

The improvement of qualitative traits is based on observation and selection of traits guided by Mendelian genetics, caused by one or a few genes with a marked effect, little affected by environmental factors. Discrete or discontinuous variation of the trait is obtained in progeny, whereas with quantitative traits the variation of trait progeny is continuous, being distributed according to the normal curve.

In almost all cases, the best genotypes are selected as progenitors. In specific situations, which require the inclusion of one or more traits in the resulting progeny, the progenitor is chosen for its gene or genes responsible for the desired trait.

As a rule, one should try to improve few traits at a time, not several at the same time, to achieve the goal proposed.

The soybean (*Glycine max* (L) Merrill) is a species of the Fabaceae family, domesticated from the wild species *Glycine soybean* Sieb. and Zucc., possibly with an intermediate species, *Glycine ussuriensis* Regel and Maack. Its genotype consists of $2n = 40$ chromosomes, compatible with the *Glycine soybean* and with a huge genetic diversity. It is an autogamous plant with less than 1% of allogamy.

More information about the center for diversity and genetic resources, as well as its vegetative and reproductive morphology, is detailed in Chaps. 3 and 4 of this book.

Root and Nodulation

The roots of soybean plants may exhibit fluorescence when subjected to ultraviolet radiation. This trait can be used in the separation of genotypes. Five *loci* are responsible for their expression: *Fr1 fr1*, *Fr2 fr2*, *Fr3 fr3*, *Fr4 fr4*, and *Fr5 fr5* (Fehr and Giese 1971; Dellaney and Palmer 1982; Sawada and Palmer 1987).

The roots of developing plants, in general, establish a symbiotic process with nitrifying bacteria of at least three genera and five species: *Bradyrhizobium japonicum*, *B. elkanii*, *B. lianingense*, *Sinorhizobium fredii*, and *Mesorhizobium tianshanense* (Kuykendall et al. 2000; Kuykendall 2005). The symbiosis involves the formation of nodules in the roots by the bacteria, which absorb N_2 from the atmosphere and transfer it to the plants. The N is a key element for nutrition and productivity of plants.

There is a set of genes that interfere in the process of symbiosis in various ways (Vernetti 1983; Pracht et al. 1993; Devine 1985):

rj1 = absence of nodulation or formation of few nodules
Rj2 = ineffective nodulation by strains *b7*, *b14*, and *b122*
Rj3 = ineffective nodulation by strain *33*
Rj4 = ineffective nodulation by strain *61*
Rj5 and *Rj6* = absence of nodulation
rj7 = supernodulation
Rfg1 = ineffective nodulation by strain 205 of *Sinorhizobium fredii*

The leghemoglobin protein imparts a pink color to the interior of the nodule; it is similar to the hemoglobin of animals. More than one leghemoglobin is found in nodules encoded by more than one gene: *Lba*, *Lbc1*, *Lbc2*, and *Lbc3* (Morrison et al. mentioned by Sadasivam and Krishnaveni 2002).

The selection and characterization of efficient nitrifying bacteria strains are important tasks for soybean breeding. Commercial inoculants are composed of mixtures of strains of bacteria selected for their efficiency, competitiveness, and/or aggressiveness.

Stem

The type of soybean stem growth is one of the most relevant qualitative traits in the breeding of the species. It has one of three different types of growth: determined, indeterminate, and semi-determined. And rarely, it shows one of two mutations: fasciated habit and miniature plant. In addition to these, it eventually displays changes to the length of the internode or branches of the side branches. All of them affect the architecture of the plants in different ways.

The three most common types (Bernard 1972) have the following characteristics:

1. Determined—dense distribution of legumes in the main stem, dispersion in side branches, and abrupt interruption of stem growth giving rise to flattened apex.
2. Indeterminate—sparse distribution of legumes on the main stem and branches and decreasing frequency of pods on the main stem upward.
3. Semi-determined—growth of the main stem interrupted almost as abruptly as in the determined type; the long stem is gradually thinner to the apex, but thicker and shorter than in the indeterminate, end inflorescence with five to ten legumes or more, as in the determinate.

The flowering, fruiting, and filling of grains are differentiated:

1. Determined—begins when the terminal node of the main stem completed the development; growth of pods and grains occur at the same time along the stem.
2. Indeterminate—begins when almost half of the main stem nodes have developed; therefore, vegetative and reproductive development is concomitant almost to maturity; pods and grains grow upward as the new nodes arise, but all grains reach maturity together.
3. Semi-determined—begins when half of the main stem nodes have developed as in the determined type; the growth of the main stem ends earlier in relation to what happens in the indeterminate type.

The genotypes and phenotypes are detailed below:

Dt1Dt1 dt2dt2 = indeterminate
Dt1Dt1 Dt2Dt2 = semi-determined
Dt1Dt1 Dt2dt2 = semi-determined (*Dt2* dominant on *dt2*)
Dt1dt1 dt2dt2 = semi-determined (effect of *Dt1 dt1*)
Dt1dt1 Dt2dt2 = semi-determined
Dt1dt1 dt2dt2 = semi-determined (effect of *Dt1 dt1*)
dt1dt1 Dt2Dt2 = determined (*dt1dt1* epistatic on *Dt2Dt2*)
dt1dt1 Dt2dt2 = determined (*dt1dt1* epistatic on *Dt2dt2*)
dt1dt1 dt2dt2 = determined

Other genes that affect the growth of the stem are *dt1-t* effect similar to that of *dt1*, but significantly slows the expression of determined habit (Thompson et al. 1997); *SS* reduces the height of plant in cultivars Harosoy and Clark, causing shortening of the internodes and slight reduction in the number of nodes; *ss* height of plant, internode length and number of nodes normal; $s^t s^t$ increases the length of the stem (size) of Clark, gradually extending the length of the internodes, but with negligible effect on the number of nodes (Bernard 1975d).

Brachytic or zigzag main stem is determined by two pairs of alleles (Kilen and Hartwig 1975; Kilen 1977; Boerma and Jones 1976, 1978):

Sb1 Sb2; *Sb1 sb2 sb2*; *sb1 sb1 Sb2* // normal stem
sb1 sb1 sb2 sb2 // brachytic stem, short internodes, and height of a plant much smaller than normal

The distribution of the branches along the main stem is modified by two pairs of alleles *Br1 br1* and *Br2 br2* (Nelson 1996). Its recessive combination *br1 br1 br2 br2* presents few ramifications at the lower nodes of the main stem, impairing the yield of the plants.

The type of growth of the soybean plants has outstanding importance in the breeding of the species. As with other plant species, the plant height has to some extent significant participation in the productivity of grain crops. On soybeans, for example, it is considered that it should have a minimum of 70 cm and a maximum of 1.20 m high and the absence of lodging, to achieve maximum productivity.

Several factors influence on the length of the biological cycle of the soybean: genotype, photoperiod, temperature (minimum, medium, and maximum), availability and distribution of rain precipitations, proper nutrition (primary, secondary chemical elements and microelements), nodulation, and plant population (between and within rows).

The height of plants, in turn, is directly related to the time of insertion of the pods close to the ground (which may bring losses at harvest if below the height of the cutting bar).

Numerous scenarios are presented when considering the type of growth facing genetic improvement. For example, if the biological cycle begins about 30 days after the spring equinox, through the summer solstice, and ends at the autumn equinox, the variation of the photoperiod and temperatures will have the average of occurrences in the corresponding latitude range. The cultivar adapted to this succession of climate subperiods will be the one to provide the best productivity under such conditions. If the pre-flowering period is short, it should be indeterminate to continue growing until the end of the cycle and reaching the plant height that guarantee good productivity. If it is a cultivar with a long juvenile period, it will grow sufficiently before flowering, and it may be of a determined type. Considering these two examples, we can imagine the presence of differentiated scenarios in one or several factors which, given due consideration in their effects on the growth and development of cultivars, will dictate the objective to reach at plant height (type of growth) in the breeding process.

Two mutations that rarely occur with this kind of growth are (a) miniature plant, which has small primary leaves, very short internodes, tiny curved and trifoliolate leaves, cleistogamic flowers, and sometimes a seed pod, and (b) fasciation of the stem, with normal growth beginning until the first trifoliolate leaves appear, when the meristem growth is divided in such a way that several stems undergo anastomoses and remain united; two or three trifoliolate leaves then appear per node, and at the stem apex, many floral clusters develop, and consequent compact clusters of pods.

Both traits have monogenic inheritance (Dellaney and Palmer 1984; Woodworth 1932, 1933):

$Mn-=$ normal plant // $mn\ mn$: miniature plant
$F-=$ normal stem // ff: stem with fasciation

Multiple recessive mutants producing small plants weak and usually twisted have been reported. A mutant was found in the Habaro cultivar, which produced fusiform, light-green leaves with few seeds (gene *df*). Later, the occurrence of three dwarf mutants was reported; designated *df2*, *df3*, and *df4*; and found in the Lincoln, Adams, and Hark cultivars (Porter and Weiss 1948; Byth and Weber 1969; Fehr 1972a). All are stunted and produce few seeds. A similar mutant was designated *pm* (pseudomosaic) and has very twisted, brittle, and wrinkled leaves and does not produce seeds (Probst 1950). Another dwarf mutant was designated *df5* by Palmer 1984, as nonallelic and unlike the others (T 263). A dwarf mutant was recovered in cultivar C 1421, whose seed was treated with the mutagenic ethyl methanesulfonate, which is inherited as a recessive monogenic. The symbol *df6* (T 286) was assigned (Werner et al. 1987). A trait controlled by a duplicate factor producing dwarfism was designated *df7 df8*, given by the Soybean Genetics Committee (1995).

Leaf and Petiole

The length of the petiole of the soybean leaf is controlled by one pair of alleles and the short petiole with the normal pulvinus by another pair of alleles (Kilen 1983; You et al. 1998). Thus:

Lps1 ___ = normal length of the petiole
lps1 lps1 = short petiole
Lps2 ___ = normal petiole and pulvinus
lps2 lps2 = short petiole and normal pulvinus

The leaves of most cultivars are of five kinds: cotyledonary, unifoliolate (primary), trifoliolate, sessile, and bracteous.

The inheritance of the number of leaflets, shape, color, and the abscission of the leaves was revealed by several authors. The number of leaflets is thus determined (Fehr 1972b):

Lf1 Lf1 = about 40% of pentafoliolate leaves (with two additional leaflets on the base of the side leaflets)
Lf1 lf1 = few multifoliolate leaves
lf1 lf1 = trifoliate leaves
Lf2 Lf2 = trifoliate leaves
Lf2 lf2 = few or no heptafoliolate leaves
lf2 fl2 = about 35% of heptafoliolate leaves
Lf1 = partially dominant
Lf2 = shows partial or complete dominance, according to the genome of its origin

The leaflets can be longer and narrower than the normal ones, which are oval or rhomboids (Bernard and Weiss 1973):

Ln Ln = normal, oval, or rhomboid leaflet
Ln ln = leaflet of intermediate form
ln ln = narrow leaflet (lanceolate) and high number of legumes with four seeds

A mutant, narrow, and wrinkled leaf was found (Wilcox and Abney 1991):

Lnr __ = normal leaflets
lnr lnr = narrow, wrinkled leaflets

Another form of leaflet studied was the oval one (Johnson and Bernard 1963; Weiss 1970):

Lo _ = normal, oval leaflet
lo lo = oval leaflet and large number of pods with a seed

The crossbreeding between the latter (*lo lo*) and the narrow leaflet (*ln ln*) resulted in a genotype, *ln ln lo lo*, with extremely narrow leaflets (Soybean Genetics Newsletter 1975).

Another mutation presents leaflets with corrugated edge. The inheritance is digenic and linked to the presence of the pair responsible for the color of the pubescence (Rode and Bernard 1975):

lw1 lw1 *lw2 lw2* (*t t*) = leaflet with corrugated edge, gray pubescence
lw lw1 *lw2 lw2* (*TT*) = leaflet with normal edge, brown pubescence
*Lw1*__ *lw2 lw2* (*T* or *t*) = leaflet with normal edge, gray or brown pubescence
lw1 lw1 *Lw2* – (*T* or *t*) = leaflet with normal edge, gray or brown pubescence
*Lw1*__ *Lw2* – (*T* or *t*) = leaflet with normal edge, gray or brown pubescence

One mutation presents a blistered leaf, with no virus attack, which was designated as vesicular (Singh et al. 1974; Rode and Bernard 1975):

lb1 lb1 *lb2 lb2* // vesicular leaf
*Lb1*__ *lb2 lb2*; *lb1 lb1* *Lb2*__ and *Lb1*__ *Lb2*__ // normal leaf

Plant senescence is a systematic degenerative process involving the activation of genes and downregulation of other genes. Unraveling the regulatory steps of senescence and the participating genes will enable strategies to improve yield and reduce postharvest losses.

Few cultivars retain the leaf for longer than normal (maturity), a desirable trait in those that will be used in the production of hay for animal feed.

Leaf retention is usually accompanied by stems and/or green branches, increasing the humidity of the harvested grain, making mechanized harvest and pod opening difficult. Its heritage is monogenic (Probst 1950):

*Ab*__ = normal abscission
ab ab = delayed abscission (usually in black-grain cultivars)

From the foregoing, we can see that the petiole and leaf do not have significant trait for breeding, only the priority of eliminating any harmful mutants described above and the genotypes that carry them in their genomes.

Pubescence: Presence, Arrangement, Density, and Shape

Soybean stems, petioles, leaves, calyxes, and pods are usually covered with hair or trichomes called pubescence. Eventually, there are lines that are devoid of hair, generally unproductive: they are glabrous (Hartwig and Edwards 1970; Singh et al. 1971). The inheritance of trait is monogenic (Nagai and Saito 1923):

P1 – = glabrous plants
p1 p1 = pubescent plants

There is another mutant showing miniature pubescence, whose inheritance is also monogenic (Stewart and Wentz 1926):

P2 – = pubescent plants
p2 p2 = plants with miniature hair, little height, and small productivity

P1 is epistatic on *P2* and *p1*, which in turn does not influence *p2* (Woodworth and Veatch 1929).

The arrangement of hair on the stem is in vertical columns very close to each other. In the leaves, they are more numerous in the veins.

The density of the hair is genetically controlled (Bernard and Singh 1969; Bernard 1975a; Gunashinge et al. 1988):

Pd1 _ = dense pubescence
pd1 pd1 = normal pubescence
Ps _ = sparse pubescence
$ps^s ps^s$ = semi-sparse pubescence
ps ps = normal pubescence
Pd2 _ = dense pubescence
pd2 pd2 = normal pubescence
Pd1 Pd1 Pd2 Pd2 = extra-dense pubescence

The 20 populations tested showed that the yields were slightly higher in those with dense gray pubescence (*tt*). Pubescence can be beneficial in environments with irregular precipitation, by reducing the transpiration of plants.

The pubescence of the soybean may be smooth, straight, or curly, as well as erect or with different degrees of slope (decumbence). Curly pubescence is decrepit or deciduous, that is, it detaches at plant maturity (Vernetti 1983).

The position of the hair on the leaf surface may be erect, almost vertical, slightly sloping toward the margins and apex of the leaflet, or decumbent, lying on the leaf blade.

The two traits have monogenic and digenic control, respectively (Bernard and Singh 1969; Bernard 1975b):

Pc Pc = normal pubescence
Pc pc = intermediate pubescence between normal and curly
pc pc = curly and deciduous pubescence
Pa1 Pa1 Pa2 _ = erect pubescence on the leaves
pa1 pa1 Pa2 Pa2 = semi-decumbent pubescence on leaves
pa1 pa1 pa2 pa2 = decumbent pubescence on leaves

Inflorescence, Flowering, and Maturity

The reproductive stage of soybean begins with the development of axillary sprouts with groups of 2 to 35 flowers each. In the plant of indeterminate habit, the inflorescences designated racemes are axillary. In those of determined habit, there are axillary racemes and a terminal. The raceme may be peduncular or subsessile, according to the genotype (Schaik and Probst 1958):

Se _ = pedunculate raceme
se se = subsessile raceme

Soybean is a very sensitive species to the photoperiod. Sensitivity varies according to the genotype, and the degree of response to this stimulus is the main determinant of the adaptation area of different cultivars. Along with this, the thermal regime encodes the reproductive process.

A series of genes with their own characteristics controls or influences the time of flowering (Owen 1927; Bernard 1971):

E1 E1 = slows the normal flowering and maturity
E1 e1 = slows less the normal flowering and maturity
e*1 e1* = normal flowering of cultivar Clark-e
E2 E2 = typical flowering and maturity of cultivar Clark-*E2*
E2 e2 = smaller anticipation of flowering and maturity
e2 e2 = anticipates flowering and maturity

The presence of *E1* and *e2* in the same genotype causes additive effect of two genes on the times of flowering and maturity (Bernard and Weiss 1973).

E3 E3 and *E3 e3* determine late flowering, other than that caused by *E2*, besides sensitivity to fluorescent light: Harosoy 63, Lee, Hill, PI 297550, for example. (Buzzell 1971; Kilen and Hartwig 1971; Buzzell and Voldeng 1980; Bonato and Vello 1999; Cober and Voldeng 2001; Cober et al. 1996).

e3 e3 anticipate the times of flowering and maturity in the field and causes insensitivity to fluorescent light: Dorman, Arksoy, and Blackhawk, for example.

E4 E4 = determines the long cycle and sensitivity to long days: Harcor
e4 e4 = determines short cycle and insensitivity to long days: PI 297550
E5 E5 = Harosoy late cycle
e5 e5 = Harosoy early cycle
JJ = normal juvenile period
jj = long juvenile period
E6 E6 = short juvenile period; precocity of Paraná, for example
e6 e6 = long juvenile period; SS1, Paranagoiana, for example
E7 E7 = slows flowering in Harosoy and causes sensitivity to light incandescent
e7 e7 = causes precocity (short cycle) in PI 196529

It was found that under cultivation, conditions *E1*, *E2*, or *E3* delay flowering and maturity; the pair *E3 e3* is the main cause of insensitivity to long days; *E3* and *E4* should be considered in breeding for the biological cycle trait; and *e3* and *e4* are valuable in obtaining a short cycle (precocity).

One of the most important events for the genetic improvement of soybean was the discovery of the qualitative trait heritage designated "long juvenile period" (Sinclair and Hinson 1992), responsible for obtaining cultivars adapted to low tropical latitudes. Hartwig and Kihil (1979) described the "delayed flowering under short-day conditions." The trait is controlled by a recessive gene *j* (Ray et al. 1995). Thanks to it, soybean cultivation has expanded from temperate regions to tropical regions.

The time of flowering and maturity and the time of sowing are very important traits for breeding. Despite the control attributed to few genes, the photoperiod and temperature mainly affect the biological cycle of the genotypes, generating continuous but not quantitative variation. It then requires careful attention to latitude and its consequences on photoperiod and temperature, as well as times of sowing and the genome involved (behavior under different conditions).

Pigmentation: Pubescence, Flower, Leaf, Legume, Seed, and Hilum

One of the most evident characteristics of almost all soybean plants is the presence of hair or trichomes on the stem, branches, petioles, leaves, and sepals. Few genotypes are devoid of hair (glabrous) or have miniature hair.

The colors of soybean pubescence can be brown or light brown (almost gray) or gray. They are indispensable traits to the production technique of pure seeds, which shall produce uniform plants regarding hair color of the corresponding cultivar. This characteristic must be observed in adult plants, that is, at full maturity, because pubescence is always gray in young plants.

The color of the pubescence was initially said to be controlled by a pair of alleles: *TT*, *tt* (Piper and Morse 1910; Woodworth 1921; Nagai 1921; Williams 1952; Bernard 1975c). Subsequently, two pairs of alleles proved to encode:

T T *Td Td* = dark-brown pubescence
T T *Td td* = brown pubescence, intermediate tone
T T *td td* = light-brown or almost gray pubescence (e.g., Grant, Sooty, Korean, Kingwa, cultivars)
tt tt *td td* = gray pubescence
tt tt *Td Td* = gray pubescence
tt tt *Td td* = gray pubescence

It was not possible to unravel the genetic base for the difference between light brown and almost gray; the trait is very variable.

The *T* gene encodes the flavonoid 3-hydroxylase; *tt* results from the deletion of a nitrogenous base from the chromosomal DNA of *T*. The *T* allele responds by the presence in the leaves of the flavonoid glycosides quercetin and kaempterol; *tt* determines the absence of both (Toda et al. 2002).

Soybean flowers are usually white or purple (violet). The purple shade varies with the genotype of the cultivar and is also present in the hypocotyl, possibly in the plumes. During maturation under intense sunlight, the purple color formed by anthocyanin appears on the walls of the pods, on the petioles, and on the stem. The synthesis and intensity of anthocyanin are very variable but never occur in white-flowered plants.

The white and purple colors were considered to be encoded by a pair of alleles *w1w1* and *W1W1*, respectively, which also determine the green hypocotyl and purple hypocotyl of the stem (pleiotropy). Later, other genes were assigned to the encoding of the purple tones (Hartwig and Hinson 1962):

W1 _ *W3* _ *W4* _ = intense purple
W1 _ *W3* _ *w4 w4* = diluted purple (pigment only at the base of the banner)
W1 _ *w3w3W4* _ = typical purple (the most common)
W1 _ *w3w3w4w4* = purple with very diluted or almost white coloration, showing only traces of color at the base of the banner

The presence of *w1 w1* with *W3* _ *W4* _, *W3* _ *w4 w4*, *w3 w3 W4* _, or *w3 w3 w4 w4* encodes the white color of the five petals of the flower. Therefore, *W1* is essential for the production of the intense purple shade in combination with *W3*_ *W4*_and of diluted shades in the absence of *W3*_ or *W4*_, and the recessive form *w1w1* determines lack of anthocyanins.

Other mutations involving the alleles *W*1 and *w*1 and a new allele *w* were described (Buzzell et al. 1974, 1977):

W1 _ *Wm* _ = purple flower, glycosides present
W1 _ *wm wm* = magenta flower, glycosides absent
w1 w1 Wm _ = white flower
w1 w1 wm wm = white flower

Magenta is a dark-red color; *Wm wm* is not allele of the other three *W1 w3*, *W3 w3*, and *W4 w4* but linked to *W1 w1* (2.2% of recombination); it requires the presence of *W1*_ to be expressed.

The *w4-m* mutable recessive gene frequently reverts to the dominant form. It produces nearly white flowers (quite diluted purple) and purple flowers with almost white sectors.

Another *wp* mutant is pink-flowered (Stephens and Nickell 1992). A new variegated mutant came from it: pink or purple flowers or pink with purple sectors (Johnson et al. 1998).

Bleached purple color is another flower color mutant, coded by *W4p* (pale) (Xu and Palmer 2005).

Finally, a new allele of *W1*, *w1-lp*, was identified in *G. soja*, which encodes the color light purple.

Consequently, the color trait of the soybean flower can be used in the technology of the production of genetic, registered, or certified seeds. However, it requires thorough knowledge of the genetics of the trait and careful analysis of the tones of purple and white color that the progenitors of the crossbreeding show.

The leaves of soybeans typically show green shades characteristic of the color of chlorophyll. However, there are mutants for the partial absence and for retention of chlorophyll.

The partial absence of chlorophyll in the leaves is called variegation. It appears in the first trifoliolate leaf or in the newly formed ones, and also in the fully developed leaves. It is controlled by pairs of alleles *V1v1 and V2v2* (Vernetti 1983):

V1V1 = normal green leaf
v1v1 = variegated, green and yellow leaf
V2V2 = normal green leaf
V2v2 = variegated, green and yellow leaf

There were already identified 32 nuclear genes causing chlorophyll deficiency. Some are unstable and produce chlorophyll chimera; others are almost lethal yellow recessive; of these, one originates an albino plant as it grows; others form yellow-green leaves; one forms yellow leaves and is lethal; other forms yellow-greenish leaves; and another produces leaves that gradually become yellowish white and chlorotic (Vernetti and Vernetti Júnior 2009).

Chlorophyll deficiency can also have cytoplasmic inheritance. The mutant *cit Y2* produces yellow leaves that turn to yellowish green, while *cit G2* originates green leaves. The *cit Y3* mutant emits yellow leaves on very weak plants, while *cit G3* generates normal green leaves. Another five mutants were identified: *cit Y4*, *cit I5*, *cit I6*, *cit Y7*, and *cit Y8*; however, *cit G4*, *cit G5*, *cit G6*, *cit G7*, and *cit G8* form normal green leaves.

The soybean legume or pod shows variable shape and color in different cultivars. It ranges from flattened to oval, almost cylindrical, according to the shape and size of the seed that it contains. Color varies from straw yellow to brown and even black.

It is necessary to emphasize that the color of the hair surrounding the pod can make the correct observation of its color difficult. Therefore, when in doubt, you should shave the pubescence on the surface of the pod so that you can see its real color.

The genetic control of the legume color is made by the interaction of two alleles (Piper and Morse 1923; Woodworth 1923; Bernard 1967):

L1 _ *L2* _ = black-colored legume
L1 _ *l2 l2* = black-colored legume
l1 l1 L2 _ = brown-colored legume
l1 l1 l2 l2 = straw-yellow legume

Therefore, this trait can be used to prove the purity of the cultivar in production.

The colors of the integument and the seed hilum are quite variable and codified mainly by the joint action of four series of genes: inhibitor of dark color expression (*I*) generates brown and black colors (*T* and *R*) and flower color (*W1*) (Vernetti and Vernetti Júnior 2009).

The integument or shell of the seed may present one or two and possibly three colors. In general, the hilum has only one color. The colors of the integument and/or hilum are yellow, imperfect yellow, green, brown (more than one shade), black and imperfect black, or also gray.

The imperfect black color in the integument shows irregular black areas over light brown ones. Genotypes *I I T T rr* have imperfect yellow hilum.

The integument is maternal tissue, and the embryo belongs to the subsequent generation, that is, the seed integument of a genotype has an identical genome, and its internal structures may show that they are hybrid.

The black and brown colors depend on the interaction of three pairs of alleles:

R _ (*T* -) = black (brown pubescence)
R _ *tt W1* _ = imperfect black, gray pubescence, and purple flower
R _ *tt w1 w1* = light-brown, gray pubescence, and white flower
rm. - (*T* -) = black dots or stripes on brown (brown pubescence)
rr (*T* -) = brown (brown pubescence)
rr tt = light brown

The imperfect black or light-brown genotypes generally present irregular shell cracks impairing the conservation of seeds for future sowing.

The reddish-brown color which sometimes appears in the seed integument is encoded by the interaction of a pair of alleles causing the brown color (Weiss 1970):

r r T _ *O* _ = brown
r r T _ *o o* = reddish brown

The presence and distribution of the colors mentioned above, which occur in the integument or hilum of the seed, are controlled by an allelomorphic series of four alleles. Their effects are as follows:

I _ = prevents the expression of colors previously listed in the integument and hilum
i^i _ = the colors described appear only in the hilum; this is the most common
i^k _ = the colors described are seen in the hilum and spread around it until reaching half of the seed, in the form of an eyebrow or a saddle
i i = the colors related cover the integument and hilum

Therefore, in the presence of *I* or i^i, the integument and hilum are yellow (*gg*) or green (*G* -). With the alleles *ii* the hilum is black or brown. The i^k gene, in turn, determines the black, imperfect black or brown hilum, and the integument is bicolored in green or yellow in its greatest extent and the same color of the hilum around it, in the form of a saddle or eyebrow.

In some genotypes, allele *I* modifies the color of the hilum (Mahmud and Probst 1953; Bhatt and Torrie 1968):

I _ *R* _ *T* _ = gray hilum, and not yellow or green
I _ *R* _ *tt w1* _ = gray hilum, and not imperfect black
I _ *R* _ *tt w1 w1* = yellow or green hilum, and not brown

*I _ rr T _ O_* = yellow or green hilum, and not brown
I _ rr T _oo = yellow or green hilum, and not reddish brown
I _ rr tt = yellow or green hilum, and not light brown

There are other cases of variation of gray hilum in the presence of *I*, as well as discoloration of the integument.

Most of the genes that govern qualitative traits in plants are regulated at transcript levels. However, the color determination of the soybean integument is a rare example of posttranslational regulated phenotype. Although the genes responsible for integument pigmentation—chalcone synthases (CHS)—are functional and normally transcribed, their *m-RNAs* are destroyed via *RNA-silencing agent*. The result is colorless or light yellow hila. Nowadays, breeders, manufacturers, and producers of *natto* and *tofu* are focusing on obtaining cultivars with colorless or lightly pigmented hila, since anthocyanins from dark pigments can interfere with the extraction of oil and protein during the industrial process. There are nine chalcone synthases: *CHS 1*, *2*, *3*, *4*, *5*, *6*, *7*, *8*, and *9*; and *CHS 7* and *8* transcribed are abundant in cultivars carrying the mutant for reversion of gene *I*, i.e., it is directly responsible for the suppression of the pigment.

Other non-allele genes or not influenced by the allelomorphic series *I*, i^i, i^k, and *i* were genetically studied. They produce dark-colored pigments in the integument and hilum, and with saddle or eyebrow standard. These were designated: *K1K1*, *K1k1*, *k1k1*; *K2K2*, *K2k2*, *k2k2*; and *K3K3*, *K3k3*, *k3k3*.

The retention of chlorophyll in the seed integument is another characteristic encoded by three pairs of alleles (Reese and Boerma 1989):

G1 _ = green integument
g1 g1 = yellow integument
*G2_* = green integument
g2 g2 = yellow integument
*G3_* = yellow integument
g3 g3 = green integument

The cotyledons of soybean seeds under development have bright yellow or green color. Those of green color are genetic inheritance or cytoplasmic inheritance. Two pairs of alleles encode the color of cotyledons (Woodworth 1921):

D1 _ D2 _ = yellow cotyledons (most cultivars)
D1 _ d2 d2 = yellow cotyledons (some cultivars)
d1 d1 D2 _ = yellow cotyledons (some cultivars)
d1 d1 d2 d2 = green cotyledons (the minority of cultivars)

The green cotyledon cultivars have green seed integument, and their vegetative and reproductive structures remain green at maturity.

Cytoplasmic inheritance of green cotyledons is thus observed (Veatch and Woodworth 1930):

cit Y1 = embryo (cotyledons) of yellow seed
cit G1 = embryo of lighter green color than that encoded by *d1 d1 d2 d2*; carries the nuclear gene G to determine green integument

The color characteristics of the tegument, hilum, and cotyledons of the seeds and the color of the legumes are valuable auxiliaries in the process of genetic breeding of cultivars. The possible occurrence of natural crossings, mixtures of genotypes, and/or mutant plants can be confirmed by the presence of plants differing from the expected characteristics within the segregating generations of progenies. They are quite striking traits, generally easily observable or analyzable in the field or in the laboratory, giving a guarantee of genetic purity of the plants selected. Although each trait is encoded by one or a few genes, as the whole, the number of alleles involved is large enough to allow adequate evaluation of the progenies of the crossings made.

Chemical Composition of the Seed

The chemical composition of the soybeans includes proteins, lipids, carbohydrates, and minerals. Protein and oil make up about 60% of the total. Protein levels vary from 30 to 46%, and the oil from 12 to 24%, according to the genome of the cultivar and the host environment.

In soybeans, the requirements of essential amino acids in human feeding are equal to or greater than those found in the egg, except for methionine.

Most proteins are globulins soluble with the addition of salt. The extraction of soy flour followed by centrifugation results in protein 2S, 7S, 11S, and 15S.

There are 27 enzymes, but the most important ones are lipoxygenases (catalyze the oxidation of lipids forming fatty acids), hydroperoxidases (resulting in unpleasant flavors and aromas), lipoperoxidase (destroy fatty acids), lipases (catalyze the formation of free fatty acids), and α- and β-amylases, proteases, and proteinases.

Soluble and insoluble carbohydrates are present in sugars.

Proteases

The protease inhibitor proteins are very significant, like: trypsin inhibitor Kunitz and Bowman-Birk inhibitor and trypsin inhibitors, considered as antinutritional factors of cultivated soybean (*G. max*) and wild soybean (*G. soja*), whose action is to delay the digestion of casein.

The codominant genes responsible for the inhibition of trypsin are seven (Hymowitz 1973; Orf and Hymowitz 1977, 1979; Singh et al. 1969; Hymowitz et al. 1972; Zhao et al. 1995):

Ti-a = variant for mobility of the trypsin inhibitor Kunitz (Harosoy)

Ti-b = variant for mobility of the trypsin inhibitor Kunitz (Aoda)
Ti-c = variant for mobility of the trypsin inhibitor Kunitz (PI 86084)
Ti-x = variant for mobility of the trypsin inhibitor Kunitz
Ti-d = variant for mobility of the trypsin inhibitor Kunitz
Ti-e = variant for mobility of the trypsin inhibitor Kunitz
Ti-b-f = variant for mobility of the trypsin inhibitor Kunitz
ti = trypsin inhibitor Kunitz absent

Raw soybean intake causes pancreatic hypertrophy. It is believed to inhibit 30–50% growth of monogastric animals by causing disturbance in the balance between methionine and cystine in the animal organism.

Fatty Acids

There is a clear expectation that the proper use of genes determining beneficial changes in the composition of the seed will have significant influence on the quality, utility, and economic value of soybeans. Market trends, industry priorities, and government policies will influence the course of scientific efforts to genetically alter the traditional composition of the seed, but strengthening the competitive position of soybeans in the global and domestic markets will be the ultimate goal: to have always a better product.

One of the major goals is to have oil with less saturated fats and trans isomers.

Conventional soybean cultivars contain approximately 11% of palmitate, 4% of stearate, 24% of oleate, 54% of linoleate, and 7% of linolenate (Fehr 1991).

The genes involved influencing the fatty acid composition of soybean seed are:

Palmitic

Fap1, *Fap2*, *Fap3*, *Fap4*, *Fap5*, *Fap6*, and *Fap7* // Normal level of palmitic acid
fap1 (T308), *fap2* (J10), *fap3* (A 22), *fap3-nc* (N79–2077-12), *fapx* (ELLP-2), *fapx* (KK 7), *fap?* (J3) and *fap?* (ELHP-1) // Reduced level of palmitic acid
fap2 (T309), *fap2-b* (A 21), *fap4* (A 24), *fap5* (A 27), *fap6* (A 25) and *fap7* (A 30) // High level of palmitic acid

Stearate

Fas St1 St2 // normal level of stearic acid
faz (A9), *fas-a* (A 6), *fas-b* (A 10), *st1*(KK 2) and *st2*(M-25) // high level of stearic acid

Oleate

Ol // normal level of oleic acid
ol (M-23) and *ol-a* (M-11) // high level of oleic acid

Linoleate

No gene was identified for mutants of linoleate (linoleic acid). Reduction of linolenate content was obtained (65–3%) with the use of antisense oleate desaturase DNA (Yadav 1995).

Linolenate

Fan1, *Fan2*, and *Fan3* // normal level of linolenic acid

fan1 (PI 123440), *fan1* (A5, T307),*fan1* (C1640, T280), *fan1* (PI 361088B), *fan1* (M-5), *fan1-b* (RG10), *fan2* (A23), *fan3* (A26), *fanx* (KL-8) and *fanx-a* (M-24) // reduced level of linolenic acid

Flavonoids

Flavonoids are generally found in higher plants. Its greatest value is as a chemical compound employed in pharmaceuticals and food, such as genetic markers and its function in the vegetable.

Isoflavonoids are active natural products that, according to genetic and environmental factors, accumulate in soybean seeds during their development.

Two genes are responsible to control 2-hydroxisoflavone synthase (Dhaubhadel et al. 2003):

IFS 1 = high concentration in the root and seed integument
IFS 2 = high concentration in embryos and pods

A maternal effect on the control of the isoflavonoid content in the seed was found. Numerous genes control flavonoids and glycosides in the leaves (Buzzell and Buttery 1992):

T _ = quercetin and kaempferol present, brown pubescence
t t = quercetin absent, kaempferol present, gray pubescence
Wm _ = flavonoid glycosides present; magenta flower
wm wm = flavonoid glycosides absent (T235)
Fg1 _ *B (1–6)* = flavonoid glycosides present (T31)
fg1 fg1 B (1–6) = flavonoid glycosides absent (Chippewa 64)
Fg2 _ = rutinoside kaempferol normal (Ox-724)
Fg2-b _ = less rutinoside kaempferol (Ox-730)
fg2 fg2 α (1–6) = rhamnoside absent (Chippewa 64)
Fg3 _ *B (1–2)* = glycoside present (T31)
fg3 fg3 B (1–2) = glycoside absent (Chippewa 64)
Fg4 _ *α (1–2)* = rhamnoside present (T31)
fg4 fg4 α (1–2) = rhamnoside absent (AK-FC 30761)

The increased flavonoid content in the leaves can become an important objective of breeding as a possible source of its production to meet the possible increase in the demand of the pharmaceutical and food industries.

Proteins and Isozymes

Only about half of soybean proteins and isozymes was genetically analyzed. Their alleles and their phenotypic expression were studied and approved by the Soybean Genetics Committee.

Among them is β-conglycinin, a reserve protein accumulated in the seed during development, as a source of C and N. It is composed of three main subunits α', α, and β and three secondary subunits, β', γ, and δ.

Several studies have indicated that there are at least 15 different genes for *β*-conglycinin in the soybean genome. In the subunit α' are *CG1*, *CG2*, *CG3*; in β are *CG4*, *CG8*, *CG11*, *CG12*, *CG13*, *CG15*; homologues with the subunits α', α and β are *CG5*, *CG6*, *CG7*, *CG14*.

The family of genes encoding glycinin is *Gy1*, *Gy2*, *Gy3*, *Gy4-a*, *Gy4-b*, *Gy5*, *Gy6*, and *Gy7*; and *gy4* encodes the absence of glycinin.

Peroxidase

The peroxidase activity of the soybean seed integument is one of the standard tests used to identify cultivars. The genotypes can thus be separated (Buzzell and Buttery 1969):

*Ep_* = high peroxidase activity
ep ep = low peroxidase activity

The test is performed with oxidized guaiacol, which causes change in the color of the integument according to the peroxidase activity. A more accurate test was developed by Vierling and Wilcox (1996), using a monoclonal antibody to isolate the enzyme before measuring its activity.

Ureases

The solubility of the protein in soybean meal is positively associated to its digestibility and the presence of active antinutritional factors. The goal of toasting it immediately after solvent extraction and its immediate withdrawal is to inactivate trypsin inhibitors, lecithin, and other toxic components. Ruminants are able to inactivate it, but monogastric animals digest it, replacing it with much less protein quality. Overheating the meal is also harmful.

Ureases, which resemble protein inhibitors in their action, should be inactivated.

Urease isozymes are thus inherited (Buttery and Buzzell 1971; Holland et al. 1987; Palmer et al. 2004; Kloth et al. 1987; Polacco et al. 1989):

Eu1-a = urease of the embryo (Prize)
Eu1-b = urease of the embryo (Williams)
eu1-sun = urease of the embryo absent (PI 229324)
eu1-n4 = urease null—without mRNA (Williams mut. NMU)
eu1-n6 = urease mRNA present
eu1-n7 = urease null without mRNA
eu1-n8 = urease mRNA present
Eu2 = normal levels of urease
eu 2 = without urease in the leaf, present in the embryo
Eu3 = normal levels of urease
eu3-e1 = both ureases absent
Eu3-e3 = reduced levels of both ureases
Eu4 = normal levels of urease
eu4 = normal levels of urease in the embryo and no urease in leaf (Williams mut. EMS)

The presence or absence of urease in the seed is thus encoded:

Sun = urease present
sun = urease absent

Phytohemagglutinin or Lectin

Lectin or phytohemagglutinin is a glycoprotein that causes agglutination of certain red blood cells. In the soybean seed, there are four different forms of lectin. They are bound by carbohydrates to the nitrifying bacteria, that is, they participate in the recognition that leads to symbiosis. They also carry O_2 into the tissues of the nodules, where it remains. It is, therefore, a nodulin, a polypeptide specific for lymph node hosting.

The main form of lectin consists of four identical subunits. The presence or absence of soybean lectin is decoded by a gene (Orf et al. 1978):

*Le_* = presence of lectin in the seed
le le = absence of lectin in the seed

Therefore, since a large number of genotypes do not have lectin, which contributes to the success of symbiosis/nodulation and absorption of N2 from the air, it is essential to verify their presence in the genotypes involved in the breeding program.

Seed Lipoxygenase

Soybean seeds contain lipoxygenases 1, 2, and 3, and they play two enzymatic roles in the plant: participate or intervene in the hydroperoxidation of lipids and transient N storage during vegetative growth. Lipoxygenase has been considered for some time as the cause of undesirable flavors in products based on soybean oil.

The inheritance of lipoxygenase is controlled as follows (Hildebrand and Hymowitz 1981, 1982; Kitamura et al. 1983; Davies and Nielsen 1986; Pfeiffer et al. 1993):

Lx1-a = lipoxygenase 1, PI 5,85 (Century)
Lx1-b = lipoxygenase 1, PI 5,79 (PI86523)
lx1 = lipoxygenase absent
Lx2 = lipoxygenase 2 present
lx2 = lipoxygenase 2 absent
Lx3 = lipoxygenase 3 present
lx3 = lipoxygenase 3 absent (PI 417458)

Genes of the series *Lox 1* to *8* and *Lox A* and *Lox B* were considered related to lipoxygenases, but without the defined inheritance mode.

It can be concluded that the absence of lipoxygenase in soybean cultivars is hypothetically advantageous.

Phytic Acid (Inositol Hexaphosphate)

Soybeans is, along with wheat, corn, and rice, one of the crops that most removes the phosphorus available in the soil, stored as phytic acid.

In the case of soybeans, this fact may be a problem, since it forms an insoluble protein-phytate complex. These complexes tend to exert negative effects on bioavailability of phosphorus. As a result, significant amount of digestible P returns to the soil in animal manure, with a negative impact on the environment. Therefore, it is important to reduce it in the feed of chickens, for example, to improve the availability of eight amino acids.

The nutritional quality of soybeans can be improved by reducing the P of phytic acid in the seed (70% of total) (Raboy and Dickinson 1993). Raboy et al. (1984) found that many genes are involved in the control of this trait. Some mutants have lower content of phytic acid and greater inorganic P, a valuable fact if used successfully in breeding. The corresponding gene encoding was not presented.

The use of phytase introduced by genetic engineering in soybean seeds seems to degrade phytate and increase the availability of P in animal diets. The gene that changed the soybean cells to exhibit phytase was *An Phy*, found in the fungus *Aspergillus niger* (Denbow et al. 1997).

Inositol hexaphosphate is potentially useful in the prevention of human colon cancer, since it is able to reduce cell proliferation and increase differentiation and can be incorporated, modifying the signaling pathway.

Lecithin

Lecithin, a byproduct of soybean oil processing, is a phosphatide (natural emulsifier) widely used in the food industry, especially in chocolate. It has a hypophilic group and a hydrophilic group in the same cell, which allows it to promote the union of products considered noninterchangeable.

Raw soy lecithin contains about 34.5% pure lecithin (phosphatidylcholine), 23.2% commercial cephalin (phosphatidylethanol amine), 15.7% phosphatidylinositol, 7.0% phosphatidic acid, and 6.1% phosphatidylserine.

Lecithin has colloidal, emulsifying, softening, antioxidant, and physiological properties, mainly with great application in margarine, chocolate, ice cream, pasta, cakes, crackers, and biscuit industry, for example. However, it can cause food allergies resulting from the fractioning of proteins during their preparation (Xiang et al. 2008). In addition, it is one of the by-products of greatest significance in the production of soybean oil.

Final Considerations

The genetic improvement has had great success in productivity and quality of soybeans, both in the public sector and in private companies.

The selection of progenitors is very important for the success of genetic breeding, both in adapted elite cultivars and exotic germplasm from germplasm banks. Elite cultivars from distinct origin as a rule produce progenies superior to each progenitor in relation to those genetically more related.

The selection of progenitors depends on factors such as the characteristic desired, the purpose of crossbreeding, the relative importance of the traits, the genealogy of parents, and resources and time available. Testcross models are indicated for prospecting higher alleles, and the genomic analysis can assist in progenitor selection.

The selection itself may be held among plants in the first generation, or during the evaluation period of the lines. At this point, the heritability of the traits is essential.

The selection types commonly employed are genealogy; single seed descent (SSD); advancement of populations without early selection; early testing of generations, backcross breeding and recurrent selection (usually involves male-sterility); and genetic marker-assisted selection. Recurrent selection is difficult to address due to laborious attainment of crossbreeding population.

The genetic gain and the harvest index are indicative of the success of each breeding program conducted.

The elite cultivars obtained contribute to the economic gain of productive activity.

References

Bernard, R.L.: The inheritance of pod color in soybeans. J. Hered. **58**, 165–168 (1967)

Bernard, R.L.: Two major genes for time of flowering and maturity in soybeans. Crop Sci. **11**, 242–244 (1971)

Bernard, R.L.: Two genes affecting stem termination in soybeans. Crop Sci. **12**, 235–239 (1972)

Bernard, R.L.: The inheritance of semi-sparse pubescence. Soybean Genet. Newslett. **2**, 33–34 (1975a)

Bernard, R.L.: The inheritance of appressed pubescence. Soybean Genet. Newslett. **2**, 34–36 (1975b)

Bernard, R.L.: The inheritance of near-gray pubescence color. Soybean Genet. Newslett. **2**, 31–33 (1975c)

Bernard, R.L.: An allelic series affecting stem length. Soybean Genet. Newslett. **2**, 28–30 (1975d)

Bernard, R.L., Singh, B.B.: Inheritance of pubescence type in soybeans: glabrous, curly, dense, sparse and puberulent. Crop Sci. **9**, 92–97 (1969)

Bernard, R.L., Weiss, M.G.: Qualitative genetics. In: Caldwell, B.E. (ed.) Soybeans: Improvement, Production and Uses, pp. 117–154. American Society of Agronomy, Madison (1973)

Bhatt, G.M., Torrie, J.H.: Inheritance of pigment color in the soybean. Crop Sci. **8**, 617–619 (1968)

Boerma, H.R., Jones, B.G.: Evidence of a second gene controlling short internode (zigzag stem) character. Soybean Genet. Newslett. **3**, 22–23 (1976)

Boerma, H.R., Jones, B.G.: Inheritance of a second gene for brachytic stem in soybean. Crop Sci. **18**, 559–560 (1978)

Bonato, E.R., Vello, N.A.: *E6*, a dominant gene conditioning early flowering and maturity in soybeans. Genet. Mol. Biol. **22**, 229–232 (1999)

Buttery, B.R., Buzzell, R.I.: Properties and inheritance of urease isoenzymes in soybean seeds. Can. J. Bot. **49**, 1101–1105 (1971)

Buzzell, R.I.: Inheritance of a soybean flowering response to fluorescent-daylength conditions. Can. J. Genet. Cytol. **13**, 703–707 (1971)

Buzzell, R.I., Buttery, B.R.: Inheritance of peroxidase activity in soybean seed coats. Crop Sci. **9**, 387–388 (1969)

Buzzell, R.I., Buttery, B.R.: Inheritance of an anomalous flavonoid glycoside gene in soybean. Genome. **35**, 636–638 (1992)

Buzzell, R.I., Voldeng, H.D.: Inheritance of insensitivity to long day-length. Soybean Genet. Newslett. **7**, 26–29 (1980)

Buzzell, R.I., Bernard, R.L., Buttery, B.R.: Inheritance of magenta flower color. Soybean Genet. Newslett. **1**, 14–15 (1974)

Buzzell, R.I., Buttery, B.R., Bernard, R.L.: Inheritance and linkage of a magenta flower gene in soybeans. Can. J. Genet. Cytol. **19**, 749–751 (1977)

Byth, D.E., Weber, C.R.: Two mutant genes causing dwarfness in soybeans. J. Hered. **60**, 278–280 (1969)

Cober, E.R., Voldeng, H.D.: A new soybean maturity and photoperiod sensitivity locus linked to *E1* and *T*. Crop Sci. **41**, 698–701 (2001)

Cober, E.R., Tanner, J.W., Voldeng, H.D.: Soybean photoperiod sensitivity loci respond differentially to light quality. Crop Sci. **36**, 606–610 (1996)

Davies, C.S., Nielsen, N.C.: Genetic analysis of a null-allele for lipoxygenase-2 in soybean. Crop Sci. **26**, 460–463 (1986)

Dellaney, X., Palmer, R.G.: Two genes controlling root fluorescence in soybean. Crop Sci. **22**, 278–281 (1982)

Dellaney, X., Palmer, R.G.: Inheritance of a miniature mutant in soybean. Soybean Genet. Newslett. **11**, 92–93 (1984)

Denbow, D.M., Grabau, E.A., Hanlon, R.W., Hegeman, C.E., Lacy, G.H., Li, J.: Secretion of active recombinant phytase from soybean cell-suspension cultures. Plant Physiol. **114**(3), 1103–1111 (1997)

Devine, T.E.: Nodulation of soybean (*Glycine max* (L.) Merrill) plant introduction lines with the fast-growing rhizobial strain USDA 205. Crop Sci. **25**, 354–356 (1985)
Dhaubhadel, D., Mcgarvey, B.D., Williams, R., Gijzen, M.: Isoflavonoid biosynthesis and accumulation in developing soybean seeds. Plant Mol. Biol. **53**(6), 733–914 (2003)
Fehr, W.R.: Inheritance of a mutation for dwarfness in soybeans. Crop Sci. **12**, 212–213 (1972a)
Fehr, W.R.: Genetic control of leaflet number in soybeans. Crop Sci. **12**, 221–224 (1972b)
Fehr, W.R.: Principles of Cultivar Development: Theory and Technique, vol. 1, pp. 1–527. Macmillan, New York (1987)
Fehr, W.R.: A plant breeder's response to changing needs in soybean oil quality. In: Wilkinson, D. (ed.) Proceedings of the XXI Soybean Seed Research Conference, Chicago, IL, 10–-11 December 1991, pp. 79–87. American Seed Trade Association, Washington, DC (1991)
Fehr, W.R., Giese, J.H.: Genetic control of root fluorescence. Crop Sci. **11**, 771 (1971)
Gunashinge, U.B., Irwin, M.E., Kampmeir, G.E.: Soybean leaf pubescence affects aphid vector transmission and field spread of soybean mosaic virus. Ann. Appl. Biol. **112**, 259–272 (1988)
Hartwig, E.E., Edwards Jr., C.J.: Effects of morphological characteristics upon seed yield in soybeans. Agron. J. **62**, 64–65 (1970)
Hartwig, E.E., Hinson, K.: Inheritance of flower color of soybean. Crop Sci. **2**, 152–153 (1962)
Hartwig, E.E., Kiihl, R.A.S.: Identification and utilization of a delayed flowering character in soybeans for short-day conditions. Field Crop Res. **2**, 145–151 (1979)
Hildebrand, D.F., Hymowitz, T.: Two soybean genotypes lacking lipoxygenase-1. J. Am. Oil Chem. Soc. **58**, 583–586 (1981)
Hildebrand, D.F., Hymowitz, T.: Inheritance of lipoxygenase-1 activity in soybean seeds. Crop Sci. **22**, 851–853 (1982)
Holland, M.A., Griffin, J.D., Meyer-Bothling, L.E., Polacco, J.C.: Developmental genetics of soybean urease isozymes. Dev. Genet. **8**, 375–387 (1987)
Hymowitz, T.: Electrophoretic analysis of SBT-A, in the USDA soybean germplasm collection. Crop Sci. **13**, 420–421 (1973)
Hymowitz, T., Palmer, R.G., Hadley, H.R.: Seed weight, protein, oil, and fatty acid relationships within the genus *Glycine*. Trop. Agric. **49**, 245–250 (1972)
Johnson, H.W., Bernard, R.L.: Soybean genetics and breeding. In: Norman, A.G. (ed.) The Soybean, pp. 1–73. Academic, New York (1963)
Johnson, E.O.C., Stephens, P.A., Fasoula, D.A., Nickell, C.D., Vodkin, L.O.: Instability of a novel multicolored flower trait in inbred and outcrossed soybean lines. J. Hered. **89**, 508–515 (1998)
Kilen, T.C.: Inheritance of a brachytic character in soybean. Crop Sci. **17**, 853–854 (1977)
Kilen, T.C.: Inheritance of a short petiole trait in soybean. Crop Sci. **23**, 1208–1210 (1983)
Kilen, T.C., Hartwig, E.E.: Inheritance of a light quality sensitive character in soybeans. Crop Sci. **11**, 559–561 (1971)
Kilen, T.C., Hartwig, E.E.: Short internode character in soybeans and its inheritance. Crop Sci. **15**, 878 (1975)
Kitamura, K., Davies, C.S., Kaizuma, N., Nielsen, N.C.: Genetic analysis of a null-allele for lipoxygenase-3 in soybean seeds. Crop Sci. **23**, 924–927 (1983)
Kloth, R.H., Polacco, J.C., Hymowitz, T.: The inheritance of a urease-null trait in soybean. Theor. Appl. Genet. **73**, 410–418 (1987)
Kuykendall, L.D.: Order VI. Rhizobiales ord. nov. In: Brenner, D.J., Krieg, N.R., Staley, J.T., Garrity, G.M. (eds.) Bergey's Manual of Systematic Bacteriology, 2nd edn, vol. 2 (The Proteobacteria), Part C (The Alpha-, Beta-, Delta-, and Epsilonproteobacteria), p. 324. Springer, New York (2005)
Kuykendall, L.D., Hashem, F.M., Dadson, R.B., Elkan, G.H.: Nitrogen fixation. In: Lederberg, J., Alexander, M. (eds.) Encyclopedia of Microbiology, vol. 1, pp. 492–505. Academic, New York (2000)
Mahmud, I., Probst, A.H.: Inheritance of gray hilum color in soybeans. Agron. J. **45**, 59–61 (1953)
Nagai, I.: A genetic physiological study on the formation of anthocyanin and brown pigments in plants. Tokyo Univ. Coll. Agric. J. **8**, 1–92 (1921)

Nagai, I., Saito, S.: Linked factors in soybeans. Jpn. J. Bot. **1**, 121–136 (1923)

Nelson, R.L.: The inheritance of a branching type in soybean. Crop Sci. **36**, 1150–1152 (1996)

Orf, J.H., Hymowitz, T.: Inheritance of a second trypsin inhibitor variant in seed protein of soybeans. Crop Sci. **17**, 811–813 (1977)

Orf, J.H., Hymowitz, T.: Inheritance of the absence of the kunitz trypsin inhibitor in seed protein of soybeans. Crop Sci. **19**, 107–109 (1979)

Orf, J.H., Hymowitz, T., Pull, S.P., Pueppke, S.G.: Inheritance of soybean seed lectin. Crop Sci. **18**, 899–900 (1978)

Owen, F.V.: Inheritance studies in soybeans. II—Glabrousness, color of pubescence, time of maturity and linkage relations. Genetics. **12**, 519–529 (1927)

Palmer, R.G.: Genetic studies with T263. Soybean Genet. Newslett. **11**, 94–97 (1984)

Palmer, R.G., Pfeifer, T.D., Buss, G.R., Kilen, T.C.: Qualitative genetics. In: Boerma, H.R., Specht, J.E. (eds.) Soybeans: Improvement, Production, and Uses, 3rd edn, pp. 137–233. ASA Inc., CSSA Inc., SSSA Inc., Madison (2004)

Pfeiffer, T.W., Hildebrand, D.F., Orf, J.H.: Inheritance of a lipoxygenase-1 allozine in soybean. Crop Sci. **33**, 91–693 (1993)

Piper, C.V., Morse, W.J.: The Soybean: History, Varieties and Field Studies, USDA Bureau of Plant Industry Bulletin 197, p. 84. US Government Printing Office, Washington, DC (1910)

Piper, C.V., Morse, W.J.: The Soybean, 2nd edn, p. 329. McGraw-Hill, New York (1923)

Polacco, J.C., Judd, A.K., Dybing, J.K., Cianzio, S.R.: A new mutant class of soybean lacks urease in leaves but not in leaf-derived callus or in roots. Mol. Gen. Genet. **217**, 257–262 (1989)

Porter, K.B., Weiss, M.G.: The effect of polyploidy on soybean. J. Am. Soc. Agron. **40**, 710–724 (1948)

Pracht, J.E., Nickell, C.D., Harper, J.E.: Genes controlling nodulation in soybean: *Rj5* and *Rj6*. Crop Sci. **33**, 711–713 (1993)

Probst, A.H.: The inheritance of leaf abscission and other characters in soybeans. Agron. J. **42**, 35–45 (1950)

Raboy, V., Dickinson, D.B.: Phytic acid levels in seeds of *Glycine max* and *Glycine soja* as influenced by phosphorus status. Crop Sci. **33**, 1300–1305 (1993)

Raboy, V., Dickinson, D.B., Below, F.E.: Variation in seed total phosphorus, phytic acid, zinc, calcium, magnesium, and protein among lines of *Glycine max* and *Glycine soja*. Crop Sci. **24**, 431–434 (1984)

Ray, J.D., Hinson, K., Mankow, J.E.B., Malo, M.F.: Genetic control of a long-juvenile trait in soybean. Crop Sci. **35**, 1001–1006 (1995)

Reese Jr., P.F., Boerma, H.R.: Additional genes for green seed coat in soybean. J. Hered. **80**, 86–88 (1989)

Rode, M.W., Bernard, R.L.: Inheritance of wavy leaf. Soybean Genet. Newslett. **2**, 44–45 (1975)

Sadasivam, S., Krishnaveni, S.: Host gene expression in nodulation and nitrogen fixation. In: Kannaiyan, S. (ed.) Biotechnology of Fertilizers, pp. 84–97. Narosa Publishing House, New Delhi (2002)

Sawada, S., Palmer, R.G.: Genetic analysis of non-fluorescent root mutants induced by mutagenesis in soybean. Crop Sci. **27**, 62–65 (1987)

van Schaik, P.H., Probst, A.H.: The inheritance of inflorescence type, peduncle length flowers per node and percent flower shedding in soybeans. Agron. J. **50**, 98–102 (1958)

Sinclair, T.R., Hinson, K.: Soybean flowering in response to the long-juvenile trait. Crop Sci. **32**, 1242–1248 (1992)

Singh, L., Wilson, C.M., Hadley, H.H.: Genetic differences in soybean trypsin inhibitors separated by disc electrophoresis. Crop Sci. **9**, 489–491 (1969)

Singh, B.B., Hadley, H.H., Bernard, R.L.: Morphology of pubescence in soybeans and its relationship to vigor. Crop Sci. **11**, 13–16 (1971)

Singh, B.B., Gupta, S.C., Singh, B.D.: An induced crinkled leaf mutant in soybean. Soybean Genet Newslet **1**, 16–17 (1974)

Soybean Genetics Committee: Soybean genetics committee report. Soybean Genet. Newslett. **22**, 11–14 (1995)

Soybean Genetics Newsletter: Soybean isolines. Soybean Genet. Newslett. **2**, 75–76 (1975)

Stephens, P.A., Nickell, C.D.: Inheritance of pink flower in soybean. Crop Sci. **32**, 1131–1132 (1992)
Stewart, R.T., Wentz, J.B.: A recessive glabrous character in soybeans. J. Am. Soc. Agron. **18**, 997–1009 (1926)
Thompson, J.A., Bernard, R.L., Nelson, R.L.: A third allele at the soybean *dt1* locus. Crop Sci. **37**, 757–762 (1997)
Toda, K., Yang, D., Yamanaka, N., Watanabe, S., Harada, K., Takahashi, R.: A single base deletion in soybean flavonoid 3-hydroxylase gene is associated with gray pubescent color. Plant Mol. Biol. **50**, 187–196 (2002)
Veatch, C., Woodworth, C.M.: Genetic relations of cotyledon color types of soybeans. J. Am. Soc. Agron. **22**, 700–702 (1930)
Vernetti, F.de J.: Genética da Soja—Caracteres qualitativos. In: Vernetti, F.de J. (ed.) Soja: Genética e Melhoramento, vol. II, pp. 476–740. Fundação Cargill, Campinas (1983)
Vernetti, F.J., Vernetti Júnior, F.J.: Genética da soja: caracteres qualitativos e diversidade da soja. Embrapa Informação Tecnológica, Brasília (2009)
Vierling, R.A., Wilcox, J.R.: Microplate assay for soybean seed coat peroxidase activity. Seed Sci. Technol. **24**, 485–494 (1996)
Weiss, M.G.: Genetic linkage in soybeans: linkage group IV. Crop Sci. **10**, 368–370 (1970)
Werner, B.K., Wilcox, J.R., Housley, T.L.: Inheritance of an ethyl methanesulfonate-induced dwarf in soybean and analysis of leaf cell size. Crop Sci. **27**, 665–668 (1987)
Wilcox, J.R., Abney, T.S.: Inheritance of a narrow, rugose-leaf mutant in *Glycine max*. J. Hered. **82**, 421–423 (1991)
Williams, L.F.: The inheritance of certain black and brown pigments in soybean. Genetics. **37**, 208–215 (1952)
Woodworth, C.M.: Inheritance of cotyledon, seed-coat, hilum, and pubescence color in soybeans. Genetics. **6**, 487–553 (1921)
Woodworth, C.M.: Inheritance of growth habit, pod color and flower color in soybeans. J. Am. Soc. Agron. **15**, 481–495 (1923)
Woodworth, C.M.: Genetics and breeding in the improvement of the soybean. Bull. Agric. Exp. Station. **384**, 297–404 (1932)
Woodworth, C.M.: Genetic of the soybean. J. Am. Soc. Agron. **25**, 36–51 (1933)
Woodworth, C.M., Veatch, C.: Inheritance of pubescence color in soybeans and its relation to pod color. Genetics. **14**, 512–518 (1929)
Xiang, P., Baird, L., Jung, R., Zeece, M., Markwell, J., Sarath, G.: P39, a novel soybean protein allergen, belongs to a plant-specific protein family and is present in protein storage vacuoles. J. Agric. Food Chem. **56**(6), 2266–2272 (2008)
Xu, M., Palmer, R.G.: Genetic analysis and molecular mapping of a pale flower allele at the *W4* locus in soybean. Genome. **48**, 334–340 (2005)
Yadav, N.S.: Genetic modification of soybean oil quality. In: Verma, D.P.S., Schoemaker, R.C. (eds.) Soybean Biotechnology, pp. 165–188. CAB Int, London (1995)
You, M., Zhao, T., Gai, J., Yen, Y.: Genetic analysis of short petiole and abnormal pulvinus in soybean. Euphytica. **102**, 329–333 (1998)
Zhao, S., Qimin, G., Hai, W.: Inheritance of new variant of SBTi-A2 in seed protein of soybean (*Glycine max*) in China. Soybean Genet. Newslett. **22**, 85–88 (1995)

Chapter 6
Quantitative Traits in Breeding

Felipe Lopes da Silva, Marcos Deon Vilela de Resende, Willian Hytalo Ludke, and Thays Vieira Bueno

Abstract Quantitative traits are those controlled by several genes and are strongly influenced by the environment, presenting continuous distribution, for example, soybean grain yield. Most of the important economic characteristics for the soybean crop are quantitative in nature. These peculiarities make the genetic control of the traits considered complex and require the study of genetic parameters that are used to increase the accuracy of the process of selection of superior genotypes. This chapter aims to present QLTs associated with some quantitative traits of importance in soybean breeding programs. It also presents new breeding strategies for quantitative traits based on new developments in genomics and quantitative genetics.

Keywords *Glycine max* (L.) Merrill • Selection strategies • QTL • GWS

Introduction

Most economically important traits in various agricultural crops are quantitative nature. Quantitative traits are, therefore, those controlled by several genes and strongly influenced by the environment, with continuous distribution, such as soybean grain yield (Fig. 6.1). These peculiarities cause the genetic control of these traits to be considered complex.

The genes or chromosomal loci that control a certain quantitative trait are called QTLs (quantitative trait loci) (Möller 2010). The identification of QTLs associated with the traits of interest for soybean breeding and the knowledge about genetic

F.L. da Silva, M.S., D.S. (✉) • W.H. Ludke, M.S. • T.V. Bueno, M.S.
Federal University of Viçosa, Viçosa, Minas Gerais 36570-900, Brazil
e-mail: felipe.silva@ufv.br; willianludke@gmail.com; thaysvieirabueno@gmail.com

M.D.V. de Resende, M.S., D.S.
Brazilian Agricultural Research Corporation/Federal University of Viçosa, Viçosa, Minas Gerais 36570-000, Brazil
e-mail: marcos.deon@gmail.com

F. Lopes da Silva et al. (eds.), *Soybean Breeding*,
DOI 10.1007/978-3-319-57433-2_6

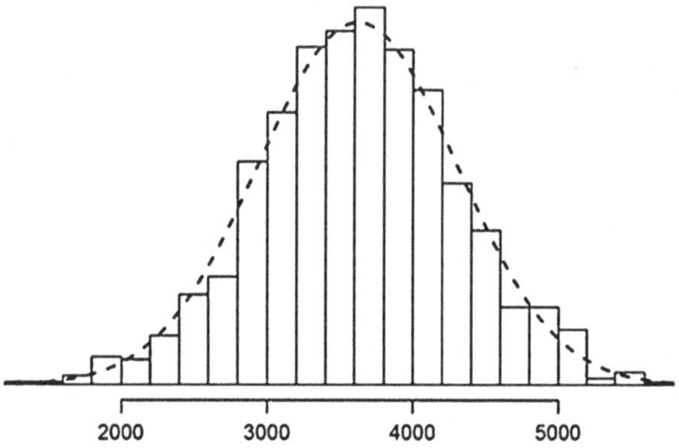

Fig. 6.1 Continuous graphical distribution of an F_2 population for soybean grain yield traits (kg/ha)

diversity, among genotypes, for these loci is of great importance, inasmuch as it facilitates the process of choice of parents and strategies for obtaining elite lines in the breeding program.

The genetic studies of the quantitative traits are carried out adopting the basic model that defines the phenotypic value as a result of the effect of the genotype and the environment (Barona et al. 2009). Hence, the study of the inheritance of these traits takes into account information based on the population, as the information based on a single individual has little value, due to the random effect of the environment (Cruz 2012).

Given their complex nature, quantitative traits require the study of genetic parameters, such as heritability, which are used to increase the accuracy of superior genotype selection process. Therefore, the application of quantitative genetics is necessary, and a broader approach of the application of this science in the soybean breeding program will be made in Chap. 7 of this book.

In quantitative traits, as they suffer great environmental influence, the estimates of heritability values are low. Thus, the genetic gains with the selection of these characteristics, in general, are smaller, and demand adequate breeding strategies to optimize the selection process. These strategies include recurrent selection and genomic selection, which will be discussed in Chaps. 10 and 14 of this book, respectively.

Certain quantitative traits may be correlated, which allows the indirect selection, essential to breeding programs, which is based on the selection of a trait correlated with the one of interest, whose genetic gain is greater or with easier visual selection than that of interest (Carpentieri-Pípolo et al. 2005). For example, in soybean farming, there is a negative correlation between oil and protein levels, and often between protein content and yield. Conversely, the relationship between oil content and productivity is positive, which makes it possible to maintain productivity when selecting for high oil content (Rodrigues et al. 2013).

The purpose of this chapter was to present QTLs associated with some quantitative traits that are important to the soybean breeding program, as well as new breeding strategies for the quantitative traits, given the new developments in genomics and quantitative genetics.

QTLs Associated with Quantitative Traits

In soybean breeding there are numerous important agronomic characteristics with quantitative trait. In this topic we will demonstrate the QTLs of some quantitative traits of importance to soybean breeding. Such information was taken from SoyBase, the USDA-ARS genetic soybean database (Grant et al. 2009). To access SoyBase, go to http://soybase.org.

As examples, we will demonstrate QTLs associated with the following traits: seed production (Table 6.1); number of nodes (Table 6.2) and pods (Table 6.3); number of pods per node (Table 6.4); seed weight (Table 6.5); oil content (Table 6.6); protein content (Table 6.7) in seeds; tolerance to aluminum, drought, flooding, and salinity; and water use efficiency (Table 6.8). The information contained in the tables allows for greater efficiency of breeding programs for these traits, favoring also genomic selection and assisted selection processes.

Table 6.1 Binding group, initial and final position, in centimorgan (cM), in the quantitative trait loci chromosome associated with the seed production characteristic in soybean

Group	Start	End	Reference	Group	Start	End	Reference	Group	Start	End	Reference
A1	25.42	27.42	Kabelka et al. (2004)	C2	44.80	46.80	Orf et al. (1999a)	C2	115.70	117.70	Mansur et al. (1993b)
A1	28.95	31.30	Rossi et al. (2013)	C2	97.07	99.07	Kabelka et al. (2004)	C2	116.72	131.75	Gai et al. (2007)
A1	29.93	31.93	Guzman et al. (2007)	C2	101.75	111.68	Du et al. (2009)	C2	116.87	118.87	Reinprecht et al. (2006)
A1	64.73	78.44	Du et al. (2009)	C2	106.58	108.58	Reinprecht et al. (2006)	C2	116.87	118.87	Orf et al. (1999a)
A1	93.19	95.19	Li et al. (2008a)	C2	106.58	108.58	Palomeque et al. (2009a)	C2	116.87	118.87	Mansur et al. (1996)
A1	94.16	96.16	Kabelka et al. (2004)	C2	106.58	108.58	Rossi et al. (2013)	C2	126.66	128.66	Reinprecht et al. (2006)
A2	107.78	109.78	Orf et al. (1999a)	C2	106.60	108.60	Orf et al. (1999a)	D1a	63.52	65.52	Orf et al. (1999b)
B1	4.07	6.07	Orf et al. (1999a)	C2	107.58	109.26	Zhang et al. (2004)	D1a	66.79	68.79	Kabelka et al. (2004)
B1	36.48	46.38	Du et al. (2009)	C2	109.26	113.40	Zhang et al. (2004)	D1b	24.60	26.60	Orf et al. (1999a)
B2	54.20	56.20	Kabelka et al. (2004)	C2	111.19	113.19	Zhang et al. (2004)	D1b	74.00	76.00	Rossi et al. (2013)
B2	72.20	74.20	Guzman et al. (2007)	C2	111.19	113.19	Guzman et al. (2007)	D1b	74.01	87.19	Du et al. (2009)
B2	74.34	76.34	Guzman et al. (2007)	C2	111.20	113.20	Specht et al. (2001)	D2	46.73	48.73	Reinprecht et al. (2006)
B2	77.84	79.84	Orf et al. (1999a)	C2	111.84	113.84	Wang et al. (2004)	D2	46.73	48.73	Orf et al. (1999a)
C1	11.51	16.61	Kim et al. (2012)	C2	111.84	113.84	Du et al. (2009)	D2	52.85	54.85	Orf et al. (1999a)
C1	20.85	26.35	Palomeque et al. (2009a)	C2	112.40	114.40	Orf et al. (1999a)	D2	85.69	89.19	Kazi et al. (2010)
C1	20.89	27.09	Kim et al. (2012)	C2	112.45	112.80	Wang et al. (2004)	D2	104.15	106.15	Kabelka et al. (2004)
C1	71.45	82.51	Palomeque et al. (2009a)	C2	112.83	112.84	Du et al. (2009)	E	2.30	4.30	Wang et al. (2004)

C1	73.45	75.45	Rossi et al. (2013)	C2	112.95	114.95	Palomeque et al. (2009a)	E	2.30	4.30	Wang et al. (2004)
C1	75.23	77.23	Guzman et al. (2007)	C2	112.95	114.95	Rossi et al. (2013)	E	32.47	34.47	Wang et al. (2004)
C1	77.65	79.65	Yuan et al. (2002)	C2	113.41	113.62	Du et al. (2009)	E	43.75	45.75	Reinprecht et al. (2006)
C2	29.46	31.46	Guzman et al. (2007)	C2	114.78	116.72	Mansur et al. (1993a)	E	44.00	46.00	Rossi et al. (2013)
C2	39.30	41.30	Specht et al. (2001)	C2	115.20	116.72	Gai et al. (2007)	F	2.63	4.63	Gai et al. (2007)
F	70.42	72.42	Orf et al. (1999a)	J	11.41	13.41	Guzman et al. (2007)	L	8.40	10.40	Mansur et al. (1996)
F	74.97	76.97	Rossi et al. (2013)	J	11.41	13.41	Orf et al. (1999a)	L	33.54	35.54	Guzman et al. (2007)
F	101.08	103.08	Orf et al. (1999a)	J	19.80	21.80	Mansur et al. (1996)	L	60.34	62.34	Panthee et al. (2007)
F	141.40	143.40	Specht et al. (2001)	J	28.38	39.18	Eskandari et al. (2013b)	L	65.50	67.50	Orf et al. (1999a)
G	1.20	3.20	Reinprecht et al. (2006)	J	36.04	38.04	Guzman et al. (2007)	L	69.36	71.36	Orf et al. (1999a)
G	37.47	39.47	Reinprecht et al. (2006)	J	40.90	42.90	Li et al. (2008a)	L	70.44	72.44	Orf et al. (1999b)
G	61.63	76.76	Du et al. (2009)	J	41.25	43.25	Guzman et al. (2007)	L	85.60	87.60	Orf et al. (1999a)
G	68.66	76.76	Du et al. (2009)	J	43.08	45.08	Guzman et al. (2007)	L	88.10	90.10	Specht et al. (2001)
G	94.50	105.50	Kim et al. (2012)	J	66.79	68.79	Guzman et al. (2007)	L	91.00	93.00	Mansur et al. (1996)
G	95.47	97.47	Kabelka et al. (2004)	J	72.74	74.74	Specht et al. (2001)	L	94.40	96.40	Orf et al. (1999a)
H	68.12	70.12	Specht et al. (2001)	J	77.57	79.57	Reinprecht et al. (2006)	L	106.24	108.24	Orf et al. (1999a)
H	85.49	87.49	Kabelka et al. (2004)	K	0.80	2.80	Yuan et al. (2002)	L	107.23	115.07	Du et al. (2009)
H	86.48	89.51	Du et al. (2009)	K	35.99	37.99	Wang et al. (2004)	M	6.84	8.84	Zhang et al. (2004)
I	31.40	33.40	Sebolt et al. (2000)	K	36.99	40.86	Wang et al. (2004)	M	17.60	19.60	Specht et al. (2001)
I	34.35	36.35	Sebolt et al. (2000)	K	42.35	44.35	Kabelka et al. (2004)	M	17.60	19.60	Orf et al. (1999a)
I	35.40	37.40	Nichols et al. (2006)	K	44.59	46.59	Guzman et al. (2007)	M	18.58	40.35	Zhang et al. (2004)

(continued)

Table 6.1 (continued)

Group	Start	End	Reference	Group	Start	End	Reference	Group	Start	End	Reference
I	36.40	36.94	Chung et al. (2003)	K	45.74	47.38	Yuan et al. (2002)	M	33.40	50.10	Wang et al. (2004)
I	45.22	47.22	Reinprecht et al. (2006)	K	47.38	49.53	Yuan et al. (2002)	M	33.47	50.09	Wang et al. (2004)
I	85.73	87.73	Palomeque et al. (2009a)	K	50.56	56.61	Eskandari et al. (2013b)	M	34.85	36.85	Orf et al. (1999a)
I	85.73	87.73	Rossi et al. (2013)	K	55.61	57.61	Rossi et al. (2013)	M	37.98	39.98	Mansur et al. (1996)
I	97.11	99.11	Du et al. (2009)	K	58.00	61.66	Du et al. (2009)	M	38.98	58.50	Mansur et al. (1993a)
I	111.70	113.70	Yuan et al. (2002)	K	107.19	109.19	Palomeque et al. (2009a)	M	39.35	41.35	Mansur et al. (1993b)
I	111.70	113.70	Kabelka et al. (2004)	K	107.19	109.19	Rossi et al. (2013)	M	53.54	63.93	Du et al. (2009)
M	60.47	63.93	Du et al. (2009)	N	135.00	137.00	Hnetkovsky et al. (1996)	O	66.93	68.93	Rossi et al. (2013)
M	129.76	131.76	Kabelka et al. (2004)	O	0.00	8.75	Du et al. (2009)	O	81.08	83.08	Guzman et al. (2007)
M	139.69	141.69	Du et al. (2009)	O	4.44	6.44	Kabelka et al. (2004)				
N	52.25	54.25	Kabelka et al. (2004)	O	19.45	21.45	Guzman et al. (2007)				
N	74.91	74.91	Kabelka et al. (2004)	O	38.82	40.82	Rossi et al. (2013)				
N	90.55	94.55	Kim et al. (2012)	O	39.82	55.81	Palomeque et al. (2009a)				

Source: *SoyBase*

Table 6.2 Binding group, initial and final position, in centimorgan (cM), in the quantitative trait loci chromosome associated with the characteristic of number of nodes in soybean

Group	Start	End	Reference	Group	Start	End	Reference	Group	Start	End	Reference
A1	70.38	86.38	Chen et al. (2007)	C2	112.40	114.40	Zhang et al. (2004)	F	2.63	4.63	Gai et al. (2007)
B1	27.33	49.33	Chen et al. (2007)	C2	112.95	114.95	Liu et al. (2011)	F	11.37	16.08	Zhang et al. (2004)
B1	28.33	32.50	Chen et al. (2007)	C2	115.20	116.71	Gai et al. (2007)	F	76.69	78.69	Gai et al. (2007)
B1	32.50	46.38	Chen et al. (2007)	C2	115.20	116.72	Zhang et al. (2004)	F	83.50	85.50	Gai et al. (2007)
B1	36.48	46.38	Chen et al. (2007)	C2	116.71	131.75	Gai et al. (2007)				
B1	58.10	60.10	Gai et al. (2007)	D1a	48.13	49.84	Chen et al. (2007)				
B2	72.00	74.00	Gai et al. (2007)	D1b	83.04	85.04	Liu et al. (2011)				
C2	67.98	82.23	Moongkanna et al. (2011)	F	0.00	11.37	Gai et al. (2007)				

Source: *SoyBase*

Table 6.3 Binding group, initial and final position, in centimorgan (cM), on the quantitative trait loci chromosome associated with the characteristic of number of pods in soybean

Group	Start	End	Reference	Group	Start	End	Reference	Group	Start	End	Reference
A1	14.37	17.16	Zhang et al. (2010)	C2	106.58	111.80	Sun et al. (2006)	I	98.05	98.38	Zhang et al. (2010)
A2	8.14	10.14	Liu et al. (2011)	C2	106.58	108.58	Palomeque et al. (2009b)	J	39.63	42.25	Zhang et al. (2010)
A2	35.68	53.75	Zhang et al. (2010)	C2	112.95	114.95	Liu et al. (2011)	K	54.50	65.60	Palomeque et al. (2009b)
B1	28.33	40.33	Sun et al. (2006)	C2	112.95	114.95	Palomeque et al. (2009b)	K	66.43	90.93	Vieira et al. (2006)
B1	32.50	47.60	Sun et al. (2006)	C2	116.76	121.76	Chen et al. (2007)	L	0.00	107.23	Zhang et al. (2010)
B1	36.48	46.38	Zhang et al. (2010)	D1b	25.60	30.74	Zhang et al. (2010)	N	101.75	112.20	Zhang et al. (2010)
C1	74.50	85.70	Vieira et al. (2006)	E	64.18	70.23	Zhang et al. (2010)	O	58.40	106.02	Chen et al. (2007)
C2	106.58	111.60	Sun et al. (2006)	G	55.99	57.32	Zhang et al. (2010)				

Source: *SoyBase*

Table 6.4 Binding group, initial and final position, in centimorgan (cM), in the quantitative trait loci chromosome associated with the characteristic of number of pods per node in soybean

Group	Start	End	Reference	Group	Start	End	Reference	Group	Start	End	Reference
A1	48.29	50.29	Gai et al. (2007)	C2	112.95	114.95	Palomeque et al. (2009b)	I	85.73	87.73	Palomeque et al. (2009b)
A1	48.29	50.29	Zhang et al. (2004)	C2	114.20	116.20	Zhang et al. (2004)	K	107.19	109.19	Palomeque et al. (2009b)
C1	25.35	27.35	Zhang et al. (2004)	C2	115.20	116.42	Gai et al. (2007)	M	49.09	51.09	Gai et al. (2007)
C2	106.58	108.58	Palomeque et al. (2009b)	C2	115.20	116.72	Zhang et al. (2004)				
C2	108.26	110.26	Gai et al. (2007)	C2	116.72	131.75	Gai et al. (2007)				

Source: *SoyBase*

Table 6.5 Binding group, initial and final position, in centimorgan (cM), in the quantitative trait loci chromosome associated with the seed weight characteristic in soybean

Group	Start	End	Reference	Group	Start	End	Reference	Group	Start	End	Reference
A1	16.16	19.16	Chen et al. (2007)	A2	100.08	159.63	Han et al. (2012)	B2	64.55	66.55	Li et al. (2008a)
A1	16.16	21.66	Chen et al. (2007)	A2	100.48	119.18	Sun et al. (2012)	B2	64.60	66.60	Hoeck et al. (2003)
A1	26.42	95.95	Han et al. (2012)	A2	107.78	109.78	Orf et al. (1999a)	B2	71.51	73.51	Specht et al. (2001)
A1	26.42	28.08	Han et al. (2012)	A2	115.73	117.73	Specht et al. (2001)	B2	71.80	73.80	Hoeck et al. (2003)
A1	26.78	28.78	Orf et al. (1999a)	A2	118.09	120.09	Mansur et al. (1996)	C1	17.62	19.62	Mian et al. (1996b)
A1	27.66	37.86	Sun et al. (2012)	B1	4.07	6.07	Orf et al. (1999a)	C1	32.30	34.30	Orf et al. (1999a)
A1	27.78	28.08	Han et al. (2012)	B1	11.10	53.90	Sun et al. (2012)	C1	43.08	45.08	Yang et al. (2011)
A1	29.93	36.93	Chen et al. (2007)	B1	31.50	33.50	Gai et al. (2007)	C1	72.17	74.17	Orf et al. (1999a)
A1	87.58	89.58	Orf et al. (1999a)	B1	31.50	33.50	Zhang et al. (2004)	C1	72.84	74.84	Liang et al. (2010b)
A1	91.89	93.89	Specht et al. (2001)	B1	32.50	46.38	Chen et al. (2007)	C1	73.32	84.80	Han et al. (2012)
A1	92.59	94.59	Mansur et al. (1996)	B1	53.41	57.91	Liang et al. (2010b)	C1	82.51	84.80	Chen et al. (2007)
A2	8.14	10.14	Liu et al. (2011)	B1	56.91	58.91	Liang et al. (2010b)	C2	22.34	24.34	Specht et al. (2001)
A2	9.14	60.59	Han et al. (2012)	B1	57.91	59.91	Maughan et al. (1996)	C2	22.34	24.34	Orf et al. (1999a)
A2	36.77	100.08	Han et al. (2012)	B1	58.66	60.66	Lee et al. (2001)	C2	24.07	26.07	Mansur et al. (1996)
A2	36.77	60.59	Han et al. (2012)	B1	64.91	123.95	Li et al. (2008b)	C2	45.75	61.98	Moongkanna et al. (2011)
A2	38.06	145.57	Han et al. (2012)	B1	79.19	89.19	Han et al. (2012)	C2	94.61	96.61	Mian et al. (1996b)
A2	49.40	51.40	Maughan et al. (1996)	B1	80.31	82.31	Specht et al. (2001)	C2	97.83	107.58	Han et al. (2012)
A2	53.90	55.90	Hoeck et al. (2003)	B1	84.19	100.87	Han et al. (2012)	C2	105.00	112.45	Funatsuki et al. (2005)
A2	53.91	55.91	Kim et al. (2010)	B1	100.87	123.95	Han et al. (2012)	C2	106.58	108.58	Palomeque et al. (2009b)
A2	53.92	55.92	Orf et al. (1999a)	B2	5.05	7.05	Liu et al. (2011)	C2	106.58	108.58	Rossi et al. (2013)

A2	56.13	58.13	Yang et al. (2011)	B2	39.63	72.80	Han et al. (2012)	C2	106.60	108.60	Orf et al. (1999a)
A2	95.97	97.97	Zhang et al. (2004)	B2	54.18	56.18	Mian et al. (1996b)	C2	107.58	121.26	Han et al. (2012)
C2	107.60	119.80	Hyten et al. (2004)	D2	21.97	23.97	Zhang et al. (2004)	F	70.42	72.42	Orf et al. (1999a)
C2	109.00	116.00	Funatsuki et al. (2005)	D2	22.97	24.92	Zhang et al. (2004)	F	71.41	77.69	Eskandari et al. (2013b)
C2	112.19	113.38	Han et al. (2012)	D2	28.56	30.56	Chapman et al. (2003)	F	73.70	75.70	Lee et al. (2001)
C2	112.95	114.95	Palomeque et al. (2009b)	D2	38.34	40.34	Kim et al. (2010)	F	74.97	76.97	Rossi et al. (2013)
C2	112.95	114.95	Rossi et al. (2013)	D2	46.73	48.73	Panthee et al. (2005)	G	33.10	47.40	Hyten et al. (2004)
C2	113.95	117.76	Han et al. (2012)	D2	56.10	58.10	Hoeck et al. (2003)	G	44.40	46.40	Fasoula et al. (2004)
C2	117.76	126.23	Han et al. (2012)	D2	91.20	93.70	Han et al. (2012)	G	51.60	53.60	Mian et al. (1996b)
C2	120.26	122.26	Reinprecht et al. (2006)	E	22.70	24.70	Lee et al. (2001)	G	52.00	58.40	Vieira et al. (2006)
D1a	16.52	18.52	Panthee et al. (2005)	E	31.27	33.27	Liu et al. (2011)	G	52.42	54.42	Fasoula et al. (2004)
D1a	17.52	107.89	Panthee et al. (2005)	E	35.79	41.24	Moongkanna et al. (2011)	G	55.81	57.81	Yang et al. (2011)
D1a	42.00	88.20	Hyten et al. (2004)	E	39.16	45.15	Li et al. (2008b)	G	62.16	85.56	Sun et al. (2012)
D1a	49.84	56.57	Han et al. (2012)	E	44.27	45.40	Han et al. (2012)	G	66.53	68.53	Maughan et al. (1996)
D1a	56.57	69.91	Han et al. (2012)	E	45.30	47.30	Mian et al. (1996b)	G	68.87	70.87	Specht et al. (2001)
D1a	59.00	64.70	Sun et al. (2012)	E	45.40	46.63	Han et al. (2012)	G	96.24	98.24	Lee et al. (2001)
D1a	63.62	73.62	Han et al. (2012)	E	46.57	48.57	Orf et al. (1999a)	G	96.24	98.24	Mian et al. (1996b)
D1a	74.30	76.30	Orf et al. (1999a)	F	2.04	4.04	Specht et al. (2001)	H	1.00	23.00	Funatsuki et al. (2005)
D1a	107.88	109.88	Liu et al. (2011)	F	14.33	17.83	Sun et al. (2012)	H	37.90	39.90	Hoeck et al. (2003)
D1a	107.88	109.88	Kim et al. (2010)	F	46.63	48.63	Mian et al. (1996b)	H	52.34	54.34	Li et al. (2008a)
D1b	75.94	77.34	Han et al. (2012)	F	50.50	52.50	Specht et al. (2001)	H	67.16	81.04	Han et al. (2012)

(continued)

Table 6.5 (continued)

Group	Start	End	Reference	Group	Start	End	Reference	Group	Start	End	Reference
D1b	76.04	77.34	Han et al. (2012)	F	51.20	77.70	Hyten et al. (2004)	H	80.00	82.00	Hoeck et al. (2003)
D1b	116.34	128.04	Sun et al. (2012)	F	55.45	62.45	Funatsuki et al. (2005)	H	81.04	91.20	Han et al. (2012)
D2	9.35	11.35	Mian et al. (1996b)	F	58.77	60.77	Mian et al. (1996b)	H	81.04	89.08	Han et al. (2012)
D2	21.97	23.97	Gai et al. (2007)	F	62.70	64.70	Hoeck et al. (2003)	H	81.04	89.51	Eskandari et al. (2013b)
I	18.50	46.22	Han et al. (2012)	K	55.61	57.61	Rossi et al. (2013)	M	79.02	81.02	Csanádi et al. (2001)
I	21.84	23.84	Csanádi et al. (2001)	K	66.43	90.93	Vieira et al. (2006)	M	111.08	113.08	Csanádi et al. (2001)
I	21.90	46.22	Han et al. (2012)	K	66.73	68.73	Gai et al. (2007)	M	132.80	134.80	Hoeck et al. (2003)
I	26.98	28.98	Reinprecht et al. (2006)	K	66.73	68.73	Mian et al. (1996b)	N	31.84	33.84	Kim et al. (2010)
I	31.42	33.42	Sebolt et al. (2000)	K	83.54	85.54	Mian et al. (1996b)	N	32.84	34.52	Li et al. (2008b)
I	35.40	37.40	Nichols et al. (2006)	L	3.10	5.10	Mian et al. (1996b)	N	64.45	66.45	Kim et al. (2010)
I	43.00	45.00	Sebolt et al. (2000)	L	29.20	31.20	Hoeck et al. (2003)	N	78.51	84.51	Chen et al. (2007)
I	45.22	47.22	Reinprecht et al. (2006)	L	33.54	35.54	Csanádi et al. (2001)	N	78.51	83.01	Chen et al. (2007)
I	45.22	47.22	Liu et al. (2011)	L	35.70	37.70	Maughan et al. (1996)	O	4.44	6.44	Csanádi et al. (2001)
I	46.22	112.69	Han et al. (2012)	L	59.52	70.36	Han et al. (2012)	O	54.20	118.30	Han et al. (2012)
I	60.53	65.13	Sun et al. (2012)	L	62.20	70.50	Hyten et al. (2004)	O	58.40	93.37	Chen et al. (2007)
I	82.78	100.78	Hyten et al. (2004)	L	65.26	85.26	Stombaugh et al. (2004)	O	58.40	60.40	Vieira et al. (2006)
I	85.73	87.73	Palomeque et al. (2009b)	L	69.36	71.36	Orf et al. (1999a)	O	58.40	71.90	Sun et al. (2012)
J	26.63	28.63	Mian et al. (1996b)	L	70.00	72.00	Han et al. (2012)	O	63.00	65.00	Csanádi et al. (2001)
J	27.13	29.13	Maughan et al. (1996)	L	70.36	71.44	Han et al. (2012)	O	81.09	83.09	Csanádi et al. (2001)

J	42.00	44.00	Han et al. (2012)	L	77.23	79.23	Orf et al. (1999a)	O	113.30	123.30	Han et al. (2012)
J	43.68	45.68	Kim et al. (2010)	L	88.10	90.10	Mian et al. (1996b)	O	118.30	127.50	Han et al. (2012)
K	15.59	42.70	Han et al. (2012)	L	91.00	93.00	Hoeck et al. (2003)	O	128.30	130.30	Liu et al. (2011)
K	29.28	31.28	Specht et al. (2001)	L	92.89	94.89	Csanádi et al. (2001)				
K	41.00	43.00	Han et al. (2012)	L	100.30	102.30	Maughan et al. (1996)				
K	42.70	45.59	Han et al. (2012)	M	6.84	8.84	Specht et al. (2001)				
K	45.20	47.20	Kim et al. (2010)	M	13.56	18.58	Han et al. (2012)				
K	46.63	56.62	Hyten et al. (2004)	M	17.60	19.60	Orf et al. (1999a)				

Source: *SoyBase*

Table 6.6 Binding group, initial and final position, in centimorgan (cM), in the quantitative trait loci chromosome associated with the oil content characteristic in soybean

Group	Start	End	Reference	Group	Start	End	Reference	Group	Start	End	Reference
A1	0.00	4.95	Wang et al. (2012)	B1	28.17	30.17	Brummer et al. (1997)	C2	106.58	108.58	Reinprecht et al. (2006)
A1	28.00	30.00	Rossi et al. (2013)	B1	36.48	46.38	Qi et al. (2011)	C2	106.58	108.58	Palomeque et al. (2009b)
A1	28.08	30.28	Qi et al. (2011)	B1	80.30	82.30	Qi et al. (2011)	C2	107.59	117.77	Hyten et al. (2004)
A1	29.28	31.28	Mansur et al. (1996)	B2	32.13	34.13	Diers et al. (1992)	C2	112.38	114.38	Reinprecht et al. (2006)
A1	52.36	54.36	Brummer et al. (1997)	B2	68.98	78.66	Qi et al. (2011)	C2	112.95	114.95	Kim et al. (2010)
A1	74.41	76.41	Brummer et al. (1997)	B2	71.10	73.10	Csanádi et al. (2001)	C2	112.95	114.95	Palomeque et al. (2009b)
A1	85.58	96.58	Liang et al. (2010a)	B2	72.20	75.40	Chen et al. (2007)	C2	112.95	114.95	Rossi et al. (2013)
A1	87.58	89.58	Orf et al. (1999a)	B2	72.80	89.90	Liang et al. (2010a)	D1a	17.52	56.20	Hyten et al. (2004)
A1	91.24	94.19	Qi et al. (2011)	B2	77.83	79.83	Eskandari et al. (2013a)	D1a	56.43	59.00	Qi et al. (2011)
A1	91.30	93.30	Brummer et al. (1997)	B2	96.38	104.73	Tajuddin et al. (2003)	D1a	68.62	72.26	Qi et al. (2011)
A1	92.59	94.59	Orf et al. (1999a)	B2	99.10	110.30	Tajuddin et al. (2003)	D1a	68.90	70.90	Specht et al. (2001)
A1	92.59	94.59	Mansur et al. (1996)	B2	115.23	117.23	Gai et al. (2007)	D1a	68.91	70.91	Qi et al. (2011)
A1	92.88	101.57	Moongkanna et al. (2011)	C1	5.36	15.67	Qi et al. (2011)	D1a	85.48	91.26	Qi et al. (2011)
A1	93.92	95.92	Specht et al. (2001)	C1	9.34	11.34	Orf et al. (1999a)	D1a	108.66	110.66	Eskandari et al. (2013a)
A1	95.54	101.57	Moongkanna et al. (2011)	C1	61.88	65.08	Li et al. (2011)	D1b	7.63	9.63	Zhang et al. (2004)
A2	13.99	15.99	Qi et al. (2011)	C1	89.70	91.70	Lee et al. (1996)	D1b	36.07	38.07	Kabelka et al. (2004)

A2	43.90	67.33	Mansur et al. (1993a)	C1	89.72	91.72	Fasoula et al. (2004)	D1b	74.29	76.29	Rossi et al. (2013)
A2	50.36	53.36	Liang et al. (2010a)	C1	122.79	124.79	Kabelka et al. (2004)	D1b	86.19	88.19	Kim et al. (2010)
A2	54.91	78.21	Liang et al. (2010a)	C2	4.22	8.34	Wang et al. (2012)	D1b	115.35	117.35	Panthee et al. (2005)
A2	54.91	79.75	Tajuddin et al. (2003)	C2	37.05	39.05	Orf et al. (1999a)	D1b	116.34	137.05	Qi et al. (2011)
A2	63.33	71.33	Tajuddin et al. (2003)	C2	96.20	98.20	Brummer et al. (1997)	D2	24.52	57.07	Hyten et al. (2004)
A2	131.31	133.31	Brummer et al. (1997)	C2	100.78	107.58	Liang et al. (2010a)	D2	29.79	31.79	Eskandari et al. (2013a)
D2	72.50	74.50	Lee et al. (1996)	F	70.41	72.41	Specht et al. (2001)	I	26.98	28.98	Reinprecht et al. (2006)
D2	75.29	80.19	Qi et al. (2011)	F	74.97	76.97	Rossi et al. (2013)	I	27.53	35.16	Tajuddin et al. (2003)
D2	78.40	80.40	Lee et al. (1996)	F	81.97	97.97	Eskandari et al. (2013a)	I	30.56	34.66	Qi et al. (2011)
D2	78.60	80.60	Lee et al. (1996)	G	47.79	49.79	Reinprecht et al. (2006)	I	30.78	42.02	Shibata et al. (2008)
D2	84.18	92.12	Qi et al. (2011)	G	58.00	63.00	Qi et al. (2011)	I	31.42	33.42	Sebolt et al. (2000)
D2	105.44	113.74	Liang et al. (2010a)	G	64.56	66.56	Brummer et al. (1997)	I	34.22	36.22	Specht et al. (2001)
E	5.30	7.30	Diers et al. (1992)	G	65.56	73.41	Qi et al. (2011)	I	34.35	36.35	Tajuddin et al. (2003)
E	12.60	14.60	Diers et al. (1992)	G	66.53	68.53	Brummer et al. (1997)	I	36.02	49.34	Qi et al. (2011)
E	15.72	22.86	Shibata et al. (2008)	G	66.70	68.70	Brummer et al. (1997)	I	36.40	36.94	Chung et al. (2003)
E	22.70	24.70	Lee et al. (1996)	G	94.40	96.57	Qi et al. (2011)	I	37.06	39.06	Diers et al. (1992)
E	22.84	28.27	Qi et al. (2011)	G	94.83	99.78	Wang et al. (2012)	I	38.40	40.40	Diers et al. (1992)
E	27.27	29.27	Diers et al. (1992)	G	96.69	98.69	Lee et al. (1996)	I	49.11	51.11	Qi et al. (2011)
E	28.27	35.79	Qi et al. (2011)	G	98.30	100.30	Lee et al. (1996)	I	49.11	51.11	Reinprecht et al. (2006)

(continued)

Table 6.6 (continued)

Group	Start	End	Reference	Group	Start	End	Reference	Group	Start	End	Reference
E	29.90	31.90	Diers et al. (1992)	H	32.17	34.17	Brummer et al. (1997)	I	57.82	59.82	Reinprecht et al. (2006)
E	31.27	33.27	Reinprecht et al. (2006)	H	45.95	47.95	Shibata et al. (2008)	I	85.73	87.73	Reinprecht et al. (2006)
E	33.62	35.62	Diers et al. (1992)	H	81.04	89.51	Eskandari et al. (2013a)	I	85.73	87.73	Palomeque et al. (2009b)
E	45.09	53.34	Wang et al. (2012)	H	85.80	87.80	Fasoula et al. (2004)	I	118.09	120.09	Specht et al. (2001)
E	53.10	57.70	Li et al. (2011)	H	85.80	87.80	Lee et al. (1996)	J	22.85	33.41	Tajuddin et al. (2003)
F	1.09	3.04	Qi et al. (2011)	H	86.48	90.25	Qi et al. (2011)	J	24.51	26.51	Kabelka et al. (2004)
F	11.37	16.08	Qi et al. (2011)	H	88.52	90.52	Panthee et al. (2005)	J	27.13	29.13	Tajuddin et al. (2003)
F	25.58	32.32	Wang et al. (2012)	H	123.05	125.05	Qiu et al. (1999)	J	41.63	46.53	Eskandari et al. (2013a)
F	50.69	77.69	Eskandari et al. (2013a)	I	17.50	19.50	Reinprecht et al. (2006)	J	56.20	58.20	Lee et al. (1996)
F	59.59	65.19	Wang et al. (2012)	I	21.84	23.84	Csanádi et al. (2001)	K	2.40	28.72	Mansur et al. (1993a)
K	42.38	45.68	Wang et al. (2012)	L	33.70	36.04	Qi et al. (2011)	M	102.49	112.89	Tajuddin et al. (2003)
K	50.56	56.61	Eskandari et al. (2013a)	L	35.70	37.70	Diers et al. (1992)	M	106.70	108.70	Tajuddin et al. (2003)
K	93.63	99.53	Wang et al. (2012)	L	35.70	37.70	Lee et al. (1996)	M	130.75	140.18	Tajuddin et al. (2003)
K	97.13	106.52	Qi et al. (2011)	L	66.51	89.13	Hyten et al. (2004)	N	28.52	32.84	Qi et al. (2011)
K	97.87	99.87	Brummer et al. (1997)	L	67.26	69.26	Qi et al. (2011)	N	31.84	33.84	Qi et al. (2011)
K	102.59	104.59	Eskandari et al. (2013a)	L	90.34	95.37	Qi et al. (2011)	N	91.55	102.55	Chen et al. (2007)
K	103.79	105.79	Csanádi et al. (2001)	L	91.00	93.00	Mansur et al. (1996)	N	92.55	102.05	Qi et al. (2011)
L	13.03	15.03	Reinprecht et al. (2006)	L	91.10	101.90	Hyten et al. (2004)	N	97.92	102.05	Wang et al. (2012)

L	13.98	27.92	Eskandari et al. (2013a)	L	94.40	96.40	Orf et al. (1999a)	N	101.05	115.05	Chen et al. (2007)
L	29.93	31.93	Kim et al. (2010)	M	35.85	49.90	Hyten et al. (2004)	O	4.44	6.44	Qi et al. (2011)
L	30.58	34.54	Fasoula et al. (2004)	M	39.35	41.35	Gai et al. (2007)	O	49.71	54.20	Panthee et al. (2005)
L	30.58	31.21	Qi et al. (2011)	M	50.09	60.04	Eskandari et al. (2013a)	O	121.04	127.50	Li et al. (2011)

Source: *SoyBase*

Table 6.7 Binding group, initial and final position, in centimorgan (cM), in the quantitative trait loci chromosome associated with soybean seed protein content characteristic

Group	Start	End	Reference	Group	Start	End	Reference	Group	Start	End	Reference
A1	29.28	31.28	Mansur et al. (1996)	B2	42.60	45.60	Lee et al. (1996)	C2	112.95	114.95	Palomeque et al. (2009b)
A1	90.30	94.30	Tajuddin et al. (2003)	B2	54.20	56.20	Kabelka et al. (2004)	C2	112.95	114.95	Rossi et al. (2013)
A1	92.45	95.95	Lu et al. (2013)	B2	77.38	94.48	Liang et al. (2010a)	C2	117.80	121.30	Hyten et al. (2004)
A1	92.59	94.59	Mansur et al. (1996)	B2	92.48	94.48	Reinprecht et al. (2006)	C2	121.02	123.02	Csanádi et al. (2001)
A1	92.59	94.59	Orf et al. (1999a)	B2	96.91	98.91	Zhang et al. (2004)	D1a	5.43	7.43	Brummer et al. (1997)
A1	93.92	95.92	Specht et al. (2001)	C1	9.34	11.34	Orf et al. (1999a)	D1a	39.80	41.80	Brummer et al. (1997)
A2	42.98	54.71	Tajuddin et al. (2003)	C1	20.04	22.04	Lee et al. (1996)	D1a	76.48	78.48	Pandurangan et al. (2012)
A2	48.50	49.50	Tajuddin et al. (2003)	C1	32.30	34.30	Specht et al. (2001)	D1a	76.50	78.50	Csanádi et al. (2001)
A2	52.57	81.01	Lu et al. (2013)	C1	47.08	75.08	Stombaugh et al. (2004)	D1b	7.63	9.63	Reinprecht et al. (2006)
A2	52.57	54.57	Lu et al. (2013)	C1	64.08	66.08	Orf et al. (1999a)	D1b	36.07	38.07	Kabelka et al. (2004)
A2	53.91	55.91	Reinprecht et al. (2006)	C1	89.70	91.70	Brummer et al. (1997)	D1b	70.65	75.66	Chen et al. (2007)
A2	131.31	133.31	Brummer et al. (1997)	C1	96.00	98.00	Lee et al. (1996)	D2	78.23	80.23	Reinprecht et al. (2006)
A2	144.57	146.57	Kabelka et al. (2004)	C1	120.00	122.00	Lee et al. (1996)	D2	99.02	115.64	Tajuddin et al. (2003)
A2	149.00	151.00	Vollmann et al. (2002)	C1	122.79	124.79	Kabelka et al. (2004)	D2	105.49	109.49	Tajuddin et al. (2003)
B1	24.66	35.00	Lu et al. (2013)	C1	125.50	127.50	Lee et al. (1996)	E	2.30	4.30	Pandurangan et al. (2012)

B1	28.17	30.17	Brummer et al. (1997)	C2	29.92	50.65	Tajuddin et al. (2003)	E	5.30	7.30	Diers et al. (1992)
B1	35.48	37.48	Chapman et al. (2003)	C2	41.36	43.36	Pandurangan et al. (2012)	E	13.73	24.85	Tajuddin et al. (2003)
B1	46.38	58.91	Gai et al. (2007)	C2	69.69	90.93	Lu et al. (2013)	E	18.80	20.00	Tajuddin et al. (2003)
B1	58.91	59.10	Gai et al. (2007)	C2	97.07	99.07	Kabelka et al. (2004)	E	25.02	27.02	Lee et al. (1996)
B1	99.87	101.87	Reinprecht et al. (2006)	C2	106.58	108.58	Reinprecht et al. (2006)	E	26.30	28.30	Lee et al. (1996)
B2	28.19	30.19	Lee et al. (1996)	C2	106.58	108.58	Rossi et al. (2013)	E	29.88	31.88	Brummer et al. (1997)
B2	32.13	34.13	Diers et al. (1992)	C2	112.90	114.90	Liang et al. (2010a)	E	29.89	31.89	Fasoula et al. (2004)
E	29.90	31.90	Lee et al. (1996)	H	32.17	34.17	Brummer et al. (1997)	I	85.73	87.73	Reinprecht et al. (2006)
E	44.00	46.00	Lee et al. (1996)	H	53.34	58.91	Lu et al. (2013)	J	26.63	28.63	Lee et al. (1996)
E	44.93	45.40	Chen et al. (2007)	H	81.04	89.51	Eskandari et al. (2013b)	K	5.80	25.80	Hyten et al. (2004)
E	44.93	45.77	Chen et al. (2007)	H	85.49	86.49	Kabelka et al. (2004)	K	30.75	32.75	Lee et al. (1996)
F	2.35	4.35	Reinprecht et al. (2006)	H	85.80	87.80	Lee et al. (1996)	K	39.86	41.86	Specht et al. (2001)
F	46.63	48.63	Brummer et al. (1997)	H	123.05	125.05	Qiu et al. (1999)	K	52.90	54.90	Lee et al. (1996)
F	70.41	72.41	Kabelka et al. (2004)	I	17.50	19.50	Reinprecht et al. (2006)	K	55.61	57.61	Rossi et al. (2013)
F	71.41	77.69	Eskandari et al. (2013b)	I	20.90	22.90	Reinprecht et al. (2006)	K	67.90	80.68	Lu et al. (2013)
F	79.70	99.70	Hyten et al. (2004)	I	25.68	48.18	Tajuddin et al. (2003)	K	71.41	77.69	Eskandari et al. (2013b)
F	104.84	106.84	Qiu et al. (1999)	I	31.40	33.40	Diers et al. (1992)	K	103.79	105.79	Csanádi et al. (2001)
F	129.63	131.63	Reinprecht et al. (2006)	I	31.42	33.42	Sebolt et al. (2000)	L	14.03	27.92	Eskandari et al. (2013b)

(continued)

Table 6.7 (continued)

Group	Start	End	Reference	Group	Start	End	Reference	Group	Start	End	Reference
G	1.84	17.64	Liang et al. (2010a)	I	31.42	33.42	Brummer et al. (1997)	L	29.19	33.19	Lu et al. (2013)
G	3.53	5.53	Reinprecht et al. (2006)	I	31.42	33.42	Diers et al. (1992)	L	35.70	37.70	Diers et al. (1992)
G	11.74	13.74	Panthee et al. (2005)	I	34.35	36.35	Sebolt et al. (2000)	L	51.02	56.13	Tajuddin et al. (2003)
G	42.38	44.38	Reinprecht et al. (2006)	I	35.16	50.11	Lu et al. (2013)	L	52.14	56.14	Tajuddin et al. (2003)
G	56.18	61.41	Lu et al. (2013)	I	35.40	37.40	Nichols et al. (2006)	L	65.50	67.50	Orf et al. (1999a)
G	62.16	67.16	Liang et al. (2010a)	I	35.40	37.40	Pandurangan et al. (2012)	L	91.00	93.00	Mansur et al. (1996)
G	66.53	68.53	Brummer et al. (1997)	I	35.94	37.94	Tajuddin et al. (2003)	L	106.24	108.24	Chapman et al. (2003)
G	66.70	68.70	Brummer et al. (1997)	I	36.40	36.94	Chung et al. (2003)	M	32.50	34.50	Specht et al. (2001)
G	88.97	90.97	Diers et al. (1992)	I	37.06	39.06	Diers et al. (1992)	M	32.50	34.50	Csanádi et al. (2001)
G	96.24	98.24	Brummer et al. (1997)	I	38.40	40.40	Diers et al. (1992)	M	35.85	50.10	Hyten et al. (2004)
G	106.62	112.34	Tajuddin et al. (2003)	I	49.11	51.11	Reinprecht et al. (2006)	M	37.98	39.98	Orf et al. (1999a)
H	30.95	46.95	Liang et al. (2010a)	I	83.73	85.73	Rossi et al. (2013)	M	50.09	60.07	Eskandari et al. (2013b)
M	129.76	131.76	Kabelka et al. (2004)	N	33.40	35.40	Lee et al. (1996)	O	4.44	6.44	Kabelka et al. (2004)
M	130.75	136.75	Lu et al. (2013)	N	74.91	76.91	Kabelka et al. (2004)	O	58.40	106.02	Chen et al. (2007)
N	29.35	31.35	Lee et al. (1996)	N	91.55	114.55	Chen et al. (2007)	O	70.10	72.10	Specht et al. (2001)

Source: *SoyBase*

Table 6.8 Binding group, initial and final position, in centimorgan (cM), on the quantitative trait loci chromosome associated with the aluminum tolerance, drought tolerance, flood tolerance, salinity tolerance, and water use efficiency characteristics in soybean

Group	Start	End	Reference	Group	Start	End	Reference	Group	Start	End	Reference
Aluminum tolerance											
A2	117.50	119.50	Bianchi-Hall et al. (2000)	D1b	38.04	42.04	Korir et al. (2011)	J	56.20	58.20	Bianchi-Hall et al. (2000)
B1	19.00	21.00	Bianchi-Hall et al. (2000)	D2	86.42	95.42	Korir et al. (2011)	K	60.32	62.32	Korir et al. (2011)
B1	39.10	41.10	Bianchi-Hall et al. (2000)	D2	86.42	95.55	Korir et al. (2011)	L	3.10	5.10	Bianchi-Hall et al. (2000)
B1	46.38	53.41	Korir et al. (2011)	F	16.08	33.18	Sharma et al. (2011)	L	34.54	38.16	Korir et al. (2011)
C2	126.23	145.47	Sharma et al. (2011)	F	66.20	68.20	Bianchi-Hall et al. (2000)	O	68.97	74.05	Korir et al. (2011)
Drought tolerance											
A1	14.08	41.82	Carpentieri-Pípolo et al. (2012)	N	70.43	81.39	Carpentieri-Pípolo et al. (2012)	O	60.00	88.90	Carpentieri-Pípolo et al. (2012)
H	80.42	89.51	Carpentieri-Pípolo et al. (2012)								
Flood tolerance											
A1	64.73	67.58	Cornelious et al. (2005)	C2	107.58	123.37	Githiri et al. (2006)	G	107.70	109.70	VanToai et al. (2001)
A2	54.91	67.86	Sayama et al. (2009)	C2	112.62	114.62	Githiri et al. (2006)	H	83.19	99.50	Sayama et al. (2009)
A2	86.83	100.83	Rizal and Karki (2011)	D1a	16.52	31.52	Rizal and Karki (2011)	K	66.01	73.01	Rizal and Karki (2011)
B1	56.60	61.60	Githiri et al. (2006)	D1b	3.79	20.61	Sayama et al. (2009)	L	27.27	39.27	Rizal and Karki (2011)
B2	11.82	22.82	Githiri et al. (2006)	D1b	51.61	53.61	Githiri et al. (2006)	L	91.37	96.37	Githiri et al. (2006)
C1	123.79	132.46	Sayama et al. (2009)	D1b	75.09	77.09	Githiri et al. (2006)	M	72.37	74.37	Githiri et al. (2006)

(continued)

Table 6.8 (continued)

Group	Start	End	Reference	Group	Start	End	Reference	Group	Start	End	Reference
C1	131.46	133.46	Githiri et al. (2006)	F	16.08	23.56	Cornelious et al. (2005)	O	81.08	85.08	Githiri et al. (2006)
C2	101.75	103.75	Githiri et al. (2006)	F	36.47	41.47	Rizal and Karki (2011)	O	106.02	108.02	Githiri et al. (2006)
Salinity tolerance											
N	73.91	77.91	Hamwieh and Xu (2008)	N	75.91	78.21	Hamwieh and Xu (2008)	N	78.51	76.51	Hamwieh et al. (2011)
N	74.98	76.98	Ha et al. (2013)	N	78.51	80.51	Lee et al. (2004)				
Water use efficiency											
C1	89.70	91.70	Mian et al. (1998)	H	32.64	34.64	Mian et al. (1996a)	J	87.90	89.90	Mian et al. (1996a)
G	51.60	53.60	Mian et al. (1996a)	J	67.30	69.30	Mian et al. (1996a)	L	94.40	96.40	Mian et al. (1998)
H	10.25	12.25	Mian et al. (1996a)	J	82.15	84.15	Mian et al. (1996a)				

Source: *SoyBase*

Breeding Strategies in View of New Developments in Genomics and Quantitative Genetics

Usually, several populations from different crosses are evaluated in self-pollinated plants, and several simple F_1 hybrids and their corresponding F_2 generations are generated. Plants from each F_2 population are taken for harvesting of seeds, and endogamous generations are advanced until lines are obtained, mainly by the method bulk within progenies (BDP, the acronym in Portuguese) (Ramalho et al. 2012a, b). With the advent of genome selection, new breeding and selection strategies for quantitative traits can be coupled to the BDP method.

Wide genome selection (WGS) is defined as the simultaneous selection for thousands of markers, which densely covers the genome, so that all genes of a quantitative trait are in linkage disequilibrium with at least a part of the markers. These markers on linkage disequilibrium with QTLs, of both major and minor effects, will explain almost the entire genetic variation of a quantitative trait. The number of SNPs (single-nucleotide polymorphisms) is of such magnitude that the probability of finding a QTL in linkage disequilibrium with at least one marker is very high.

The markers will have their genetic effects estimated from a sample of at least 1000 genotyped and phenotyped individuals, i.e., based on at least 1000 experimental replicates of each locus. Thus, although the heritability of each effective marker (the one that accurately identifies one of the polygenes) is very low, with 1000 replicates such heritability becomes high. In other words, the environment effect will be minimized by using a very high number of repetitions. This is the same evaluation and selection of quantitative philosophy of traits based on phenotypes in field experiments, implanted with large numbers of replicates.

In autogamous species, the benefit of WGS is due to three factors: (a) increased accuracy, (b) increased selection intensity in F_2, and (c) reduced selective cycle. The estimation of the effects of marks can be based on genotypes of F_2 generation plants and $F_{2:3}$ or $F_{2:4}$ generation phenotypes. This can also be performed based on genotypes and phenotypes (on repetition average level) of parent lines.

Tables 6.9, 6.10, 6.11, and 6.12 present the types of information used and the selection methods corresponding to four breeding strategies combining genomic, phenomic, and genealogical information. The four alternative breeding strategies are (1) genomic + phenomic selection at all stages with recombination in $F_{2:6}$ generation for recurrent selection, (2) genomic + phenomic selection at all stages with recombination in $F_{2:4}$ generation for recurrent selection, (3) genomic + phenomic selection at all stages with recombination in the F_2 generation for recurrent selection, and (4) purely genomic selection at some stages with recombination in the F_2 generation for recurrent selection.

Good alleles for quantitative traits of interest can be identified with WGS in F_2 generation, by directing the cross between the best plants, making intrapopulation recurrent selection in autogamous plants according to strategy 3 (Table 6.11). This allows to capitalize on the three benefits of WGS and, consequently, increases

Table 6.9 Source of information and combined genomic and phenomic selection methods associated with breeding strategy 1—recombination for recurrent selection in $F_{2:6}$ generation

Stage Selection method	Information sources used
Selection of F_0 parents and F_1 crosses	Phenotypes and genotypes of parent lines; catalog of marking effects
Genomic selection through prediction of genotypic values of crosses	
Selection of populations and plants in F_2	F_2 plant genotypes; catalog of estimated marking effects in F_0
Genomic selection through prediction of plant genetic values	
Selection of progenies/populations in $F_{2:3}$	Phenotypes of $F_{2:3}$ progenies/populations in $F_{2:3}$; genealogy; catalog of estimated marking effects in $F_2/F_{2:3}$
Genomic and phenomic combined selection through prediction of genetic additive values of the progenies	
Selection of progenies/populations in $F_{2:4}$	Phenotypes of progenies/$F_{2:3}$ populations; phenotypes of progenies/$F_{2:4}$ populations evaluated at various sites; genealogy; catalog of estimated marking effects in $F_2/F_{2:3}$
Genomic and phenomic combined selection based on multigeneration index (BLUP) via $F_{2:3} + F_{2:4}$ with additive effects of progenies and populations	
Selection of progenies/populations in $F_{2:5}$	Phenotypes of progenies/$F_{2:5}$ populations; evaluated at various sites; genealogy; catalog of estimated marking effects
Genomic and phenomic combined selection based on multigeneration index (BLUP) via $F_{2:3} + F_{2:4} + F_{2:5}$ with additive effects of progenies and populations	
Selection of progenies/populations in $F_{2:6}$	Phenotypes of progenies/$F_{2:6}$ populations; evaluated at various sites; genealogy; catalog of estimated marking effects
Genomic and phenomic combined selection based on multigeneration index (BLUP) via $F_{2:3} + F_{2:4} + F_{2:5} + F_{2:6}$ with additive effects of progenies and populations	
Selection of F_0 parents and F1 crosses—selection cycle 2 of the recurrent selection program; line selection	Phenotypes and genotypes of parent lines; catalog of marking effects; line tests
Genomic selection through prediction of genotypic values of crosses; selection based on line genotypic values	

Table 6.10 Source of information and combined genomic and phenomic selection methods associated with breeding strategy 2—recombination for recurrent selection in $F_{2:4}$ generation

Stage Selection method	Information sources used
Selection of F_0 parents and F_1 crosses	Phenotypes and genotypes of parent lines; catalog of marking effects
Genomic selection through prediction of genotypic values of crosses	
Selection of populations and plants in F_2	F_2 plant genotypes; catalog of estimated marking effects in F_0
Genomic selection through prediction of plant genetic values	
Selection of progenies/populations in $F_{2:3}$	Phenotypes of progenies/$F_{2:3}$ populations; genealogy; catalog of estimated marking effects in F_2/$F_{2:3}$
Genomic and phenomic combined selection through prediction of genetic additive values of the progenies	
Selection of progenies/populations in $F_{2:4}$; selection of F_0 parents and F_1 crosses—selective cycle 2 of the recurrent selection program	Phenotypes of progenies/$F_{2:3}$ populations; phenotypes of progenies/$F_{2:4}$ populations evaluated at various sites; genealogy; catalog of estimated marking effects in F_2/$F_{2:3}$
Genomic and phenomic combined selection based on multigeneration index (BLUP) via $F_{2:3} + F_{2:4}$ with additive effects of progenies and populations	
Selection of progenies/populations in $F_{2:5}$	Phenotypes of progenies/$F_{2:5}$ populations; evaluated at various sites; genealogy; catalog of estimated marking effects
Genomic and phenomic combined selection based on multigeneration index (BLUP) via $F_{2:3} + F_{2:4} + F_{2:5}$ with additive effects of progenies and populations	
Selection of progenies/populations in $F_{2:6}$	Phenotypes of progenies/$F_{2:6}$ populations; evaluated at various sites; genealogy; catalog of estimated marking effects
Genomic and phenomic combined selection based on multigeneration index (BLUP) via $F_{2:3} + F_{2:4} + F_{2:5} + F_{2:6}$ with additive effects of progenies and populations	
Line selection	Line tests
Selection based on genotypic values of lines	

Table 6.11 Source of information and combined genomic and phenomic selection methods associated with breeding strategy 3—recombination for recurrent selection in F_2 generation

Stage Selection method	Information sources used
Selection of F_0 parents and F_1 crosses	Phenotypes and genotypes of parent lines; catalog of marking effects
Genomic selection through prediction of genotypic values of crosses	
Selection of populations and plants in F_2; selection of F_0 parents and F_1 crosses—selective cycle 2 of the recurrent selection program	F_2 plant genotypes; catalog of estimated marking effects in F_0
Genomic selection through prediction of genetic additive values of plants; genomic selection via prediction of the genotypic values of crosses	
Selection of progenies/populations in $F_{2:3}$	Phenotypes of progenies/$F_{2:3}$ populations; genealogy; catalog of estimated marking effects in F_2/$F_{2:3}$
Genomic and phenomic combined selection through prediction of genetic additive values of the progenies	
Selection of progenies/populations in $F_{2:4}$	Phenotypes of progenies/$F_{2:3}$ populations; phenotypes of progenies/$F_{2:4}$ populations evaluated at various sites; genealogy; catalog of estimated marking effects in F_2/$F_{2:3}$
Genomic and phenomic combined selection based on multigeneration index (BLUP) via $F_{2:3} + F_{2:4}$ with additive effects of progenies and populations	
Selection of progenies/populations in $F_{2:5}$	Phenotypes of progenies/$F_{2:5}$ populations; evaluated at various sites; genealogy; catalog of estimated marking effects
Genomic and phenomic combined selection based on multigeneration index (BLUP) via $F_{2:3} + F_{2:4} + F_{2:5}$ with additive effects of progenies and populations	
Selection of progenies/populations in $F_{2:6}$	Phenotypes of progenies/$F_{2:6}$ populations; evaluated at various sites; genealogy; catalog of estimated marking effects
Genomic and phenomic combined selection based on multigeneration index (BLUP) via $F_{2:3} + F_{2:4} + F_{2:5} + F_{2:6}$ with additive effects of progenies and populations	
Line selection	Line tests
Selection based on genotypic values of lines	

Table 6.12 Source of information and combined genomic and phenomic selection methods associated with breeding strategy 4—recombination for recurrent selection in F_2 generation and purely genomic selection up to $F_{2:6}$

Stage	Information sources used	Selection method
Selection of F_0 parents and F_1 crosses	Phenotypes and genotypes of parent lines; catalog of marking effects	Genomic selection through prediction of genotypic values of crosses
Selection of populations and plants in F_2; selection of F_0 parents and F_1 crosses—selective cycle 2 of the recurrent selection program	F_2 plant genotypes; catalog of estimated marking effects in F_0	Genomic selection through prediction of genetic additive values of plants; genomic selection via prediction of the genotypic values of crosses
Advance of progenies in bulk within progenies until $F_{2:3}$	None	No selection
Advance of progenies in bulk within progenies until $F_{2:4}$	None	No selection
Advance of progenies in bulk within progenies until $F_{2:5}$	None	No selection
Selection of progenies/ populations in $F_{2:6}$	Phenotypes of progenies/$F_{2:6}$ populations; evaluated at various sites; genealogy; catalog of estimated marking effects	Genomic and phenomic combined selection based on $F_{2:6}$ with additive effects of progenies and populations
Line selection	Line tests	Selection based on genotypic values of lines

genetic gain. The potential increase in selection intensity is remarkable, since it is possible to evaluate a much higher number of F_2 plants via genotyping.

Now, to advance F_2 generation plants to homozygous lines, one can perform early selection via WGS in each generation, with no need to try progeny, according to strategy 4 (Table 6.12), thus maximizing the genetic gain per unit for quantitative traits.

In summary, strategy 4 (Table 6.12) presents the shorter interval between breeding cycles as the main trait, both for recurrent selection and for the new cultivar recommendation. However, it demands accurate catalog of genetic effects of markers. Strategy 3 (Table 6.11) is fast for recurrent and orthodox selection in the case of recommendation of new cultivars. Strategy 2 (Table 6.10) is intermediate for recurrent and orthodox selection for recommendations of new cultivars, but it has the advantage of being more accurate than strategy 3 for recurrent selection, for it uses more precise information (closer to F_∞) and in higher number (from experimentation in several locations). In turn, strategy 1 (Table 6.9) does not reduce the selective cycle, but provides greater selective accuracy by using all available phenotyping and genotyping information from multiple sites, multiple generations, and the effects both of progenies and populations.

Final Considerations

The main traits explored by breeding programs are quantitative nature. Thus, the breeding strategies for these traits, in view of new developments in genomics and quantitative genetics, were presented in this chapter. These strategies prioritize the use of WGS associated with phenomics to increase efficiency in predicting genomic values and, consequently, to improve selective accuracy.

References

Barona, M.A.A., Colombari Filho, J.M., Geraldi, I.O.: Epistasia para a produção de grãos em soja. Bragantia. **68**(2), 313–318 (2009)

Bianchi-Hall, C.M., Carter Jr., T.E., Bailey, M.A., Mian, M.A.R., Rufty, T.W., Ashley, D.A., Boerma, H.R., Arellano, C., Hussey, R.S., Parrott, W.A.: Aluminum tolerance associated with quantitative trait loci derived from soybean PI 416937 in hydroponics. Crop Sci. **40**(2), 538–545 (2000)

Brummer, E.C., Graef, G.L., Orf, J., Wilcox, J.R., Shoemaker, R.C.: Mapping QTL for seed protein and oil content in eight soybean populations. Crop Sci. **37**(2), 370–378 (1997)

Carpentieri-Pípolo, V., Gastaldi, L.F., Pipolo, A.E.: Correlações fenotípicas entre caracteres quantitativos em soja. Semina Ciênc. Agrár. **26**(1), 11–16 (2005)

Carpentieri-Pípolo, V., Pipolo, A.E., Abdel-Haleem, H., Boerma, H.R., Sinclair, T.R.: Identification of QTLs associated with limited leaf hydraulic conductance in soybean. Euphytica. **186** (3), 679–686 (2012)

Chapman, A., Pantalone, V.R., Ustun, A., Allen, F.L., Landau-Ellis, D., Trigiano, R.N., Gresshoff, P.M.: Quantitative trait loci for agronomic and seed quality traits in an F2 and F4:6 soybean population. Euphytica. **129**(3), 387–393 (2003)

Chen, Q., Zhang, Z., Liu, C., Xin, D., Qiu, H., Shan, D., Shan, C., Hu, G.: QTL analysis of major agronomic traits in soybean. Agric. Sci. China. **6**(4), 399–405 (2007)

Chung, J., Babka, H.L., Graef, G.L., Staswick, P.E., Lee, D.J., Cregan, P.B., Shoemaker, R.C., Specht, J.E.: The seed protein, oil, and yield QTL on soybean linkage group I. Crop Sci. **43**(3), 1053–1067 (2003)

Cornelious, B., Chen, P., Chen, Y., de Leon, N., Shannon, J., Wang, D.: Identification of QTLs underlying water-logging tolerance in soybean. Mol. Breed. **16**(2), 103–112 (2005)

Cruz, C.D.: Princípios de genética quantitativa, p. 394. Ed. UFV, Viçosa (2012)

Csanádi, G., Vollmann, J., Stift, G., Lelley, T.: Seed quality QTLs identified in a molecular map of early maturing soybean. Theor. Appl. Genet. **103**(6–7), 912–919 (2001)

Diers, B.W., Keim, P., Fehr, W.R., Shoemaker, R.C.: RFLP analysis of soybean seed protein and oil content. Theor. Appl. Genet. **83**(5), 608–612 (1992)

Du, W., Wang, M., Fu, S., Yu, D.: Mapping QTLs for seed yield and drought susceptibility index in soybean (*Glycine max* L.) across different environments. J. Genet. Genomics. **36**(12), 721–731 (2009)

Eskandari, M., Cober, E.R., Rajcan, I.: Genetic control of soybean seed oil: I. QTL and genes associated with seed oil concentration in RIL populations derived from crossing moderately high-oil parents. Theor. Appl. Genet. **126**(2), 483–495 (2013a)

Eskandari, M., Cober, E.R., Rajcan, I.: Genetic control of soybean seed oil: II. QTL and genes that increase oil concentration without decreasing protein or with increased seed yield. Theor. Appl. Genet. **126**(6), 1677–1687 (2013b)

Fasoula, V.A., Harris, D.K., Boerma, H.R.: Validation and designation of quantitative trait loci for seed protein, seed oil, and seed weight from two soybean populations. Crop Sci. **44**(4), 1218–1225 (2004)

Funatsuki, H., Kawaguchi, K., Matsuba, S., Sato, Y., Ishimoto, M.: Mapping of QTL associated with chilling tolerance during reproductive growth in soybean. Theor. Appl. Genet. **111**(5), 851–861 (2005)

Gai, J., Wang, Y., Wu, X., Chen, S.: A comparative study on segregation analysis and QTL mapping of quantitative traits in plants-with a case in soybean. Front. Agric. China. **1**(1), 1–7 (2007)

Githiri, S., Watanabe, S., Harada, K., Takahashi, R.: QTL analysis of flooding tolerance in soybean at an early vegetative growth stage. Plant Breed. **125**(6), 613–618 (2006)

Grant, D., Nelson, R.T., Cannon, S.B., Shoemaker, R.C.: SoyBase, the USDA-ARS soybean genetics and genomics database. Nucleic Acids Res. **38**, D843–D846 (2009). doi:10.1093/nar/gkp798

Guzman, P.S., Diers, B.W., Neece, D.J., St Martin, S.K., Leroy, A.R., Grau, C.R., Hughes, T., Nelson, R.L.: QTL associated with yield in three backcross-derived populations of soybean. Crop Sci. **47**(1), 111–122 (2007)

Ha, B.K., Vuong, T.D., Velusamy, V., Nguyen, H.T., Shannon, J.G., Lee, J.D.: Genetic mapping of quantitative trait loci conditioning salt tolerance in wild soybean (*Glycine soja*) PI 483463. Euphytica. **193**(1), 79–88 (2013)

Hamwieh, A., Xu, D.: Conserved salt tolerance quantitative trait locus (QTL) in wild and cultivated soybeans. Breed. Sci. **58**(4), 355–359 (2008)

Hamwieh, A., Tuyen, D.D., Cong, H., Benitez, E.R., Takahashi, R., Xu, D.H.: Identification and validation of a major QTL for salt tolerance in soybean. Euphytica. **179**(3), 451–459 (2011)

Han, Y., Li, D., Zhu, D., Li, H., Li, X., Teng, W., Li, W.: QTL analysis of soybean seed weight across multi-genetic backgrounds and environments. Theor. Appl. Genet. **125**(4), 671–683 (2012)

Hnetkovsky, N., Chang, S.J.C., Doubler, T.W., Gibson, P.T., Lightfott, D.A.: Genetic mapping of loci underlying field resistance to soybean sudden death syndrome (SDS). Crop Sci. **36**(2), 393–400 (1996)

Hoeck, J.A., Fehr, W.R., Shoemaker, R.C., Welke, G.A., Johnson, S.L., Cianzio, S.R.: Molecular marker analysis of seed size in soybean. Crop Sci. **43**(1), 68–74 (2003)

Hyten, D.L., Pantalone, V.R., Sams, C.E., Saxton, A.M., Landau-Ellis, D., Stefaniak, T.R., Schmidt, M.E.: Seed quality QTL in a prominent soybean population. Theor. Appl. Genet. **109**(3), 552–561 (2004)

Kabelka, E.A., Diers, B.W., Fehr, W.R., Leroy, A.R., Baianu, I.C., You, T., Neece, D.J., Nelson, R.L.: Putative alleles for increased yield from soybean plant introductions. Crop Sci. **44**(3), 784–791 (2004)

Kazi, S., Shultz, J., Afzal, J., Hashmi, R., Jasim, M., Bond, J., Arelli, P., Lightfoot, D.A.: Iso-lines and inbred-lines confirmed loci that underlie resistance from cultivar 'Hartwig' to three soybean cyst nematode populations. Theor. Appl. Genet. **120**(3), 633–644 (2010)

Kim, H.K., Kim, Y.C., Kim, S.T., Son, B.G., Choi, Y.W., Kang, J.S., Park, Y., Cho, Y., Choi, I.S.: Analysis of quantitative trait loci (QTLs) for seed size and fatty acid composition using recombinant inbred lines in soybean. J. Life Sci. **20**(8), 1186–1192 (2010)

Kim, K.S., Diers, B.W., Hyten, D.L., Mian, M.R., Shannon, J.G., Nelson, R.L.: Identification of positive yield QTL alleles from exotic soybean germplasm in two backcross populations. Theor. Appl. Genet. **125**(6), 1353–1369 (2012)

Korir, P., Qi, B., Wang, Y., Zhao, T., Yu, D., Chen, S., Gai, J.: A study on relative importance of additive, epistasis and unmapped QTL for aluminum tolerance at seedling stage in soybean. Plant Breed. **130**(5), 551–562 (2011)

Lee, S.H., Bailey, M.A., Mian, M.A.R., Carter Jr., T.E., Shipe, E.R., Ashley, D.A., Parrott, W.A., Hussey, R.S., Boerma, H.R.: RFLP loci associated with soybean seed protein and oil content across populations and locations. Theor. Appl. Genet. **93**(5–6), 649–657 (1996)

Lee, S.H., Park, K.Y., Lee, H.S., Park, E.H., Boerma, H.R.: Genetic mapping of QTLs conditioning soybean sprout yield and quality. Theor. Appl. Genet. **103**(5), 702–709 (2001)

Lee, G.J., Boerma, H.R., Villagarcia, M.R., Zhou, X., Carter Jr., T.E., Li, Z., Gibbs, M.O.: A major QTL conditioning salt tolerance in S-100 soybean and descendent cultivars. Theor. Appl. Genet. **109**(8), 1610–1619 (2004)

Li, D., Pfeiffer, T.W., Cornelius, P.L.: Soybean QTL for yield and yield components associated with alleles. Crop Sci. **48**(2), 571–581 (2008a)

Li, W., Zheng, D.H., Van, K., Lee, S.H.: QTL mapping for major agronomic traits across two years in soybean (*Glycine max* L. Merr.) J. Crop. Sci. Biotechnol. **11**, 171–190 (2008b)

Li, H., Zhao, T., Wang, Y., Yu, D., Chen, S., Zhou, R., Gai, J.: Genetic structure composed of additive QTL, epistatic QTL pairs and collective unmapped minor QTL conferring oil content and fatty acid components of soybeans. Euphytica. **182**(1), 117–132 (2011)

Liang, H.Z., Yu, Y.L., Wang, S.F., Yun, L.I.A.N., Wang, T.F., Wei, Y.L., Gong, P., Liu, X., Fang, X., Zhang, M.C.: QTL mapping of isoflavone, oil and protein contents in soybean (*Glycine max* L. Merr.) Agric. Sci. China. **9**(8), 1108–1116 (2010a)

Liang, Q., Cheng, X., Mei, M., Yan, X., Liao, H.: QTL analysis of root traits as related to phosphorus efficiency in soybean. Ann. Bot. **106**(1), 223–234 (2010b)

Liu, W., Kim, M., Kang, Y., Van, K., Lee, Y., Srinives, P., Yuan, D., Lee, S.: QTL identification of flowering time at three different latitudes reveals homeologous genomic regions that control flowering in soybean. Theor. Appl. Genet. **123**(4), 545–553 (2011)

Lu, W., Wen, Z., Li, H., Yuan, D., Li, J., Zhang, H., Huang, Z., Cui, S., Du, W.: Identification of the quantitative trait loci (QTL) underlying water soluble protein content in soybean. Theor. Appl. Genet. **126**(2), 425–433 (2013)

Mansur, L.M., Lark, K.G., Kross, H., Oliveira, A.: Interval mapping of quantitative trait loci for reproductive, morphological, and seed traits of soybean (*Glycine max* L.) Theor. Appl. Genet. **86**(8), 907–913 (1993a)

Mansur, L.M., Orf, J., Lark, K.G.: Determining the linkage of quantitative trait loci to RFLP markers using extreme phenotypes of recombinant inbreds of soybean (*Glycine max* L. Merr.) Theor. Appl. Genet. **86**(8), 914–918 (1993b)

Mansur, L.M., Orf, J.H., Chase, K., Jarvik, T., Cregan, P.B., Lark, K.G.: Genetic mapping of agronomic traits using recombinant inbred lines of soybean. Crop Sci. **36**(5), 1327–1336 (1996)

Maughan, P.J., Maroof, M.S., Buss, G.R.: Molecular-marker analysis of seed-weight: genomic locations, gene action, and evidence for orthologous evolution among three legume species. Theor. Appl. Genet. **93**(4), 574–579 (1996)

Mian, M.A.R., Bailey, M.A., Ashley, D.A., Wells, R., Carter Jr., T.E., Parrott, W.A., Boerma, H. R.: Molecular markers associated with water use efficiency and leaf ash in soybean. Crop Sci. **36**(5), 1252–1257 (1996a)

Mian, M.A.R., Bailey, M.A., Tamulonis, J.P., Shipe, E.R., Carter Jr., T.E., Parrott, W.A., Ashley, D.A., Hussey, R.S., Boerma, H.R.: Molecular markers associated with seed weight in two soybean populations. Theor. Appl. Genet. **93**(7), 1011–1016 (1996b)

Mian, M.A.R., Ashley, D.A., Boerma, H.R.: An additional QTL for water use efficiency in soybean. Crop Sci. **38**(2), 390–393 (1998)

Möller, M.: Mapeamento de locos de resistência quantitativa da soja ao complexo de percevejos. 87 f. Tese de D.S., Escola Superior de Agricultura "Luiz de Queiroz", Universidade de São Paulo, Piracicaba (2010)

Moongkanna, J., Nakasathien, S., Novitzky, W., Kwanyen, P., Sinchaisri, P., Srinives, P.: SSR markers linking to seed traits and total oil content in soybean. Thai J. Agric. Sci. **44**(4), 233–241 (2011)

Nichols, D.M., Glover, K.D., Carlson, S.R., Specht, J.E., Diers, B.W.: Fine mapping of a seed protein QTL on soybean linkage group I and its correlated effects on agronomic traits. Crop Sci. **46**(2), 834–839 (2006)

Orf, J.H., Chase, K., Jarvik, T., Mansur, L.M., Cregan, P.B., Adler, F.R., Lark, K.G.: Genetics of soybean agronomic traits: I. Comparison of three related recombinant inbred populations. Crop Sci. **39**(6), 1642–1651 (1999a)

Orf, J.H., Chase, K., Adler, F.R., Mansur, L.M., Lark, K.G.: Genetics of soybean agronomic traits: II. Interactions between yield quantitative trait loci in soybean. Crop Sci. **39**(6), 1652–1657 (1999b)

Palomeque, L., Li-Jun, L., Li, W., Hedges, B., Cober, E.R., Rajcan, I.: QTL in mega-environments: I. Universal and specific seed yield QTL detected in a population derived from a cross of high-yielding adapted × high-yielding exotic soybean lines. Theor. Appl. Genet. **119**(3), 417–427 (2009a)

Palomeque, L., Li-Jun, L., Li, W., Hedges, B., Cober, E.R., Rajcan, I.: QTL in mega-environments: II. Agronomic trait QTL co-localized with seed yield QTL detected in a population derived from a cross of high-yielding adapted x high-yielding exotic soybean lines. Theor. Appl. Genet. **119**(3), 429–436 (2009b)

Pandurangan, S., Pajak, A., Molnar, S.J., Cober, E.R., Dhaubhadel, S., Hernández-Sebastià, C., Kaiser, W., Nelson, R., Huber, S., Marsolais, F.: Relationship between asparagine metabolism and protein concentration in soybean seed. J. Exp. Bot. **63**(8), 3173–3184 (2012). doi:10.1093/jxb/ers039

Panthee, D.R., Pantalone, V.R., West, D.R., Saxton, A.M., Sams, C.E.: Quantitative trait loci for seed protein and oil concentration, and seed size in soybean. Crop Sci. **45**(5), 2015–2022 (2005)

Panthee, D.R., Pantalone, V.R., Saxton, A.M., West, D.R., Sams, C.E.: Quantitative trait loci for agronomic traits in soybean. Plant Breed. **126**(1), 51–57 (2007)

Qi, Z.M., Wu, Q., Han, X., Sun, Y.N., Du, X.Y., Liu, C.Y., Jiang, H., Hu, G., Chen, Q.S.: Soybean oil content QTL mapping and integrating with meta-analysis method for mining genes. Euphytica. **179**(3), 499–514 (2011)

Qiu, B.X., Arelli, P.R., Sleper, D.A.: RFLP markers associated with soybean cyst nematode resistance and seed composition in a ‘Peking’ x ‘Essex’ population. Theor. Appl. Genet. **98** (3–4), 356–364 (1999)

Ramalho, M.A.P., Abreu, A.F., Santos, J.B., Nunes, J.A.: Aplicações da genética quantitativa no melhoramento de plantas autógamas, p. 522. Ed. UFLA, Lavras (2012a)

Ramalho, M.A.P., Carvalho, B.L., Nunes, J.A.R.: Perspectives for the use of quantitative genetics in breeding of autogamous plants. ISRN Genet. **2013**, 1–6 (2012b)

Reinprecht, Y., Poysa, V.W., Yu, K., Rajcan, I., Ablett, G.R., Pauls, K.P.: Seed and agronomic QTL in low linolenic acid, lipoxygenase-free soybean (*Glycine max* (L.) Merrill) germplasm. Genome. **49**(12), 1510–1527 (2006)

Rizal, G., Karki, S.: Alcohol dehydrogenase (ADH) activity in soybean (*Glycine max* [L.] Merr.) under flooding stress. Electron. J. Plant Breed. **2**(1), 50–57 (2011)

Rodrigues, J.I., Arruda, K.M.A., Cruz, C.D., Piovesan, N.D., Barros, E.G., Moreira, M.A.: Associação de marcadores microssatélites com teores de óleo e proteína em soja. Pesq. Agrop. Bras. **48**(3), 255–262 (2013)

Rossi, M.E., Orf, J.H., Liu, L.J., Dong, Z., Rajcan, I.: Genetic basis of soybean adaptation to North American vs. Asian mega-environments in two independent populations from Canadian x Chinese crosses. Theor. Appl. Genet. **126**(7), 1809–1823 (2013)

Sayama, T., Nakazaki, T., Ishikawa, G., Yagasaki, K., Yamada, N., Hirota, N., Hirata, K., Yoshikawa, T., Saito, H., Teraishi, M., Okumoto, Y., Tsukiyama, T., Tanisaka, T.: QTL analysis of seed-flooding tolerance in soybean (*Glycine max* [L.] Merr.) Plant Sci. **176**(4), 514–521 (2009)

Sebolt, A.M., Shoemaker, R.C., Diers, B.W.: Analysis of a quantitative trait locus allele from wild soybean that increases seed protein concentration in soybean. Crop Sci. **40**(5), 1438–1444 (2000)

Sharma, A., Sharma, H., Lightfoot, D.: The genetic control of tolerance to aluminum toxicity in the 'Essex' by 'Forrest' recombinant inbred line population. Theor. Appl. Genet. **122**(4), 687–694 (2011)

Shibata, M., Takayama, K., Ujiie, A., Yamada, T., Abe, J., Kitamura, K.: Genetic relationship between lipid content and linolenic acid concentration in soybean seeds. Breed. Sci. **58**(4), 361–366 (2008)

Specht, J.E., Chase, K., Macrander, M., Graef, G.L., Chung, J., Markwell, J.P., Germann, M., Orf, J.H., Lark, K.G.: Soybean response to water. Crop Sci. **41**(2), 493–509 (2001)

Stombaugh, S.K., Orf, J.H., Jung, H.G., Chase, K., Lark, K.G., Somers, D.A.: Quantitative trait loci associated with cell wall polysaccharides in soybean seed. Crop Sci. **44**(6), 2101–2106 (2004)

Sun, D., Li, W., Zhang, Z., Chen, Q., Ning, H., Qiu, L., Sun, G.: Quantitative trait loci analysis for the developmental behavior of soybean (*Glycine max* L. Merr.) Theor. Appl. Genet. **112**(4), 665–673 (2006)

Sun, Y.N., Pan, J.B., Shi, X.L., Du, X.Y., Wu, Q., Qi, Z.M., Jiang, H., Xin, D., Liu, C., Hu, G., Chen, Q.S.: Multi-environment mapping and meta-analysis of 100-seed weight in soybean. Mol. Biol. Rep. **39**(10), 9435–9443 (2012)

Tajuddin, T., Watanabe, S., Yamanaka, N., Harada, K.: Analysis of quantitative trait loci for protein and lipid contents in soybean seeds using recombinant inbred lines. Breed. Sci. **53**(2), 133–140 (2003)

Vantoai, T.T., St Martin, S.K., Chase, K., Boru, G., Schnipke, V., Schmitthenner, A.F., Lark, K.G.: Identification of a QTL associated with tolerance of soybean to soil waterlogging. Crop Sci. **41** (4), 1247–1252 (2001)

Vieira, A., de Oliveira, A., Soares, T., Schuster, I., Piovesan, N., Martinez, C., de Barros, E., Moreira, M.: Use of the QTL approach to the study of soybean trait relationships in two populations of recombinant inbred lines at the F7 and F8 generations. Braz. J. Plant Physiol. **18** (2), 281–290 (2006)

Vollmann, J., Schausberger, H., Bistrich, H., Lelley, T.: The presence or absence of the soybean Kunitz trypsin inhibitor as a quantitative trait locus for seed protein content. Plant Breed. **121** (3), 272–274 (2002)

Wang, D., Graef, G.L., Procopiuk, A.M., Diers, B.W.: Identification of putative QTL that underlie yield in interspecific soybean backcross populations. Theor. Appl. Genet. **108**(3), 458–467 (2004)

Wang, X., Jiang, G.L., Green, M., Scott, R.A., Hyten, D.L., Cregan, P.B.: Quantitative trait locus analysis of saturated fatty acids in a population of recombinant inbred lines of soybean. Mol. Breed. **30**(2), 1163–1179 (2012)

Yang, K., Moon, J.K., Jeong, N., Chun, H.K., Kang, S.T., Back, K., Jeong, S.C.: Novel major quantitative trait loci regulating the content of isoflavone in soybean seeds. Genes Genomics. **33**(6), 685–692 (2011)

Yuan, J., Njiti, V.N., Meksem, K., Iqbal, M.J., Triwitayakorn, K., Kassem, M.A., Davis, G.T., Schmidt, M.E., Lightfoot, D.A.: Quantitative trait loci in two soybean recombinant inbred line populations segregating for yield and disease resistance. Crop Sci. **42**(1), 271–277 (2002)

Zhang, W., Wang, Y., Luo, G., Zhang, J., He, C., Wu, X., Gai, J., Chen, S.: QTL mapping of ten agronomic traits on the soybean (*Glycine max* L. Merr) genetic map and their association with EST markers. Theor. Appl. Genet. **108**(6), 1131–1139 (2004)

Zhang, D., Cheng, H., Wang, H., Hengyou, Z., Liu, C., Yu, D.: Identification of genomic regions determining flower and pod numbers development in soybean (*Glycine max* L). J. Genet. Genomics. **37**(8), 545–556 (2010)

Chapter 7
Quantitative Genetics in Breeding

Magno Antonio Patto Ramalho, Adriano Teodoro Bruzi, and Rita de Kássia Siqueira Teixeira

Abstract Plant breeding has improved the world soybean yield average substantially over the last decades. Its challenge is to continue to increase it. To accomplish that, it is necessary to use all tools available in the plant breeding. Normally, the traits that soybean breeders work are controlled by several genes and are also greatly affected by the environment. These traits are the object of quantitative genetics. In this chapter, some of the most relevant strategies using quantitative genetics, such as genetic components of means and variance, which can be useful in breeding, are presented. The implications of the genotype × environment interactions in soybean breeding are discussed. Some alternatives that may be used to quantify the G × E interaction and to identify lines that are well adapted and stable are presented. It is expected that quantitative genetics continue to help breeders in the process of decision-making that usually take place daily in soybean breeding.

Keywords Soybean breeding • Quantitative genetics • Genotype × environment interaction • Genetic mean and variance components

Introduction

Most of the time, the goal of soybean breeders is to improve the phenotypic expression of several traits whose segregating populations for these traits have continuous distribution. This continuous distribution will depend on the environmental effect and the presence of several genes involved in the genetic control of said traits. Since they do not form distinct classes, quantitative genetics tools should be used to obtain information about genetic control.

Ultimately, quantitative genetics helps breeders in decision-making while conducting breeding programs. These decisions are related among other factors to the choosing the genitors to obtain the populations to be bred, conducting segregating populations, mitigating the effects of the interaction of genotypes × environments, and making the breeding programs more efficient.

M.A.P. Ramalho, M.S., D.S. (✉) • A.T. Bruzi, M.S., D.S
R. de Kássia Siqueira Teixeira, M.S.
Federal University of Lavras, Lavras, MG 37200-000, Brazil
e-mail: magnoapr@dbi.ufla.br; adrianobruzi@dag.ufla.br; ritadekassiast@gmail.com

F. Lopes da Silva et al. (eds.), *Soybean Breeding*,
DOI 10.1007/978-3-319-57433-2_7

The use of averages or variances can help obtain information about the genetic control of the quantitative traits, and in this chapter both will be considered. However, the focus will be more on information to be used than on how the genetic components are estimated. More details on this can be found in some textbooks, such as Bernardo (2010), Cruz and Carneiro (2003), and Ramalho et al. (2012).

Genetic Structure of Autogamous Plant Populations

Understanding what happens to segregating populations of autogamous plants is key to understanding the quantitative genetics of this group of plants. As an example, consider gene B, with alleles B^1 and B^2, in the frequencies p and q, respectively. What will be said is valid for any allelic frequency, but we will consider the most common case in the soybean crop, which is to obtain segregating populations from the crossing of two lines, that is, $p = q = 1/2$.

$$
\begin{array}{ccc}
L_1 & \text{x} & L_2 \\
B^2B^2 & \downarrow & B^1B^1 \\
 & F_1 & \\
 & B^1B^2 & \\
 & \downarrow & \\
 & F_2\text{: } 1/4\ B^1B^1 & \\
 & 2/4\ B^1B^2 & \\
 & 1/4\ B^2B^2 &
\end{array}
$$

What happens to the other generations with successive self-fecundations is shown in Table 7.1. Note that the frequency of heterozygotes is reduced by half at each generation. So, in a given t generation, the frequency of the locus with the genotype B^1B^2 will be $(1/2)^{t-1}$ In this condition, the frequency of loci $B^1B^1 + B^2B^2$ is thus obtained:

Frequency $B^1B^1 + B^2B^2 = 1 - (1/2)^{t-1}$, being half of the frequency of each genotype, i.e., $\text{Freq.}B^1B^1 = \text{Freq.}B^2B^2 = \frac{1-(1/2)^{t-1}}{2} = \frac{2^{t-1}-1}{2^t}$. Considering that the

Table 7.1 Expected genotype frequencies from the crossing of two lines with genotypes B^2B^2 x B^1B^1

	Genotypes		
	B^2B^2	B^1B^2	B^1B^1
F_1	0.0000	1.0000	0.0000
F_2	0.2500	0.5000	0.2500

(continued)

favorable allele is B^1, the frequency of genotypes B^1B^1 + B^1B^2 will be $\frac{2^{t-1}-1}{2^t} + (1/2)^{t-1} = \frac{2^{t-1}+1}{2^t}$.

The expression that provides the coefficient of inbreeding (I_t) in any generation from t generations of self-fecundation is shown by Ramalho et al. (2012). It is appropriate to point out that as the population of reference is F_2, our interest is inbreeding occurred from this generation on. In this situation, the coefficient of inbreeding is given by the estimator:

$$I_t = 1 - (1/2)^{t-2}.$$

Genetic Component of Averages

Genetic models employed in the early twentieth century continue to be used today, despite the development in molecular biology. This is because there are still many questions surrounding the concept of the gene and, above all, how they act to produce the phenotypes, especially of quantitative traits.

Disregarding the effect of the environment and initially considering only locus B with alleles B^1 and B^2, the genotypic value can be represented in the following ideogram.

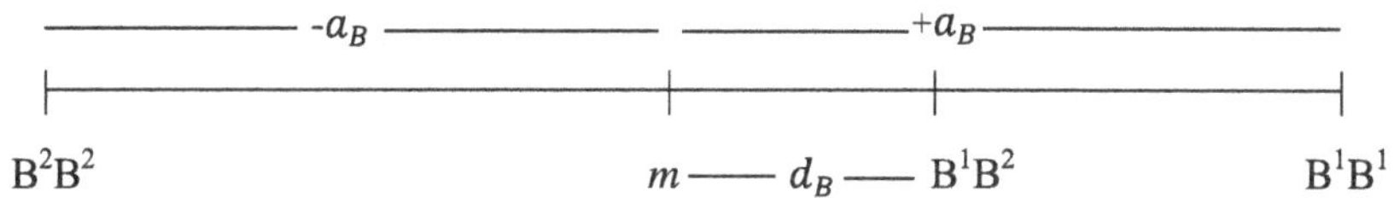

where μ corresponds to the mean value between the two homozygous genotypes, a_B is the homozygous deviation from the mean, and d_B the deviation of the heterozygous relative to the mean. Note that the relation d_B and a_B allows inferring about the type of allelic interaction, i.e., if:

$d_B = 0$, dominance is not observed in the locus considered.
$d_B = a_B$ there is complete dominance.
$0 < d_B < a_B$ there is partial dominance.
$d_B > a_B$ shows the presence of overdominance.

In a Hardy-Weinberg equilibrium population, in which the allele B^1 is in the frequency of p and B^2 of q and the genotypes $B^1B^1 = p^2$, $B^1\,B^2 = 2pq$, and $B^2B^2 = q^2$, the population mean($\bar{P}$) is thus obtained:

$$\bar{P} = \mu + (p - q)a_B + 2pqd_B.$$

Note that the average consists of two parts: one depending on the locus in homozygous $[(p - q)a_B]$ and the other in heterozygous ($2pqd_B$). Also note that the mean is a function of the allele frequency p or q and the type of allelic interaction.

If we consider k locus as in the quantitative trait, we have:

$$P = m + \sum (p - q)a_k + \sum (2pqd_k)$$

where the k-index refers to the effect and frequency of the allele of any of the k locus that are segregating and m the midpoint between the two homozygotes of the k genes. To facilitate what will be discussed below, the expression $\sum(p - q)a_k$ will be replaced by $\boldsymbol{a}$ and $\sum(2pqd_k)$ by $\boldsymbol{d}$. Thus, the average population for k locus becomes

$$\bar{\boldsymbol{P}} = \boldsymbol{m} + \boldsymbol{a} + \boldsymbol{d}$$

When considering the crossing of two lines, for example

	L_1		L_2
Genotypes	ddEEffggHH	×	DDeeffGGHH

the average of line 1 will be $\bar{L}_1 = m - a_\mathrm{D} + a_\mathrm{E} - a_\mathrm{F} - a_\mathrm{G} + a_\mathrm{H}$ and $\bar{L}_2 = m + a_\mathrm{D} - a_\mathrm{E} - a_\mathrm{F} + a_\mathrm{G} + a_\mathrm{H}$. Thus, the average of the two lines will be $\bar{L} = m + a_\mathrm{X} - a_\mathrm{Y}$ where a_X and a_Y are the algebraic sums of locus contributions in homozygous with favorable and unfavorable alleles, respectively. The letter a' will be used to represent this algebraic sum. Then the average of the two lines will be $\bar{L} = m + a'$. The generation F_1 of hybrid between them will have the constitution F_1 = DdEeffGgHH and the average $\mathrm{F}_1 = m + a' + d$, i.e., a' contribution of locus in that they are already fixed and d the algebraic sum of the contribution of the locus in heterozygous. In this situation, the average generation F_2 shall be achieved by $\mathrm{F}_2 = m + a' + 1/2d$. With successive self-fecundations, heterozygosis will be halved in each generation, and in the n lines F_∞ from the segregating population will have an average $\mathrm{F}_\infty = m + a'$. There are some alternatives to estimating $\boldsymbol{a}$ and $\boldsymbol{d}$, which are presented in the textbooks. What will be reviewed here is the estimated $m + a'$ and d, which will be of great use to the breeders.

Consider two segregating populations P_1 and P_2. The average of generation F_1 not considering the epistasis and based on what has been reviewed will be $\mathrm{F}_{1P1} = m + a'_1 + d_1$ and $\mathrm{F}_{1\mathrm{P}2} = m + a'_2 + d_2$. For the generation F_2, we have $\mathrm{F}_{2P1} = m + a'_1 + 1/2d_1$ and $\mathrm{F}_{2\mathrm{P}2} = m + a'_2 + 1/2d_2$.

Thus, if we obtain the average of the generations F_1 and F_2 simultaneously, to reduce the environmental effect, we can obtain the contrast $2\overline{\mathrm{F}_2} - \overline{\mathrm{F}_1} = m + a'$ and also $2(\mathrm{F}_1 - \mathrm{F}_2) = d$.

So, if population 1 has a much higher estimate of $m + a'$ than P_2, we can infer that P_1 will produce lines on F_∞ whose average will be higher than the P_2. Similarly, if the estimate of d of P_1 is greater than P_2, this indicates that this population has a higher number of loci in heterozygosity that P_2.

These estimates are easy to obtain, which makes it possible to eliminate non-promising segregating populations and early low estimates of $m + a'$ and d, increasing the work efficiency of the breeders. What was commented refers to a population coming from the crossing of two lines, therefore the frequency of the segregating alleles $p = q = 1/2$. If they are populations of multiple crossings where $p \neq q \neq 1/2$, then the procedure is similar. Additional details in this regard are found in Ramalho et al. (2012).

Heterosis (h) is another very important concept for quantitative traits. It is estimated to be the superiority of the hybrid (F_1 generation) in relation to the genitors' average, i.e., $h = F_1 - \frac{\overline{P_1 + \bar{P}_2}}{2}$. Falconer and Mackay (1996) showed that heterosis is $h = F_1 - \frac{\overline{P_1 + \bar{P}_2}}{2} = dY^2$, where d has already been defined previously and Y is the measure of divergence between the genitors. Thus, for heterosis to occur, d has to be nonzero, allelic dominance interaction is required, and the genitors must be divergent.

Another key concept for breeders is the average effect of allelic substitution (α) – breeding value. It is defined as the change in the average of the offspring when one allele is changed for another. Bernardo (2010) shows that the breeding value (α) for the locus B is $\alpha = a_B + d_B(q - p)$. Then it will depend on the allele frequencies and the type of allelic interaction. We call α the additive effect of the locus. Often, we estimate a, which can represent the additive effect. This inference is correct only if for the trait in question there is no dominance ($d = 0$) or the frequency of alleles are equal ($p = q$). Otherwise, this is not correct.

As an example of the estimate of average components in soybeans, we will use the work of Ribeiro et al. (2009) concerning the control of Asian soybean rust. This disease is caused by the fungus *Phakopsora pachyrhizi* Sydow & P. Syd. The genetic control may be monogenic, vertical (qualitative), or quantitative resistance or horizontal resistance. The discussion here will emphasize on the polygenic control of this trait. Five lines (A, B, C, D, E) of Embrapa soybean were used to differ in the response to the pathogen.

In an experiment conducted for 2 years, five lines were crossed two by two in a diallel scheme, with ten populations being obtained. Then, 25 treatments, the 5 genitors, and 10 populations were evaluated in generations F_2 and F_3. The severity of the disease, obtained by means of a percentage of the leaf area infected by the pathogen, was also evaluated. The evaluations were performed at 30 and 37 days after the first blast. We estimated $m + a'$ by the estimator 2 $F_3 - F_2$ and $d = 4(F_2 - F_3)$. The average results of the two evaluations and the two harvests are presented in Table 7.2.

The estimates of $m + a'$ and d are given in Table 7.3. Note that the estimate of $m + a'$ varied among the populations. Considering that in this case it is desirable that the lines to be selected show less severity of the disease, the most promising population is the A × D, with the lowest estimate of $m + a'$. Note also that the dominance was not expressive, even with the lack of constancy in the direction of dominance. Although no test was applied, it is assumed that the negative estimates of d are not different from zero. Note the estimates of h: although there was no

Table 7.2 Average results (%) of two evaluations of the severity of soybean rust

Genitors	A	B	C	D	E
A	38.2	45.4	43.4	45.2	45.1
B	45.8	52.7	49.4	49.0	50.7
C	42.3	49.2	44	49.2	48.7
D	42.3	49.2	47.8	51.7	50.5
E	43.6	49.3	47.5	48.9	46.3

Source: Adapted from Ribeiro et al. 2009
Data above the diagonal, population F_2; below, population F_3; and in the diagonal, the genitors

Table 7.3 Estimates of $m + a'$, d and h for the severity of soybean rust, using the average of two harvests

Hybrids	$m + a'$	d	h
A × B	46.2	−1.6	−0.1
A × C	41.2	4.4	4.6
A × D	39.4	11.6	0.5
A × E	42.1	6.0	5.7
B × C	49.0	0.8	2.1
B × D	49.4	−0.8	−6.4
B × E	47.9	5.6	2.4
C × D	46.4	5.6	2.7
C × E	46.3	−2.4	7.1
D × E	47.4	6.3	3.0

Source: Adapted from Ribeiro et al. 2009

coincidence in the magnitudes of the estimates of d, they had the same tendency, except in the case of the crossing C × E. Even if there is dominance, the contribution of the loci in homozygous is much more expressive than the loci in heterozygous. As in that case the allele frequency is 1/2, we can say that for this trait there is predominance of additive effects.

Employment of Variances in the Study of Quantitative Traits

The use of variance provides useful information to breeders, as it makes it possible to estimate the heritability and the expected gain with the selection, which are two very useful estimates for the breeders. In addition, it also has the advantage of obtaining estimates without canceling the genetic deviations of opposite signals, as it happens with the components of averages, since these are squared. The big problem in the use of variances is the error associated, which is typically high (Wricke and Weber 1986; Ramalho et al. 2012).

In a given environment, the phenotypic variance (V_F) contains the genetic variance (V_G) and the environmental variance (V_E). What interests us at the moment is the genetic variance. Fisher (1918), decomposed V_G into additive genetic variance (V_A) + genetic variance of dominance (V_D) + epistatic genetic variance (V_I).

This situation is valid for the populations in Hardy-Weinberg equilibrium or inbreeding populations with a frequency of $p = q = 1/2$. When the population is inbreeding and $p \neq q \neq 1/2$, other components occur to explain the genetic variation. These components will not be presented in this publication; further details are found in Souza-Júnior (1989).

The additive variance measures the variation that is transmitted to the offspring by means of the gametes. It is of key importance in predicting the changes in population average with the selection (Bernardo 2010). It can be shown that V_A is the variance of the breeding values. We have already commented that the breeding value is symbolized by $\alpha.V_A$ corresponds to $2pq\alpha^2$. As $\alpha = [a + d\,(q - p)]$, then $V_A = 2pq\alpha^2 = 2pq\,[a + d\,(q - p)]^2$ (Bernardo 2010).

It is apparent that V_A is a function of the allelic frequency, which varies between populations and the contribution of a and d, which are properties of the allelic interaction of each of the loci involved in the trait expression.

"The term additive variance is somewhat misleading, because the meaning of additive can be vague. Within a locus, the term additive may imply that the alleles act in a purely additive fashion - that the value of the heterozygote is the means of the values of the two homozygotes. In other words additive may imply that V_A exists only at the loci where dominance is absent. However, the expression of V_A indicates that any segregating locus with either no dominance ($d = 0$), partial dominance ($0 < d < a$) or complete ($d = a$) or even overdominance ($d > a$), can contribute to V_A. It is clear, then, that the presence of V_A in the population does not necessarily imply that only the homozygous alleles contribute to V_A" (Bernardo 2010).

The variance of dominance can be obtained by the estimator $V_D = (2pqd)^2$. It is easy to see that V_D is a function of the allele frequencies and the degree of dominance. For those loci where d is nonzero, the genetic variance contains V_D. Only when $d = 0$, the genetic variance will depend only of V_A. In this condition, V_A will be maximum when $p = q = 1/2$. When $d = a$, V_A will be maximum for $p = 1/4$. An important aspect emphasized by Bernardo (2010) is that it is not justified for a population to show that V_A is bigger that V_D, as it has been frequent in the literature. He argues that, as the method of least squares tends to minimize the deviations from the regression line, V_D by definition is considered as the least variation possible between the genotypic values. Thus, a priori, V_A will always be greater than V_D, unless the allelic frequencies are extreme or if there is overdominance.

Another component of the genetic variance is the epistatic variance, which corresponds to the variance of the deviations of the interaction between two or more loci. The epistatic variance can also be decomposed as a function of the interactions between the different types of genetic effects. Considering only two loci, the interaction between genetic values, in both loci, gives an additive x additive epistasis (V_{AA}). The interaction between the genetic values of one locus and the deviations of dominance of another locus gives the additive x dominance epistasis (V_{AD}), and the interaction between dominance deviations of two loci gives the dominance x dominance epistasis (V_{DD}), i.e., the variance of interaction is given by $V_I = V_{AA} + V_{AD} + V_{DD}$.

In order to estimate these components of genetic variance, there are innumerable procedures that are found in some publications (Hallauer et al. 2010; Kearsey and Pooni 1998). At the moment, the greatest interest is to comment what happens to the components of genetic variance of a soybean segregating population in successive generations of inbreeding. In this species, as in any other self-pollination plant, it is possible to predict what occurs with the genotypic constitution of the segregating population in any generation of inbreeding.

The work of Cockerham (1963) shows an easy way to estimate the components of V_G. The methodology is based on the covariance between closely related individuals. To exemplify we will consider the bulk method within F_2 progenies. According to Horner and Weber (1956), the genotypic variances and covariance of a segregating population of related plants can be expressed as a linear function of $COV_{tgg'}$, to simplify, $C_{tgg'}$, where, in the example, t is the generation of F_2 plant, g can be the $F_{2:3}$ progeny, and g' can be $F_{2:4}$ progeny, or any other progeny in the hierarchy. According to the notation of Cockerham (1963), this covariance among related plants can be thus expressed, disregarding the occurrence of epistasis:

$$C_{tgg'} = zV_A + wV_D.$$

z and w are obtained by the expressions

$$z = (1 + I_t)$$
$$w = \frac{1 + I_t}{1 - I_t}(1 - I_g)(1 - I_{g'})$$

where I_t is the coefficient of inbreeding in the genitor generation and I_g and $I_{g'}$ the coefficient of endogamy in the generation of the descendants according to the hierarchy.

What is within in the genetic variance among plants F_2 is $C_{ttt} = C_{222}$. This is because, in this case, the population will be F_2 (*t generation*). Thus, $C_{222} = zV_A + wV_D$.

In this case, $I_t = I_2 = 1 - \left(\frac{1}{2}\right)^{t-2} = 0$. As previously mentioned, F_2 generation is considered the reference population with zero inbreeding. Thus:

$$z = (1 + I_t) = (1 + 0) = 1$$
$$w = \frac{1 + I_t}{1 - I_t}(1 - I_t)(1 - I_t) = 1.$$

Thus, $V_{G_{F_2}} = C_{222} = 1V_A + 1V_D$. The total genetic variance between F_3 plants is provided by:

$$V_{G_{F_3}} = C_{333} = zV_A + wV_D$$

$$z = 1 + I_3, \text{as } I_3 = \frac{1}{2} \text{ then :}$$

$$z = 1 + \frac{1}{2} = \frac{3}{2}$$

$$w = \frac{1 + I_3}{1 - I_3}(1 - I_3)(1 - I_3) = \frac{1 + \frac{1}{2}}{1 - \frac{1}{2}}\left(1 - \frac{1}{2}\right)^2 = \frac{3}{4}.$$

It appears that:

$$V_{G_{F_3}} = \frac{3}{2}V_A + \frac{3}{4}V_D.$$

This procedure can be used, as already mentioned, for all schemes of segregating populations of self-pollination plants. Further details on this subject can be obtained on Ramalho and Vencovsky (1978) and Souza-Júnior (1989).

For example, the following scheme of conducting a segregating population is considered, which is very common among soybean breeders:

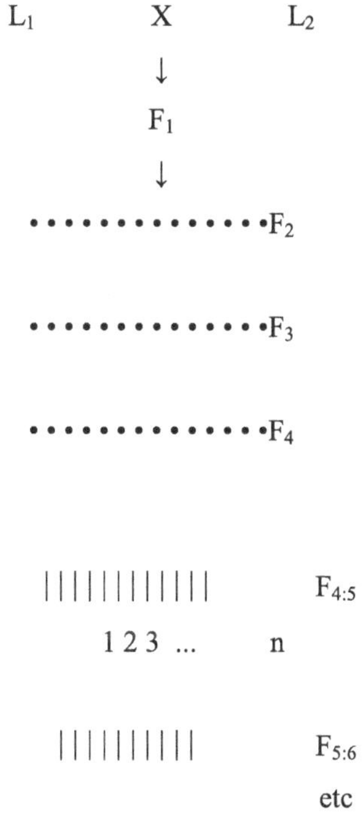

In this scheme, V_G on F_4 generation will have:

$$V_{G_{F_4}} = C_{ttt} = C_{444} = zV_A + wV_D.\ \text{As}\, z = 1 + I_t\, \text{and}\, w = \frac{1 + I_t}{1 - I_t}(1 - I_g)(1 - I_g).$$

In this case, as the covariance is C_{ttt}, then $I_t = I_g = I_{g'} = I_4 = 1 - 1/2^{4-2} = 3/4$. In this case, $V_{G_{F_4}} = \left(1 + \frac{3}{4}\right)V_A + \left(\frac{1+3/4}{1-3/4}\right)(1 - 3/4)^2 V_D = 7/4V_A + 7/16V_D$.

Thus, between plants of the F_4 generation occurs 7/4, that is, 1.75 times the V_A in the F_2 population and 0.4375 of V_D of F_2. This is the main reason for soybean breeders to drive the population in *bulk* to F_4 generation. If the *bulk* opening is postponed to F_5generation, 1.875 of V_A of F_2 will be explored. Therefore, the wait for one more generation does not compensate, because the increase in the proportion of the additive variance from F_4 to F_5 is only 0.125.

What will the composition of the genetic variation between and within $F_{4:5}$ progenies be? That is, the first generation that the breeders have offspring to be evaluated. Using the notation of Cockerham (1963): $V_{P_{F_{4:5}}} = zV_A + wV_D = C_{455}$. Thus, $I_t = I_4$ and $I_g = I_{g'} = I_5$. Note that $I_4 = 3/4$ and $I_5 = 1 - (1/2)^{5-2} = 7/8$. Then:

$$V_{P_{F_{4:5}}} = C_{455} = 1,75V_A + \left[\left(\frac{1 + 3/4}{1 - 3/4}\right)(1 - 7/8)^2\right]V_D = 1,75V_A + 7/64V_D.$$

The genetic variance within progenies $F_{4:5}$ $\left(V_{Gw_{F_{4:5}}}\right)$ will be $V_{Gw_{F_{4:5}}} = C_{555} - C_{455}$, that is, the total genetic variance between plants of F_5 generation minus the variation between the $F_{4:5}$ progenies. Using the same logic as above, we have $V_{Gw_{F_{4:5}}} = C_{555} - C_{455} = 1/8V_A + 8/64V_D$.

In the schema usually adopted by soybean breeders, they identify the best plants in the best progenies $F_{4:5}$ and then $F_{5:6}$ progenies. Note that the amount of variation within progenies $F_{4:5}$ is very small and only explored $1/8 = 0.125$ of the additive genetic variance present in F_2. From $F_{5:6}$, usually the selection is not performed within it, and thus there are in the next generation basically pure lines to be evaluated in value for cultivation and use trials (VCU).

When working with variances, it is possible to obtain estimates of heritability (h^2), which is one of the most useful parameters for breeders. The h^2 estimates the reliability of phenotypic value as an indicator of the breeding value (Bernardo 2010). Therefore, it is always present in the decisions of the breeders in their selection programs.

The h^2 can be in the broad sense $\left(h_a^2\right)$ or in the strict sense $\left(h_r^2\right)$. The difference between them is only in the numerator of the expression. In the case of h_r^2, the numerator of the estimator contains only V_A and not all the genetic variation among individuals or progenies of populations, i.e.:

$$h_a^2 = \frac{V_G}{V_F} = \frac{V_A + V_D}{V_A + V_D + V_E} \qquad h_r^2 = \frac{V_A}{V_F} = \frac{V_A}{V_A + V_D + V_E}$$

where:

V_F is the phenotypic variance in the selective unit, in the example, among plants.

There are several procedures to estimate the h^2 (Nyquist 1991; Hallauer et al. 2010; Ramalho et al. 2012), but they will not be described here. At the moment, it is necessary to mention that the h^2 is not an immutable estimate, since it depends on the genetic variance, that is, it varies according to the segregating population used; it depends on the precision with which individuals and/or progenies were obtained; and it also varies according to the unit of evaluation, that is, according to individuals or the progeny mean in one environment or in various environments. Therefore, to infer whether a trait has high heritability or not, estimates must be accumulated, and only then we can say that the trait is normally h^2 high or not.

The greater importance of heritability is that it takes part in the expression of the expected gain on the selection (GS). The estimate of expected gain with the selection enables breeders to evaluate their selection options before they are carried out. The basic expression of gain with the selection is

$$\mathrm{GS} = \mathrm{ds} \cdot h^2$$

where: ds is the selection differential, that is, the difference between the average of the individuals selected (ms) minus the average of the original population (ds = ms − mo).

The selection may be truncated or not. If the selection is performed simultaneously in some traits, for example, grain type, disease resistance, and grain production, it is not possible to specify a specific point of discard, and then a non-truncated selection is obtained. If the selection criterion involves only one of the traits, for example, grain production, it is possible to discard all individuals and/or progenies below a certain value, and then we have what is called a truncated selection.

When the selection is truncated, the ds can be expressed in units of phenotypic standardized deviations (i), being also called standard or standardized selection intensity. Thus, the data is adjusted to a normal distribution, and i may be determined directly from the properties of a normal distribution and applicable to any situation. Falconer and Mackay (1996) state that:

$$i = \frac{\mathrm{ds}}{\sqrt{V_{\mathrm{X}}}} = \vartheta/f,$$

where ϑ is the height in the standardized normal curve at the truncation point and f is the selected proportion.

The intensity of the standardized selection measures the number of standard deviations in which the average of the selected individuals exceeds the average of the original population. For example, consider any trait whose population standard deviation $\left(\sqrt{V_{\mathrm{X}}}\right)$ is 4 and the selection differential (ds) is 10 units. In this case, $i = \mathrm{ds}/\sqrt{V_{\mathrm{X}}} = 10/4 = 2,5$.This value of i indicates that the average of the selected population is 2.5 standardized standard deviations higher than the average of the population. The advantage of using i is that this value is found in tables, not being

Table 7.4 Estimates of the severity variance of the occurrence of soybean rust

Genitors	A	B	C	D	E
A	72.95	116.15	124.01	128.85	127.66
B	130.35	99.25	114.37	126.75	122.03
C	193.21	128.04	101.37	133.99	129.25
D	152.99	140.90	141.76	88.68	121.55
E	143.19	130.10	151.50	131.70	109.08

Data from Ribeiro et al. 2009

Data above the diagonal, population F_2; below, population F_3; and in the diagonal, the genitors

dependent on data, and can then be used to predict the gain, giving greater flexibility to the breeder in the comparisons of selection methods.

As an example, the study on genetic control of resistance to soybean rust previously mentioned (Ribeiro et al. 2009) will again be considered. As mentioned in this study, the genitors and the F_2 and F_3 generations were evaluated simultaneously. Data was obtained on the percentage of leaf tissue of the affected host of 50 plants of each genitor, 160 from F_2 population, and 200 from F_3 population. Table 7.4 shows the phenotypic variances obtained from the genitors and F_2 and F_3 generations.

In this situation, the phenotypic variance (V_F) of the genitors, due to being lines, is all environmental $V_{F_{genitores}} = V_E$. Based on the comment and using the Cockerham methodology, the genetic variance of the F_2 generation contains $V_{G_{F_2}} = 1V_A + V_D$ and $V_{F_{F_2}} = V_A + V_D + V_E$ and $V_{F_{F_3}} = 1,5V_A + 0,75V_D + 1V_E$. Using the estimates in Table 7.4, we have for the crossing A × B:

$$V_E = V_{F_{genitores}} = \frac{V_{F_A} + V_{F_B}}{2} = \frac{72,95 + 99,25}{2} = 86,10$$

$$V_{F_{F_2}} = 116,15 = 1V_A + 1V_D + 1V_E$$

$$V_{F_{F_3}} = 130,35 = 1,5V_A + 0,75V_D + 1V_E.$$

Solving it, we have:

$$V_E = 86,10; V_{G_{F_2}} = 29,55; V_{G_{F_3}} = 43,75; V_A = 28,78; V_D = 0,77.$$

It should be pointed out that the estimates shown in Table 7.5 for V_A and V_D, presented by Ribeiro et al. (2009), differ somewhat from those obtained by the algebraic method, because the methodology used by the authors was the weighted least square method (Ramalho et al. 2012) and not the algebraic one, as done at the time. However, the difference does not change the interpretations to be obtained.

The use of quantitative genetics in the choice of genitors/segregating populations has been little explored in the soybean crop. For example, Bernardo (2014) shows a typical soybean breeding program of the American seed companies. He mentions that these companies obtain 200 populations annually and evaluate

Table 7.5 Estimates of components of additive genetic variance (V_A), of dominance (V_D), environment (V_E), and of heritability $\left(h_r^2\right)$ concerning the evaluation of percentage of leaf area infected by the pathogen

Hybrids	V_A	V_D	V_E	h_r^2
A × B	29.85	2.91	83.39	41.05
A × C	38.06	−0.72	86.67	45.43
A × D	48.14	−0.25	80.96	55.83
A × E	32.58	3.01	92.07	40.63
B × C	23.77	−7.12	97.73	25.51
B × D	30.67	4.65	91.40	39.98
B × E	17.36	2.44	102.22	23.12
C × D	23.43	15.80	94.76	37.76
C × E	37.60	−13.95	105.56	33.32
D × E	20.92	1.24	99.39	26.75

Adapted from Ribeiro et al. 2009
Data obtained from population evaluations F_2 and F_3, in the two harvests

them using markers, resistance to some pathogens, and cyst nematodes. The populations are carried in *bulk*, and from 200 to 500 $F_{4:5}$ progenies are removed, evaluating approximately 70,000 progenies in experiments without repetition at one or two sites. After this evaluation, there are 5000 $F_{4:6}$ progenies to be selected to continue the process. The soybean breeding programs in Brazil follow the same line. Hundreds of populations are obtained annually. They are conducted in *bulk* up to F_4, with more than 70,000 progenies, which is drastically reduced to 2000–5000 progenies, usually using visual selection. In other words, both in the United States and Brazil, no discard process of the segregating populations is applied.

The proposal is to evaluate the *bulks* and use some of the various methods available to eliminate populations with little likelihood of success. The focus is only on the best and, thus, in the evaluation of fewer $F_{4:5}$ progenies, but only from promising populations to produce grains.

A good alternative of choice is the methodology proposed by Jinks and Pooni (1976). The basis of this methodology is that for a trait controlled by at least six genes, the distribution of lines in F_∞ from the crossing of two lines is normal.

This procedure was also used by Ribeiro et al. (2009) aiming at the resistance to Asian soy rust, as previously mentioned. The probability of a population producing lines above a given standard corresponds to:

$$Z = \frac{\bar{L} - \bar{X}}{s}$$

where $\bar{L}$ is the average of the line used as control, that is, the average of the standard; $\bar{X}$ is the average of the lines in generation F_∞. In a nondominant model, that is, $m + a$ and s is the phenotypic standard deviation between lines, i.e., $s = \sqrt{V_F}$.

What should be estimated is the phenotypic variance (V_F) of lines in F_∞. Note that it should contain $2V_A + V_E$. Thus, V_A e V_E already estimated can be used for each population, as seen in Table 7.5. In the case of Ribeiro et al. (2009), $\bar{L}$ was

Table 7.6 Estimates of $m + a'$, genetic and environmental variance, and genetic potential related to the evaluations of percentage in the leaf area infected by the pathogen

Population	$m + a'$	V_A	V_E	Genetic potential (%)
A × B	46.09	29.85	83.39	9.01
A × C	41.12	38.06	86.67	18.67
A × D	39.3	48.14	80.96	23.89
A × E	42.16	32.58	92.07	16.60
B × C	48.93	23.77	97.73	5.05
B × D	49.31	30.67	91.40	6.06
B × E	47.94	17.36	102.22	6.30
C × D	46.51	23.43	94.76	9.34
C × E	46.27	37.60	105.56	9.85
D × E	47.35	20.92	99.39	7.08

Adapted from Ribeiro et al. 2009
Data obtained in the evaluation of two harvests

considered the average minus a standard deviation of the genitor; $\bar{X}$ was the estimate of $m + a'$, the average estimate of lines in F_∞.

The estimates obtained in Table 7.6 confirm that the A × D population has a higher probability of obtaining lines to exceed the standard. Note that if the B × C population was maintained, the probability of exceeding the standard would be four times lower than the A × D population, with a great loss of time and resource at the end of the selection process.

The Interaction Between Genotypes × Environments

Soybeans in Brazil are grown in more than 30 million hectares. In this condition, the expectation is of enormous variation in the environmental conditions of cultivation. Although soybean farmers are characterized by the use of high technology among and within growing regions, there is variation in crop management. Considering the wide environmental variation and also among the lines, great interaction between lines × environments is expected to occur. That is, the behavior of the lines is not always coincident in the different environments.

The nature of the interaction should be attributed to physiological and/or biochemical factors of each cultivar. In genetic terms, interaction occurs when the contribution of the alleles of different genes that control the trait and/or their level of expression differs among the environments. This is because gene expression is influenced and/or regulated by the environment (Kang and Gauch 1996).

Interaction is the main complicator of the work of the breeders, both to select the progenies and recommend new cultivars. In fact, the greatest challenge of breeders is to select and/or recommend a cultivar at the experimental stations in a few years,

Table 7.7 Result of the analysis of variance of six cultivars of soybeans evaluated at three sites and 2 years

SV	df	Mean Square	R^2 (%)
Blocks (years*sites)	12	163435.92*	–
Cultivars (C)	5	3343186.62**	31.49
Years (Y)	1	278973.34	2.63
Sites (S)	2	733627.69**	6.91
C × Y	5	2073446.85**	19.53
C × S	10	1131917.03**	10.66
Y × S	2	2094654.01**	19.73
C × Y × S	10	959703.85**	9.04
Error	60	81815.35	–
Total	107		

** and * significant, respectively, at 1 and 5% probability by the F test.

Table 7.8 Estimates of contribution of each line to the interaction and the risk of their adoption by farmers

Cultivars	Average (kg/ha)	Contribution for interaction[a]	Risk of adoption[b]
A	2818	18.2	86.6
B	2795	14.6	86.1
C	2937	12.3	90.3
D	3720	11.8	116.0
E	3367	10.3	105.2
F	3704	32.7	115.9

[a]Estimation of the contribution to the interaction by the Wricke (1965)
[b]Estimation of the risk of adoption by the Annicchiarico (1992), Annicchiarico et al. 1995) and $\alpha = 0.20$

aiming at future use for different management systems of farmers in subsequent years.

Since more than one line and more than one environment are evaluated, we can estimate the interaction and its implications. As an example, the evaluation data of six soybean cultivars (A, B, C, D, E, and F) will be used in six environments. The summary of the analysis of variance is presented in Table 7.7. Note that all interactions were significant ($P \leq 0.01$). The contribution of each one to explain the sum of squares associated to the cultivar and environments shows the importance of each type. Note that the interaction between cultivars x years ($R^2 = 19.53\%$) was more important than cultivars × sites ($R^2 = 10.66\%$).

There is a multitude of methodologies to select the most adapted and stable cultivars. However, the most relevant for breeders is to identify the cultivars and/or environments that contribute the most to the interaction and the risk of adopting a given cultivar. The relative contribution of each line or environment to the interaction was estimated. Note that cultivars D and E contributed the least to the interaction (Table 7.8). In addition, cultivar D showed the lowest risk of adoption, that is, at worst, this cultivar would perform 16% higher than the general average of each environment. Cultivars A and B are the ones with the highest risks, as they are likely to perform below the environment average.

What are the main alternatives to mitigate the effect of the interaction? The first would be the zoning of breeding programs, as proposed by Embrapa for soybean VCU trials (Kaster and Farias 2012). This strategy is already used by companies that subdivide their breeding programs by region. The second is to keep in mind that no cultivar shall be recommended without being widely evaluated in multiple sites and years. It is common for some soybean breeding programs in the VCU to eliminate large numbers of lines during the first year. This is not advisable because the interaction per year is usually large and consequently a hasty elimination can neglect great lines.

A third option, which is widely used and that should be stimulated, is the evaluation of the lines before the recommendation, in the properties of the farmers. In this case, macroplots are used, and a smaller number of lines are observed, but in a large number of properties.

One last alternative would be the annual recommendation of a greater number of lines/cultivars. After 1 or 2 years of intensive cultivation with the support of the farmers, the cultivar that would have intensified seed production and its dissemination would be identified.

Final Considerations

The success of soybean breeding in Brazil is undeniable. However, it is also clear that the continuity of this genetic progress will be more difficult, as the differences to be detected among the lines available will be smaller and smaller. Moreover, the farming areas will certainly be able to migrate to marginal regions where conditions will not necessarily be favorable. In this scenario, the breeders should use all possible strategies to be successful.

The purpose of quantitative genetics is mainly to provide the basis, so that the breeders can be more efficient in their daily activities. This efficiency refers in particular to the rationalization of work, cost reduction, and increased genetic progress.

References

Annicchiarico, P.: Cultivar adaptation and recommendation from alfalfa trials in Northern Italy. J. Genet. Breed. **46**(1), 269–278 (1992)
Annicchiarico, P., Bertolini, M., Mazzinelli, G.: Analysis of genotype-environment interactions for maize hybrids in Italy. J. Genet. Breed. **49**, 61–68 (1995)
Bernardo, R.: Quantitative Traits in Plant Breeding. Stemma Press, Woodbury (2010)
Bernardo, R.: Essentials of Plant Breeding, p. 260. Stemma Press, Woodbury (2014)

Cockerham, C.C.: Estimation of genetic variance. In: Hanson, W.D., Robinson, H.F. (eds.) Statistical Genetics and Plant Breeding, pp. 53–93. The National Academies Press, Washington (1963)
Cruz, C.D., Carneiro, P.C.S.: Modelos Biométricos aplicados ao Melhoramento Genético, vol. 2, p. 585. UFV, Viçosa (2003)
Falconer, D.S., Mackay, T.F.C.: Introduction to Quantitative Genetics, 4a edn, p. 463. Longman, London (1996)
Fisher, R.A.: The correlation between relatives on the supposition of mendelian inheritance. Trans. Royal Soc. **5**, 399–433 (1918)
Hallauer, A.R., Carena, M.J., Miranda Filho, J.B.: Quantitative Genetics in Maize Breeding. Springer, New York (2010)
Horner, T.W., Weber, C.R.: Theoretical and experimental study of self-fertilized populations. Biometrics. **12**, 404–414 (1956)
Jinks, J.L., Pooni, H.S.: Predicting the properties of recombinant inbred lines derived by single seed descendent. Heredity. **36**(2), 243–266 (1976)
Kang, M.S., Gauch Jr., H.G.: Genotype-by-Environment Interaction. CRC, Boca Raton, FL (1996)
Kaster, M., Farias, J.R.B.: Regionalização dos testes de valor de cultivo e uso e da indicação de cultivares de soja – Terceira Aproximação, p. 69. EmbrapaSoja, Londrina (2012)
Kearsey, M.J., Pooni, H.S.: The Genetical Analysis of Quantitative Traits, p. 381. Chapman & Hall, London (1998)
Nyquist, W.E.: Estimation of heritability and prediction of selection response in plant populations. Crit. Rev. Plant Sci. **10**(3), 235–322 (1991)
Ramalho, M.A.P., Vencovsky, R.: Estimação dos componentes da variância genética em plantas autógamas. Ciência e Prática. **2**(2), 117–140 (1978)
Ramalho, M.A.P., Abreu, A.F.B., Santos, J.D., Nunes, J.A.R.: Aplicações da genética quantitativa no melhoramento de plantas autógamas, p. 522. UFLA, Lavras (2012)
Ribeiro, A.S., de Toledo, J.F.F., Ramalho, M.A.P.: Selection strategies of segregant soybean populations for resistance to Asian rust. Pesq. Agrop. Brasileira. **44**, 1452–1459 (2009)
Souza-Júnior, C.L.: Componentes de Variância genética e suas implicações nomelhoramento vegetal. FEALQ, Piracicaba (1989)
Wricke, G.: ZurBerechnung der ÖkovalenzbeiSommerweizen und Hafer. ZeitschriftfürPflanzenzüchtung. **52**, 127–138 (1965)
Wricke, G., Weber, W.E.: Quantitative Genetics and Selection in Plant Breeding, p. 406. Walter de Gruyter, New York (1986)

Chapter 8
Parental Selection

Leonardo Lopes Bhering, Leonardo de Azevedo Peixoto, and Cosme Damião Cruz

Abstract In this chapter, we will describe several biometrical and statistical concepts that may help researches and breeders to analyze their experimental data and generate useful information to select superior genitors more accurate in a soybean breeding program. Methodologies presented here aim the formation of base population with large genetic variability, suitable adaptability and repeatability, and with high average for the most important agronomics traits. Diallel analysis and genetic diversity analysis stand out among methodologies. Diallel analysis aims to evaluate general and specific combining ability for each genitor, and then superior genitors are selected based on these parameters. Moreover, diallel analysis allows for estimating additive, dominance, maternal, and reciprocal effects, and this information will help breeders to decide what strategy is better to be used for the target trait. On the other hand, genetic diversity analysis aims to evaluate how close or distant are the genotypes and then select superior genotypes based on breeders' objectives. Several methods are described to evaluate the genetic diversity such as principal component analysis, canonical correspondence analysis, and clustering analysis. These analyses should be used together in a soybean breeding due to while breeders will select genitors with high average based on diallel analysis, they also will select genitors with large genetic variability based on genetic diversity analysis. Therefore, the use of biometric information during genitors' selection stage in a soybean breeding may be a useful tool to help breeders make better decision and allow them to obtain the maximum information about evaluated experiments. It is the researcher's responsibility to use all tools available based on the desired objectives, experience with soybean breeding, and available biometrical techniques.

Keywords Biometry • Diallel • Genetic diversity • Clustering analysis • Principal component analysis

L.L. Bhering, M.S., D.S. (✉) • L. de Azevedo Peixoto, M.S., D.S. • C.D. Cruz, M.S., D.S.
The Federal University of Viçosa, Viçosa, Minas Gerais, Brazil
e-mail: leonardo.bhering@ufv.br; leoazevedop@gmail.com; cdcruz@ufv.br

F. Lopes da Silva et al. (eds.), *Soybean Breeding*,
DOI 10.1007/978-3-319-57433-2_8

Introduction

This chapter addresses the various biometric and statistical principles that enable researchers to analyze their experimental data and generate useful information to select the parents, in the best way possible, in a soybean breeding program. The purpose of the procedures presented herein is to form a base population of broad genetic variability, with characteristics of agronomic interest and good adaptability.

Formation of Population-Base

One of the main stages of the breeding program is the choice of the parents which, after crossbreeding, will form the base population, on which researchers will invest their efforts in the search for productive genetic material, with quality and good adaptation. Several criteria can be adopted for the choice of the parents, with highlight to the performance as regards the traits of agronomic interest, the combinatorial ability, and the adaptability. Another fundamental aspect is the diversity among the parental group, evaluated with the purpose of identifying the hybrid combinations of greater heterotic effect and greater heterozygosity, so that, in its segregant generations, there is greater possibility of recovery of superior genotypes. The formation of the base population is also of great importance in the context of the management and conservation of soybean germplasm, inasmuch as it provides information on the available resources and assists the localization and exchange of these resources.

The genetic diversity has been evaluated by means of biometric techniques based on the quantification of heterosis or by predictive processes. Among the methods based on biometric models, which are intended to evaluate the parental diversity, we should mention the diallelic analyses (Cruz 2005).

Diallelic Analysis

The term diallel has been used to express a set of $p(p-1)/2$ hybrids, resulting from the mating between p parents (lines, varieties, clones etc.), and may include, in addition to the respective parents, the reciprocal hybrids and, or, other related generations, such as F_2's, backcrossings, etc.

The methodologies of diallel analysis have the purpose of analyzing the genetic outline, providing estimates of useful parameters in the selection of parents for hybridization and in the understanding of the genetic effects involved in the determination of the traits.

In general, there are various types of diallel, for example, balanced diallels (which include the F_1's hybrids among all pairs of parental combinations, and may additionally include the parents, and their reciprocal hybrids); partial diallels (involving two groups of parents and their crosses); and circulating diallels (the parents are represented in the outlining by the same number of crosses, but below $p - 1$, as for the balanced ones. Circulating diallels allow obtaining information about the parents with a smaller number of crosses, although there is loss of information regarding certain hybrid combinations, which are absent). Other types of diallels are incomplete (when parents are represented by a variable number of crosses) and the unbalanced ones (when hybrid combinations and other generations, reciprocal, etc., are represented, but in variable frequency, due to the unequal number of replications per treatment). For more details on diallel types and different methodologies, please refer to Cruz et al. (2012).

Diallelic analyses are also intended to quantify the trait genetic variability and to evaluate the genetic value of the parents as well as the specific capacity and heterosis manifested in specific crosses. Diallels require the evaluation of hybrid combinations among parents. Diallel analysis has been routinely used in breeding to evaluate small numbers of parents (around 10). However, when a large number of potential parents are available to be used in crossing for the formation of a base population, obtaining the experimental material may be impracticable, and the study may be unfeasible.

For diallels that only include hybrid combinations, it has been adopted the following statistical model proposed by Griffing (1956):

$$Y_{ij} = m + g_i + g_j + s_{ij} + \bar{\varepsilon}_{ij},$$

where

Y_{ij}: hybrid average value ij $(i, j = 1, 2, ... p, i < j)$
m: general average
g_i, g_j: effects of the general combining ability (GCA) of the ith and the jth parent, respectively
s_{ij}: effect of the specific combining ability (SCA) for crosses between the i and j parents
$\bar{\varepsilon}_{ij}$: mean experimental error

Consider, as an illustration, a design involving four parents. The diallel scheme is shown in Table 8.1.

The effects will be estimated using the following formulas:

$$\hat{m} = \frac{2}{p(p-1)}Y$$

$$\hat{g}_i = \frac{1}{(p-2)}[Y_i - (p-1)\hat{m}] = \frac{1}{(p-2)}\left[Y_i - \frac{2}{p}Y\right] = \frac{1}{p(p-2)}[pY_i - 2Y]$$

Table 8.1 Diallel scheme including the F_1 hybrids of four parents

Parent	1	2	3	4
1	–	Y_{12}	Y_{13}	Y_{14}
2		–	Y_{23}	Y_{24}
3			–	Y_{34}
4				–

Table 8.2 Variance analysis for balanced diallels involving only the F_1 hybrids, according to the methodology proposed by Griffing (1956)

			F		E(MS)	
SV	DF	MS	Fixed	Random	Fixed	Random
GCA	$p-1$	GMS	GMS/RMS	GMS/QMS	$\sigma_\varepsilon^2+(p-2)\varphi_g$	$\sigma_\varepsilon^2+\sigma_s^2+(p-2)\sigma_g^2$
SCA	$p(p-3)/2$	QMS	QMS/RMS	QMS/RMS	$\sigma_\varepsilon^2+\varphi_s$	$\sigma_\varepsilon^2+\sigma_s^2$
Residue	f	RMS			σ_ε^2	σ_ε^2

GCA general combining ability, *SCA* specific combining ability

$$\widehat{s}_{ij}=Y_{ij}-\left(\widehat{m}+\widehat{g}_i+\widehat{g}_j\right)=Y_{ij}-\frac{1}{(p-2)}(Y_i+Y_j)+\frac{2}{(p-1)(p-2)}Y$$

and the variance analysis will be performed as shown in Table 8.2.
The sum of squares (SS) will be given by

$$\text{SQ(GCA)}=\sum\widehat{g}_iY_i=\frac{1}{(p-2)}\sum Y_i^2-\frac{4}{p(p-2)}Y^2e$$

$$\text{SS(SCA)}=\sum_i\sum_{<j}\widehat{s}_{ij}Y_{ij}$$

$$=\sum_i\sum_{<j}Y_{ij}^2-\frac{1}{(p-2)}\sum_i\sum_{<j}Y_{ij}(Y_i+Y_j)+\frac{2}{(p-1)(p-2)}Y^2.$$

Based on this analysis, it is possible to assess the importance of the additive genetic effects, expressed by the effects associated with GCA as well as the effects due to dominance deviations associated with SCA. This information is useful to establish the best breeding strategy. When additive effects are pronounced, gains of greater magnitude are the anticipated, even when using the simplest breeding strategies. The results are also useful to indicate the best performance and genetic complementarity of parents to be intercrossed.

Dialectical crosses can still be used to obtain the heterodox potential of certain crosses. Gravina et al. (2003) crossed seven soybean cultivars with different levels of resistance to *Cercospora sojina* using Griffing methodology, in order to aim to identify genes in breeding programs aimed at obtaining soybean cultivars with resistance to frogeye leaf spot. These authors identified three cultivars that presented high GCA and SCA estimates, with the purpose of reducing the disease, so they had potential for the desired purpose. Zorzetto et al. (2008) used three soybean cultivars and their F1 hybrids to identify promising parents and combinations to generate segregating populations that attend a soybean breeding program in the state of São Paulo. These authors identified cultivar A-7002 and a promising hybrid combination to achieve the desired goal. Maphosa et al. (2012) assessed eight soybean lines that were crossed in balanced diallel, in order to obtain hybrids resistant to soybean rust. They assessed F_2 and F_3. Griffing model was used to estimate general and specific combination capabilities. Through these analyses, they could identify that genotypes UG5 and G7955 showed good tolerance to soybean rust and could be used in future crosses.

Genetic Diversity

When a large number of generic elements are available, previous predictive studies have been recommended, which indicate the number and type of crosses on which most effort should be concentrated to obtain hybrids. As previous hybrid combinations are not required, the predictive methods of diversity among the parents have gained considerable emphasis. Predictive methods are the ones based on morphological, physiological, and other differences—presented by the parents in the diversity determination, which is usually quantified by a measure of dissimilarity (e.g., Euclidean distance and Mahalanobis distance). Inference based on ecogeographic diversity is also an example of a predictive heterosis method.

It is important to evaluate the diversity found in soybean germplasm banks, insofar as commercial cultivars are increasingly related. Thus, diversity must be analyzed and quantified so that it can be made available for breeding programs in order to maintain the diversity within such programs and have some of these accessions selected as parents. Statistical and genetic tools are used to evaluate and quantify this diversity.

In genetic diversity prediction, several multivariate methods can be applied. Among them, the analysis by principal components and by canonical variables and the agglomerative methods are mentioned. The most appropriate method has been chosen based on the precision desired by the researcher, the ease of analysis, and how data were obtained.

Methods based on major components or on canonical variables allow for the study of diversity in scatter plots, in which two Cartesian axes are generally considered. In these studies, several characteristics are evaluated in a set of genotypes that, by statistical procedures, are summarized in few components (or canonical variables) and given by linear combinations of the original traits, independent of each other and with descending discrimination capacity, so that the first components (or canonical variables) explain the maximum variation in the original data. The choice of one or the other depends on the experimental type evaluated. In experiments without replicate, the main components are used, whereas in experiments with statistical outlines with replicate, one must make use of canonical variables.

Agglomerative methods differ from the others, for they essentially depend on previously estimated dissimilarity measures, such as Euclidean distance or the generalized Mahalanobis distance, among others.

Cluster Analysis

The cluster analysis aims to gather, by means of some classification criteria, the parents (or any other type of sample unit) in several groups, so that there is homogeneity within the group and heterogeneity among groups. Alternatively, the cluster analysis techniques also have the purpose of dividing an original group of observations into several groups, according to some similarity or dissimilarity criteria (Cruz et al. 2012).

In cluster analysis, several questions arise. Thus, the final number of desired groups, the adequacy of the partition obtained, and the type of similarity measure to be used are questioned. With regard to the number of groups desired, the most commonly done is to employ several numbers of groups and, by some optimization criteria, to select the most convenient. In order to assess the adequacy of the partition, it is common to use discriminant analysis. With regard to similarity measures, many are mentioned; however, the most used in breeding are the Euclidean and Mahalanobis distances for quantitative variables and Jaccard or Nei and Li indices for binary variables resulting from molecular marker studies.

The cluster process basically involves two steps: the first one relates to the estimation of the similarity measure (or dissimilarity) between the parents and the second one with the adoption of cluster technique to form groups.

Dissimilarity Measures

Genetic diversity studies aimed at the identification of parents for hybridization have been developed based on quantitative traits information or molecular markers. In the case of quantitative traits, it has been used to express the genetic diversity the mean Euclidean distance $\left(d_{ii'}\right)$ or the generalized Mahalanobis distance $\left(D^2_{ii'}\right)$. This latter is preferred; nevertheless, it can only be estimated when the residual covariance matrix is available, based on experimental trials that have replicates.

In general, if X_{ij} is the observation in the ith parent ($i = 1, 2, ..., p$), with reference to the jth ($j = 1, 2, ..., n$) under study, the Euclidean distance between two parents I and i' is defined by means of the expression

$$d_{ii'} = \sqrt{\sum_j \left(X_{ij} - X_{i'j}\right)^2}.$$

The generalized Mahalanobis distance is thus defined:

$$D_{ii'}^2 = \delta' \Sigma^{-1} \delta,$$

where

δ: deviation vector ($v \times 1$) among the average values of the parents in relation to the variables studied

Σ: residual variance-covariance matrix ($v \times v$), obtained from previous analyses, according to an appropriate statistical model

Additionally, it is possible to obtain the 2D projection, in which the dissimilarity measures are converted into scores relative to two variables (X and Y) which, when represented in scatter plots, will reflect, in the two-dimensional space, the distances originally obtained from the v-dimensional space (v = number of traits used to obtain distances). Thus, it is graphically possible to verify the distances between the genotypes. The feasibility of use of this technique is evaluated by the correlation between the original distances and those that will be represented in the scatter plot or by alternative measures such as the degree of distortion and the stress coefficient. In general, it is assumed that the graphic representation is satisfactory when the correlation value between the measures of original and graphic distances is greater than 0.9 and the values of distortion and stress are less than 20%. Figure 8.1 depicts the 2D graphic dispersion obtained on the basis of Mahalanobis distance resulting from the evaluation of 12 cultivars, with respect to eight traits (*NDF* number of days for flowering, *NDM* number of days for maturation, *A1V* height of the first pod insertion, *AP* height of plants, *NVP* number of pods per plants, *NSV* number of seeds per pod, *PCS* weight of 100 seeds, *PRO* production per plant), whose data are presented in Table 8.3.

In order to study genetic diversity, it is necessary, first of all, to estimate the residual variance-covariance matrix by means of variance analysis. With the mean values and the matrix Σ, it is then obtained the 12×12 dissimilarity matrix, which must be submitted to cluster analysis. Hence, it will be possible to make inference about the groups and determine which ones are similar or dissimilar.

In the case of molecular data, different dissimilarity measures can be obtained for later cluster analysis. Specifically in the selection of parents, it is important to point out that many authors do not advise the single use of molecular markers to identify the genotypes to form their base population, as markers are able to capture any genome variation, and not only referring to traits of interest in soybeans. We will present here, in a summarized form, some alternatives to obtain similarity measures based on molecular markers. Dissimilarity measures can be considered as multivariate statistics for reducing data or information or just a way of comparing pairs of populations or the basis to construct the evolutionary history of the

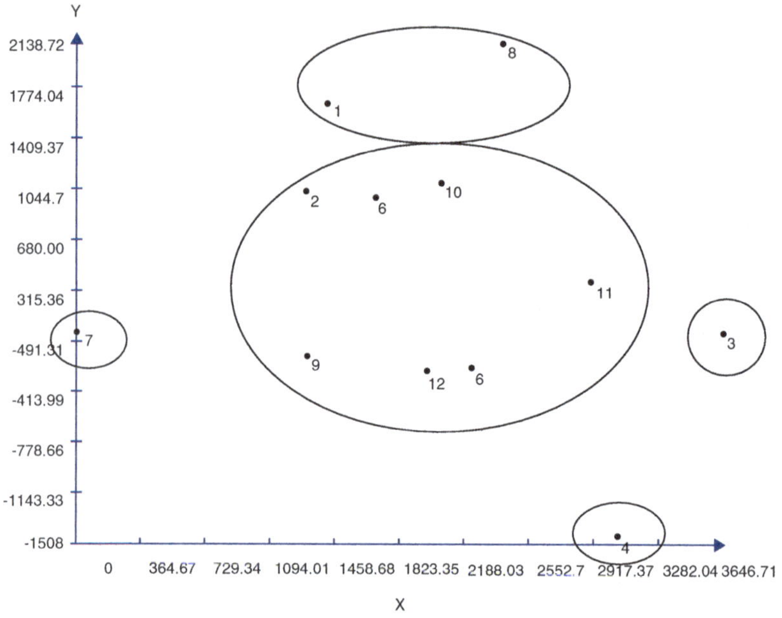

Fig. 8.1 2D projection obtained from the Mahalanobis dissimilarity measure of 12 soybean genotypes evaluated for eight phenotypic traits

populations (Weir 1996). Distance estimates obtained pairwise between populations generate a matrix capable of providing objective and stable classification, as much as possible, of the populations studied (Dias 1998). Some of these measures compile the information in binary form, whose data are encoded in ones and zeros, representing, respectively, the presence and absence of a given allele or mark. These measures have been the most used for information from dominant markers, since it is not possible to distinguish the dominant homozygous genotype from the heterozygote, and the calculation of allele frequencies for this type of marker is only possible under special conditions, that is, when populations are in equilibrium with respect to the reproductive system and the natural rate of cross-fertilization is known (Robinson 1998). Nonetheless, other dissimilarity measures are based on allelic or genotypic frequencies, the calculation of \vis based on individual loci, and the final distance is represented by the mean distance of each loci.

Dominant markers, for example, RAPD and AFLP, are identified by a binary code, code 1 being used to represent the occurrence of bands on the gel and code 0 to characterize the absence thereof. The similarity coefficients (S) between genotypes pairs are obtained taking into account:

a: Number of coincidences of 1-1 type for each pair of populations
b: Number of dislocations of 1-0 type for each pair of populations
c: Number of discordances of 0-1 type for each pair of populations

Table 8.3 Evaluation of 12 soybean cultivars in relation to eight phenotypic traits

Genotypes	Blocks	NDF	NDM	A1V	AP	NVP	NSV	PCS	PRO
1	1	38.01	100.21	21.53	51.28	33.22	2.17	12.08	2689.66
1	2	36.12	94.76	13.58	59.75	23.60	1.93	9.40	2737.54
1	3	38.80	96.17	21.18	65.43	37.23	1.99	11.97	2639.15
2	1	38.92	93.89	17.09	60.78	33.23	1.97	10.67	2168.66
2	2	38.84	95.50	20.02	79.84	32.84	2.04	11.65	2971.08
2	3	39.86	98.29	23.1	66.82	38.04	2.16	12.69	2950.25
3	1	38.05	89.63	17.43	59.89	33.00	1.87	16.97	1903.93
3	2	37.28	89.41	14.75	76.32	29.11	1.86	16.07	2005.19
3	3	39.06	96.90	19.00	65.68	38.13	2.18	17.50	3353.03
4	1	40.89	96.80	14.47	38.55	36.29	2.05	15.35	3341.27
4	2	39.32	98.22	20.83	79.15	28.33	2.11	17.50	3623.4
4	3	40.01	97.77	18.23	59.53	31.81	2.09	16.62	3770.82
5	1	38.25	92.79	15.37	80.33	31.09	2.02	11.06	2480.42
5	2	39.65	92.79	17.72	72.60	38.17	2.02	11.85	3018.13
5	3	37.83	88.42	12.24	84.16	28.93	1.84	10.01	3091.96
6	1	40.69	90.20	18.38	51.04	38.67	1.85	14.05	2478.45
6	2	39.30	95.75	20.59	60.56	31.62	2.09	14.80	3063.1
6	3	41.97	94.40	17.41	66.88	45.14	2.03	13.72	2812.12
7	1	44.05	71.17	12.72	71.34	45.71	1.61	6.84	1933.01
7	2	43.76	70.43	17.33	63.22	44.25	1.58	8.39	2106.67
7	3	43.29	81.39	14.34	78.02	41.88	2.05	7.39	2238.72
8	1	36.25	91.08	21.62	75.11	30.82	1.95	14.26	2617.33
8	2	32.09	88.19	25.85	79.49	9.65	1.83	15.68	3003.3
8	3	35.17	93.36	17.16	95.19	25.31	2.05	12.75	2560.75
9	1	40.81	86.48	15.89	68.50	35.09	1.91	11.02	2807.96
9	2	42.25	89.85	12.88	78.87	42.43	2.05	10.01	2063.29
9	3	41.27	85.20	19.91	75.29	37.42	1.85	12.38	2468.53
10	1	36.49	92.94	19.60	59.28	24.55	1.97	13.30	2992.11
10	2	38.20	94.27	19.48	63.50	33.23	2.03	13.26	2174.83
10	3	38.28	94.05	21.61	68.21	33.62	2.02	13.98	2400.25
11	1	36.53	102.12	14.03	55.60	25.22	2.01	13.75	2283.91
11	2	36.51	105.50	15.78	68.89	25.14	2.15	14.34	1806.03
11	3	35.62	111.05	18.12	64.98	20.57	2.39	15.13	2035.59
12	1	39.40	100.77	16.51	63.47	30.83	2.28	12.82	3318.72
12	2	39.86	93.66	20.09	68.11	33.17	1.98	14.03	2982.71
12	3	41.34	94.60	22.15	72.35	40.66	2.02	14.72	2760.14

NDF number of days for flowering, *NDM* number of days to maturity, *A1V* height of first pod insertion, *AP* height of plants, *NVP* number of pods per plant, *NSV* number of seeds per pod, *PCS* weight of 100 seeds, *PRO* production per plant

d: Number of coincidences of 0-0 type for each pair of populations

In practice, the similarity and dissimilarity coefficients most commonly used in genetic diversity studies have been those of simple coincidence (Sneath and Sokal 1973), Jaccard's (Jaccard 1908), and Nei and Li's (Nei and Li 1979). For further information on similarity measures, we recommend reading Cruz et al. (2011). The simple coincidence coefficient has, in its arithmetical complement, the advantage of being identical to the average Euclidean distance square ($D^2_{ii'}$). In contrast with the simple coefficient of coincidence, the coefficients of Jaccard and Nei and Li feature the advantage of not considering 0-0-type coincidences. For example, if there is a high probability of non-amplification of bands and their absence in both populations cannot be construed as a common trait, i.e., it does not necessarily mean identical regions of DNA, it is more appropriate to apply coefficients that exclude negative co-occurrence.

Thus, the estimators of the said similarity coefficients are:

Simple coincidence

$$S_{\mathrm{CS}} = \frac{a+d}{a+b+c+d}$$

Jaccard

$$S_{\mathrm{J}} = \frac{a}{a+b+c}$$

Nei and Li

$$S_{\mathrm{SD}} = \frac{2a}{2a+b+c}$$

As this is focused on similarity measures, we recommend, in cluster analysis, using dissimilarity measures, as defined: $D = 1 - S$.

Codominant markers, for example, RFLP and microsatellites, allow the identification of heterozygous and homozygous genotypes, generating more genetic information. In this situation, the allele frequency information is given based on the population, so that the allele frequencies are quantified based on the sampled genotypes of each population. The similarity can be obtained by:

$$S_{ii'} = \frac{1}{2}\sum_{j=1}^{L} p_j c_j$$

where

$p_j = \frac{a_j}{A}$: weight associated with locus j, determined by:

a_j: total number of locus j alleles

A: total number of alleles studied

L: total number of locus studied

being $\sum_{j=1}^{L} p_j = 1$

c_j: number of common alleles between pairs of accessions i and i'

The number of common alleles between two accessions of genotypes A_iA_i and A_iA_i (or A_iA_j and A_iA_j) is two; between A_iA_i and A_iA_j (or A_iA_j and A_iA_k), one; and between A_iA_j and A_kA_l, zero. The specified index varies from 0 to 1, and, as it is a matter of similarity measures, in cluster analysis it is recommended the use of dissimilarity measures, defined by $D = 1 - S$. For information on other indexes, refer to Cruz et al. (2011).

Clustering Techniques

Given the fact that in the clustering process it is desirable to have information about each pair of parents, the number of dissimilarity measures estimates is relatively large, and this makes it impracticable to recognize homogeneous groups by the simple visual examination of those estimates. To accomplish such task, clustering methods are used.

Among the clustering methods most commonly used in plant breeding, we mention hierarchical and optimization methods, which will be presented below.

Hierarchical Methods

In hierarchical methods, parents are grouped by a process that is repeated in several levels, until the dendrogram or the tree diagram is established. In this case, there is no concern about the optimal number of groups, since the greatest interest is in the "tree" and in the branches obtained. Delimitations can be established by visual examination of the dendrogram, through which points of high-level change are evaluated, generally taking them as delimiters of the number of parents for a given group.

Hierarchical methods are divided into agglomerative and divisive. Agglomerative ones include single linkage method; complete linkage method; average linkage or unweighted pair group method with arithmetic mean (UPGMA), which can be weighted or not, centroid, also weighted or not; and the method proposed by Ward (1963). Among the divisive methods, the best known is that of Edwards and Cavalli-Sforza (1965). For the example under consideration, clustering was performed through UPGMA, based on the generalized distance of Mahalanobis, and the result obtained is shown in Fig. 8.2.

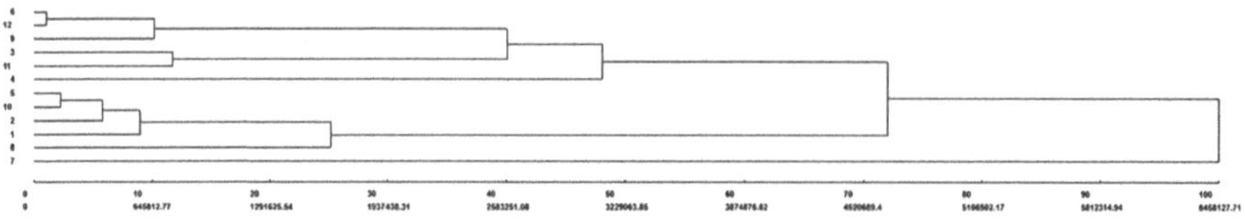

Fig. 8.2 Dendrogram generated by the UPGMA clustering method, based on the generalized distance of Mahalanobis

If a cut is made in the dendrogram above at the 70% melt level or the cut based on Mojena (1977), three groups will be formed: group A (6, 12, 9, 3, 11, 4), group B (5, 10, 2, 1, 8), and group C (7).

Santos et al. (2011) evaluated the genetic divergence among 48 soybean genotypes, grown in irrigated floodplain in the State of Tocantins, in order to identify the most promising combinations to produce superior recombinations, both of genotypes intended for oil and meal production and those of the special group intended for human consumption. The authors verified the existence of genetic variability among the genotypes tested and identified promising combinations for the production of soybeans for oil and meal as well as promising crosses for soybean genotypes for human consumption. It is worth mentioning that these authors used the UPGMA clustering method, Tocher method, and canonical variables analysis, the latter two of which are discussed later in this chapter. It is also noteworthy that, according to the authors, the three methods were concordant and, through them, four distinct groups of genotypes were detected.

In another study, Vieira et al. (2009) evaluated the genetic diversity of 23 soybean cultivars with 283 microsatellite markers. Subsequently, 53 markers that showed easily detectable polymorphism in agarose gels were used in the characterization of 53 cultivars using the UPGMA hierarchical method. The authors concluded that it is possible to detect significant variability in the Brazilian soybean germplasm evaluated, even among elite cultivars, when using microsatellite markers selected due to the information provided.

Min et al. (2010), aimed at evaluating the genetic diversity among commercial cultivars, local varieties, and strains from the soybean germplasm bank, used 40 brands of microsatellite type, and, from these brands, they estimated the diversity by means of the UPGMA method. They verified that local varieties and lines have far superior genetic diversity than the cultivars, and these must be preserved and used in soybean breeding programs, in order to maintain the variability of the population. The use of these local varieties can increase the productivity of cultivars and deliver resistance to various diseases and pests, as many alleles found in these strains are not seen in commercial cultivars. These alleles can be transferred to commercial cultivars by the backcrossing technique.

Table 8.4 Groups formed by the Tocher methodology in evaluation of 12 soybean cultivars

Group	Cultivars
I	6, 12, 9, 11, 5, 10, 2
II	1, 8
III	3
IV	7
V	4

Optimization Methods

In optimization methods, the set of parents is divided into nonempty and mutually exclusive subgroups, by maximizing or minimizing some preestablished measure. One of the optimization methods most commonly used in genetic improvement is the one proposed by Tocher (quoted by Rao 1952).

In the Tocher method, the criterion adopted is that the mean dissimilarity measures within each group should be less than the mean distances between any groups. The method requires the dissimilarity matrix, on which the most similar pair of parents is identified. These parents will form the initial group. From this point, the possibility of inclusion of new parents is evaluated, by adopting the previously mentioned criterion.

For the example under consideration, the clustering was performed using Tocher's proposal, based on the generalized distance of Mahalanobis, and the result obtained is shown in Table 8.4.

It has been observed the formation of five groups and that cultivars 3, 7, and 4 are the most divergent in relation to the others. For the selection of cultivars to be intercrossed, the potential of evaluated traits and diversity must be taken into account. In this case, it is recommended cross cultivars with good agronomic potential, that is, that belong to different groups.

Júnior et al. (2015) evaluated the agronomic performance of a group of advanced and superior soybean lines using genetic diversity based on Mahalanobis distance and Tocher's optimization method. According to the authors, it was possible to identify the presence of genetic variability among the lines evaluated.

Analysis of Canonical Variables

This type of analysis, reported by Rao (1952), requires more refined knowledge about multivariate statistical procedures but is easy to interpret and useful in genetic diversity studies. The analysis of canonical variables allows the data set simplification, summarizing the information, originally contained in a group of variables, in a few variables. They hold the maximum of the originally available variation and are independent of each other. However, this analysis is based on information between and within genotypes (or between individuals of each genotype); therefore, data is required at accession level, with replicates.

This is an alternative analysis to evaluate the degree of genetic similarity between genotypes, which takes into account both the residual covariance matrix and the phenotypic covariance matrix among the traits evaluated. It is based on the generation of new variables (called canonical variables), in which the information of the original variables measured is represented. These canonical variables are independent of each other and are estimated in order to retain the maximum variation in descending order.

Thus, for the example under consideration, analyses allow to conclude that with only two canonical variables (VC_1 and VC_2), it is possible to explain 99.72% of the variation found in the original data. These variables are defined by the following equations:

$$VC_1 = 16108.12\,NDF - 19377.60\,NDM - 1910.7\,A1V + 9.92\,AP + 29.41\,NVP + 19.19\,NSV + 11.53\,PCS - 5.9648\,PRO$$

$$VC_2 = 16501.49\,NDF - 19898.49\,NDM - 1900.46\,A1V + 10.59\,AP + 30.68\,NVP + 20.01\,NSV + 12.48\,PCS - 5.97\,PRO$$

The graphical analysis of the cultivar scores can be made based on Fig. 8.3. Again, it is observed that cultivars 3, 7, and 4 are the most divergent in relation to the others.

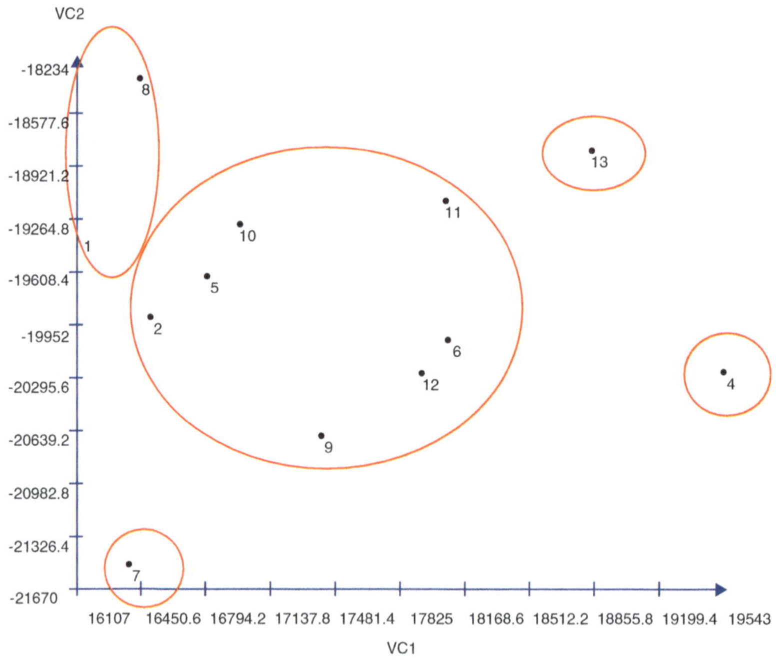

Fig. 8.3 Graphic dispersion of canonical variable scores obtained from the linear combination of eight traits evaluated in 12 soybean cultivars

Analysis of Performance Per Se

For a long time, plant breeding was only performed as an art, through which people selected the best behavior genotypes and used them in crosses for the formation of the next crossing cycle and, consequently, of a new population. Even with all techniques, breeders use this performance information for important and interesting traits at the time of selecting the parents in their breeding program. Therefore, it is up to the breeder, whenever possible, to try to select as parents the genotypes that present good performance and genetic diversity and should, therefore, always cross the "good and divergent" ones. Accordingly, the graphical analysis of the traits of interest is interesting. The disadvantage of this representation is the high number of possible graphic combinations. Considering the data set, represented herein with eight variables, it would be possible to obtain 28 graphs with paired combinations of each variable ($v(v - 1)/2$). Hence the importance of using dissimilarity measures and multivariate techniques. In any case, the projection for the main traits at the moment of making the decision regarding the future parents is interesting. Figure 8.4 shows the graphic dispersion, with the performance of each genotype for two traits of interest, the production (PRO), and the weight of 100 seeds (PCS). It is found, for example, that genotype 4 has satisfactory performance for the two characters. In contrast, genotype 7 has poor performance for the two characters evaluated.

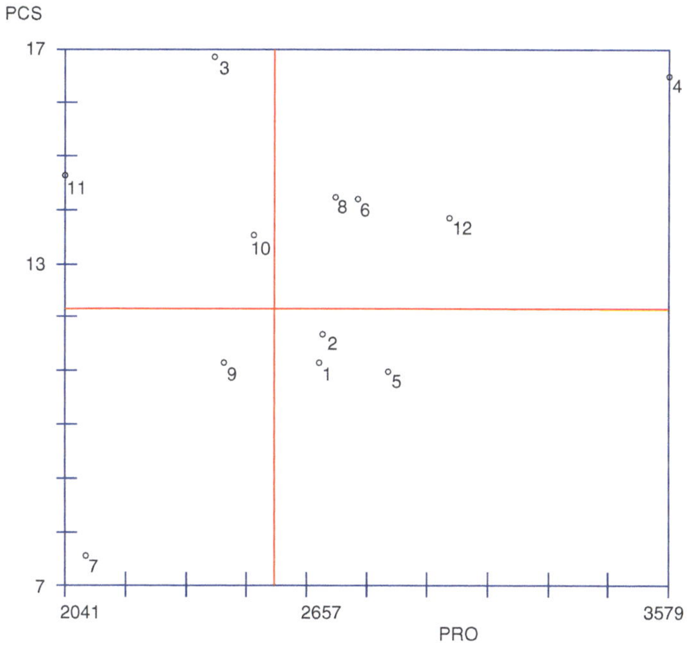

Fig. 8.4 Graphic dispersion on the average of 12 soybean cultivars for yield and weight traits of 100 seeds

Using the information generated by the analyses used, it is possible to see that genotypes 4 and 12 would be ideal in the selection process. From Fig. 8.4 we see that they are in the upper right quadrant; therefore, they have high productive capacity, and by observing the genetic diversity information (Figs. 8.1, 8.2, 8.3 and Table 8.4), these genotypes are divergent. Thus, the researchers are able to select good and divergent materials, increasing the chance of success when starting their programs, with parents selected in the best possible way.

The success of a soybean breeding program, aimed at conservation, evaluation, or use of accessions such as parents, depends exclusively on the knowledge and understanding of the population structure of these accessions, including diversity and parentage. Therefore, using the correct techniques to assess genetic diversity is essential to the success of the future breeding program. The first step is the correct choice of the parents, based on diversity and, later, diallel analysis. Li et al. (2011) evaluated the genetic diversity of 150 soybean accessions from China using seven characteristics of economic importance: (a) resistance to soybean cyst nematode, (b) resistance to soybean mosaic virus, (c) salinity, (d) tolerance to cold temperatures, (e) drought tolerance, (f) oil concentration, and (g) protein concentration. They also used 55 microsatellites distributed throughout the soybean genome and evaluated the genetic diversity using the main components method using STRUCTURE software. Soybean accessions were divided into four main groups that clearly indicated the four different regions from which the accesses were derived.

Final Considerations

The correct choice of parents can represent a major genetic progress, saving of time, and financial resources. Due to the existence of genetic diversity in soybeans, it is possible to select and recombine the better adapted, with better quality and more efficient genetic forms.

Several strategies can be used by researchers at this time of choice, as they must evaluate the performance of parents in traits of interest, genetic diversity for a set of traits, and complementary molecular information when existing.

The use of biometrics during this parent selection stage in a soybean breeding program is, therefore, an extremely useful decision-making tool for researchers, for it allows obtaining the maximum amount of information about the experiments evaluated, planning as accurately as possible the decision to be made. Thus, the correct selection of the genetic material to be used in the initial crosses of the breeding program is essential, and it is up to the researcher to make use of all the available tools to this end, taking into account the desired intent, the experience with the crops, and the several biometric techniques synthesized herein.

References

Cruz, C.D.: Princípios de Genética Quantitativa, 1st edn, p. 391. UFV, Viçosa (2005)
Cruz, C.D., Ferreira, F.M., Pessoni, L.A.: Biometria Aplicada ao Estudo da Diversidade Genética, 1st edn, p. 620. UFV, Viçosa (2011)
Cruz, C.D., Regazzi, A.J., Carneiro, P.C.S.: Modelos Biométricos Aplicados ao Melhoramento Genético, 4th edn, p. 514. UFV, Viçosa (2012)
Dias, L.A.S.: Análises multidimensionais. In: Alfenas, A.C. (ed.) Eletroforese de Isoenzimas e Proteínas Afins, pp. 405–475. UFV, Viçosa (1998)
Edwards, A.W.F., Cavalli-Sforza, L.L.: A method for cluster analysis. Biometrics. **21**, 362–375 (1965)
Gravina, G.A., Sediyama, C.S., Filho, S.M., Moreira, M.A., Barros, E.G.: Análise dialélica da resistência à mancha olho-de-rã em soja. Pesquisa agropecuária brasileira. **38**(6), 673–680 (2003)
Griffing, B.: Concept of general and specific combining ability in relation to diallel crossing systems. Aust. J. Biol. Sci. **9**, 463–493 (1956)
Jaccard, P.: Nouvelles recherches sur la distribution florale. Bull. Soc. Vaud. Sci. Nat. **44**, 223–270 (1908)
Júnior, J.A.F., Unêda-Trevisoli, S.H., Espíndola, S.M.C.G., Vianna, V.F., Mauro, A.O.: Diversidade genética em linhagens avançadas de soja oriundas de cruzamentos biparentais, quádruplos e óctuplos. Rev. Ciênc. Agron. **46**(2), 339–351 (2015)
Li, Y., Smulders, M.J.M., Chang, R., Qiu, L.: Genetic diversity and association mapping in a collection of selected Chinese soybean accessions based on SSR marker analysis. Conserv. Genet. **12**, 1145–1157 (2011)
Maphosa, M., Talwana, H., Gibson, P., Tukamuhabwa, P.: Combining ability for resistance to soybean rust in F_2 and F_3 soybean population. Field Crop Res. **130**, 1–7 (2012)
Min, W., Run-Zhi, L., Wan-Ming, Y., Wei-Jun, D.: Assessing the genetic diversity of cultivars and wild soybeans using SSR markers. Afr. J. Biotechnol. **9**(31), 4857–4866 (2010)
Mojena, R.: Hierarchical grouping methods and stopping rules: an evaluation. Comput. J. **20**, 359–363 (1977)
Nei, M., Li, W.H.: Mathematical model for studying genetic variation in terms of restriction endonucleases. Proc. Natl. Acad. Sci. **76**(10), 5269–5273 (1979)
Rao, C.R.: Advanced Statistical Methods in Biometric Research, p. 400. Wiley, New York (1952)
Robinson, I.P.: Aloenzimas na Genética de Populações de Plantas. In: Alfenas, A.C. (ed.) Eletroforese de Isoenzimas e Proteínas Afins: Fundamentos e Aplicações em Plantas e Microrganismos, pp. 329–380. UFV, Viçosa (1998)
Santos, E.R., Barros, H.B., Ferraz, E.C., Cella, A.J.S., Capone, A., Santos, A.F., Fidelis, R.R.: Divergência entre genótipos de soja, cultivados em várzea irrigada. Revista Ceres. **58**(6), 755–764 (2011)
Sneath, P.H., Sokal, R.R.: Numerical Taxonomy: The Principles and Practice of Numerical Classification, p. 573. W.H. Freeman, San Francisco (1973)
Vieira, E.S.N., Schuster, I., Silva, R.B., Oliveira, M.A.R.: Variabilidade genética em cultivares de soja determinada com marcadores microssatélites em gel de agarose. Pesq. Agrop. Brasileira. **44**(11), 1460–1466 (2009)
Ward, J.H.: Hierarchical grouping to optimize an objective function. J. Am. Stat. Assoc. **58**, 236–244 (1963)
Weir, B.S.: Genetic Data Analysis II, p. 445. Sinauer Associates, Sunderland, MA (1996)
Zorzetto, M.M., Motta, F.C., Morais, L.K., Kiihl, T.A.M., Silva, R.M.: Análise dialélica da capacidade combinatória em soja. Revista Biociências. **14**(2), 105–109 (2008)

Chapter 9
Methods for Advancing Segregating Populations

Felipe Lopes da Silva, Willian Hytalo Ludke, Murilo Viotto Del Conte, Thays Vieira Bueno, and Alisson Santos Lopes da Silva

Abstract The use of hybridization method in soybean breeding is extremely important for the development of cultivars with desired agronomic characteristics. Segregating populations from these crosses must be conducted through successive generations of self-pollination, and the choice of the appropriate method of conduction favors to increase the efficiency of selective process of superior lineages. In this sense, the present chapter aims to describe the main methods for advancing soybean segregating populations and to conduct a comparative discussion among them.

Keywords *Glycine max* (L.) Merrill • Advancing generations • Methods for obtaining lines

Introduction

The accumulated benefits of soybean breeding are undoubtedly the main contributors toward achieving the current levels of crop productivity. In Brazil, soybean yield increased from 1644 kg ha^{-1} in the 1970s and 1980s to the current 3000 kg ha^{-1} (Conab 2015). In the 2014/2015 harvest, production of soybean grains in Brazil was approximately 96 million tons of beans produced on 32 million hectares. In Argentina, in the same harvest, production of soybean grains was 60.8 million tons produced on a total area of 19.3 million hectares (USDA 2015). In the United States of America, in the 2014/2015 harvest, approximately 108 million tons of beans were produced on an area of 33.6 million hectares (USDA 2015).

The genetic differences to be detected between soybean cultivars already launched are becoming ever smaller, thus requiring greater efficiency from breeding programs. One of the factors that has been contributing to this reduction in

F.L. da Silva, M.S., D.S. (✉) • W.H. Ludke, M.S. • M.V. Del Conte, M.S.
T.V. Bueno, M.S. • A.S.L. da Silva
Federal University of Viçosa, Viçosa, Minas Gerais 36570-900, Brazil
e-mail: felipe.silva@ufv.br; willianludke@gmail.com; muriloviottodelconte@hotmail.com; thaysvieirabueno@gmail.com; alissonufv@gmail.com

F. Lopes da Silva et al. (eds.), *Soybean Breeding*,
DOI 10.1007/978-3-319-57433-2_9

differences is the narrowing of the genetic base, which has arisen from adopting few genotypes when developing new soybean cultivars. The base is so narrow that the genetic base of the material cultivated in Brazil in 1985 was represented by only 26 ancestors, and only 11 of these contributed 89% of the genetic set of Brazilian cultivars (Hiromoto and Vello 1986). Furthermore, six of these ancestors are also the most frequent in the genealogies of North American cultivars. Vello (1985) suggests that 25% of exotic germplasm selected can be introduced gradually into the cultivated germplasm in triple crosses or into populations with a broad genetic base.

Thus, the use of the hybridization method in soybean breeding becomes increasingly important. Borém and Miranda (2013) define hybridization as the fusion of genetically different gametes coming from different parents, which results in hybrid heterozygote individuals for one or more loci. The different methods for selecting parents were set out in Chap. 8 of this book.

After hybridization, the objective of the soybean breeding program is to obtain superior homozygous individuals (lines) for breeding the target trait through successive generations of self-fertilization. Individuals that have desirable traits from both parents are selected in the segregant populations, and the lines arising from selected individuals are evaluated in comparative productivity tests, and those that prove to be superior are launched as cultivars (Borém and Miranda 2013).

In this respect, choosing a suitable method for conducting the segregant populations in the soybean breeding program increases the efficiency of the selection process for superior lines. The methods for conducting population segregation were proposed in the early twentieth century, and some modifications took place in the 1950s and 1960s (Raposo et al. 2000).

The aim of this chapter is to describe the main methods for conducting segregant populations and to make a brief comparison between them.

Methods for Conducting Segregant Populations

Species of plants that are predominantly autogamous are those that have a natural crossing rate of less than 5%. Different authors have shown that the natural crossing rate in soybean is less than 1% (Miyasaka 1958; Sediyama et al. 1970; Vernetti et al. 1972). In Chap. 7 of this book, the genetic structure of autogamous populations and the use of variances in the study of quantitative traits were described and so were their implications in breeding.

Before presenting the methods for conducting segregant populations, it is worth pointing out, in a simplified way, what happens to the genetic variance while the generations of endogamy go by in autogamous populations, as presented by Ramalho et al. (2012) (Table 9.1).

In one segregant F_2 population with allelic frequencies in the segregant loci of 0.5, which came from across between two lines without any epistatic effects on genetic variance, (V_G) is given as $V_G = 1V_A + 1V_D$. With endogamy, the

Table 9.1 Coefficients of genetic variance both additive (V_A) and dominance (V_D) in the total variation between and within the progenies in the successive generations of endogamy

Generations	Variance	V_A	V_D
F_2	Total	1.00	1.00
$F_{2:3}$	Between	1.00	0.25
	Within	0.50	0.25
	Total	1.50	0.50
$F_{3:4}$	Between	1.500	0.125
	Within	0.250	0.125
	Total	1.750	0.250
$F_{4:5}$	Between	1.7500	0.0625
	Within	0.1250	0.0625
	Total	1.8750	0.1250
$F_{5:6}$	Between	1.87500	0.03125
	Within	0.06250	0.03125
	Total	1.93750	0.06250
$F_{6:7}$	Between	1.937500	0.015625
	Within	0.031250	0.015625
	Total	1.968750	0.031250
F_∞	Between	2.00	0.00
	Within	0.00	0.00
	Total	2.00	0.00

Source: Ramalho et al. (2012)

contribution from the additive variance (V_A) increases, and the dominance variance (V_D) decreases (Ramalho et al. 2012), as shown in Table 9.1. This occurs because of the reduction of loci in heterozygotes and, consequently, the increase of loci in homozygotes.

It can be observed that the genetic variance between progenies $F_{2:3}$ contains 1.0 V_A between and 0.5 V_A within. In $F_{4:5}$, there is 1.75 V_A between and 0.125 V_A within. One can understand, therefore, that the V_A within is dissipated endogamy (Ramalho et al. 2012).

Table 9.1 contains the information needed for conducting segregant populations and shows which one is the best generation for selecting individual plants from, so that they can be conducted in family structures or lines.

In this respect, the methods for conducting segregant populations are classified into methods that do not separate the endogamy phase from the selection phase (genealogical method, mass method, and early generation test) and the methods that separate these two phases (population method, bulk method within families, descendent method from a single seed (SSD), and descendent method from a single pod (SPD).

Genealogical Method

The genealogical method, also known as the pedigree method, was proposed at the end of the XIX century by Hjalman Nilsson, at the *Swedish Seed Association* in Svalof, Sweden. This method is based on selecting individuals with higher phenotypes in the segregant populations by evaluating each progeny individually and by keeping up the data on the kinships between the parents and their descendants. The first description of the method was presented by Love (1927).

This method was started and developed by crossing two parents that had been selected in accordance with the objective of the program. In F_1 generation, the highest possible number of seeds must be obtained.

F_2 generation must be conducted under conditions that are representative of cultivation, except for the spacing which must be slightly wider so that visually identifying the superior individuals is made easier. In soybean, the features evaluated in F_2 are bean productivity, plant height, resistance to lodging, number of days for flowering, number of days for ripening, number of pods per plant, growth type and behavior, resistance to root diseases and airborne diseases, and visual quality of seeds, among other factors. In this generation, the plants that have desirable features will be selected and picked individually (Fig. 9.1).

In generation $F_{2:3}$, each plant selected in F_2 will make up a line of 2.0–3.0 m in length, spaced at 0.60–0.70 m apart and with a density of 15–20 plants m^{-1}. At this stage, selection of the agronomically superior lines is started, and the best plants within each line, which will later be threshed individually, are selected.

The plants selected in $F_{2:3}$ will compose the $F_{3:4}$ families and will be conducted in the same way as the previous generation. In generation $F_{3:4,}$ it is possible to observe that there is a greater number of families with desirable features and, once again, selection is carried out between and within the families. This procedure is followed until reaching generation $F_{5:6}$, since in this generation the homozygosity level is high and the variance within the family is practically nonexistent (Table 9.1). This generation must be conducted in regions and at sowing times that are representative of the environment into which the new cultivar will be recommended.

In generation $F_{5:6}$, the promising families are picked in masse and sent off for the preliminary line trial (PLT), which is the first comparative productivity test. The PLT must be conducted under conditions that are similar to those of cultivation and replicated twice, and this is when the superior lines will be kept and the inferior ones discarded.

The intermediary line trial (ILT) will be conducted with the lines that have been outstanding in the PLT. The ILT are conducted in two or three places, with outlining of random blocks, and are replicated three times. Generally, lines of 5 m in length are used with spacing of 0.5 m between rows and with a density of 15–20 seeds m^{-1}. Besides the lines, controls are sown, which are usually cultivars that have adapted to the cultivation region and which have been outstanding

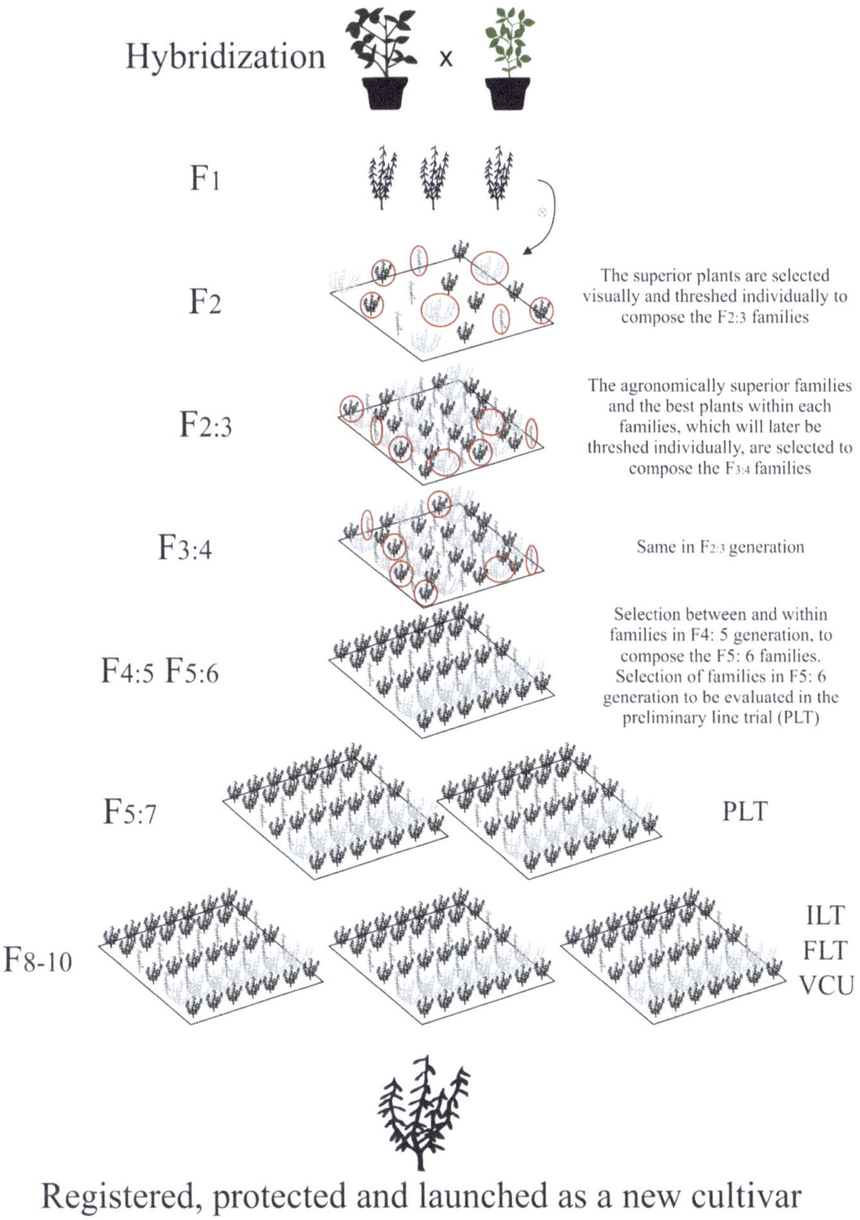

Fig. 9.1 Scheme of genealogical method

because they have been well accepted by the farmers. Superior lines are kept and inferior ones discarded.

The superior lines in the ILT will be conducted in the final line trial (FLT). The FLT are conducted in five or more places that are representative of the region into which the future cultivar will be recommended. The experimental layout and the details regarding the plots and traits being evaluated are similar to the ILT. In this generation, the value for cultivation and use (VCU) lines are also carried out. After evaluating the lines in comparative productivity tests for a period of 2 years and having established the VCU, the superior lines are registered, protected, and launched as cultivars.

The principle of the genealogical method is that selecting with a progeny test and knowing the lines selected allow maximization of selection efficiency and establish the degree of kinship of the genealogy between the lines selected.

Among the advantages of the method, the main ones are controlling the degree of kinship between the selections, discarding the inferior individuals in early generations, and using the data obtained for genetic studies. Some disadvantages are the following: the method only allows the advance of one generation per year, which makes it slow; it requires a large labor force and large experimental area; and it requires qualified professionals for selecting the superior individuals.

Mass Method

The mass method is carried out by selecting a large number of plants that have desirable phenotypical traits, which are then picked all together so that the next generation may be sown. The central idea of the method is to choose plants that have the best phenotypical expression, thus increasing the general average of the population by bringing the superior individuals together.

The method can be conducted in two ways, either by positive or negative mass selection. In positive mass selection, the plants with desirable phenotypes are selected, and in the negative one, the plants that are considered to be inferior are eliminated from the population.

Through this methodology, the phenotypical selection of plants in generation F_2 is made, and this must be conducted with wider spacing so as to make visual selection easier. The plants selected in F_2 are picked and the seeds mixed. A sample of the seeds is planted and these give rise to the plants belonging to generation F_3. This procedure is replicated until the desired level of homozygosity is reached (e.g., F_5), and, at the end of this generation, superior plants are selected for composing the lines of the following generation (Fig. 9.2).

Generally, in generation $F_{5:6}$, the best ones are selected and go on for evaluation in the PLT, ILT, FLT, and VCU trials, like the genealogical method. The best lines, after being evaluated in these trials, can be registered, protected, and launched as cultivars.

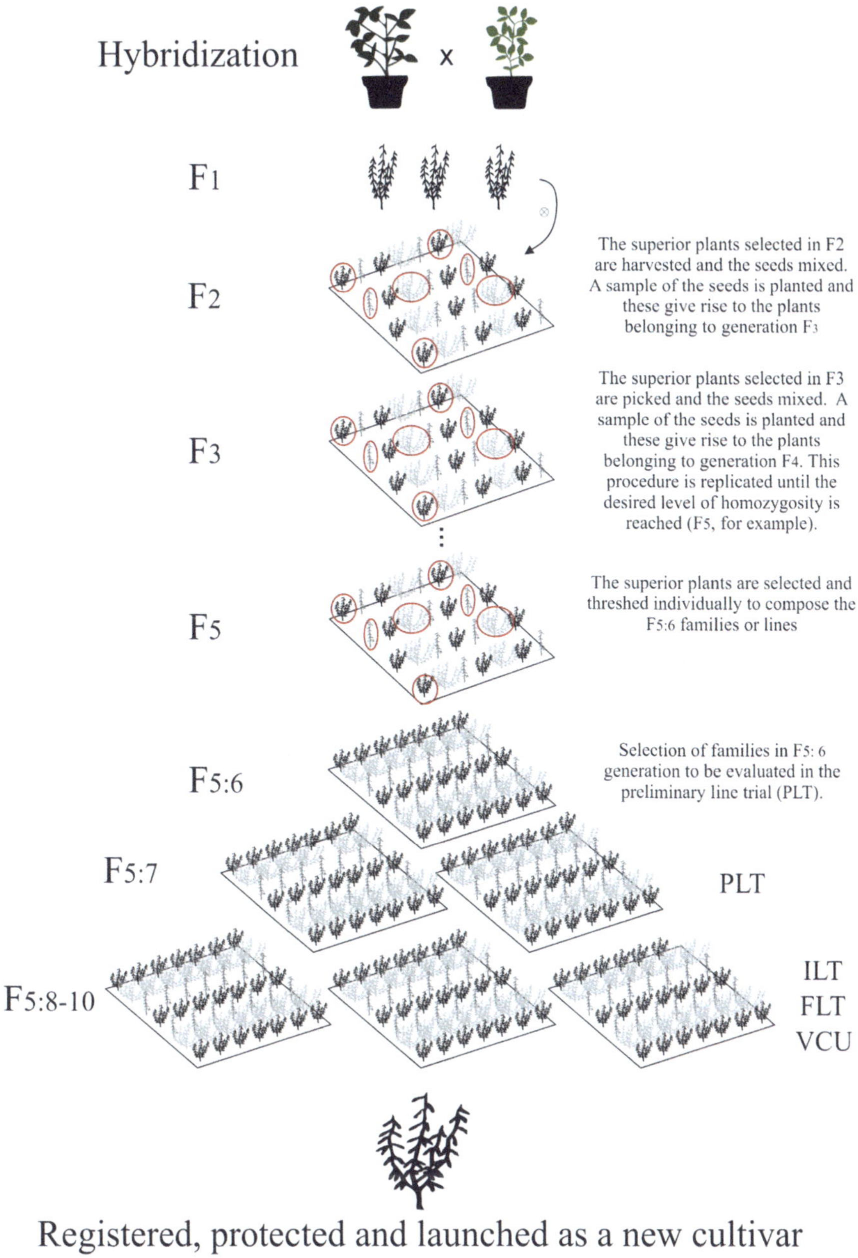

Fig. 9.2 Scheme of mass method

The mass method is efficient when used for high heritability traits, but when there are low heritability traits, the intensity of selection should be reduced, as this avoids losing favorable alleles in the initial generations of selection.

For soybean crops, some of the advantages shown by this method are the ease with which the segregant populations are conducted and the low operational cost. However, as we are dealing with a method in which visual selection is fundamental, there is the need for a qualified workforce to carry out this selecting. It is worth underlining that visual selection is not efficient enough for low heritability traits or, rather, quantitative ones. For these traits, the breeder is faced with a strong environmental influence, which makes success impossible when selecting superior plants. Another disadvantage of the method is that one needs to conduct the generations under environmental conditions that are favorable for cultivating soybean, and this makes it impossible to make many generations per year.

Early Generation Test

The early generation test was proposed by Immer (1941) as a method for identifying crosses between parents that have the possibility of generating progenies in early generations and greater potential for bean productivity. In other words, in this test one is trying to increase the percentage of lines arising from families with high productivity that were evaluated in early generations. Populations coming from crosses with low potential are discarded, and the breeding program can concentrate its resources solely on evaluating promising families, thus increasing the efficiency of the whole process. Fehr (1978) proposed using the early generation test for producing soybean grains. According to the author, the method consists of sowing F_2 seeds arising from F_1 plants, arising from F_2 individuals (Fig. 9.3).

F_3 seeds from each F_2 plant that have been selected are sown in separate lines, and these make up the $F_{2:3}$ families. Families that show inferior agronomical traits are discarded. The families from the $F_{2:4}$ generation are formed from the seeds of plants from the previous generation, and they are replicatedly used in productivity tests.

The $F_{2:4}$ families that have been identified as superior in the competition trial are cultivated in the $F_{2:5}$ generation, and, after having ripened, the superior plants are selected for composing the families of the $F_{5:6}$ generation, in which those that show the best uniformity are picked in bulk and sent on to the PLT, ILT, FLT, and VCU trials. Finally, the superior lines are ready to be registered, protected, and launched as cultivars.

This method has the advantage of being able to discard crosses and inferior families in the early generations, thus concentrating the program's resources on conducting superior genotypes. However, it does commit part of the resources to evaluate the populations in early generations when there is not much guarantee of the procedure being efficient, limits the number of generations that can be advanced per year, and shows lack of efficiency with regard to low heritability traits.

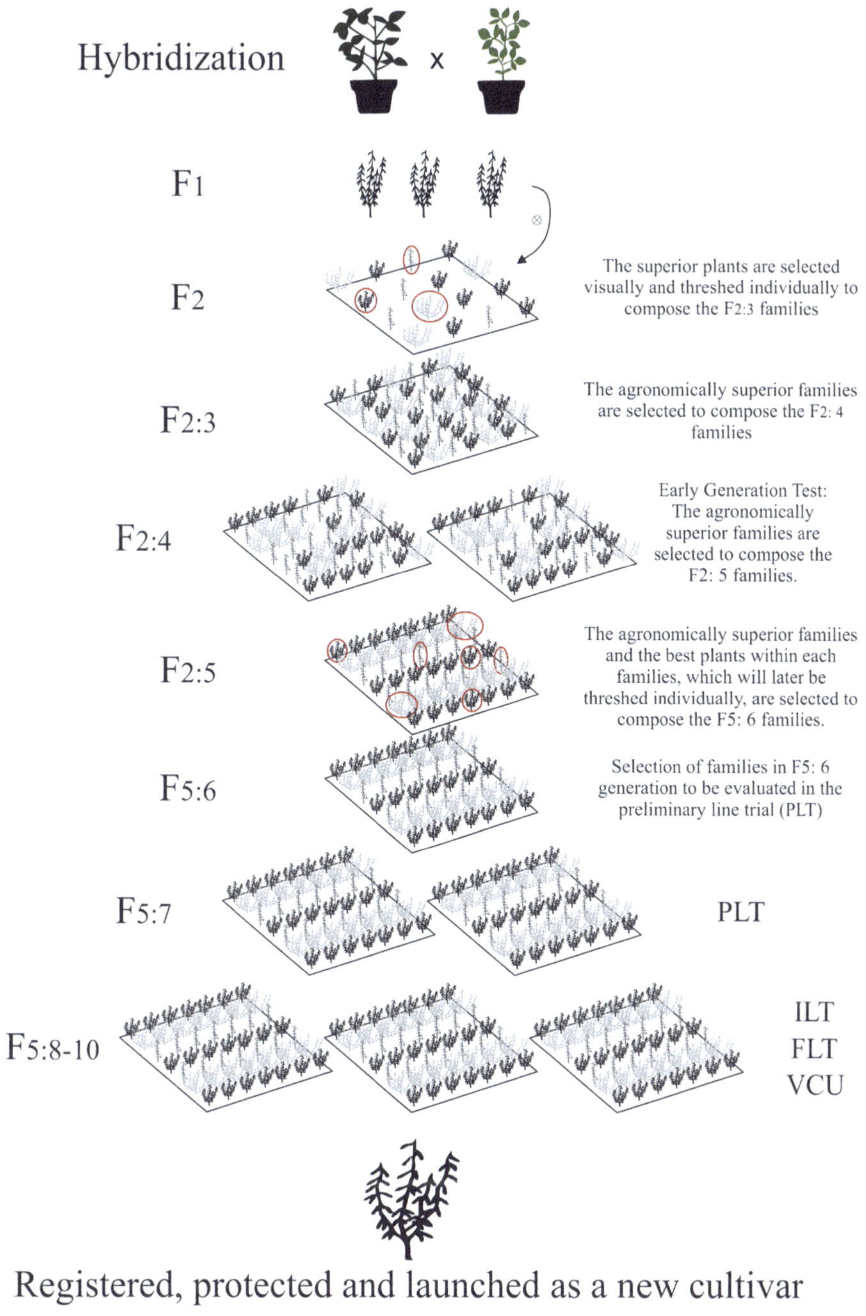

Fig. 9.3 Scheme of early generation test, proposed by Fehr (1978)

Population Method

The population method, also known as the bulk method, was proposed in 1908 by Nilsson-Ehle, at the Svalof Institute, Sweden, and was described in 1912 by Newman (Florell 1929). This is an economical method and one that is easy to conduct for obtaining homozygote lines from F_2 plants. After the hybridization of two parents, the F_1 seeds are conducted in the field or in greenhouses, and all the plants are picked together, and then one sample of the seeds will form the F_2 generation (Fig. 9.4).

Generation F_2 must be conducted under conditions that are representative of the regions into which the new cultivar will be recommended, and representative agricultural practices must also be used (15–20 plants m^{-1} and space between rows of 0.6 and 0.7 m). After ripening, all the plants are picked.

From the seeds obtained in generation F_2, a sample that will form generation F_3 is removed, and this procedure is replicated until the desired level of homozygosity has been obtained, normally up to F_5, but this can vary depending on the degree of genetic divergence of the parents that were chosen.

Generation F_6, following the last generation picked in bulk, is conducted with reduced plant density, generally 10 plants m^{-1}, thus reducing the number of progenies to be tested. It is in this phase when selecting superior individuals for the target traits of the breeding program starts. The individuals are picked and threshed separately, and the seeds from each plant will compose a line that is to be evaluated in the PLT. The best lines are sent on for the ILT, FLT, and VCU trials, until the superior ones are registered, protected, and launched as cultivars. The principle of the population method is that natural selection favors the superiority of plants that have a greater capacity to leave descendants for the next generation. In other words, it is based on the survival capacity of the individual. In this respect, individuals that produce more seeds in one generation have a better chance of being sampled for the later generation.

Some of the advantages of the method are saving on labor force when conducting the segregant population, conducting a large number of populations more easily, and, as stated above, benefitting from the process of natural selection which encourages the prevalence of more productive plants in the population.

Some of the disadvantages are not being able to use greenhouses or conduct more than one generation per year, loss of genotypes that have a low capacity for competition, necessity to select from a large number of plants in the more advanced generations, and low efficiency in natural selection for traits like resistance to diseases, to lodging, and to the dehiscence of pods, seeing as breeders do not practice selection during the advanced generation stage. Thus, unwanted plants are able to leave descendants for the next generation.

Owing to the high number of limitations that the population method presents, various modifications to it have been proposed.

The Harrington method (Harrington 1937) brought the advantages of the population method and the genealogical method together by conducting the segregant populations through the population method until favorable environmental

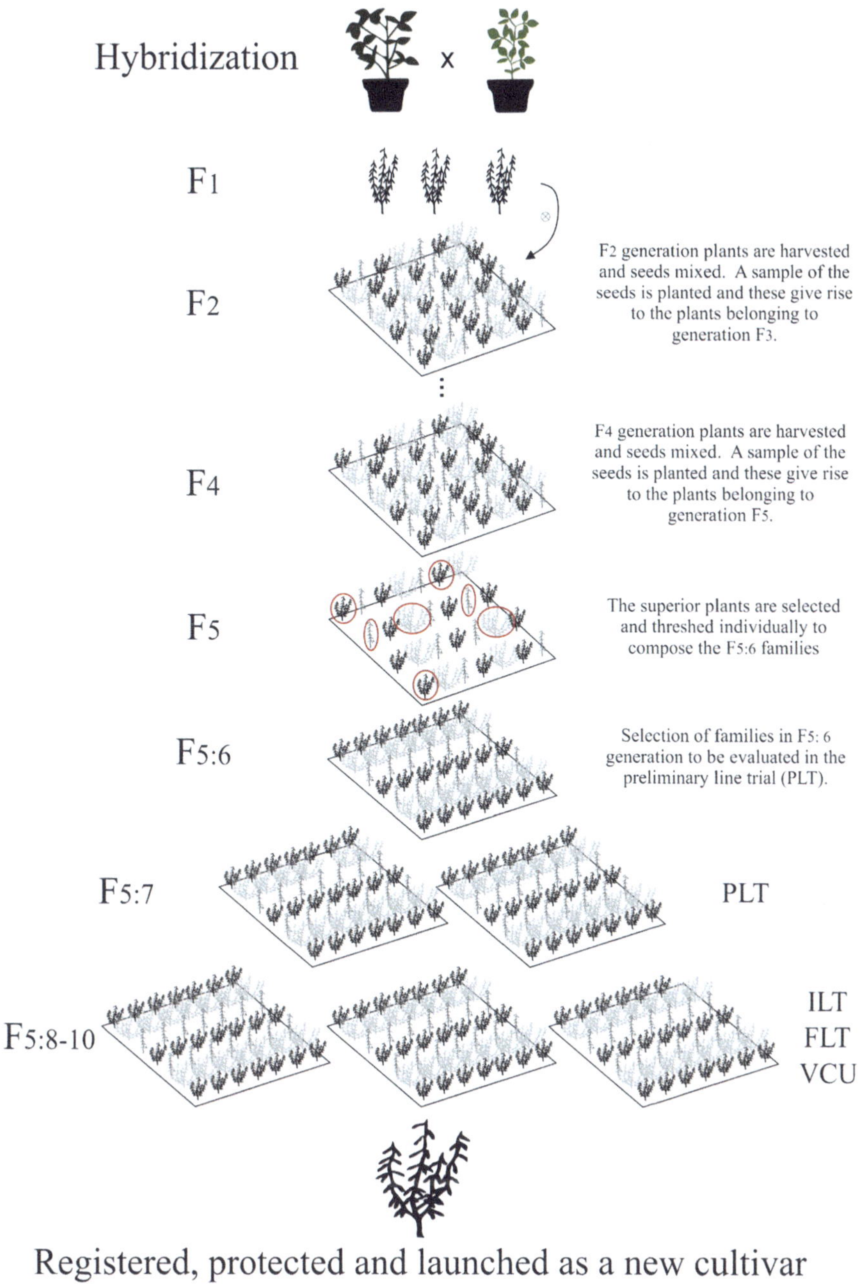

Fig. 9.4 Scheme of population method

conditions enabled the traits of interest to be phenotypically expressed and, in this case, superior plants to be selected. After selection, the genotypes are conducted through the genealogical method. This method has the advantage of being flexible, and the progeny test is used efficiently when the environmental conditions are right for selection. What is more, the method enables a larger number of genotypes to be evaluated.

The Empig and Fehr method (Empig and Fehr 1971) consists of subdividing the segregant population into groups that have different ripening periods, and each subpopulation is conducted through the original procedure until generation F_5, when arbitrary lines are removed for the PLT. The aim of this method is to eliminate competition between individuals that have different cycles.

The Van der Kley method (van der Kley 1955) is based on gradual negative mass selection, that is to say, eliminating the undesirable traits in the population. In this method, light selection pressure is used in generation F_2, and this is gradually increased until generation F_4 or F_5. The method increases the probability of success while lines are being opened up in advance generations.

Bulk Method Within Families

The bulk method within families was proposed by Frey (1954), and it makes a better sampling of the segregant population in the initial generations possible (Fig. 9.5).

The method starts with selecting plants in the F_2 generation that will form the $F_{2:3}$ families, which are sown in lines.

The seeds coming from each family are picked in bulk and used for obtaining $F_{2:4}$ families. The effect of natural selection occurs only within families, and the existing variation between the F_2 plants is maintained.

In generation $F_{2:4,}$ plants are selected, within the families, that stand out in accordance with the breeding program's trait of interest, and these will be sown in lines in the following generation $F_{4:5}$. In generation $F_{4:5}$ and $F_{4:6}$, evaluation of the lines is carried out, and this is when the inferior ones are discarded and the superior ones are kept for starting the PLT, ILT, FLT, and VCU trials. Finally, the superior lines can be registered, protected, and launched as cultivars.

Descendent Method from a Single Seed

The descendent method from a single seed, commonly known as SSD (single seed descent), was proposed by Goulden (1939), and its aim is to reduce the time needed to reach a high proportion of loci in homozygosity by advancing the generations outside the normal sowing time and maximizing the number of lines in homozygosity descendants from different individuals from generation $F_{2.}$ This is why it is one of the methods that is most used by the soybean breeders.

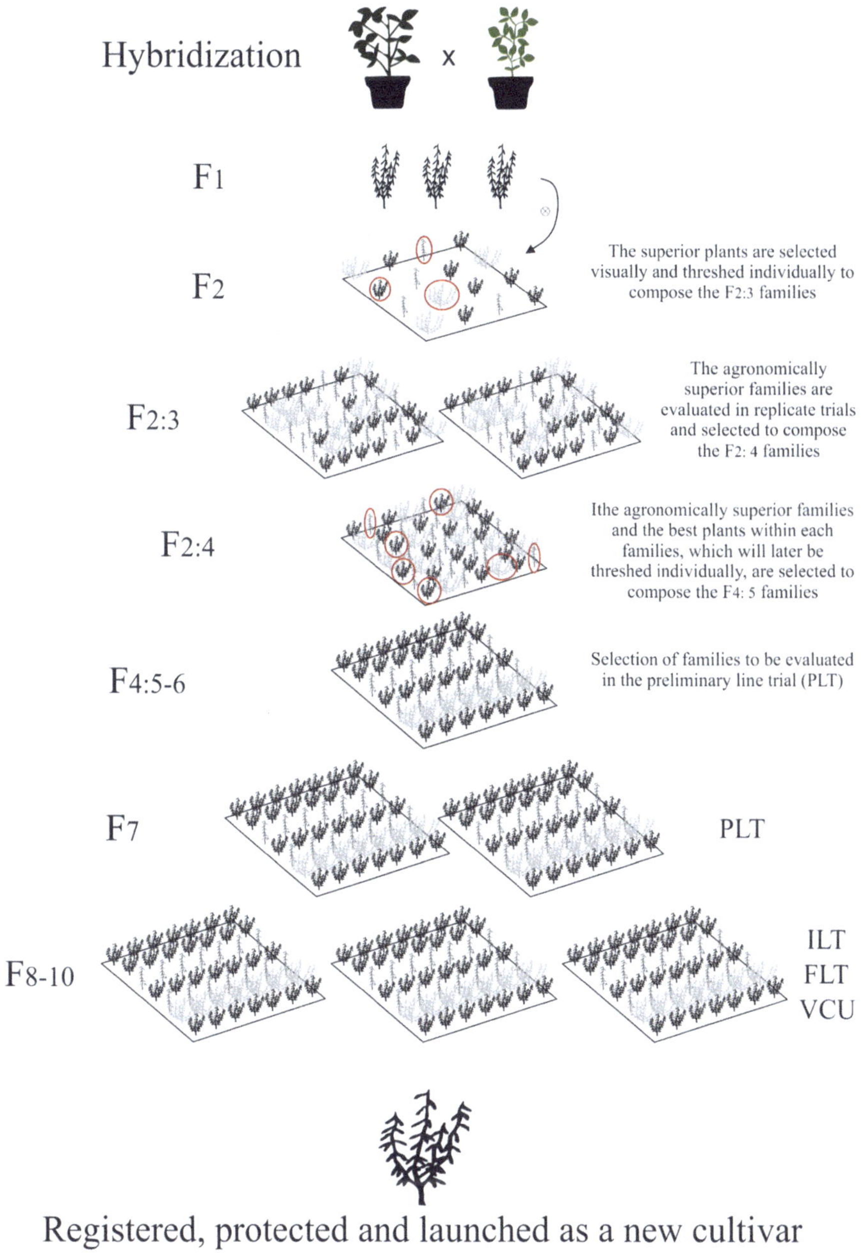

Fig. 9.5 Scheme of bulk method within families deriving from F_2

The SSD method ensures that each homozygote line in the end population corresponds to an F_2 plant. However, this does not mean that all the F_2 plants will be represented in the end population, because it is possible for some seeds to be incapable of germinating and for some plants not to complete the cycle.

In generation $F_{2,}$ one seed for sowing is collected from each plant, and this forms the population in generation F_3 (Fig. 9.6).

Thus, the number of plants in generation F_3 is the same as in generation F_2, disregarding the factors stated above like the non-germination of seeds that have been collected. In generation $F_{3,}$ one picks again, randomly, one seed to compose the population of the next generation. This procedure is replicated until generation F_5 or until the population reaches the desirable degree of homozygosity.

After having attained the desired homozygosity, the plants are picked individually and threshed, which forms the lines in generation F_6. In this generation, verification of the phenotypical homogeneity and the superiority for the traits being selected is made. The superior lines are picked separately, in bulk, to be evaluated in the PLT, ILT, FLT, and VCU trials, so that later the best line can be registered, protected, and launched as a new cultivar.

The principle of the SSD method is to separate the phase of advancement of generations from the selection phase. By doing this, segregant populations do not need to be conducted in environments that are similar to the one into which the future cultivar will be recommended. This method is best suited for greenhouses and enables more than one generation to be obtained annually.

The advantages of the method are as follows: providing a wide genetic variance between lines in the end population, seeing as the genetic variability of generation F2 can be maintained throughout the generation-advancing process; ability to reach the suitable level of homozygosity quickly; ease of conduction; no need to register genealogies; and requires a small experimental area and workforce. However, it is important to point out that selecting only one seed per plant, mainly in initial generations, does not represent all the genetic variability within progenies.

Descendent Method from a Single Pod

The descendent method from a single pod, commonly known as SPD (single pod descent), is a variation of the SSD method that was implemented with the aim of increasing the representativeness of variation within progenies, in initial generations F_2, F_3, and F_4, and this is compromised by selecting only one seed in the SSD method.

It is currently the method that is most used in soybean breeding, under greenhouse conditions, because of the ease in conducting the segregant population until the desired levels of homozygosity are reached.

In generation $F_{2,}$ one pod from each plant is picked to compose generation F_3. The pods that are picked are put in a group and threshed together. The seeds obtained are used for planting generation F_3. This procedure is replicated until

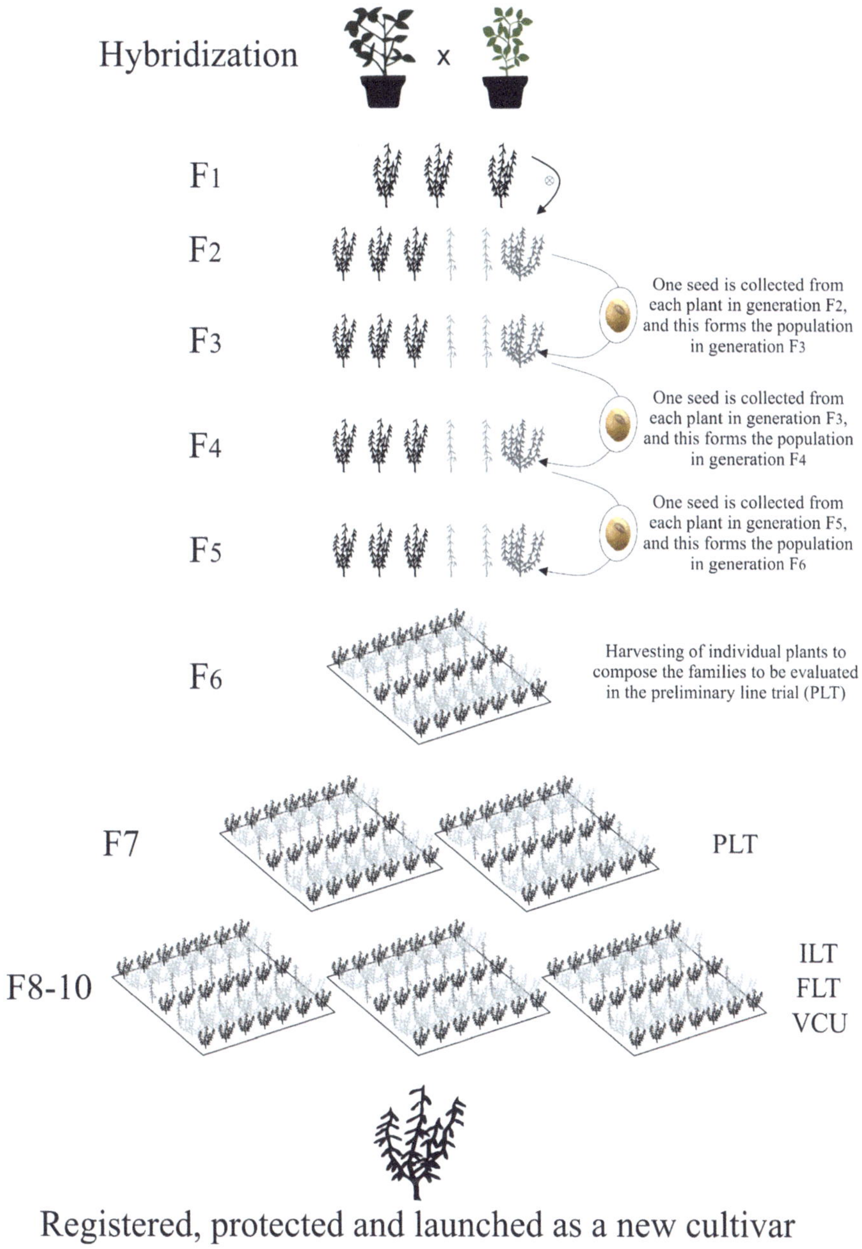

Fig. 9.6 Scheme of SSD method

the desired level of homozygosity (F_6) is obtained. Later, the method becomes similar to the SSD one.

Comparison of the Methods for Conducting Segregant Populations

The various methods for conducting segregant populations found in the literature were created in response to the disadvantages or limitations noticed in the methods described initially (Borém and Miranda 2013). It is the responsibility of the breeder to adopt a conducting strategy that provides him with the greatest genetic gains in the shortest space of time and at the lowest costs (Vencovsky 1987).

In general, the conduction methods differ from each other in the following aspects: absence or presence of selection during the endogamy phases; number of generations per year; requirement for human, physical, and financial resources; genetic variance exploited at the time of selection; and obtaining genetic gains.

Estimating the genetic gains is important in the breeding programs because it helps when choosing the most appropriate method for conducting segregant populations (Cruz 2005), and it is directly related to heritability, selection index, and selection differential.

In general, when comparing the genealogical methods, the population, the SSD, and the SPD, genetic gains and heritability, for most of the traits evaluated in soybean, are greater when segregant populations are conducted by the former. However, there are types of crosses in which it is evident that the SSD method results in greater genetic gains (Table 9.2).

Normally, the population method shows lower genetic gains than the SSD, SPD, and genealogical methods (Empig 1969; Miladinović et al. 2011). This fact occurs due to the constant reduction in genetic variability with each generation in the population method, which is due to the insufficient size of the seed sample to be planted in the following generation and to natural selection (Empig and Fehr 1971; Miladinović et al. 2011). What is more, the SSD and SPD methods enable more than one population per year to be conducted in greenhouses, and this provides greater genetic gain per year (Empig 1969).

Many comparisons have been carried out between the population method and the SSD and SPD methods. The SSD method offers the possibility to exploit the maximum genetic variability at the time of selection. However, in the population method, the same will possibly not be noticed. In the population method, the reduction in variability occurs because of the sampling, as individuals that have a higher capacity to survive and to produce seeds will be more representative within the seed sample for the next generation. In this respect, in order to benefit from natural selection, the breeder must choose an environment that is favorable for expressing the alleles relating to the target traits of the breeding program (Empig and Fehr 1971; Borém and Miranda 2013).

Table 9.2 Estimates of heritability (h^2) and gain per selection (GS) for productivity of beans in different methods for conducting segregant populations of soybean

		Sordi (2010)		Miladinović et al. (2011)	
Cross	Method	h^2 (%)	GS (%)	h^2 (%)	GS (%)
Cross 1	Population	–	–	18.60	2.42
	Genealogical	36.07	24.10	55.90	8.15
	SSD	1.64	−1.73	10.10	1.27
	SPD	12.61	1.46	–	–
Cross 2	Population	–	–	42.70	4.37
	Genealogical	13.60	10.98	67.00	8.86
	SSD	5.38	15.42	78.80	11.80
	SPD	20.53	20.44	–	–
Cross 3	Population	–	–	6.33	0.48
	Genealogical	26.46	23.17	40.50	5.97
	SSD	26.15	11.69	53.80	5.53
	SPD	13.68	4.15	–	–

Source: Adapted from Sordi (2010) and Miladinović et al. (2011)
Sordi (2010), estimates of populations in F_7; "Cross 1" corresponds to a set of nine biparental crosses; "Cross 2" corresponds to a set of nine cross quadruplet crosses; "Cross 3" corresponds to a set of nine octuplet crosses. Miladinović et al. (2011), estimates of the populations in F_6; "Cross 1" corresponds to the biparental cross (Gordana × Kolubara); "Cross 2" corresponds to the biparental cross (Krajina × Gadir); "Cross 3" corresponds to the biparental cross (Balkan × 9273). h^2, estimate of heritability. GS, estimate of genetic gain per selection

Funada et al. (2013) compared the conduction of segregant populations of soybean using the following methods: population, SSD, and SPD. SSR markers were made use of for genotyping 100 individuals in generation $F_{5:6}$ within the populations conducted by each of the methods. The results show that the numbers of unique lines found (lines that were not paired with any other line at the genetically similar level of 0.875) were, in each of the methods, 38, 49, and 44 for the population, SSD, and SPD methods, respectively. The data does not present any significant differences. The same occurred when the level of genetic similarity of 0.75 was considered. The authors have concluded that, in order to choose the method for conducting segregant populations, the time requirement to obtain the cultivar must be taken into consideration.

The genealogical, bulk within F_2, and SSD methods were compared by Boerma and Cooper (1975a). Taking into consideration the characteristic yield of beans, there was no significant difference between the methods compared. However, regarding the precocity trait, the SSD and genealogical methods were shown to be more efficient. Among the methods for conducting segregant populations presented in this chapter, the SSD one is shown to be an excellent alternative in cases which the time needed for obtaining homozygote lines is short, seeing as evaluations are not carried out in the initial generations, and this enables, in the case of soybean cultivation, the advance of three or four generations per year.

As described previously, some modifications to the population method have been proposed, among them the bulk one within progenies. Ramalho et al. (2012) point out that, despite involving more work, these modifications to the population method reduce the number of errors that have occurred when doing sampling, above all, when the number of individuals within each progeny is more numerous. The authors also stress that if the progenies are obtained in F_3, the exploited additive variance is 1.5 times higher compared to what is obtained from progenies in generation F_2, as shown in Table 9.1. As the self-fertilized generations go by, the V_A between increases, while the V_A within gradually becomes less; this fact is because, according to Borém and Miranda (2013), selecting individuals within rows must be carried out only in the first generations.

The genealogical method allows control of the degree of kinship between progenies (Borém and Miranda 2013), which can be advantageous when the kinship information under the mixed model approach is used for helping to select the best progenies (Nunes et al. 2008, quoted by Ramalho et al. 2012). However, the notations and management of all this information, added to the demand for field workforce, technically qualified personnel, and an experimental area, make it unfeasible to conduct many populations simultaneously through this methodology (Borém and Miranda 2013; Miladinović et al. 2015).

Some authors report that in the genealogical method, when using visual selection between and within families, there is an increment in the frequency of lines with desirable traits, and this increases the amount of genetic gain (Miladinović et al. 2011; Borém and Miranda 2013). In this respect, the genealogical method is a good strategy for conducting populations that arise from crosses between divergent parents, since it is possible to identify and eliminate low-standard progeny in early generations (Miladinović et al. 2015).

The early generation test was proposed with the aim of evaluating and discarding $F_{2:3}$ progenies that perform unsatisfactorily for the traits being evaluated. This enables the size of the population to be reduced and therefore allows evaluating more promising progenies to be concentrated on (Borém and Miranda 2013). Cooper (1990) states that the largest proportion of genetic variance for bean productivity in soybean cultivation is of an additive nature and suggests that F_2 plants that perform well for the traits in question would provide lines with high bean productivity. This fact encourages breeders to start selection, even for quantitative traits, in early generations, the aim of which is to obtain high genetic gains in future generations.

The early generation test can be efficient when selection indexes are used, by considering a set of traits simultaneously in the selection process (Carvalho 2008). However, the early generation test is recommended and is superior to the other methods with regard to obtaining genetic gains, when working with simple inheritance traits and high heritability traits (Boerma and Cooper 1975b), in which the selection process favors obtaining high selective accuracy.

When comparing the bulk, genealogical, and early generation test methods, it was observed that there is no significant difference between the methods with regard to the heritability estimate (Luedders et al. 1973). Borém and Miranda

(2013) state as possible causes for unfavorable results when using the early generation test the limited number of seeds in the first generations, which prevents implementing the trials in more than one place, and the genotype × environment interaction.

Most of the work that has been carried out to make comparisons between the methods of conducting segregant populations in autogamy shows there is no significant difference regarding efficiency in obtaining superior lines (Empig and Fehr 1971; Luedders et al. 1973; Boerma and Cooper 1975b). However, Miladinović et al. (2011) report that the choice of methods for conducting segregant population depends on the aim of the breeding program, as well as on the genetic variability available to be exploited through the generations, on the availability of program infrastructure, and on the size and abilities of the breeding team.

Final Considerations

The methods for conducting segregant populations, ever since they were first described, have been undergoing modifications proposed by various authors in order to correct some unfavorable aspects. Nevertheless, there are still many ongoing discussions regarding the efficiency of using each of these methods. What is more, there are infinite possibilities for breeders to adapt the method chosen to suit their objectives and available resources.

Currently, many companies share the soybean cultivar market in Brazil and around the world. Thus, it is essential to be as fast as possible when launching cultivars that are being evermore adapted, and this involves using methods that require less time and enable undesirable genotypes to be eliminated early on in the first-generation branches. On the other hand, for research and study purposes, the option for using more laborious and time-consuming methods could be chosen.

However, the choice of methods for conducting segregant populations depends on the objective of the breeding program, as well as on the genetic variability available to be exploited, throughout the generations, the availability of program infrastructure, and the size and abilities of the breeding team.

References

Boerma, H.R., Cooper, R.L.: Effectiveness of early generation yield selection of heterogeneous lines in soybean. Crop Sci. **15**(3), 313–315 (1975a)

Boerma, H.R., Cooper, R.L.: Comparison of three selection procedures for yield in soybeans. Crop Sci. **15**(2), 225–229 (1975b)

Borém, A., Miranda, G.V. (eds.): Melhoramento de plantas, 6ª edn. Rev. e Ampl., p 523. Viçosa, Ed. UFV (2013)

Carvalho, A.D.F.: Avaliação da eficiência do teste precoce no melhoramento genético de soja. 103 f. Tese de D.S.—Escola Superior de Agricultura "Luiz de Queiroz". Universidade de São Paulo, Piracicaba (2008)
Companhia Nacional De Abastecimento—CONAB. http://www.conab.gov.br/conteudos.php?a=1252&t=2 (2015). Accessed 3 Nov 2015
Cooper, R.L.: Modified early generation testing procedure for yield selection in soybean. Crop Sci. **30**(2), 417–419 (1990)
Cruz, C.D.: Princípios de genética quantitativa, p. 394. Viçosa, Ed. UFV (2005)
Empig, L.T.: Comparisons of means and variances among bulk soybean populations. 74 f. Retrospective Theses and Dissertations, Iowa State University, Michigan (1969)
Empig, L.T., Fehr, W.R.: Evaluation of methods for generation advance in bulk hybrid soybean populations. Crop Sci. **11**(1), 51–54 (1971)
Fehr, X.R.: Breeding. In: Norma, A.G. (ed.) Soybean Physiology Agronomy and Utilization, pp. 119–155. Academic Press, New York (1978)
Florell, V.H.: Bulked-population method of handling cereal hybrids. J. Am. Soc. Agron. **21**, 718–724 (1929)
Frey, K.J.: The use of F2 lines in predicting the performance of F3 selections in two barley crosses. Agron. J. **46**, 541–544 (1954)
Funada, M., Helms, T.C., Hammond, J.J., Hossain, K., Doetkott, C.: Single-seed descent, single-pod, and bulk sampling methods for soybean. Euphytica. **192**(2), 217–226 (2013)
Goulden, C.H.: Problems in plant selection. In: Burnettm R.C. (ed.) International Genetics Congress, 7, Edinburgh. Proceedings..., pp. 132–133. Cambridge University Press, Edinburgh (1939)
Harrington, J.B.: The mass-pedigree method in the hybridization improvement of cereals. J. Am. Soc. Agron. **29**, 379–384 (1937)
Hiromoto, D.M., Vello, N.A.: The genetic base of Brazilian soybean (*Glycine max* (L.) Merrill) cultivars. Revista Brasileira de Genética. **9**, 295–306 (1986)
Immer, F.R.: Relation between yielding ability and homozygosity in barley crosses. J. Am. Soc. Agron. **33**, 200–206 (1941)
van der Kley, F.K.: The efficiency of some selection methods in populations resulting from crosses between self-fertilizing plants. Euphytica. **4**(1), 58–66 (1955)
Love, H.H.: A program for selection and testing small grain in successive generations following hybridization. J. Am. Soc. Agron. **19**, 705–712 (1927)
Luedders, V.D., Duclos, L.A., Matson, A.L.: Bulk, pedigree, and early generation testing breeding methods compared in soybeans. Crop Sci. **13**, 363–364 (1973)
Miladinović, J., Burton, J.W., Balešević Tubić, S., Miladinović, D., Djordjević, V., Djukić, V.: Soybean breeding: comparison of the efficiency of different. Turk. J. Agric. For. **35**, 469–480 (2011)
Miladinović, J., Vidić, M., Đorđević, V., Balešević-Tubić, S.: New trends in plant breeding-example of soybean. Genetika. **47**(1), 131–142 (2015)
Miyasaka, S.: Contribuição para o melhoramento da soja no Estado de São Paulo. 47 f. Tese de D.S.—Escola Superior de Agricultura "Luiz de Queiroz". Universidade de São Paulo, Piracicaba (1958)
Nunes, J.A.R., Ramalho, M.A.P., Ferreira, D.F.: Inclusion of genetic relationship information in the pedigree selection method using mixed models. Genet. Mol. Biol. **31**(1), 73–78 (2008)
Ramalho, M.A.P., Abreu, A.F.B., Santos, J.B., Nunes, J.A.R.: Aplicações da genética quantitativa no melhoramento de plantas autógamas, p. 522. Lavras, Ed. UFLA (2012)
Raposo, F.V., Ramalho, M.A.P., Abreu, A.F.B.: Comparação de métodos de condução de populações segregantes do feijoeiro. Pesq. Agrop. Brasileira. **35**(10), 1991–1997 (2000)
Sediyama, T., Cardoso, A.A., Vieira, C., Andrade, D.: Taxa de hibridação natural em soja, em Viçosa e em Capinópolis, Minas Gerais. Revista Ceres. **17**, 229–331 (1970)
Sordi, D: Parâmetros genéticos em populações de soja derivadas de cruzamentos simples e múltiplos, conduzidas por três diferentes métodos de derivadas de cruzamentos simples e

múltiplos. 76 f. Dissertação de M.S.—Faculdade de Ciências Agrárias e Veterinárias. Universidade Estadual Paulista "Julio de Mesquita Filho", Jaboticabal (2010)
United States Department of Agriculture (USDA): http://www.usda.gov/wps/portal/usda/usdahome (2015). Accessed 3 Nov 2015
Vello, N.A.: Efeitos da introdução de germoplasma exótico sobre a produtividade e relações com a base genética das cultivares de soja (*Glycine max* (L.) Merrill). 91 f. Tese (Livre-Docência)—Escola Superior de Agricultura "Luiz de Queiroz". Universidade de São Paulo, Piracicaba (1985)
Vencovsky, R.: Herança quantitativa. In: Paterniani, E., Viegas, G.P. (eds.) Melhoramento e produção do milho, 2nd edn, pp. 137–214. Fundação Cargill, Campinas (1987)
Vernetti, F.J., Bonatto, E.R., Terazawa, F., Gastal, M.F.: Observações sobre a taxa de cruzamentos naturais em soja, em Pelotas e Sertão, RS e Ponta Grossa, PR. Ciência e Cultura. **24**, 36–41 (1972)

Chapter 10
Recurrent Selection

Natal Antonio Vello and Felipe Maniero Nazato

Abstract In self-pollinated species such as soybean, the use of recurrent selection began in the 1960s, aiming to increase the frequencies of favorable alleles to the cumulative genetic progress in the desired traits over several cycles, that is, to increase additive and epistatic additive × additive genetic variances in the populations used in breeding programs. Currently, in species such as soybean, the use of the term recurrent selection is restricted to the systems that involve (a) three or more parents; (b) two or three generations of recombination, through biparental, quadruple, or octuple crosses; and (c) two or three generations of inbreeding, usually by using the single seed descent (SSD) method or a method derived from it (SHD, SHDT, MSD), in order to reduce the time of each cycle. It is also possible to reduce the time of each step by using molecular markers linked to genes that control important agronomic traits, such as seed yield and resistance to diseases and insect pests. The obtained progenies are evaluated in tests of agronomic performance, selected and used as parents in new cycle and/or released as new cultivars. In most programs, with or without the use of male sterility, it is possible to anticipate the progeny evaluation for the F_3 generation, in which the plants have 75% of the loci in homozygous and 25% in heterozygosis, on average. The largest genetic gains in seed yield have been obtained from the largest numbers of genetically divergent F_1 crosses. In order to enhance the genetic base, exotic and transgenic genotypes can also be used as new parents in each cycle. Although recurrent selection consumes high amount of time and labor, it should be considered as a central methodology in well-established and well-consolidated breeding programs, because it serves as the base for the creation of genetic variability to be explored in the long-term selection for various traits. In addition, it also allows the enhancement of the genetic base of the germplasm of the cultivated species and the general organization of the breeding programs in progress.

Keywords Minimum three parents • Multiple crosses • F_3 or F_4 progenies • Genetic base enhancement

---

N.A. Vello, M.S., D.S. (✉) • F.M. Nazato, M.S.
Department of Genetics, College of Agriculture "Luiz de Queiroz", University of São Paulo, Piracicaba, São Paulo, Brazil
e-mail: natal.vello@usp.br; felipe.nazato@usp.br

F. Lopes da Silva et al. (eds.), *Soybean Breeding*,
DOI 10.1007/978-3-319-57433-2_10

Introduction

Recurrent selection is a methodology applied to plant breeding programs to obtain genetically heterogeneous populations, and, from them, progenies and/or lines are derived, evaluated, selected, and cyclically recombined in an orderly, dynamic, and continuous process which aims the increasing of the allele frequencies in favor of cumulative genetic progress in the desired trait(s) over several cycles.

The term "recurrent selection" was first used in 1940 by M. T. Jenkins to designate an intrapopulation breeding method in maize and later described in 1945 by F. H. Hull for the improvement of maize populations with tester use; since that, this methodology has been widely shared and used in the breeding of open-pollinated species (Hallauer et al. 2010).

In self-pollinated species, the use of recurrent selection began in the 1960s, first through the suggestion of E. C. Gilmore Jr., in 1964, of a system using male sterility, and in 1968, by W. A. Compton's proposal of a system involving less amount of intermating; in the 1970s, the first schemes of recurrent selection in soybean appeared, with the aid of male sterility (Brim and Stuber 1973).

From the publications of these studies and with the growing concern to increase the genetic variability available in breeding programs, aiming at overcoming productivity levels and minimizing possible genetic vulnerability events to biotic and abiotic stresses, there was an increase in interest in using this powerful tool, the recurrent selection, in breeding programs of self-pollinated species, mainly in soybean (Lewers and Palmer 1997; Posadas et al. 2014).

In Brazil, initial reports on the use of recurrent selection in soybean were made by Vello (1992) with the application of complete and partial diallels for intermating, aiming to obtain genetic variability in populations of breeding programs to increase seed and oil yield, respectively. The first results obtained involved multiple crosses with the use of genetic male sterility for seed size (weight of 100 seeds) and oil content (Miranda 1994) and multiple manual crosses for seed yield (Lopes 1996; Unêda-Trevisoli 1999; Alliprandini and Vello 2004) and for oil yield (Hamawaki et al. 2000).

Justifications for the Use of Recurrent Selection

One way to measure genetic diversity among lines (cultivars) is through their genetic base, which can be conceptualized as the foundation that contains all the genetic stocks on which the lines have developed. In Brazil, Wysmierski and Vello (2013), based on the pedigrees of 444 cultivars of soybean of public and private origin, estimated the parentage coefficient of Malécot in order to measure the relative genetic contribution of ancestral soybean strains for the current cultivars, that is, an estimate of the genetic base of the cultivated germplasm; these authors found a total of 60 ancestors, including 14 ancestors representing 92.4% of the genetic base and only four (CNS, S-100, Roanoke and Tokyo) responsible for 55.3% of the genetic base, evidencing a narrow genetic base of Brazilian soybean cultivars. The expansion of the genetic base can be done through recurrent selection programs, since this methodology allows an increasing enhancement of the genetic variability.

Most of the traits of importance for plant breeding are quantitative, with polygenic inheritance involving dozens or hundreds of genes. As a consequence, these traits present continuous variation, including an expressive number of different genotypes and phenotypes. This fact makes it difficult to obtain a plant containing all the genes of interest, since the probability of this occurring in a single step in a population coming from a cross is very small; the use of recurrent selection may increase the chances of obtaining this type of plant, since it is a methodology performed in stages (cycles), with recombination of selected higher plants, a fact that allows the complementation and gradual accumulation of favorable alleles, generating rising increases in genetic variability (additive and epistatic additive × additive variances) on which selection is practiced.

Stages of Recurrent Selection in Soybean

Although all processes involving successive cycles of choice of parents, recombination, and selection may be called recurrent selection, in self-pollinated species such as soybean, their use has been restricted to systems involving:

(a) Three or more parents
(b) Up to three generations of recombination, obtaining different types of base populations, formed from biparental, quadruple, and octuple crosses (Fig. 10.1)
(c) Three generations of inbreeding, usually through the SSD (*single seed descent*) method or a method derived therefrom, for example, SHD, *single hill descent*; SHDT, *single hill descent thinned*, in order to leave a single plant in the hill; and MSD, *multiple seed descent*, commonly by harvesting a pod with three seeds of each plant

The progenies obtained in each population are evaluated in agronomic performance tests (seed yield and other traits), selected and used as parents in new cycle, and/or released as new cultivars.

A general scheme of recurrent selection with potential use in soybean crop is presented in Fig. 10.2.

In Fig. 10.2, the first recombination involves biparental crosses and obtains simple hybrids (F_1); in the second recombination, multiple crosses between simple hybrids ($F_1 \times F_1$) originate quadruple hybrids; and in the third recombination, multiple crosses between quadruple hybrids $[(F_{1[4]} \times F_{1[4]}) \times (F_{1[4]} \times F_{1[4]})]$ result in octuple hybrids, which can combine two-to-two up to eight different alleles of each gene; a multiple cross is represented by the letter M. At each cycle, new

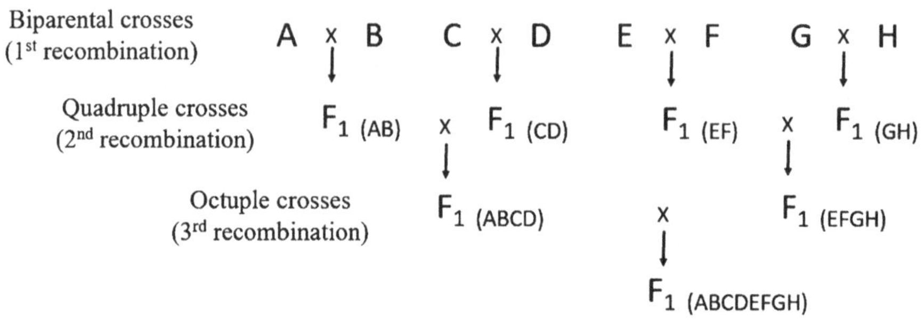

Fig. 10.1 Types of crosses most common in recurrent selection programs

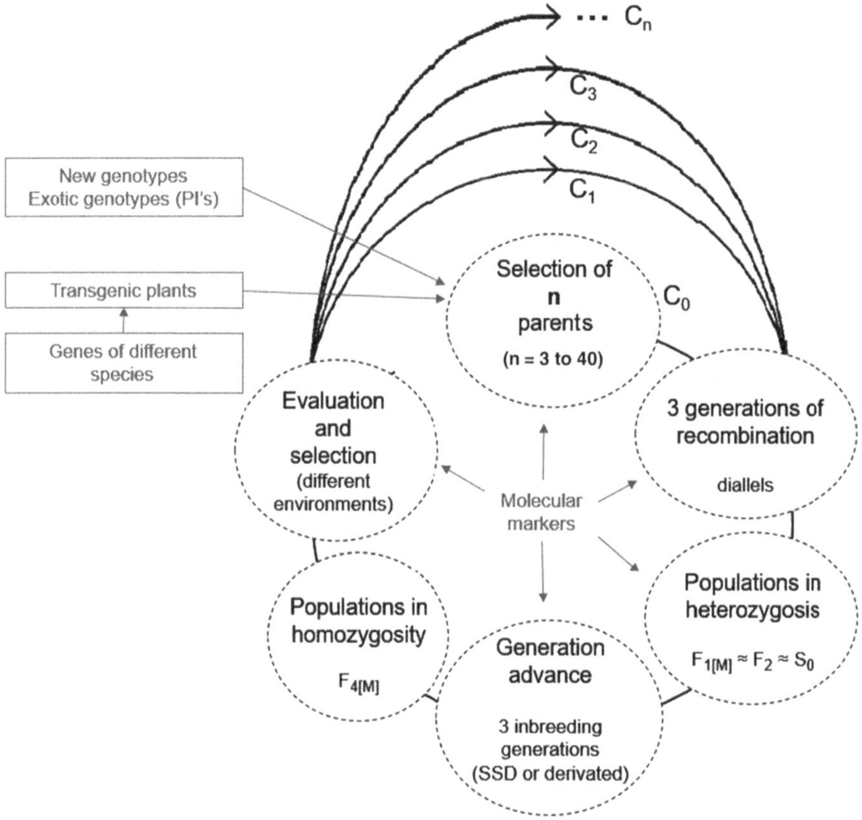

Fig. 10.2 Potential recurrent selection scheme in soybean

parents are usually added to the system. Molecular markers can contribute to improve efficiency and decrease the time of each step of the system.

The use of recurrent selection can present a number of advantages:

- To increase the number of alleles (multiples) of each gene in the synthesized base populations.
- To favor the breaking of gene blocks.
- To increase genetic variance (additive and epistatic additive × additive).
- To test a large number of crosses in each cycle.
- To accumulate genetic gains in successive cycles.
- To expand the genetic base of the germplasm.
- To allow the organization of the breeding program in the short, medium, and long term.
- The synthesized base populations can be used to provide parents to the following cycles and/or strains for commercial use as new cultivars; examples of cultivars grown from base populations obtained from recurrent selection programs were reported by Fehr (1987) and J. E. Specht and G. L. Graef in 1992 (Orf et al. 2004).

The recurrent selection may have the following disadvantages:

- It is a laborious method that requires a great number of artificial hybridizations, which in the case of soybean can become a great difficulty, since the flowers are small, hermaphrodite, and cleistogamic; this difficulty may be increased by the need to perform more hybridizations to compensate the small number of seeds (one to three) obtained at each cross.
- It is a method that requires a long time, several years per cycle, with slow renewal of parents, which may even become obsoletes.
- In the additional recombination {second ($F_1 \times F_1$) and third [($F_1 \times F_1$) × ($F_1 \times F_1$)]}, which are performed in artificial environments (e.g., greenhouses), it is not always possible to select the best plants to be hybridized.
- It is difficult to evaluate many plants at each cross.

Base Populations

The first stage of recurrent selection is the formation of base populations (initial populations in heterozygosity), which focus on the choice of the parents. It is sought at this stage to create a wide initial genetic variability with the possibility of favorable allelic combinations, which will be the basis for the beginning of the recurrent selection and will dimension the potential of the methodology. The choice and crossing of divergent and complementary parents with favorable traits is a very effective strategy for generating variability with a focus on obtaining promising genotypes.

Some important traits of soybean parents are:

- High seed yield
- Resistance/tolerance genes to biotic (diseases and insect pests) and abiotic (mineral deficiencies, drought, salinity, over- or suboptimal temperatures) factors
- Genes for late flowering in cultivation under shorter days (long juvenile period)
- Appropriate cycle
- Adaptative traits: resistance to lodging, adequate height of plant, adequate height of first pod insertion, no dehiscence of pods, absence of leaf retention, and green stem

The estimation of genetic diversity among the parents can be performed by means of estimates of parentage coefficient and/or with the aid of markers (morphological, biochemical, and molecular) and/or estimates of combining ability.

Potential parents for recurrent selection programs are preselected elite cultivars, exotic genotypes (where necessary to introduce some important traits not found in adapted sources or when there is a depletion of variability in the populations of the program), in addition to the progenies and/or superior lines selected in different cycles. Next, the chosen parents are crossed to obtain the genotypic combinations that will form the base populations.

The greater the number of different crosses performed, the greater the variability generated, and the higher the gain with recurrent selection. In some cases, for specific purposes, a smaller number of crosses may be performed, but when this occurs, it is important to increase heterozygous population sizes (F_2 generation), evaluating more plants or progenies at each cross. As will be seen later, this number depends on the effective size used to make the recombination, with a consensus that it should be equal to or greater than 30 F_2 plants per cross.

The crosses can be realized in several schemes, such as complete diallels, partial diallels, and circular chain crosses, with similar efficiencies, so that the breeder can choose the scheme most appropriate to his needs. For example, in a scheme of circular chain crosses with 20 parents, 20 crosses are performed, each parent being involved in only two crosses of the chain; with this, the number of crosses to be realized (in comparison to a complete diallel) is significantly reduced, and it may even facilitate the incorporation of exotic genotypes into this crossing scheme.

Evaluation and Selection of Soybean Progenies for Recurrent Selection

According to Lewers and Palmer (1997), in recurrent selection programs of self-pollinated species, the selection of superior genotypes may occur phenotypically at the level of individual plants (mass selection) or at the level of progenies (families) evaluated in experiments with replications.

For high heritability traits, phenotypic or mass selection, which is based on the phenotypic visual selection of individual plants, accelerates the recurrent selection process, since it is of easy management, besides presenting low operational cost. However, for traits with low heritability, the phenotypic selection is not effective, mainly because of the high environmental influence in the expression of these traits.

For traits with low heritability, the evaluation of progenies in experiments with replications is recommended, if possible conducted in different environments, because the genotypic values will be better estimated, due to the better local control and the reduction of the environmental effects on the estimates of the progeny means.

In soybean recurrent selection programs, the progeny evaluations are most commonly performed in F_3, F_4, and F_5 generations. For seed yield in soybean, W. R. Fehr and L. B. Ortiz, in 1975, found greater selection gains per year with assessments in the F_3 (or S_1) generation, but the greatest genetic variability was found among F_6 (or S_4) progenies, due the inbreeding contributions; in addition, Sumarno and W. R. Fehr, in 1982, found in F_4 and F_5 generations, the plants obtained were more consistent, making it easier to obtain lines and cultivars (Lewers and Palmer 1997).

Recombination of Selected Materials

After being evaluated and selected, the materials (plants or progenies) must be recombined (intermated) to continue the cycles of recurrent selection. The recombination of genotypes aims at creating variability for the next selection cycle. According to Lewers and Palmer (1997), both manual recombination (without the aid of male sterility) and natural recombination (using pollinator insects and with the aid of male sterility genes) are successfully used in recurrent soybean selection programs.

Recombination is responsible for new gene combinations that increase the accumulation of favorable alleles in the final population, allowing for the progress of the method. Very small samples decrease the chances of selection of superior genotypes, while large samples cause increases in labor and costs; furthermore, smaller samples would allow to increase the number of crosses involved in recurrent selection, a factor of great importance to increase the genetic gain of selection.

An important point in this step is to consider the effective size (N_e) of the sample, which is related to the genetic base involved in the recombination, that is, it is equivalent to the number of genetically different plants in the sample that contribute to the formation of the next generation. A sample of the population with an ideal effective size should preserve the genetic variability of the population and also allow the increase of the effective size through the recombination performed throughout the selection cycles.

Brim and Stuber (1973) recommend an effective size of 30 plants per cross for a selection intensity of 10% up to the first ten cycles in the soybean crop, which is in agreement with the results of L. H. Baker and R. N. Curnow, in 1969, as well as J. O. Rawlings's work in 1970, which obtained theoretical evidence of an effective size, respectively, of 24 (16–32) and 37.5 (30–45), respectively, to satisfy the need for preservation of genetic variability in the sample. Most of the programs conducted in Piracicaba use 72 F_2 soybean plants to represent a cross, commonly distributed in six replications of plots, each one with 12 individual plants in hills; this higher number of F_2 plants per cross takes into account the common occurrence of losses in the experimental field.

Manual Artificial Recombination (Without the Aid of Male Sterility)

Although the reduced size of soybean flowers, the low success rate of hybridization (setting rate), and the small amount of seeds obtained from each hybridization hinder the manual production of large quantities of recombined seeds, the amount of seeds needed to a recurrent selection program is entirely viable to be obtained manually. Moreover, manual artificial crosses are targeted and therefore more efficient in promoting recombination.

Lewers and Palmer (1997) point out the factors that must be taken into account for the success in the production of the hybrid seeds required by the recurrent selection program:

- Labor available to perform manual artificial hybridizations
- Daily hours actually available to carry out the hybridizations
- Number of pollinations per hour
- Success rate ("setting") of hybridizations
- Number of seeds per pod obtained
- Percentage of germination of hybrid seeds

In 1980, W. R. Fehr found that, generally, up to 6 h daily are effectively available for performing artificial hybridizations and one person skilled in doing hybridizations on soybean performs about 20 hybridizations per hour, and in 1987, this same author pointed out that the success rate of each hybridization resulted in a pod containing one to three seeds (Lewers and Palmer 1997). In Piracicaba, hybridizations with soybean are carried out from 8 to 11 o'clock a.m. and from 3 to 5 o'clock p.m., during the spring and summer; the average success rate of hybridizations is about 60% in greenhouses (controlled environments), resulting in pods with one to three seeds with a percentage of seed germination of 95%.

Natural Recombination with the Help of Male Sterility

The discovery of male sterility made it possible to obtain populations through random crosses by the action of pollinating insects in self-pollinated species. In this intermating system, a random reorganization of the hybridizations is constantly taking place, and a large number of parents can be used in a single crossing phase. With this, a more representative sample of the existing variability of the species can be used as parents.

Male sterility in soybean is a characteristic of monogenic recessive nuclear inheritance (ms ms), and, currently, there are known nine independent genetic loci (ms1 to ms9) conditioning this trait. Brim and Stuber (1973) were the first authors to propose the use of male sterility for intermating in recurrent soybean selection programs, and in 1990, J. W. Burton, E. M. K. Koinange, and C. A. Brim emphasized the great utility of the use of male sterility in programs of recurrent selection due to the facility provided in the recombination among soybean plants (Posadas et al. 2014).

In 1975, W. R. Fehr and L. B. Ortiz pointed out a disadvantage of the use of recessive male sterility in recurrent selection programs, which is due to the fact that heterozygous fertile plants (Ms ms) segregate and result in mixtures of fertile and sterile plants in the progeny, across generations; this may hinder the evaluation of progeny performance in the field (Lewers and Palmer 1997). In 1981, S. K. St. Martin recommended a progeny test by means of self-fertilization to eliminate male sterile plants from the experimental plots in seed yielding tests (Lewers and Palmer 1997).

Another negative feature of the use of male sterility is its correlation with low seed yield. In 1972, H. Doggett proposed a recycling system for sorghum male sterile plants, which could also be used in soybean, with selection for seed yield during the development of the base populations in order to reduce this effect of the association between the male sterility genes and low seed yield (Brim and Stuber 1973). The incorporation of the male sterility alleles can be done during the synthesis of the base populations, by means of backcrosses using as recurrent parents the lines of interest to the program.

Tanaka (2010) developed a recurrent selection system for possible large-scale use in self-pollinated species, which is based on male sterility (dominant allele) and transgenic plants. This author used a plant transformation cassette containing three main characteristics: male sterility, herbicide tolerance (characteristic used as marker), and genes for induction of lethality (characteristic used as marker, which is expressed only under a specific induction treatment). With specific treatments for the expression of the characteristics of the marker genes, the author proposes the separation of the transformed plants into two groups, one presenting male sterility and the other male fertility; then the plants of the two groups would be intermated by insects, resulting in plant populations that would be selected (for the characteristic of interest) and again separated into two groups, starting a new cycle.

Introduction of Exotic and New Materials

The creation of base populations via multiple crosses (Fig. 10.1) in complementarity to the recombination of divergent superior genotypes during selection cycles characterizes recurrent selection as a method that allows for a growing increase in the mean of the desired trait along with maintenance or even with the amplification of genetic variability.

Whenever a new cycle is initiated, the incorporation of new parents into the lot of crosses to expand the genetic base of the base populations becomes an interesting option, but the preference should be for a greater frequency of divergent and adapted genotypes. The introduction of exotic materials should be dosed so that the gains from selection are not diminished.

In researches with five soybean populations with a broad genetic base, obtained through intermating among 40 plant introductions (PI, exotic parents, previously selected for adaptation) and 40 elite cultivars (adapted parents with high seed yielding) and using backcrosses to obtain the relative percentages of adapted and exotic parents, Vello et al. (1984) found that percentages between 25 and 50% of exotic parents previously selected for adaptation originated populations with a very promising combination between maximum genetic variability and minimum loss of seed yield, relatively to the populations synthesized only from adapted high yielding parents.

The introduction of genes from other species into the gene pool of recurrent selection populations, via transgenesis, is also a way of amplifying genetic variability. Currently, several transgenic soybean materials have resistance/tolerance to herbicides and insect pests, as well as high seed yield and adaptability to the locations and current climate conditions of soybean cultivation in the world.

In 1990, J. E. Specht and G. L. Graef proposed a methodology with a cyclic system represented by the acronym MSFCB (Fig. 10.3), using male sterility to facilitate the incorporation of new parents, with only 1 year per cycle, involving testing and selection in the summer (Nebraska, temperate environment) and two inbreeding generations in the winter (Belize, tropical environment) (Orf et al. 2004). The system proposes the annual incorporation of a new set of ten parents (A, B, C ...), which may be commercial cultivars, improved lines, or exotic genotypes preselected for adaptation; these sets of parents are cultivated in an isolated lot intercalated with male sterile lines (S). A beehive is placed in the recombination area to transport the pollen from the new parents to male sterile plants (fertile females). At least one F_1 seed is harvested from the crosses of each male sterile plant. Then, the plants are advanced in nurseries and F_2 seeds are harvested in bulk. From this bulk, a small sample of seed is taken and advanced, segregating 75% male fertile plants and 25% male sterile plants; male fertile plants are rapidly eliminated at the flowering. The remaining male sterile plants of the process are pollinated again in an isolated lot by a new set of chosen parents with the aid of pollinating insects, and, thus, the cycles have continuity. The remaining seeds of the original F_2 bulk are advanced in further experiments in which

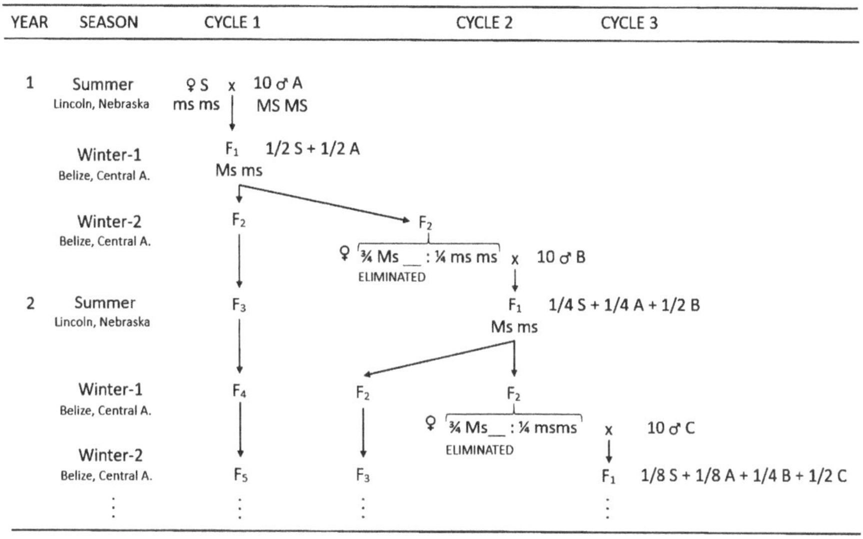

Fig. 10.3 Recurrent selection facilitated with the use of male sterility (MSFCB) based on the scheme proposed by J. E. Specht and G. L. Graef in 1990 (Orf et al. 2004)

promising genotypes or lines are evaluated, selected, and identified. In addition to the advantage of obtaining one cycle per year, this system also promotes the annual renewal of the parents, and the new parents participate in the populations obtained with larger gene contributions, compared to the parents of the previous cycles.

Recurrent Selection with the Aid of Markers

In recent years, the researches have generated a broad spectrum of markers spanning the soybean genome. RFLP (*restriction fragment length polymorphism*) markers were introduced in the late 1980s and RAPD (*random amplified polymorphic DNA*) and DAF (*DNA amplification fingerprint*) in the early 1990s; the SSR (*short sequence repeat*) or microsatellite markers were introduced in 1994 and allowed a great progress. In the early 2000s, several genetic maps, with moderate to high marker density, became available for soybean.

Recurrent marker-assisted selection utilizes the close genetic linkage between a marker (DNA segment) and a gene that affects a trait of interest, which is particularly useful for parental selection (genetic diversity studies), evaluation, and selection of promising genotypes, especially for complex inheritance characteristics (*quantitative trait loci* or QTL markers).

Priolli et al. (2013) obtained very promising results on the use of microsatellite loci allied to pedigree informations (Malécot's coefficients of parentage) for analyses of genetic diversity of Brazilian soybean germplasm; they found a nuclear collection (31 cultivars) well representative of most of the total genetic variability among 435 cultivars and established a panel of divergent genotypes, which can provide soybean breeders with efficient planning of the most promising crosses for the recurrent selection, as well as for the broadening of the germplasm genetic base. Furthermore, soybean germplasm was also classified into two divergent groups, with American and Asian origins, based on microsatellite markers for quantity (oil content) and quality (fatty acid contents) by Priolli et al. (2015).

A very efficient method to produce soybean hybrids with the aid of markers was developed by Lewers and Palmer (1997), which uses the close genetic linkage of the locus conditioning the male sterility (ms_6) with the locus w_1 which conditions the color of hypocotyl, flowers, and seed hilum. Fertile plants ($Ms_6\_$) have purple flowers and hypocotyls ($W_1\_$), while male sterile plants have white flowers and green hypocotyls ($ms_6\ ms_6\ w_1\ w_1$). This narrow genetic union facilitates recurrent selection, allowing the breeder to identify the genotype of a male sterile plant in the sowing, flowering, and maturity stages, helping to reduce the loss of nuclear and cytoplasmic genetic diversity in the population, especially in the recombination phase; it also allows to evaluate the plots containing only male fertile plants, since it is easy to identify and eliminate the male sterile plants soon after the emergence of the seedlings, through the hypocotyl color; allows the evaluation (testcross) of the parents; and enables the recurrent selection of half-sibs with a pure line as a tester, a recurrent selection method that may be useful to increase the genetic diversity of an existing breeding program.

In 2010 the complete genome of soybean was presented (Schmutz et al. 2010) with identification of approximately 46,430 genes coding for proteins distributed in 20 chromosomes (linkage groups). Information on linkage maps, genes of interest, and sequencing can be found in the SoyBase domain developed by USDA and Iowa State University (http://soybase.org/); currently, QTLs have been identified for several traits, such as seed yield, with more than 15% of the QTLs in the linkage group C2 (chromosome 6); oil and protein contents, with more than 10% of the QTLs in the linkage group I (chromosome 20); genetic resistance to *Sclerotinia sclerotiorum* (white mold) and *Fusarium solani* (soybean sudden death syndrome or SDS) with more than 10% of the QTLs in the linkage groups K (chromosome 9) and O (chromosome 10); and resistance to the soybean cyst nematode, with more than 30% of the QTLs in the linkage group G (chromosome 18); in this same linkage group G, 5–10% of the QTLs linked to the white mold and SDS resistance genes are also found.

Genetic Progress

The response to recurrent selection depends on a number of factors, such as the types of traits that are subjected to the selection process, the genetic variability of the base populations, the environmental influence on the selection unit, and the effective size of the population.

For a periodic monitoring of the selection program, the measurement of the progress or expected genetic gain through the estimates of the components of the genetic variance by the method of Cockerham (1983) is an interesting option, since it uses the progenies obtained during the implementation of the recurrent selection. In order to obtain these estimates, it is only necessary to have successive generations of inbreeding (at least three), which is already present in programs with progeny selection, that is, the expected genetic progress can be estimated using the same data of the evaluation of the progenies in each cycle. Alternative methods for estimating genetic gain in selection are discussed by Orf et al. (2004); these authors consider that the greatest gains in seed yield in recurrent selection programs have been obtained from the largest possible numbers of genetically different (F_1) crosses.

Recurring Selection for Different Traits

Seed Yield

In 1979, W. J. Kenworthy and C. A. Brim reported soybean crosses between nine exotic genotypes and an adapted line, resulting in F_1 plants that were backcrossed with the adapted line and intermated to the formation of the base populations (Posadas et al. 2014). F_3 progenies were evaluated for seed yield, productive efficiency, and an index obtained from the first two. Next, the 20 best F_3 progenies of each evaluated trait were selected, giving rise to three populations. Each population was conducted in a similar manner during three cycles (2 years per cycle) with selection of 20 F_3 progenies and one generation of interbreeding among F_4 plants derived from these progenies; at the end, significant gains of 134 ± 30 kg/ha/cycle were obtained.

In 1982, Sumarno and W. R. Fehr reported octuple crosses (Fig. 10.1) among 40 lines showing high seed yield and belonging to different maturity groups to form the base populations, which were conducted in three cycles (2 years per cycle) (Lewers and Palmer 1997). In cycle 0, the obtained progenies were advanced (MSD method) to F_5, evaluated, and subdivided into three groups with different maturity groups (early, semi-early, and late cycle); of each set ten progenies with the highest means of seed yield were selected. In cycle 1, the 30 selected progenies were recombined in a diallel scheme without reciprocals, advanced, and evaluated in the F_4 generation for seed yield; then, three sets with different maturity groups were formed. In cycle 2, the ten progenies with the highest means of seed yield from each maturity group were intermated in diallel schemes separately, giving rise to three populations, which were advanced to F_4 and evaluated for seed yield. The cycle 3 followed the same steps described for cycle 2. At the end of the three selection cycles, positive results were obtained with genetic gains of 120 ± 10 kg/ha/cycle of recurrent selection for the early cycle population, 24 ± 9 kg/ha/cycle for the late cycle population, and no significant gain for the semi-early cycle population.

Rose et al. (1992) performed octuple crosses from a diallel of 17 highly productive parents for the formation of the base population. The resulting F_3 progenies were evaluated for seed yield and selected (about 10%). After five cycles of recurrent selection, the means of seed yield of the population and selected progenies increased by 17% and 54%, respectively, compared to the average of one of the parents (Davis); the magnitude of these gains represents an evidence of the potential value of the recurrent selection as suppliers of variability for other breeding programs conducted in parallel. The average progress was estimated in 128 kg/ha/cycle, equivalent to a 5.4% increase in average productivity of the base populations.

Figure 10.4 represents the scheme proposed by Vello (1992) to obtain two-parent, quadruple, and octuple crosses in soybean, which were evaluated in the researches of Lopes (1996), Unêda-Trevisoli (1999) and Alliprandini and Vello (2004).

Lopes (1996) verified the potentiality of 60 soybean octuple crosses among the 40 genotypes belonging to four maturity groups (early, semi-early, semi-late, and late cycles), through evaluations in $F_{1[8]}$ generation for seed yield, number of days to maturity (cycle), agronomic value, and plant height at maturity. The octuple crosses presented a high genetic variability, allowing the selection of superior progenies for all evaluated traits.

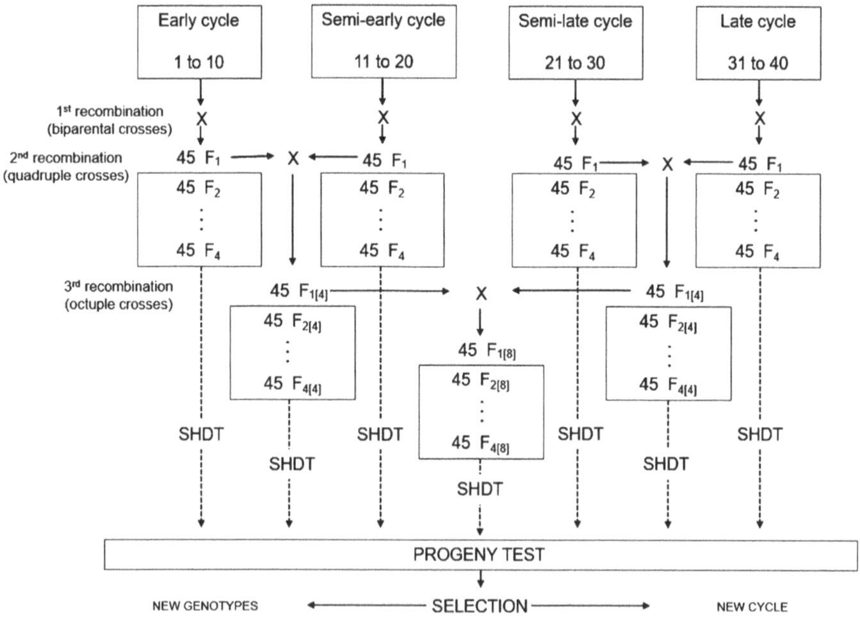

Fig. 10.4 Recurrent selection scheme for seed yield based on the Vello's proposal (1992) and used by Lopes (1996), Unêda-Trevisoli (1999) and Alliprandini and Vello (2004)

Unêda-Trevisoli (1999) evaluated the potentiality and phenotypic stability of $F_{3:4[8]}$, $F_{3:5[8]}$, and $F_{3:6[8]}$ progenies obtained from 60 soybean octuple crosses, in six different environments, for the number of days to maturity, plant height at maturity, lodging, agronomic value, soybean stem canker symptoms, and seed yield. Significant genetic variability was estimated among the progenies for all evaluated traits; the level of variability was higher than the variability presented by biparental crosses.

Alliprandini and Vello (2004) evaluated the $F_{2[4]}$ and $F_{2:3[4]}$ generations of 45 populations from quadruple crosses between plants obtained from two 10×10 complete diallels with semi-late and late maturity, aiming to estimate the potentiality of populations in terms of seed yield and other agronomic traits. The advancement of generations of inbreeding was done by the SHDT (*single hill descent thinned*) method. It was estimated high genetic variability in the populations obtained, allowing the selection of superior progenies for seed yield and agronomic value.

Posadas et al. (2014) developed base populations (UP2Y) from octuple crosses obtained in three stages of recombination (Fig. 10.5), beginning by using as parents seven exotic (PI) soybean genotypes showing high protein content and diverse geographical origins and obtaining and evaluating 100 $F_{5:7}$ progenies; the generations of inbreeding were advanced by the MSD method, by sampling a single pod with two seeds per plant. The ten best $F_{5:7}$ progenies were selected for high seed yield and recombined (one generation) in a diallel scheme, obtaining 45 crosses in the first cycle (C_1). Three cycles (C_1, C_2, and C_3) were performed, following the same scheme previously described, with a single recombination at the beginning, followed by evaluation and selection of progenies derived from F_5 plants. The average duration of each cycle was 5 years, during which there were reductions in the number of days to maturity and lodging, increase in plant height, and significant average gains of 567 kg/ha/cycle in seed yield.

Seed Yield Using Male Sterility

In 1990, J. W. Burton, E. M. K. Koinange, and C. A. Brim highlighted the role of male sterility in facilitating intermating among soybean genotypes in the recurrent selection programs (Posadas et al. 2014); the authors synthesized two populations and obtained average seed yield in four selection cycles of 76 ± 23.8 and 37 ± 55.7 kg/ha/cycle; F_3 progenies were evaluated and selected, disregarding the occurrence of male sterile plants in the experimental plots, and the recombination was mediated by pollinating insects; the first population also stood out with gains per year of 2.1% in seed yield, 8.9% reduction in lodging, and 5.9% increase in plant height.

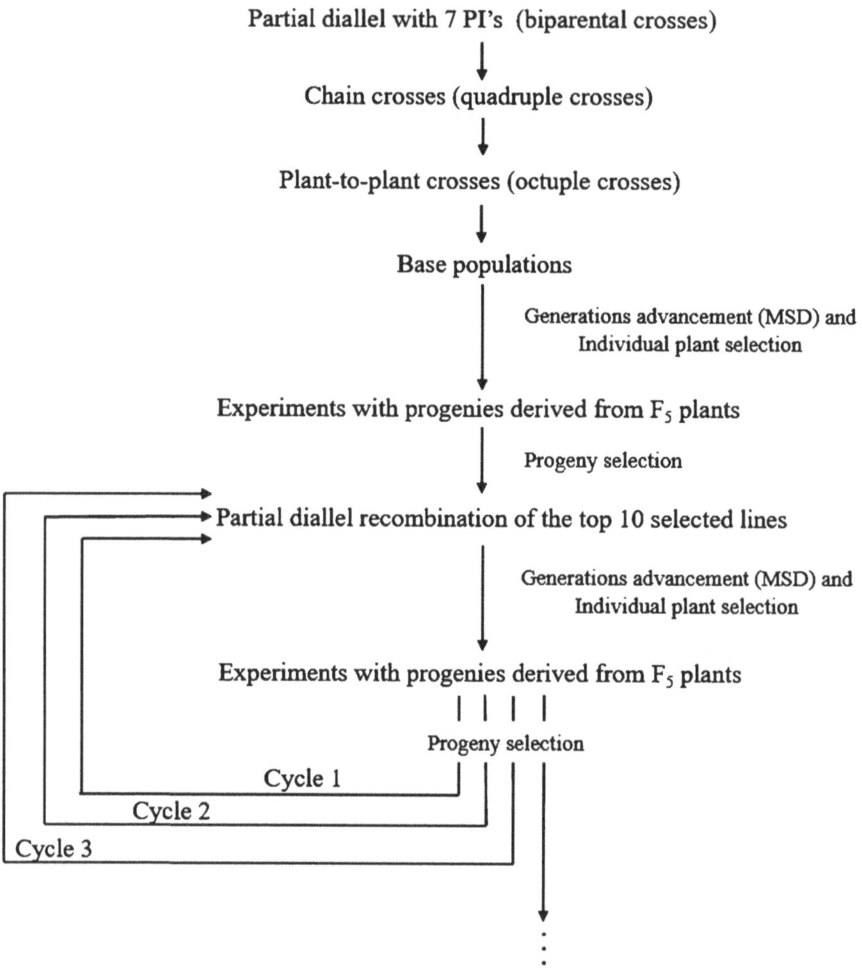

Fig. 10.5 Recurrent selection scheme proposed by Posadas et al. (2014)

Seed and Oil Yield

Hamawaki et al. (2000) evaluated $F_{3:4[8]}$ and $F_{3:5[8]}$ progenies from 45 octuple crosses synthesized from hybridizations between adapted (A) and exotic (E) parents in a circulating chain system for three generations, to obtain a group of octuple crosses with 75% of adapted genes and 25% of exotic genes (Fig. 10.6). The evaluated traits were plant height at maturity, number of days to maturity (cycle), agronomic value, lodging, seed yield, seed oil content, and oil yield. They concluded that octuple crosses resulted in superior progenies for all evaluated traits,

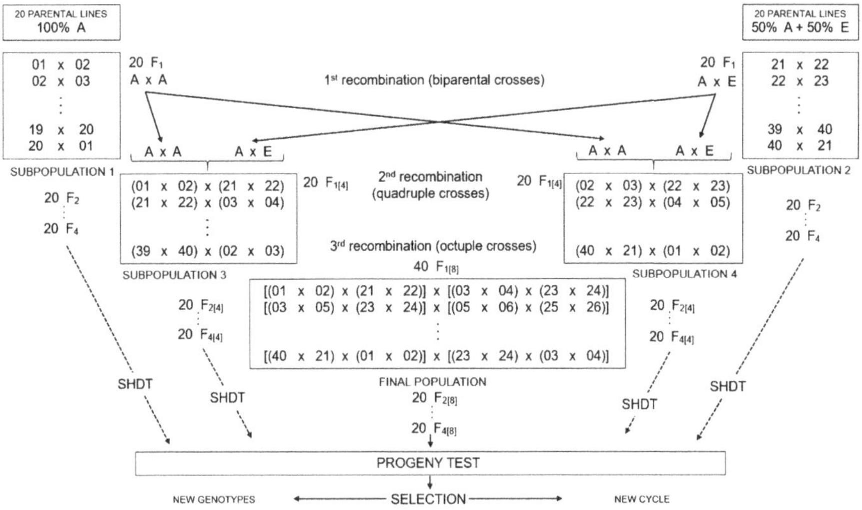

Fig. 10.6 Recurrent selection scheme for seed and oil yields proposed by Vello (1992) and used by Hamawaki et al. (2000)

mainly for seed yield; in addition, they estimated high genetic variability remaining among the selected progenies in several crosses, evidencing the possibility of obtaining additional genetic gains in advanced selection cycles.

Seed Size and Oil Content Using Male Sterility

Miranda (1994) carried out a divergent selection with the use of male sterility to seed size (weight of 100 seeds) of soybean in three stages: mass selection in male sterile plants, selection among half-sib progenies based on the average of progenies, and selection of the best plants within the selected progenies. Selection gains were relatively high, varying from 6 to 7% for seed size in four populations (two cycles for larger and smaller seed sizes). Afterward, a divergent selection was made for oil content in the seeds, reaching averages of 21.39% and 20.80%, that is, a difference of 0.59% between the oil content averages of the two selected populations; therefore, the selection increased or decreased the oil content by 0.295%, equivalent to the average of 1.4% of genetic gain in the selection. Therefore, recurrent selection was efficient in altering the two traits but with gains much higher in seed size than in oil content.

Protein Content

In 1979, J. E. Miller and W. R. Fehr applied direct selection (selecting for high protein content) and indirect selection (selecting low oil content, which has a high negative correlation with protein content) in order to increase protein content in soybean seeds (Lewers and Palmer 1997); the authors obtained base populations from octuple crosses among 12 cultivars with high protein content and 12 cultivars with high seed yield (Fig. 10.7). A random sample of 100 F_3 progenies from the base populations were evaluated in experiments with two replications and two locations, by using hills with 12 seeds each. The ten progenies with the highest protein content and the ten progenies with the lowest oil content were recombined separately in diallel schemes, giving rise to two populations in cycle 1. A total of 100 F_3 progenies of each population were evaluated in the presence of the 24 parents and six experimental checks, in two-location experiments, with a randomized block design with two replications, in the hill system. They concluded that the direct selection of plants for high protein content was efficient, but the indirect selection for lower oil content was little effective in increasing the protein content.

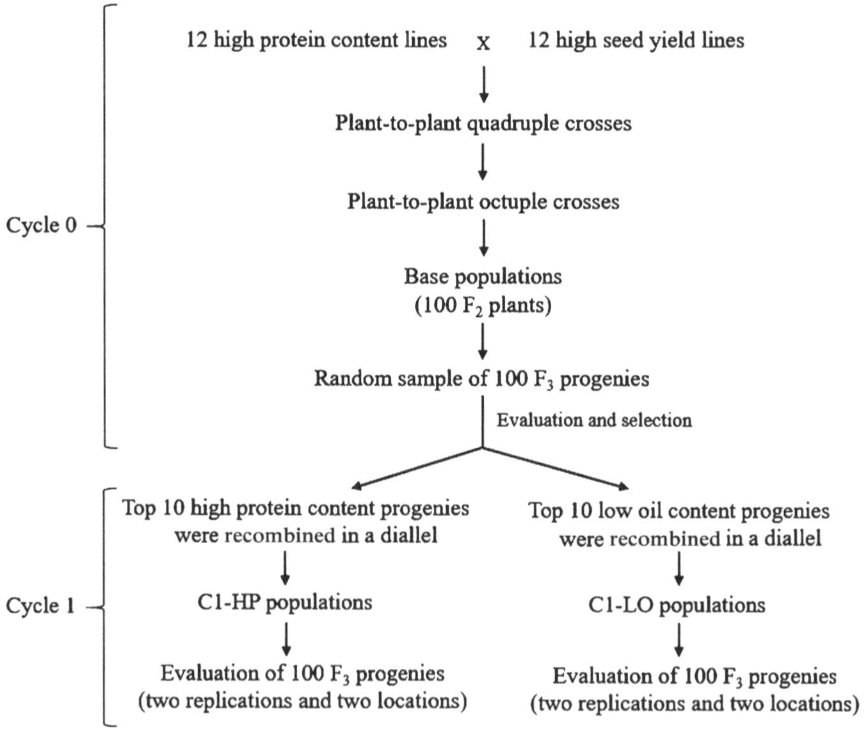

Fig. 10.7 Recurrent selection scheme proposed by J. Miller and W. R. Fehr in 1979 (Lewers and Palmer 1997)

Tolerance to Phytophthora

In 1984, A. K. Walker and A. F. Schmitthenner obtained soybean base populations through quadruple crosses, from a diallel scheme among ten parents presenting high seed yield and different levels (moderate to high) of tolerance to rot caused by *Phytophthora megasperma* (Lewers and Palmer 1997). They evaluated 100 plants in greenhouse that gave rise to 40 progenies to be cultivated in an experimental field contaminated with *Phytophthora*. The F_3 progenies were evaluated in experiments with replications, and the 20 most tolerant progenies were selected to be intermated in cycle 1. By using similar procedures, the cycles 2 and 3 were performed, and to begin the last two cycles, the ten most tolerant progenies were intermated. The gain with the selection was evaluated in experiments with field and greenhouse replications, containing the progenies selected in each cycle. They concluded that recurrent selection with progeny evaluation was efficient to increase tolerance to *Phytophthora* rot.

Resistance to Chlorosis of Iron Deficiency

In 1981, K. R. Prohaska and W. R. Fehr reported the application of recurrent selection for resistance to chlorosis of iron deficiency (Lewers and Palmer 1997); first, they obtained base populations through octuple crosses between ten resistant adapted parents and ten resistant exotic parents (PI). One hundred F_3 progenies were evaluated in experiments carried out at two locations and three replications, and the ten most resistant progenies were selected to start the cycle 1, which were recombined in diallel. Two to three plants of each cross were planted and harvested to recover 100 F_3 progenies, which were evaluated and selected, as in the previous cycle. The cycle 2 was also performed in a similar way, with recombination at the beginning of the ten most resistant progenies selected in the previous cycle, followed by evaluation and selection of the resulting F_3 progenies. Genetic progress was estimated by comparing the initial 20 parents and the parents of cycle 1 and cycle 2 (20 selected progenies, 10 of each cycle) in experiments with replications at three locations. A 9% increase in resistance to chlorosis caused by iron deficiency in soybean was observed.

Final Considerations

There is a large number of proposed methodologies and researches using recurrent selection in soybean crop. In most studies, with or without the use of male sterility, there is a predominance of progeny evaluation in the F_3 generation, in which the genotypes are, on average, 75% of the loci in homozygous and 25% of the loci in heterozygosis. The greatest genetic gains in seed yield in recurrent selection programs have been obtained from the largest possible numbers of genetically different (F_1) crosses.

In soybean, the SSD (*single seed descent*) method and their previously reported derivations are well suited for the advancement of more than one generation of inbreeding per year, even outside the ideal conditions of soil and climate, a fact that allows to reduce the time of each cycle of recurrent selection.

In general, although recurrent selection is laborious and time-consuming, it should be considered as a central methodology in well-established and consolidated soybean breeding programs, since it serves as the base for the creation of genetic variability to be explored in the long term for different traits. Moreover, it also allows the enhancement of the genetic base of the germplasm of the cultivated species and the general organization of breeding programs in progress.

References

Alliprandini, L.F., Vello, N.A.: Heritability and correlations among traits in four-way soybean crosses. Euphytica. **136**(1), 81–91 (2004)

Brim, C.A., Stuber, C.W.: Application of genetic male sterility to recurrent selection schemes in soybeans. Crop Sci. **13**, 528–530 (1973)

Cockerham, C.C.: Covariances of relatives from self-fertilization. Crop Sci. **23**(6), 1177–1180 (1983)

Fehr, W.R.: Breeding methods for cultivar development. In: Wilcox, J.R. (ed.) Soybeans: Improvement, Production and Uses, 2nd edn, pp. 249–293. American Society of Agronomy, Madison (1987)

Hallauer, A.R., Carena, M.J., Miranda Filho, J.B.: Quantitative Genetics in Maize Breeding, 3rd edn, p. 663. Springer, New York (2010)

Hamawaki, O.T., Vello, N.A., Didoné, C.A.: Improvement in genetic characteristics and oil yield of selected soybean progenies from octuple crosses. Genet. Mol. Biol. **23**(4), 855–864 (2000)

Lewers, K.S., Palmer, R.G.: Recurrent selection in soybean. In: Plant Breeding Reviews, vol. 15, pp. 275–313. Wiley, New York (1997)

Lopes, A.C.A.: Potencialidade de cruzamentos óctuplos de soja com ênfase na produtividade de grãos. Mestre dissertação, PPG-GMP, Escola Superior de Agricultura "Luiz de Queiroz", p. 102. Universidade de São Paulo, Piracicaba (1996)

Miranda, M.A.C.: Seleção recorrente divergente para peso de sementes e porcentagem de óleo em soja com o uso de machoesterilidade genética. Doutor tese, PPG-GMP, Escola Superior de Agricultura "Luiz de Queiroz", p. 112. Universidade de São Paulo, Piracicaba (1994)

Orf, J.H., Diers, B.W., Boerma, H.R.: Genetic improvement: conventional and molecular-based strategies. In: Boerma, H.R., Specht, J.E. (eds.) Soybeans: Improvement, Production, and Uses, 3rd edn, pp. 417–450. American Society of Agronomy, Madison (2004)

Posadas, L.G., et al.: Elite performance for grain yield from unadapted exotic soybean germplasm in three cycles of a recurrent selection experiment. Crop Sci. **54**(6), 2536–2546 (2014)

Priolli, R.H.G., et al.: Genetic structure and a selected core set of Brazilian soybean cultivars. Genet. Mol. Biol. **36**(3), 382–390 (2013)

Priolli, R.H.G., et al.: Association mapping of oil content and fatty components in soybean. Euphytica. **203**(1), 83–96 (2015)

Rose, J.L., Butler, D.G., Ryley, M.J.: Yield improvement in soybeans using recurrent selection. Aust. J. Agric. Res. **43**, 135–144 (1992)

Schmutz, J., et al.: Genome sequence of the palaeopolyploid soybean. Nature. **463**(7278), 178–183 (2010)

Tanaka, J.: Transgenic male sterility permits efficient recurrent selection in autogamous crops. Crop Sci. **50**(4), 1124–1127 (2010)

Unêda-Trevisoli, S.H.: Estabilidade fenotípica e potencialidade de progênies obtidas por cruzamentos óctuplos em soja. Doutor tese, PPG-GMP, Escola Superior de Agricultura "Luiz de Queiroz", p. 228. Universidade de São Paulo, Piracicaba (1999)

Vello, N.A.: Ampliação da base genética do germoplasma e melhoramento da soja na ESALQ/USP. In: I Simpósio sobre Cultura e Produtividade da Soja, 15–18 julho 1991, Piracicaba, USP/ESALQ, pp. 60–81. Editora FEALQ, Piracicaba (1992)

Vello, N.A., Fehr, W.R., Bahrenfus, J.B.: Genetic variability and agronomic performance of soybean populations developed from plant introductions. Crop Sci. **24**(3), 511–514 (1984)

Wysmierski, P.T., Vello, N.A.: The genetic base of Brazilian soybean cultivars: evolution over time and breeding implications. Genet. Mol. Biol. **36**(4), 547–555 (2013)

Chapter 11
Biometrics Applied to Soybean Breeding

Cosme Damião Cruz, Haroldo Silva Rodrigues, Renato Domiciano Silva Rosado, and Leonardo Lopes Bhering

Abstract Soybean has been an important source of protein for either food or feed; breeding program for this species has turn to a competitive business between seed companies, where the application of the right statistical method in data analysis means to lead the market. The understanding of biometrics and its use along all phases in the breeding process dictates the direction and the genetic gain that will be achieved; then, knowing the right statistical method to apply is crucial to meet the stockholders' demand. In the beginning of a breeding program, genotypes that have a good performance should be identified by genetic diversity studies and used to compose the base population, diallel crosses may be used to determine the best combinations among them, the segregant population must be right conduced, and statistical methods that fit in the adopted experimental design ought to be used so the breeder can make an accurate selection, describing the line as broad or specific adaptation and what level of stability it shows, releasing reliable information about the lines and developing elite cultivars for the farmers. This chapter shows a great number of biometrics techniques used in plant breeding as well their application in soybean breeding.

Keywords Soybean • Breeding • Biometrics • Experimental Statistics • Genetics

Introduction

In the literature we find diverse information about biometric and statistical principles, models, and procedures that allow researchers to analyze their experimental data and generate useful information for breeding programs. In this chapter we illustrate the use of these procedures in the various stages of a soybean breeding program, subdivided into three phases: *beginning*, where the purpose is to form a base population for breeding with good performance, broad variability, low genetic

---

C.D. Cruz, M.S., D.S. (✉) • H.S. Rodrigues, M.S., D.S.
R.D.S. Rosado, M.S., D.S. • L.L. Bhering, M.S., D.S.
The Federal University of Viçosa, P.H.Rolfs Avenue, Viçosa, Minas Gerais 36570-900, Brazil
e-mail: cdcruz@ufv.br; haroldo.rodrigues@ufv.br; rosado.rds@gmail.com; leonardo.bhering@ufv.br

F. Lopes da Silva et al. (eds.), *Soybean Breeding*,
DOI 10.1007/978-3-319-57433-2_11

load, and good adaptation; *middle*, where concern is related to the management of segregating families that allow maximizing direct, indirect, or simultaneous gains in important traits; and *end*, where improved genetic material is already available and one seeks to recommend it for broad or specific regions, which makes interaction studies on genotype vs. environment as well on stability and adaptability fundamental.

Base Population Formation

One of the primary phases of a breeding program is the choice of parents to be intercrossed to form a base population, on which researchers will focus their efforts in the search for genetic material that deliver higher yield, quality, and adaptation. Criteria used in the choice of parents are many, and performance is observed in relation to several aspects, the combinatorial ability and adaptation. Another major feature is diversity, and the complementarity and the variability indispensable for selection to be carried out are expected to be obtained through genetic recombination.

Thus, genetic diversity among a group of parents has been evaluated with the purpose of identifying the hybrid combinations of greater heterotic effect and greater heterozygosity, so that, in its segregating generations, there is greater possibility of recovery of superior genotypes by means of the occurrence of transgressive segregants. In the case of soybeans, initially, no emphasis is given to the search for hybrid vigor in F1 populations, considering that commercial use focuses on pure lines; however, the heterotic effect is an indication of diversity and of the genetic complementarity which, when including parents with good genetic potential per se, make it possible, in segregating generations, to obtain homozygous combinations able to exceed the limits of good performance established by these parents.

The assessment of genetic diversity, based on scientific evidence, is also of great importance in the context of species evolution, inasmuch as it provides information on available resources and assists in locating and exchanging them.

Genetic diversity has been evaluated by means of biometric techniques, based on the quantification of heterosis or by predictive processes. Among the methods based on biometric models, which are intended to assess the potential of the parents, the diallelic analysis and the predictive analysis of genetic diversity are mentioned.

Diallelic Analysis

Diallelic scheme is a very efficient genetic design, for it quantifies the genetic variability of the trait and assesses the genetic value of parents and the specific capacity and heterosis manifested in specific crosses. Some important aspects on diallelic analysis are presented in this work in Chap. 8; however, due to their relevance to the theme of this chapter, some summarized information on concept and applications will also be highlighted, considering the diallelic schemes subsequently described.

Balanced Diallels (Full or of Half Table)

They are the diallels that include F1 hybrids among all pairs of parent combinations and may additionally include the parents, their reciprocal hybrids, and, sometimes, other related generations, such as F2, backcrosses, etc.

Partial Diallels

They involve two groups of parents and their respective crosses, similar to the factorial model proposed by Comstock and Robinson (1948) (Design II). Adaptations of the Griffing (1956) and Gardner and Eberhart (1966) model, for this type of diallel, have been possible to maximize information on the groups studied with a smaller number of crosses than those required for the balanced diallel.

Circulating Diallels

In these diallels, parents are represented, in the outline, by a same number of crosses, but inferior to $p-1$, as in the balanced ones. These diallels allow obtaining information about the parents with smaller number of crosses; however, there is loss of information about certain hybrid combinations, which are not provided.

Incomplete Diallels

The diallels whose parents are represented by a variable number of crosses. In general, this type of design occurs as a consequence of treatment losses during the conduction of the tests or their absence due to various problems, such as seed insufficiency, incompatibility of crosses, etc.

Unbalanced Diallels

The diallels in which all hybrid combinations and also the other generations (parents, reciprocals, etc.) are represented, but in variable frequency, by virtue of the unequal number of repetitions per treatment.

Some scientific contributions using diallelic analysis under different approaches are presented subsequently.

Martins Filho et al. (1992) studied diallelic analysis and obtained estimates of the effects and square sums of the effects of parental combinations capacities from the Griffing (1956) model for diallels with unequal numbers of replicates per treatment. The authors report that the use of resistant cultivars and the incorporation of genes for resistance in susceptible commercial cultivars are the most efficient and economic means to control the disease, and for greater efficiency in this incorporation process, it is necessary to know the inheritance mechanism of the trait at issue. To that end, a diallel composed of five cultivars (Cristalina, Paraná, Bossier, Uberaba and Sucupira) was tested under greenhouse conditions, in order to evaluate the reaction to *Cercospora*. The authors showed that the best parents were Cristalina and Uberaba and the best hybrid combinations, Cristalina x Bossier, Cristalina x Sucupira, Bossier x Uberaba, and Uberaba x Sucupira.

Another disease that causes great concern in Brazil, and which was evaluated in a diallelic system, is the Asian rust, caused by the fungus *Phakopsora pachyrhizi*. As a result, Antunes Cruz et al. (2011) saw the opportunity to evaluate the progress of Asian rust in soybean genotypes intended for silage production, under natural infection conditions. The authors evaluated the severity of the rust over time and the combinatorial ability of five cultivars and 20 soybean F3 progenies. To evaluate the effect of the combinatorial ability of the parents and of the F3 progeny, diallelic analysis was performed according to Griffing (1956). There was a significant effect of the general and specific combination capacities for the disease progression area between the parents and progenies as well as for reciprocal crosses. The results also indicated that the resistance to soybean Asian rust can be influenced by maternal components.

Aiming at increasing productivity, Shirahige et al. (2013) estimated the effects of general and specific combining ability, identifying parents and promising combinations to generate segregating populations, in order to attend the soybean breeding program in the State of São Paulo. The 15 crosses (treatments) within F4 generation and six controls were evaluated in a randomized block experiment. The traits evaluated were grain yield (PG), days for flowering (DF), days for maturation (DM), plant height at flowering (PA), and plant height at maturity (AM). The treatment mean squares of all traits were significant by F ($P \leq 0.01$) test, as well as the mean squares of GCA and SCA. GCA was the predominant effect in the composition of the means of crosses for traits DF, DM, AF, and AM, while for PG there was preponderance of the effects of SCA. One parent (FT-14) was noted for the concentration of favorable alleles for the five characters. For PG, the best crosses involved at least one parent with high GCA or the parents with the highest SCA.

Daronch et al. (2014) analyzed a complete diallel, without reciprocals, involving six soybean genotypes (M 9350, BR/EMGOPA 314, P99R01, M 9056RR, NIDERA A 7002, P98Y70) adapted to low-latitude conditions. The characteristics studied were number of pods per plant, number of grains per plant, and grain production per plant (grams/plant). The method used to interpret the diallel was Griffing's (1956), which provided information on the general and specific combining ability of parents in artificial crosses. For all the characteristics, a significant effect was detected for GCA and SCA. The parent M 9056RR was the most promising, and the combination M 9056RR x P98Y70 was the most indicated for formation of a segregating population with superior characteristics.

Genetic Diversity

In general, soybean breeders have an interest group of lines for the formation of a breeding population in view of the political, social, or technical reality that results in demands by consumers and/or producers. Thus, it is interesting to incorporate the agronomic qualities of the parental line in a new population that will be submitted to selection cycles. However, when a large number of genetic lines is available, some predictive studies have been recommended, prior to the performance of diallelic crosses, that guide the number and type of crosses on which efforts should be concentrated to obtain hybrid combinations and, later, segregating populations of greater genetic potential. As they do not require previous hybrid combinations, parent diversity predictive methods have deserved considerable emphasis.

The predictive methods, used in an exploratory way, are based on an expression of the quantitative genetics that relates heterosis to the existence of two basic factors: dominance deviations and the gene frequency difference among the populations involved in the crosses. It is necessary to add the action of inter-allelic interactions, of epistatic and non-epistatic nature, to these two genetic factors, so as to obtain superior genotypes in F1 and subsequent generations. In practice, phenotypic differences expressing morphological, physiological, biochemical, and other traits presented by the parents in diversity determination are taken as the basis, and these are generally quantified by a measure of dissimilarity (e.g., Euclidean and Mahalanobis distances). Inference based on ecogeographic diversity is also an example of a predictive heterosis method.

Several methods that combine multiple data, known as multivariate, can be applied in the genetic diversity prediction. These methods include principal component analysis, main coordinates, canonical variables, and several clustering techniques, such as hierarchical and optimization methods. The choice of the most appropriate method has been determined by the precision intended by the researcher, ease of analysis, and how data were obtained.

The methods based on principal components or canonical variables allow the study of diversity in scatter plots, in which two Cartesian axes are generally considered. In these studies, several characteristics are evaluated in a set of genotypes that are, by statistical procedures, summarized in a few components (or canonical variables) that are given by linear combinations of the original characters, independent of each other and with decreasing discrimination capacity, so that the first components (or canonical variables) explain the maximum variation in the original data.

The clustering methods differ from the others, as they fundamentally depend on previously estimated dissimilarity measures, such as the Euclidean distance or the generalized Mahalanobis distance, among others. The cluster analysis aims to gather, by means of some classification criteria, the parents (or any other type of sample unit) in several groups, so that there is homogeneity within the group and heterogeneity among groups. Alternatively, clustering techniques also have the purpose of dividing an original group of observations into several groups, according to some similarity or dissimilarity criteria (Cruz and Regazzi 1997).

The grouping process basically involves two steps. The first one was related to the estimation of a similarity (or dissimilarity) measure between the parents and, the second, with the adoption of a clustering technique for group formation. Some of the most used dissimilarity measures are Euclidean and Mahalanobis distances, for phenotypic characteristics, and the complements of Jaccard, Nei, and Li indexes or weighted, which take into account the number of shared alleles when molecular information is available.

As it is desirable to have information about each pair of parents in the clustering process, the number of dissimilarity measure estimates is considered high, which makes the recognition of homogeneous groups unfeasible by the simple visual examination of those estimates. To this task clustering methods are used.

There is a large number of clustering methods available, and the researcher must decide which one is the most suitable for his work, since the various techniques can lead to different clustering patterns. The clustering methods most commonly used in plant breeding are hierarchical and optimization methods.

In hierarchical methods, the parents are grouped by a process that is repeated at several levels until the dendrogram or tree diagram is established. In this case, there is no concern regarding the optimal number of groups, as the greatest interest is in the "tree" and the branching obtained. Delimitations can be established by a visual examination of the dendrogram, in which points of high change of level are evaluated, generally taking them as delimiters of the number of parents for a certain group.

Hierarchical methods are also divided into agglomerative and divisive. Agglomerative ones include the single linkage method; the complete linkage method; the average linkage, weighted or not; the centroid, also weighted or not; and the one proposed by Ward (1963). Among the divisive methods, the best known is that of Edwards and Cavalli-Sforza (1965).

An example of use of the mean Euclidean distance is presented by Torres et al. (2015), in a genetic variability study of soybean cultivars, while the use of Mahalanobis distance is shown by Rigon et al. (2012). An example of genetic diversity studies with molecular marker information is found in a study by Vieira et al. (2007).

Torres et al. (2015) have adopted a procedure that has been common in many studies on diversity, which will be briefly described. The focus was on the genetic diversity study of soybean cultivars in the Brazilian *Cerrado*. Data were submitted to variance analysis, and the means were compared to Tukey's test, at 5% probability. Subsequently, the dispersion matrix was obtained, using multivariate methods, based on the single linkage method, obtaining the dendrogram, using the mean Euclidean distance as a dissimilarity measure. Singh's criterion (1981) was also applied, in order to quantify the contribution of the traits to the genetic divergence among the genotypes (Cruz et al. 2004). The authors concluded that the number of pods and height of pod insertion contributed the most to the divergence among genotypes. Other authors also applied the single linkage method to the diversity study of soybean cultivars, such as Rigon et al. (2012).

In optimization methods, the set of parents is divided into nonempty and mutually exclusive subgroups by maximizing or minimizing some preestablished measure. One of the most commonly used optimization methods in breeding is the one proposed by Tocher (cited by Rao 1952).

Tocher method adopts the criterion that the mean dissimilarity measures within each group should be less than the mean distances between any groups. The method requires obtaining the dissimilarity matrix, on which the most similar pair of parents is identified. These parents will form the initial group. From there the possibility of inclusion of new parents is evaluated by adopting the previously mentioned criterion.

The entry of a progenitor into a group always increases the mean value of the distance within the group. Thus, one can make the decision of including the parent in a group by comparing the addition of the mean distance value within the group to a maximum permissible level, which can be arbitrarily established, or adopting, as it has been generally done, the maximum dissimilarity measure value found in the set of the smaller distances involving each progenitor. The use of the Tocher method in the soybean genetic diversity study has been reported by Rigon et al. (2012).

Another important aspect when studying genetic diversity is the quantification of the relative contribution of the traits measured on the variability of the individuals or accessions measured. Through the analysis of the importance of traits, it is possible to classify variables studied according to their contribution to genetic diversity and eliminate those with lesser contribution. The genetic diversity study among a set of accessions is made from a set of information that, in some cases, requires the evaluation of many traits, demanding great manpower and cost. In these studies, it is necessary to evaluate the importance of each of them to the diversity, identifying those that contribute the least, and it is advisable to discard them in future studies. In this context, Rigon et al. (2012) studied the importance of traits for genetic diversity studies. They reported that the most intense contribution to the genetic divergence was provided by the height of insertion of the first legume, followed by the soybean physiological maturation time, which contributed 23 and 20%, respectively, while the mass variable of 100 grains had little difference among the individuals, with importance of 3.27%, similar to the data reported by Peluzio et al. (2009), with contribution of 2.8%, for Torres et al. (2015). Among the traits evaluated, the most important contribution to genetic divergence was the number of pods, followed by pod insertion height, with 25.6 and 24.2%, respectively.

Conducting Segregant Populations

Prediction of Gains by Selection

One of the great contributions of quantitative genetics is the evaluation of the gains to be obtained by a certain selection strategy. This information allows us to guide breeding programs, predict their success, choose or discard population, and concentrate efforts on the measurement of characters of greater importance and earning potential.

For the evaluation of families in soybean segregating generations, the self-fertilization process that leads to a redistribution of total genotypic variation between and within lines should be considered, as specified in Table 11.1. The amount of additive variation $\left(\sigma_a^2\right)$ available increases with inbreeding, and fixation ($F = 1$) can be twice the value that would be found in a panmictic population. The variance component assigned to dominance deviations $\left(\sigma_d^2\right)$ becomes null when inbreeding reaches its maximum limit.

With genetic variance estimates, between $\left(\widehat{\sigma}_{ge}^2\right)$ and within $\left(\widehat{\sigma}_{gd}^2\right)$, the components of additive variance and attributed to the deviations of dominance are thus estimated:

$$\widehat{\sigma}_a^2 = \frac{\widehat{\sigma}_{ge}^2 - F\widehat{\sigma}_{gd}^2}{F(1+F)}$$

and

$$\widehat{\sigma}_d^2 = \frac{2F\widehat{\sigma}_{gd}^2 - (1-F)\widehat{\sigma}_{ge}^2}{F\left(1-F^2\right)}$$

Table 11.1 Decomposition of the total variance in the evaluation of families in advanced generations of self-fertilization, with coefficient of inbreeding *F*

FV	F	$F = 0$	$F = 1$
Between lines	$2F\sigma_a^2 + F(1-F)\sigma_d^2$	0	σ_a^2
Within lines	$(1-F)\sigma_a^2 + (1-F)\sigma_d^2$	$\sigma_a^2 + \sigma_d^2$	0
Total	$(1+F)\sigma_a^2 + (1-F)\sigma_d^2$	$\sigma_a^2 + \sigma_d^2$	$2\sigma_a^2$

In this case, heritability in the restricted sense, defined as the ratio between the additive genetic variance and the phenotypic variance between the selection units, should be estimated as follows:

$$h^2_{\text{r entre}} = \frac{\widehat{\sigma}^2_{\text{a entre}}}{\widehat{\sigma}^2_{\text{f entre}}} = \frac{2F\widehat{\sigma}^2_{\text{a}}}{\widehat{\sigma}^2_{\text{f entre}}}$$

Given the importance of genetic parameters, especially of heritability, to guide strategies that maximize genetic gains, several authors have presented results that allow a relative comparison of values applicable to soybean breeding. Thus, the results of Bárbaro et al. (2007), described in Table 11.2, are shown as an illustration. There is great variation in heritability values from one characteristic to another, from one population to another, from one generation to another, and also between heritability types (broad or restricted). However, in general, the influence of environmental factors on complex characteristics such as grain yield has been determined.

$$h^2_{\text{r dentro}} = \frac{\widehat{\sigma}^2_{\text{a dentro}}}{\widehat{\sigma}^2_{\text{f dentro}}} = \frac{(1-F)\widehat{\sigma}^2_{\text{a}}}{\widehat{\sigma}^2_{\text{f dentro}}}$$

$$h^2_{\text{r total}} = \frac{\widehat{\sigma}^2_{\text{a total}}}{\widehat{\sigma}^2_{\text{f total}}} = \frac{(1+F)\widehat{\sigma}^2_{\text{a}}}{\widehat{\sigma}^2_{\text{f total}}}$$

Gains by Selection

The change produced by the selection of most interest is on the average of a population. Considering a selective scheme in which includes the test units (X), the recombination units (R), and the improved ones, the genetic values of Y can be predicted by means of the equation:

$$Y = \widehat{\beta}X$$

where:

$$\widehat{\beta} = \frac{\text{Cov}(X, Y)}{\widehat{V}(X)}$$

The prediction equation, where Y represents the genetic values in the population that improved based on the selection on the phenotypic values, can be expressed as follows:

$$\text{GS} = \widehat{\beta}\text{DS}$$

Table 11.2 Heritability estimates in the broad sense (h2ab) and (h2aw) and restricted (h2rb) and (h2rw) among and within families for 11 agronomic traits in five soybean populations in the F6 generation

	NDF	NDM	APM	AIV	NS	NV	NN	NR	PG	VA	Ac
Tracy-M x Paraná											
h2abF6	99,48	95,90	63,77	29,10	76,91	15,56	86,41	78,97	33,08	91,49	96,50
h2awF6	78,26	84,62	80,11	26,46	41,12	01,56	85,22	41,41	00,13	16,86	05,73
h2rbF6	99,09	96,27	36,86	17,88	03,53	14,79	13,36	52,76	34,04	91,10	99,20
h2rwF6	69,53	96,44	03,29	01,26	00,06	00,59	00,48	02,45	01,27	14,83	44,39
FT-Cometa x Paraná											
h2abF6	90,04	72,59	49,85	27,99	25,90	23,49	76,78	15,45	34,41	28,10	82,42
h2awF6	67,55	06,48	07,15	07,56	26,21	19,28	79,72	11,94	08,81	06,76	14,76
h2rbF6	89,15	74,77	49,09	23,94	06,17	02,89	25,83	07,20	26,08	19,09	77,51
h2rwF6	51,18	96,76	04,78	01,18	00,25	00,08	01,24	00,32	00,78	00,42	04,88
FT-Cometa x Bossier											
h2abF6	94,04	59,98	02,95	53,08	28,98	21,49	79,07	22,77	14,42	80,34	96,31
h2awF6	84,24	02,40	01,37	01,62	53,39	06,02	44,02	12,19	01,07	87,41	65,38
h2rbF6	90,14	61,92	02,16	53,36	06,55	15,44	68,81	04,40	12,46	30,07	85,88
h2rwF6	35,34	01,98	00,11	01,96	00,48	00,45	07,63	00,09	00,18	01,60	13,38
FT-Cometa x IAC-8											
h2abF6	93,66	40,81	75,97	14,57	03,92	33,45	86,00	22,71	62,79	13,30	21,83
h2awF6	71,00	73,79	36,31	04,01	04,66	08,33	77,70	03,44	58,88	11,38	17,88
h2rbF6	90,44	40,99	68,11	12,39	00,32	20,54	46,06	19,24	06,84	00,84	20,98
h2rwF6	33,16	86,02	07,74	00,66	00,01	00,39	02,70	00,51	00,22	00,02	07,78

BR-16 x IAC-11											
h2abF6	96,66	99,38	68,56	45,03	71,49	89,47	53,09	66,53	69,05	22,01	62,93
h2awF6	85,00	74,04	07,94	00,27	10,90	50,93	23,81	13,25	95,64	09,05	58,36
h2rbF6	96,11	99,60	63,51	45,89	68,32	74,49	19,89	54,97	18,40	17,78	11,68
h2rwF6	71,91	79,65	02,24	00,67	04,39	06,85	00,44	01,71	01,07	01,05	00,41

Source: Bárbaro et al. (2007)

Agricultural year 2004/2005. Jaboticabal, SP. *NDF* number of days for flowering, *NDM* number of days for maturation, *APM* plant height at ripening, *AIV* height of insertion of the first pod, *NS* number of seeds, *NV* number of pods, *NN* number of nodes, *NR* number of branches, *PG* grain yield, *VA* agronomic value, *Ac* bedding

where:
GS = selection gain in the improved population.
DS = selection differential.

$$\widehat{\beta} = \frac{\mathrm{Cov}(X, Y)}{\widehat{V}(X)}$$

where:
Cov(X, Y) = genetic covariance between related individuals X and Y.
$V(X)$ = phenotypic variance between the test units.
From the GS expression we get:

$$\mathrm{GS} = \frac{\mathrm{Cov}(X, Y)}{V(X)} \quad \mathrm{DS} = \left(\frac{2r_{XY}\widehat{\sigma}_{\mathrm{A}}^2}{\widehat{\sigma}_{\mathrm{f}}^2}\right) \quad \mathrm{DS} = h^2\mathrm{DS}$$

In the previous equation, only the inheritable fraction of the genotypic variance that makes up covariance between relatives is used for estimation of gain. The relationship between the additive variance explored in the selection and the phenotypic variance between the selection or test units was also used as the concept of heritability.

The selection gain can also be expressed as follows:

$$\mathrm{GS} = \widehat{\sigma}_{\mathrm{g}}\frac{\widehat{\sigma}_{\mathrm{g}}}{\widehat{\sigma}_{\mathrm{f}}} \quad \frac{\mathrm{DS}}{\widehat{\sigma}_{\mathrm{f}}} = \widehat{\sigma}_{\mathrm{g}}hi$$

Finally, we must also compute, for purposes of gain prediction, whether the selection occurs in one or both sexes. Thus, the more general expression is used:

$$\mathrm{GS} = p\widehat{\sigma}_{\mathrm{g}}hi$$

where:
p = parental control.
$\widehat{\sigma}_{\mathrm{g}}$ = genetic-additive standard deviation between test units. Usually, it corresponds to a fraction of the additive variance.
h = heritability square root or accuracy of the selection process.
i = selection intensity.

Gains by Selection in Self-Pollinated Populations

A situation to be considered in the selection of families, with autogamous plants as in soybeans, is the evaluation of the provisional genetic progress, obtained from generation n to n', and the definitive progress of generation n to $n' = \infty$. These developments should be predicted by the covariance between values of individuals and means of the generations considered, denoted by Y_n and $Y_{n'}$.

Selection Between Families in Generation *n*

The covariance among family means in generation n and the values of individuals in generation n' are represented as follows:

$$\mathrm{Cov}(\bar{Y}_n, Y_{n'}) = \mathrm{Cov}(k, n, n')$$

And, therefore, the provisional gain by the practice of the selection among families is given by:

$$\mathrm{GS}_{\mathrm{e.t}} = \mathrm{DS}_{\mathrm{e}} \frac{\mathrm{Cov}(k, n, n')}{\sigma^2_{\mathrm{f\,entre}}}$$

where:
DS_{e} = mean differential of selection between families, given by:

$$\mathrm{DS}_{\mathrm{e}} = \bar{Y}_{\mathrm{s}} - \bar{Y}_{\mathrm{o}}$$

where:
$\bar{Y}_{\mathrm{s}}$ = mean of selected families.
$\bar{Y}_{\mathrm{o}}$ = mean of the base population.
The covariance between the means of families in generation n and the values of individuals in generation ∞, with inbreeding coefficient $F\infty = 1$, are represented as follows:

$$\mathrm{Cov}(\bar{Y}_n, Y_{\infty}) = \mathrm{Cov}(k, n, \infty) = (1 + F_k)\sigma^2_{\mathrm{A}}$$

Thus:

$$\mathrm{GS}_{\mathrm{ep}} = \mathrm{DS}_{\mathrm{e}} \frac{\mathrm{Cov}(k, n, \infty)}{\sigma^2_{\mathrm{f\,entre}}} = \mathrm{DS}_{\mathrm{e}} \frac{(1 + F_k)\sigma^2_{\mathrm{A}}}{\sigma^2_{\mathrm{f\,entre}}}$$

Selection Within Families in Generation *n*

With the selection *within*, n' generation would be formed by individuals deriving from self-fertilization of the best plants of the superior families of generation n. The selection unit, in this case, would present values corresponding to the difference between the phenotypic value of an individual (Y_n) and the average of his family ($\bar{Y}_n$). Thus, for the prediction of interim genetic progress, to go from generation n to n', it is necessary to know the covariance:

$$\mathrm{Cov}[(Y_n - \bar{Y}_n), Y_{n'}] = \mathrm{Cov}(Y_n, Y_{n'}) - \mathrm{Cov}(\bar{Y}_n, Y_{n'})$$

where:

$$\mathrm{Cov}(Y_n, Y_{n'}) = \mathrm{Cov}(n, n, n')$$

and

$$\mathrm{Cov}(\bar{Y}_n, Y_{n'}) = \mathrm{Cov}(k, n, n')$$

Therefore:

$$\mathrm{Cov}[(Y_n - \bar{Y}_n), Y_{n'}] = (F_n - F_k)\sigma^2_{\mathrm{A}} + 2(1 - F_{n'})\frac{F_n - F_k}{1 - F_k}\sigma^2_{\mathrm{D}}$$

Once the percentage of plants *within* families to be selected has been defined, the selection gain *within* can be estimated by the expression:

$$\mathrm{GS_{dt}} = \mathrm{DS_d}\frac{\mathrm{Cov}(n, n, n') - \mathrm{Cov}(k, n, n')}{\sigma^2_{\mathrm{f\ dentro}}}$$

where:

$\mathrm{DS_d}$ = differential of mean selection of plants within families.

It is also interesting to calculate the progress expected with the selection within superior families (generation n) until pure lines are obtained. Such progress will be a function of the covariance between plants of generation n and $n' =$. Thus, one has:

$$\mathrm{Cov}[(Y_n - \bar{Y}_n), Y_\infty] = \mathrm{Cov}(Y_n, Y_\infty) - \mathrm{Cov}(\bar{Y}_n, Y_\infty)$$

Given by:

$$\mathrm{Cov}[(Y_n - \bar{Y}_n), Y_\infty] = (F_n - F_k)\sigma^2_{\mathrm{A}}$$

Hence, the permanent gain is predicted by means of the expression:

$$\mathrm{GS_{dp}} = \mathrm{DS_d}\frac{\mathrm{Cov}(n, n, \infty) - \mathrm{Cov}(k, n, \infty)}{\sigma^2_{\mathrm{f\ dentro}}} = \mathrm{DS_d}\frac{(F_n - F_k)\sigma^2_{\mathrm{A}}}{\sigma^2_{\mathrm{f\ dentro}}}$$

Massal Selection in Generation *n*

In this situation the selection of the best individuals is made, independently of the families to which they belong. The selection unit, in this case, would present values corresponding to the phenotypic value of an individual (Y_n), so that the prediction of the provisional genetic progress, when going from generation n to $n' = \infty$, would be dependent on covariance Cov(n, n, n') given by:

$$\mathrm{Cov}(n,n,n') = (1+F_n)\sigma_A^2 + (1+F_n)(1-F_{n'})\sigma_D^2$$

Once the percentage of plants to be selected has been defined, the massal selection gain can be estimated by the following expression:

$$\mathrm{GS_{mt}} = \mathrm{DS_m}\frac{\mathrm{Cov}(n,n,n')}{\sigma^2_{\mathrm{f\,total}}}$$

where:

$\mathrm{DS_m}$ = selection differential, considering the phenotypic values of the plants evaluated in the test.

In an analogous way, the expected progress is calculated with the massal selection until pure lines are obtained. This progress will be a function of the covariance between plants of generation n and $n' = \infty$, given below:

$$\mathrm{Cov}(Y_n, Y_\infty) = \mathrm{Cov}(n,n,\infty) = (1+F_n)\sigma_A^2$$

The permanent gain is predicted by the expression:

$$\mathrm{GS_{mp}} = \mathrm{DS_m}\frac{\mathrm{Cov}(n,n,\infty))}{\sigma^2_{\mathrm{f\,total}}} = \mathrm{DSm_d}\frac{(1+F_n)\sigma_A^2}{\sigma^2_{\mathrm{f\,total}}}$$

Factors Affecting Selection Gain

With gain of selection quantified by GS = $phi\hat{\sigma}_g$, it is possible to highlight some factors that directly affect selection gain and that must be observed by breeders.

Selection Differential and Intensity of Selection

One way to obtain greater selection gain is to increase the intensity of selection, but, in very intense selection, the population may present problems inherent to inbreeding, which is a consequence of the cross between related individuals and is closely related to the reduced sample size.

Another factor that raises concern is the effective selection differential, as, in practice, parents do not equally contribute to the next generation, due to, among other factors, fertility, genetic constitution, sampling, etc. Thus, in the process of recombination, one must be aware of the number of females that will contribute to the next generation and also the pollination process in which the participation of males is influenced and therefore determines the contribution of this gender.

When selecting variables with normal distribution, the intensity of selection (i) can be estimated from the percentage of individuals to be selected, according to the expression.

$$i = \frac{\mathrm{DS}}{\widehat{\sigma}_{\mathrm{f}}} = \frac{z}{c}$$

where:

z = height of the normal curve ordinate at the truncation point established from the percentage selected.

c = percentage of individuals selected.

Parental Control

Parental control defines a similarity between test units and improved units, and, consequently, it alters the genetic covariance and selection gain.

Eberhart (1970) reports that parental control in a recurrent selection process can be defined as a function of the parental relationship between a selection unit used for an identification of superior individuals and a recombination unit used to obtain the improved population.

The coefficient that expresses the parental control (p) can admit values equal to 1/2, 1, and 2, as described below.

The selection unit is equal to the recombination unit.

(a) If the recombination units are pollinated by both selected and unselected genotypes, the p value is equal to 1/2.
(b) If the recombination units are pollinated only by the genotypes selected, then p will be equal to 1.

The selection unit is different from the recombination unit.

(a) If the recombination units are pollinated by both selected and unselected genotypes, the value of p will be equal to 1.
(b) If the recombination units are pollinated only by the selected genotypes, then p will be equal to 2.

Genetic Variability

The success of a breeding program depends on the variability in a population. There is some concern as to guaranteeing broad variability in the work population, by means of the choice of divergent parents that are used in cross-linking for a formulation of the base population and high combination ability.

Genetic variability is maintained by appropriate mating and appropriate sampling, so that the effective population size is not reduced.

Environmental Variance

The phenotypic variance will be close to the genotypic variance when the environmental variations are minimal. Thus, an environmental variance influences one of the main factors that determine the selection gain, which is the heritability. This coefficient is proportional to the additive genetic variable available in the population and inversely proportional to the phenotypic variance.

As an illustration we can consider the inheritance between family means estimated in an evaluation experiment of **g** genotypes in **r** blocks at random, which is given by:

$$h^2 = \frac{\widehat{\sigma}_g^2}{\widehat{\sigma}_g^2 + \frac{\widehat{\sigma}^2}{r}}$$

This same heritability, if estimated in repeated experiments in several environments, is given by:

$$h^2 = \frac{\widehat{\sigma}_g^2}{\widehat{\sigma}_g^2 + \frac{\widehat{\sigma}_{ga}^2}{a} + \frac{\widehat{\sigma}^2}{ar}}$$

where:

σ_g^2 = component of genetic variance among family mean.

$\widehat{\sigma}^2$ = component of the environmental variance.

$\widehat{\sigma}_{ga}^2$ = component of the variance of the genotype x environment interaction.

r = number of replications (or blocks).

a = number of environments.

Therefore, the influence of the environmental component on the heritability of a characteristic can be reduced by the adoption of appropriate statistical design, the greater number of replications, greater environment control, and more careful conduction of the experiment.

Genotype x Environment Interaction

The genotype x environment interaction influences the gain to be obtained by selection. When the breeding program is restricted to a certain environmental condition, the interaction between genotypes and environments is capitalized, and, consequently, the σ_g^2 fraction used to predict the gain is the one confused with the σ_{ga}^2. interaction.

Direct and Indirect Trait Selection

The success of a breeding program is primarily based on the existence of genetic variability, which enables the breeders to select and, consequently, obtain superior genetic material. The rapid and efficient use of this variability is essential, and correlation studies are one of the ways to save time and reduce efforts.

The estimation of the correlations is important in the establishment of more suitable strategies for a breeding program condition and evaluation of indirect answers in traits of low heritability or with identification and measurement problems.

Phenotypic Correlations

The phenotypic correlations are those obtained from the averages of the traits evaluated. Considering the evaluation of two traits, X and Y, in g genotypes evaluated in b blocks at random, we have the statistical model.

$$X_{ij}\,\mathrm{ou}\,Y_{ij} = \mu + g_i + b_j + \varepsilon_{ij}$$

The phenotypic correlation is obtained as follows:

$$r_{\mathrm{f}} = \frac{\mathrm{Cov}(\bar{X}, \bar{Y})}{\sqrt{V(\bar{X})\,V(\bar{Y})}}$$

where $\bar{X}$ and $\bar{Y}$ are variables that express the genotypes means in relation to traits X and Y, respectively.

In soybean crop, an example of quantification of the trait correlation is presented by Giorgenon et al. (2014) and described in Table 11.3.

Genotypic Correlations

To estimate the environmental and genotypic correlations, one must carry out the analysis of variances and obtain the mean square values (or variances). The mean products (covariance) have been calculated using the estimates of the mean squares obtained from the variance analysis of the sum of the values of X and Y, given by:

$$Z_{ij} = X_{ij} + Y_{ij}$$

The variance analyses of variables X, Y, and $X + Y$ with the mean squares, indispensable for the calculation of correlations, are presented in Table 11.4.

Table 11.3 Phenotypic correlations among the characters evaluated in soybean progeny F7, in Jaboticabal-SP

	AIV	NN	NR	VA	Ac	NV	NS	OS
APM	0.5802**	0.4200*	−0.2986 ns	−0.0194 ns	0.2127 ns	−0.1760 ns	−0.2724 ns	−0.2724 ns
AIV		0.3895*	−0.2617 ns	−0.1801 ns	0.1162 ns	−0.2664 ns	−0.2722 ns	−0.3056 ns
NN			0.1351 ns	−0.0342 ns	0.0765 ns	−0.1120 ns	−0.0788 ns	0.0159 ns
NR				0.5676**	−0.1997 ns	0.5277**	0.6633**	0.6375**
VA					-0.0355 ns	0.8036**	0.8225**	0.8329**
Ac						−0.027 ns	−0.0084 ns	−0.0617 ns
NV							0.9179**	0.8573**
NS								0.8942**

Source: Giorgenon et al. 2014

Traits: plant height at maturity (APM), height of insertion of the first pod (AIV), number of nodes (NN), number of branches (NR), agronomic value (VA), lodging (Ac), number of pods per plant (NV), number of seeds per plant (NS), and seed production (PS) in grams/plant

*, ** Statistically significant by t test with 5% and 1% of probability, respectively

Table 11.4 Variance analysis of traits *X*, *Y*, and the sum *X* + *Y* evaluated in blocks, at random, involving g genotypes

		X		Y		$X+Y$	
FV	GL	QM	E(QM)	QM	E(QM)	QM	E(QM)
Blocks	$r-1$						
Genotypes	$g-1$	QMG_x	$\sigma_x^2+\sigma_{gx}^2$	QMG_y	$\sigma_y^2+r\sigma_{gy}^2$	QMG_{x+y}	$\sigma_{x+y}^2+\sigma_{g(x+y)}^2$
Residue	$(r-1)(g-1)$	QMR_x	σ_x^2	QMR_y	σ_y^2	QMR_{x+y}	σ_{x+y}^2

The mean products are obtained considering that:

$$V(X+Y) = V(X) + V(Y) + 2\text{COV}(X+Y)$$

So that:

$$\text{PMG}_{x,y} = \frac{\text{QMG}_{(x+y)} - \text{QMG}_x - \text{QMG}_y}{2}$$

$$\text{PMR}_{x,y} = \frac{\text{QMR}_{(x+y)} - \text{QMR}_x - \text{QMR}_y}{2}$$

The correlations can then be obtained as follows:
Environmental correlation

$$r_a = \frac{\text{PMR}_{xy}}{\sqrt{\text{QMR}_x\text{QMR}_y}}$$

Genotypic correlation

$$r_g = \frac{\widehat{\sigma}_{g(x,y)}}{\sqrt{\widehat{\sigma}_{gx}^2\widehat{\sigma}_{gy}^2}}$$

where:

$$\widehat{\sigma}_{g(x,y)} = \frac{\text{PMG}_{x,y} - \text{PMR}_{x,y}}{r}$$

$$\widehat{\sigma}_{gx}^2 = \frac{\text{QMG}_x - \text{QMR}_x}{r}$$

$$\widehat{\sigma}_{gx}^2 = \frac{\text{QMG}_y - \text{QMR}_y}{r}$$

Phenotypic correlation

The phenotypic correlation can also be obtained from the mean squares:

$$r_{\mathrm{f}} = \frac{\mathrm{PMG}_{x,y}}{\sqrt{\mathrm{QMG}_x \mathrm{QMG}_y}}$$

Relation Among Phenotypic, Genotypic, and Environmental Correlations

The directly observed correlations are phenotypic, as it is necessary to distinguish its two causes: genetic and environmental. Thus, considering two traits X and Y, it can be said that:

$$F_x = G_x + A_x$$
$$F_y = G_y + A_y$$

where:

F, G, and A are the phenotypic, genotypic, and environmental values, respectively, attributed to traits X and Y. It is determined that:

$$V(F_x) = V(G_x) + V(A_x)$$
$$V(F_y) = V(G_y) + V(A_y)$$
$$\mathrm{Cov}(F_x, F_y) = \mathrm{Cov}(G_x + A_x, G_y + A_y) = \mathrm{Cov}(G_x, G_y) + \mathrm{Cov}(A_x, A_y)$$

The phenotypic correlation coefficient is given by:

$$r_{\mathrm{f}} = \frac{\mathrm{Cov}(F_x, F_y)}{\sqrt{V(F_x)V(F_y)}}$$

The genotypic correlation coefficient is given by:

$$r_{\mathrm{g}} = \frac{\mathrm{Cov}(G_x, G_y)}{\sqrt{V(G_x)V(G_y)}}$$

The environmental correlation coefficient is given by:

$$r_a = \frac{Cov(A_x, A_y)}{\sqrt{V(A_x)V(A_y)}}$$

By defining:

$$h_x^2 = \frac{V(G_x)}{V(F_x)} \quad \text{e} \quad \text{e}_x^2 = \frac{V(A_x)}{V(F_x)}$$

$$h_y^2 = \frac{V(G_y)}{V(F_y)} \quad \text{e} \quad \text{e}_y^2 = \frac{V(A_y)}{V(F_y)}$$

It is determined that:

$$r_{\text{f}} = h_x h_y r_{\text{g}} + \text{e}_x \text{e}_y r_a$$

The expression above demonstrates that the causes of genetic and environmental correlations combine to determine the phenotypic correlation.

Causes of Genetic Correlations

Genetic correlations are mainly due to pleiotropism and to the genetic connections in situations of imbalance. Pleiotropism is the phenomenon whereby a gene simultaneously affects two or more traits, so that if it is segregating it will cause simultaneous variation in the characteristics involved. The resulting correlation of pleiotropism expresses the total effect of all genes in segregation. Some pleiotropic effects may lead to increased traits, while others may decrease them. In other cases, the effects may contribute to the increase of some traits, decreasing others, so pleiotropism does not necessarily cause a correlation that can be detected.

Gene linkage disequilibrium is a transient cause of correlation and may be altered in advanced generations by the breakdown of gene pools by the exchanges.

Interpretation of Correlation Coefficients

To interpret the correlations, one must consider its magnitude and its signal. The high magnitude indicates the existence of a linear relationship between the studied traits. As for the signal, negative correlation values are interpreted to indicate that the genetic, phenotypic, or environmental factor favors one trait to the detriment of the other, and positive values show that the two traits are either benefited or harmed by the same causes of variations. In general, genetic and environmental correlations have the same sign; however, in cases where this does not occur, there is an indication that the causes of genetic and environmental variation influence the characters through different physiological mechanisms (Falconer 1981). The signs of the phenotypic and genotypic correlation coefficients may eventually be different, and the fact is generally attributed to sampling errors (Cruz and Regazzi, 1997).

Correlation studies among soybean agronomic traits were performed. Lopes et al. (2002) present papers with information on correlation estimates, as shown in Table 11.5.

Table 11.5 Genotypic (r_F) correlations, genotypic (r_G) and environmental (r_E) correlations among the number of days for flowering (NDF), plant height at flowering (APF), number of days at maturity (NDM), plant height at maturity (APM), agronomic value (VA), grain yield (PG), oil percentage (% OL), and oil productivity (PO), for 15 diallelic crosses at F_2 (above the diagonal) and six parental ones (below the diagonal). Soybean, sowing on Nov/09/98, at ESALQ, Piracicaba—SP

Trait		NDF	APF	NDM	APM	VA[a]	PG	%OL	PO
NDF	r_F		0.933**	0.863**	0.423	0.581**	0.711**	−0.610*	0.645**
	r_G		0.950	0.884	0.427	0.57	0.742	0.715	0.68
	r_E		0.330	0.590	0.269	0.225	0.191	0.011	0.2
APF	r_F	0.965**		0.904**	0.533**	0.644**	0.803**	−0.500	0.747**
	r_G	0.970		0.930	0.555	0.654	0.843	−0.545	0.79
	r_E	−0.107		0.300	0.543	0.392	382	0.001	0.375
NDM	r_F	0.973**	0.911*		0.243	0.465	0.630*	−0.528*	0.563*
	r_G	0.973	0.913		0.261	0.46	0.653	−0.620	0.59
	r_E	0.170	0.064		0.341	0.229	0.184	−0.003	0.18
APM	r_F	0.420	0.517	0.322		0.887**	0.801	0.070	0.820**
	r_G	0.453	0.546	0.360		0.914	0.854	0.173	0.877
	r_E	0.100	0.619	0.152		0.548	0.500	0.047	0.498
VA[a]	r_F	0.505	0.600	0.349	0.939**		0.913**	−0.011	0.917**
	r_G	0.540	0.631	0.388	0.948		0.950	0.072	0.957
	r_E	0.090	0.135	−0.027	0.065		0.609	0.015	0.604
PG	r_F	0.569	0.643	0.423	0.851*	0.950**		−0.077	0.994**
	r_G	0.591	0.669	0.445	0.891	0.973		−0.008	0.996
	r_E	−0.051	0.221	0.008	0.250	0.160		−0.008	0.980

(continued)

Table 11.5 (continued)

Trait		NDF	APF	NDM	APM	VA[a]	PG	%OL	PO
%OL	r_F	0.251	−0.360	0.305	0.272	0.260	0.090		0.032
	r_G	−0.257	−0.362	0.310	0.320	0.288	0.010		0.083
	r_E	0.021	0.090	0.060	0.060	0.051	0.017		0.163
PO	r_F	0.550	0.613	0.403	0.860*	0.955**	0.957**	0.165	
	r_G	0.574	0.642	0.430	0.902	0.983	0.998	0.165	
	r_E	0.046	0.235	0.020	0.260	0.151	0.970	0.245	

*, ** Significant by test *F*, at 5 and 1% respectively

[a]Transformed value for $\sqrt{x+0.5}$

In general, the knowledge of the degrees of association between traits makes it possible to identify variables that can be used in the indirect selection on another variable, such as soybean grain yield, especially when the heritability of the main character is low (Iqbal et al. 2003; Costa et al. 2004). A more refined analysis of the correlation studies can also be performed by the unfolding of correlations between the variables, in direct and indirect effects, through track analysis, and, therefore, the results allow to identify the traits to be better weighted as selection criteria for productivity, as verified by Bizeti et al. (2004), Costa et al. (2004), Arshad et al. (2006), Alcantara Neto et al. (2011), and Santos et al. (2011).

Correlation studies among soybean agronomic traits were also performed by Rigon et al. (2012), which reported a correlation with grain yield of −0.20, with physiological maturation of −0.08, with lodging index of −0.19, with plant height (AP) of −0.44, with height of insertion for the first legume of 0.54, and with mass of 100 grains. In the track analysis, the determination coefficient was 64% and concluded that, except for the mass of 100 grains, the other traits had little influence on grain yield.

Response Correlated to Selection

The genetic correlation between traits means that selection for one trait causes changes in others. The evaluation of the meaning and magnitude of these changes is fundamental to obtain, at the end of a breeding program, genetic material with superior behavior for a series of characteristics.

The expected response on trait Y, when the selection is applied on trait X, can be estimated by the expression:

$$\mathrm{RY}(X) = \widehat{\beta}_{\mathrm{g}} \quad \mathrm{RX} = \widehat{\beta}_{\mathrm{g}}\widehat{\beta}\mathrm{DS}_X$$

where:

RX = direct response on trait X, given by $\widehat{\beta}\mathrm{DS}_X$.

$\widehat{\beta}_{\mathrm{g}}$ = regression coefficient that measures the variation in the genetic values of trait Y, with changes of one unit in the genetic values of trait Y.

$$\widehat{\beta}_{\mathrm{g}} = \frac{\mathrm{Cov}(X,Y)}{\sigma^2_{gx}} = r_{\mathrm{g}}\frac{\widehat{\sigma}_{gy}}{\widehat{\sigma}_{gx}}$$

$\widehat{\beta}$ = regression coefficient that measures the variation in the genetic values of the improved population, in relation to the trait X, with changes provoked by the phenotypic selection in the test units.

$$\widehat{\beta} = \frac{\mathrm{Cov_g}(\mathrm{UM}_x, \mathrm{UT}_x)}{\sigma^2_{\mathrm{f}x}}$$

UM and UT are units or individuals of the improved and test population, respectively.

In relation to the trait Y, direct gains can be estimated, by means of:

$$\mathrm{RY} = pi_y h_y \widehat{\sigma}_{\mathrm{gy}}$$

And the indirect gains, by means of:

$$\mathrm{RY(X)} = pi_x h_x \widehat{\sigma}_{\mathrm{gy}} r_{\mathrm{gxy}}$$

Therefore, the indirect selection efficiency in relation to the direct one is given by the ratio:

$$\frac{\mathrm{RY}(X)}{\mathrm{RY}} = \frac{pi_x h_x \widehat{\sigma}_{\mathrm{gy}} r_{\mathrm{g}xy}}{pi_y h_y \widehat{\sigma}_{\mathrm{gy}}}$$

Considering the same selection intensity for traits X and Y, we have:

$$\frac{\mathrm{RY}(X)}{\mathrm{RY}} = \frac{h_x r_{\mathrm{g}xy}}{h_y}$$

The conclusion is that $\mathrm{RY}(X) > \mathrm{RY}$ se $r_\mathrm{g} h_x > h_y$. Thus, the indirect selection response will be compensatory when the main character (Y) is of low heritability and, often, with measurement and identification problems and if there is an easy-to-measure auxiliary trait (X) with high heritability and high correlation with the main trait.

Estimation of Correlated Response

The correlated response is estimated by means of the expressions:

$$\mathrm{RY}(X) = i_x \cdot h_x \cdot \sigma_{\mathrm{gy}} \cdot r_{\mathrm{g}xy}$$

or by:

$$\mathrm{RY}(X) = h_y^2 \cdot \mathrm{DSY}(X)$$

where

$\mathrm{DSY}(X)$ is the indirect selection differential, obtained by the difference between the mean of the selected individuals, whose superiority was identified by the good performance in relation to the trait X and the original average of the population in relation to the trait X.

Example of estimates of gains through direct and indirect selection in soybean crop is presented by Giorgenon et al. (2014) and described in Table 11.6.

Simultaneous Trait Selection

In order to obtain truly superior genetic material, it is necessary to select those that, simultaneously, contain a series of favorable attributes that grant them a comparatively higher yield and meet the consumer's requirements. Thus, one way of increasing the chance of success of a breeding program is by simultaneously selecting an economically important trait set using the selection index that constitutes an additional trait, established by the linear combination of the traits evaluated, which allows optimizing selection gains. The classical index, proposed by Smith (1936) and Hazel (1943), is estimated in order to maximize the correlation between the index (I) and the genotypic aggregate (H).

The index I is given by:

$$I = b_{1x1} + b_{2x2} + \ldots b_{nxn}$$

And the genotypic aggregate H is given by:

$$H = a_1 g_1 + a_2 g_2 + \ldots a_n g_n$$

The correlation between I and H is given by:

$$r_{HI} = \frac{\mathrm{Cov}(I, H)}{\sqrt{V(I)\ V(H)}} = \frac{b'Ga}{\sqrt{(b'Pb)(a'Ga)}}$$

where:

G = matrix of genotypic variances and covariances.

P = matrix of phenotypic variances and covariances.

a = vector of economic weights of the traits studied, established by the breeder.

$a' = [a_1, a_2 \ldots a_n]$.

b = vector of coefficients of the estimated index, so that the correlation r_{HI} is maximum.

$b' = [b_1, b_2 \ldots b_n]$.

As previously mentioned, in the selection index theory, the statistical problem consists of obtaining the vector **b**, so that r_{HI} is maximized. The solution of this maximization is given by:

$$Pb = Ga$$

Table 11.6 Estimates of the selection gain (GS%) between families by the direct and indirect selection method, based on the agronomic characters evaluated in soybean progenies F7, in Jaboticabal-SP

Selection at	APM	AIV	NN	NR	VA	Ac	NV	NS	PS	Total
APM	11.33	6.43	6.75	−1.36	−4.81	−2.56	−5.34	−4.26	−3.49	2.69
AIV	13.94	21.03	14.32	−1.70	−7.21	−3.99	−7.92	−7.50	−7.70	13.27
NN	3.52	1.98	5.39	0.98	−0.59	−1.63	−0.90	−1.05	−0.79	6.91
NR	−8.47	−15.26	0.06	26.76	21.23	−2.31	18.07	21.86	21.71	83.65
VA	0.01	−1.16	−0.15	1.13	3.57	−0.21	2.64	3.34	3.18	12.35
Ac	0.17	0.19	0.17	−0.08	−0.09	−1.06	0.11	0.23	0.14	−0.22
NV	−5.36	−9.27	−5.52	10.53	19.84	1.36	23.57	21.9	21.52	78.57
NS	−6.73	−9.61	−5.82	13.31	20.79	−1.79	21.05	22.67	21.72	75.59
PS	−5.43	−12.59	−3.06	10.96	26.09	−3.59	22.45	25.43	26.14	86.4

Source: Giorgenon et al. (2014)
Traits: plant height at maturity (APM), height of insertion of the first pod (AIV), number of nodes (NN), number of branches (NR), agronomic value (VA), lodging (Ac), number of pods per plant (NV), number of seeds per plant (NS), and seed production (PS) in grams/plant

Thus:

$$b = P^{-1}Ga$$

When the selection index is obtained, gains are obtained by means of the expression:

$$R_i(I) = h_i^2 \, \mathrm{DS}_i(I)$$

The establishment of the economic value or the genetic merit of each trait has been one of the greatest concerns of breeders for the establishment of the selection index. Cruz (1990) reports that the values of the **a** vector elements can be established interactively, by means of interference of the breeder, with computational resources that allow to evaluate the selection groups and the selective units with the established values. It also suggests that the weights equivalent to the genetic variation coefficient of each trait are adopted as economic values.

Cultivar Recommendation

In a given setting, the phenotypic manifestation is the result of the action of a genotype under the influence of the environment. However, when considering a series of settings, in addition to genetic and environmental effects, an additional effect, provided by their interaction, is detected.

The evaluation of the genotype x environment interaction is of great importance in the breeding process, because, in its existence, it is possible that the best genotype in one environment is not the best one in the other. This fact influences the selection gain and hinders the recommendation of wide adaptability cultivars. Due to the importance of this interaction, it is up to the breeder to evaluate its magnitude and significance, quantify its effects on breeding techniques and technology diffusion strategies, and provide subsidies that allow for the adoption of procedures to minimize and/or exploit such interaction.

Studies on the genotype x environment interaction, although of great importance for the breeding process, do not provide detailed information on the behavior of each genotype in the face of environmental variations. For this purpose, adaptability and stability analyses are performed, through which it is possible to identify cultivars of predictable behavior and that are responsive to environmental variations under specific or broad conditions.

Currently, there are more than a dozen adaptability and stability analysis methodologies for the evaluation of a group of genotype materials tested in a number of environments. These methodologies are based on the existence of interactions and are distinguished from the concepts of stability adopted and certain statistical principles employed. The choice of an analysis method depends on the experimental data, especially those related to the number of environments available, on the required precision, and on the type of information desired. It should also be considered that some methods are alternative, while others are complementary and can be used together.

Silva and Duarte (2006) evaluated the statistical methods of genotype x environment interaction analysis (GxA), emphasizing adaptability and phenotypic stability. Data from soybean yield was used based on seven experiments in Goiás, testing 28 genotypes, of which four commercial cultivars. The Traditional, Plaisted and Peterson, Wricke, Finlay and Wilkinson, Eberhart and Russell, Verma, Chahal and Murty, Toler, AMMI (additive main effect and multiplicative interaction), Hühn, Annicchiarico, and Lin and Binns methods were evaluated. A strong association was observed between those of Plaisted and Peterson and Wricke, whose concomitant use was contraindicated. The same conclusion is attributed to the Annicchiarico and Lin and Binns methods, also strongly associated, which implies very similar phenotypic classifications. The use of one of them, however, is recommended. Methods based exclusively on regression coefficients should be used in association with another, based on the variance of the GxA interaction or on statistical measures such as variance of regression deviations. The combined use of the Eberhart and Russell and AMMI method is another indication, because of its significant correlations with most other methods and a relatively weak association between them.

Although there are many methodologies, the proposal by Eberhart and Russell (1966) has still been one of the most directly used ones in soybean breeding or as a reference for others involving different approaches. Thus, due to their relevance, some aspects of this methodology will be detailed, highlighting that the authors suggest to make the adaptability and stability analysis from a simple linear regression model, i.e., the following equation is adopted:

$$Y_{ij} = \beta_{0i} + \beta_{1i}I_j + \delta_{ij} + \bar{\varepsilon}_{ij}$$

where:

Y_{ij} = genotype mean i in environment j.
β_{0i} = general mean of genotype i.
β_{1i} = linear regression coefficient, which measures the response of the ith genotype to the mean variation.

I_j = codified environmental index $\left(\sum_j I_j = 0\right)$.

δ_{ij} = regression deviation.
$\bar{\varepsilon}_{ij}$ = mean experimental error.

By this methodology, both the regression coefficients of the phenotypic values of each genotype in relation to the environmental index and the deviations of this regression would provide estimates of stability and adaptability parameters. The concepts involved in this methodology, which are easier to understand, are the following:

Adaptability: it refers to the ability of genotypes to advantageously benefit from the environment stimulus. As regards adaptability, they are classified as:

a. Genotypes with general or broad adaptability: these are those with β_{1i} equal to 1.
b. Genotypes with specific adaptability to favorable environments: these are those with β_{1i} greater than 1.
c. Genotypes with specific adaptability to unfavorable environments: these are those with β_{1i} lower than 1.

Stability: it refers to the ability of the genotypes to show a highly predictable behavior, due to the environment stimulus. It is evaluated by the variance component assigned to the regression deviations $\left(\sigma_{\mathrm{di}}^2\right)$, and the following types of genotype are verified:

a. Genotypes with high stability or predictability: these are those with σ_{di}^2 equal to 0.
b. Genotypes with low stability or predictability: these are those with σ_{di}^2 greater than 0.

Eberhart and Russell (1966) consider as ideal genotype the one with high average yield, regression coefficient equal to 1.0, and regression deviations as small as possible.

Estimate of Stability and Adaptability Parameters

The β_{0i} and β_{1i} parameters are estimated by means of the following equations:

$$\widehat{\beta}_{0i} = \bar{Y}_i \text{ and } \widehat{V}\left(\widehat{\beta}_{0i}\right) = \frac{1}{a}\widehat{\sigma}_\varepsilon^2$$

$$\widehat{\beta}_{1i} = \frac{\sum_j Y_{ij} I_j}{\sum_j I_j^2} \text{ and } \widehat{V}\left(\widehat{\beta}_{1i}\right) = \frac{1}{\sum_j I_j^2}\widehat{\sigma}_\varepsilon^2$$

where:

$$I_j = \frac{1}{g}\sum_i Y_{ij} - \frac{1}{ag}\sum_i\sum_j Y_{ij}$$

$$\widehat{\sigma}_\varepsilon^2 = \frac{1}{r}\widehat{\sigma}^2 = \frac{\text{QMR}}{r}$$

The hypothesis $H_o : \beta_{1i} = 1$ versus $H_a : \beta_{1i} \neq 1$ is valid by the statistics t, given by:

$$t = \frac{\widehat{\beta}_{1i} - 1}{\sqrt{\widehat{V}\left(\widehat{\beta}_{1i}\right)}}$$

The stability parameter $\left(\sigma^2_{\text{di}}\right)$ is estimated by the variance analysis method, from the regression deviation mean square of each genotype (QMD$_i$) and the residual mean square, that is:

$$\widehat{\sigma}^2_{\text{di}} = \sum_j \widehat{\delta}^2_{ij}/(a-2) = \frac{\text{QMD}_i - \text{QMR}}{r}$$

where:

$$\text{QMD}_i = \frac{r}{a-2}\left[\sum_j Y^2_{ij} - \frac{Y^2_i}{a} - \frac{\left(\sum_j Y_{ij}I_j\right)^2}{\sum_j I^2_j}\right] \quad (\text{valid for any } i)$$

Sometimes many genotypes with a higher average yield present σ^2_{di} statistically different from zero; however, it may be necessary to select some genotypes in the group where stability (or predictability) is low. In these cases, an auxiliary measure of comparison between these genotypes is the coefficient of determination R^2_i, given by:

$$R^2_i = \frac{\text{SQ(Regressão linear)}_i}{\text{SQ}(A/G_i)} \times 100$$

An example of use of the Eberhart and Russel method (1966) in the soybean crop is presented by Marques et al. (2011) in Table 11.7. The author concluded that the cultivar UFUS Xavante stood out as being of specific adaptability for favorable environment and of high phenotypic stability.

Table 11.7 Estimates of the adaptability and stability parameters of seven soybean cultivars for grain yield, based on Eberhart and Russel (1966)

Cultivars	Mean	B1	S2di	R2(%)
UFUS Xavante	3017.12	1.47**	62.47	99.27
Msoy 8001	2569.81	0.95	−10162.68	98.69
Msoy 8411	2427.36	0.94	−37057.35	99.84
UFUS Milionária	2403.74	0.8	251001.5*	84.96
UFUS Guarani	2377.66	0.88	6573.55	97.68
UFUS Riqueza	2165.54	1.06	−39934.39	99.97
Msoy 8914	2030.1	0.88	−33174.047	99.62

* Significantly diverent from 0 by F test with 5% of probability
** Significantly diverent from 1 by t test with 1% of probability

Final Considerations

In order to obtain superior genetic material, the varieties need to feature, simultaneously, a number of favorable attributes that allow it to have comparatively higher yield and that meet consumer's requirements. Thus, one way to increase the chance of success of a soybean breeding program is to perform reliable experiments, from which a large volume of experimental data is obtained. The genetic parameters are estimated and the biological phenomena are interpreted based on the proper processing of these data. The knowledge on the different biometric techniques available for use by soybean breeders is essential in all stages of experimentation, involving the implantation and conduction of tests and, later, the analysis and interpretation of results.

In this context, this chapter sought to present concepts and examples on various methodologies that can be used in important themes that influence soybean breeding in its different stages. Biometry, based on the Mendelian principles, quantitative and population genetics, has been a science whose knowledge is indispensable for it allows the analysis and processing of data, as well as the interpretation of parameters and biological phenomena generally influenced by the environment.

References

Alcantara Neto, F., Amaral Gravina, G., de Sousa Monteiro, M.M., Morais, F.B., Petter, F.A., Albuquerque, J.A.A.: Análise de trilha do rendimento de grãos de soja na microrregião do Alto Médio Gurguéia. Comunicata Scientiae. **2**(2), 107–112 (2011)

Arshad, M., Ali, N., Ghafoor, A.: Character correlation and path coefficient in soybean *Glycine max* (L.) Merrill. Pak. J. Bot. **38**(1), 121 (2006)

Bárbaro, I.M., da Cruz Centurion, M.A.P., di Mauro, A.O., Unêda-Trevisoli, S.H., Costa, M.M.: Comparação de estratégias de seleção no melhoramento de populações f 5 de soja. Ceres. **54** (313), (2007)

Bizeti, H.S., et al.: Path analysis under multicollinearity in soybean. Braz. Arch. Biol. Technol. **47** (5), 669–676 (2004)

Comstock, R.E., Robinson, H.F.: The components of genetic variance in populations of biparental progenies and their use in estimating the average degree of dominance. Biometrics. **4**(4), 254–266 (1948)

Costa, M.M., di Mauro, A.O., Unêda-Trevisoli, S.H., Arriel, N.H.C., Bárbaro, I.M., Muniz, F.R.S.: Ganho genético por diferentes critérios de seleção em populações segregantes de soja. Pesq. Agrop. Brasileira. **39**(11), 1095–1102 (2004)

Cruz, C.D.: Aplicação de algumas técnicas multivariadas no melhoramento de plantas. In: de Tese, D.S. (ed.) Escola Superior de Agricultura "Luiz de Queiroz". Universidade de São Paulo, Piraicaba (1990)

Cruz, C.D., Regazzi, A.J.: Modelos biométricos aplicados ao melhoramento genético, 668 p. Ed. UFV, Viçosa (1997)

Cruz, C.D., Regazzi, O.J., Carneiro, P.C.S.: Modelos biométricos aplicados ao melhoramento genético, 480p. Imprensa Universitária, Viçosa (2004)

Cruz, M.F.A., Souza, G.A., Rodrigues, F.A., Sediyama, C.S., Barros, E.G.: Notas Científicas Reação de genótipos de soja à infecção natural por ferrugem asiática. Pesq. Agrop. Brasileira. **46**(2), 215–218 (2011)

Daronch, D.J., Peluzio, J.M., Afferri, F.S.: Capacidade combinatória de cultivares de soja em f2, sob condições de cerrado tocantinense. Combining ability of soybean cultivars in f2, under conditions of cerrado tocantinense. Biosci. J. **30**(5), (2014)

Eberhart, S.A.: Factors affecting efficiencies of breeding methods. African Soils Paris. **15**, 669–680 (1970)

Eberhart, S.A., Russell, W.A.: Stability parameters for comparing varieties. Crop Sci. **6**(1), 36–40 (1966)

Edwards, A.W.F., Cavalli-Sforza, L.L.A.: Method for cluster analysis. Biometrics. 362–375 (1965)

Falconer, D.S.: Introduction to Quantitative Genetics, 2nd edn, 340p. Longhan, London (1981)

Gardner, C.O., Eberhart, S.A.: Analysis and interpretation of the variety cross diallel and related populations. Biometrics. 439–452 (1966)

Giorgenon, C.H.B., Buzinaro, R., di Mauro, A.O.: Parâmetros genéticos em linhagens f7 de soja com fonte de resistência a ferrugem asiática. Ciência Tecnologia. **6**(esp.), (2014)

Griffing, B.: Concept of general and specific combining ability in relation to diallel crossing systems. Aust. J. Biol. Sci. **9**(4), 463–493 (1956)

Hazel, L.N.: The genetic basis for constructing selection indexes. Genetics. **28**(6), 476–490 (1943)

Iqbal, S., Mahmood, T., Tahira, M., Ali, M., Anwar, M., Sarwar, M.: Path coefficient analysis in different genotypes of soybean (Glycine max (L) Merril). Pak. J. Biol. Sci. **6**(12), 1085–1087 (2003)

Lopes, A.C.A., Vello, N.A., Pandini, F., Rocha, M.M., Yujitsutumi, C.: Variabilidade e correlações entre caracteres em cruzamentos de soja. Sci. Agric. **59**(2), 341–348 (2002)

Marques, M.C., Hamawaki, O.T., Sediyama, T., Bueno, M.R., Reis, M.S., Cruz, C.D., Nogueira, A.P.O.: Adaptability and stability of soybean genotypes under different times of sowing. Biosci. J. **27**(1), 59–69 (2011)

Martins Filho, S., Sediyama, C.S., Cruz, C.D., Sediyama, T., Gomes, J.L.L.: Análise dialélica da resistência da soja (*Glycine max* (L.) Merrill) à Cercospora sojina Hara. Ceres. **39**(221), 51–59 (1992)

Peluzio, J.M., Vaz-De-Melo, A., Afférri, F.S., Silva, R.R., Barros, H.B., Nascimento, I.R., Fidelis, R.R.: Variabilidade genética entre cultivares de soja, sob diferentes condições edafoclimáticas. Pesquisa Aplicada Agrotecnologia. **2**, 21–29 (2009)

Rao, C.R.: Advanced Statistical Methods in Biometric Research, 390p. Wiley, New York (1952)

Rigon, J.P.G., et al.: Dissimilaridade genética e análise de trilha de cultivares de soja avaliada por meio de descritores quantitativos. Revista Ceres. **59**(2), 233–240 (2012)

Santos, E.R., Barros, H.B., Ferraz, E.C., Capone, A., Santos, A.F., Fidelis, R.R.: Divergência entre genótipos de soja, cultivados em várzea irrigada. Revista Ceres. **58**(6), 755–764 (2011)

Shirahige, F.H., Fumes, L.A.A., de Castro, L.P., Farias, G.J., Colombari Filho, J.M., Geraldi, I.O.: Avaliação de Cruzamentos Dialélicos para Produção de Grãos e Caracteres Agronômicos, e Perspectivas de Melhoramento em Soja. In: Embrapa Arroz e Feijão-Artigo em anais de congresso (ALICE). CONGRESSO BRASILEIRO DE MELHORAMENTO DE PLANTAS, vol. 7, 2013, Uberlândia. Variedade melhorada: a força da nossa agricultura: anais. SBMP, Viçosa, MG (2013)

Silva, W.C.J., Duarte, J.B.: Statistical methods to study phenotypic adaptability and stability in soybean. Pesq. Agrop. Brasileira. **41**(1), 23–30 (2006)

Singh, D.: The relative importance of characters affecting genetic divergence. Indian J Genet Plant Breeding. **41**(2), 237–245 (1981)

Smith, H.F.: A discriminant function for plant selection. Ann. Eugenics. **7**(3), 240–250 (1936)

Torres, F.E., David, G.V., Teodoro, P.E., Ribeiro, L.P., Correa, C.G., Júnior, R.A.L.: Desempenho agronómico e dissimilaridade genética entre genótipos de soja. Revista de Ciências Agrárias. **38**(1), 111–117 (2015)

Vieira, E.A., Carvalho, F.I.F., Bertran, I., Kopp, M.M., Zimmer, P.D., Benin, G., Silva, J.A., Hartwig, I., Malone, G., Oliveira, A.C.: Association between genetic distance in wheat (Triticum aestivum L.) as estimated by AFLP and morphological markers. Genet. Mol. Biol. **30**, 392–399 (2007)

Ward Jr., J.H.: Hierarchical grouping to optimize an objective function. J. Am. Stat. Assoc. **58** (301), 236–244 (1963)

Chapter 12
BLUP in the Genetic Evaluation of Parents, Generations, Populations, and Progenies

Marcos Deon Vilela de Resende, Magno Antonio Patto Ramalho, José Aírton Rodrigues Nunes, Felipe Lopes da Silva, and Pedro Crescêncio Souza Carneiro

Abstract Autogamous plant breeders obtain numerous populations annually, and in the progeny selection process, the merit of the parents can also be considered. Thus, it would be important to have a selection index (BLUP) for progenies that includes not only the effects of progenies in the different generations but also the effects of populations in all the generations and the data from parents and the F_1 and F_2 generations simultaneously. The main objective of this chapter is to present a selection method that encompasses the entire structure of an autogamous plant breeding program, including all the data as of the parents, the F_1 and F_2 generations, and also both the progeny and population effects in the F_3 up to F_6 generations. Estimators are presented for computation of the contribution (through indirect heritabilities and selective accuracies) of each source of information for prediction of the additive genetic value of the lines to be obtained at the end of the selection process. The role of BLUP aiming at improving selection methods in soybean breeding is detailed.

Keywords Accuracy • Precision • BLUP • Index selection

M.D.V. de Resende, M.S., D.S. (✉)
Brazilian Agricultural Research Corporation/Federal University of Viçosa, Viçosa, Minas Gerais, Brazil
e-mail: marcos.deon@gmail.com

M.A.P. Ramalho, M.S., D.S. • J.A.R. Nunes, M.S., D.S.
Federal University of Lavras, Lavras, Minas Gerais, Brazil
e-mail: mapr@ufla.br

F.L. da Silva, M.S., D.S. • P.C.S. Carneiro, M.S., D.S.
Federal University of Viçosa, Viçosa, Minas Gerais 36570-900, Brazil
e-mail: felipe.silva@ufv.br; carneiro@ufv.br

F. Lopes da Silva et al. (eds.), *Soybean Breeding*,
DOI 10.1007/978-3-319-57433-2_12

Introduction

The soybean seed market in Brazil is very competitive; several national and international companies are on the market and make heavy investments to obtain new cultivars. These new cultivars will only succeed if they are better, in one or more attributes, than those of competing companies.

In order to overcome competition, the team of breeders of companies obtains hundreds of populations annually; conducts these populations, initially in bulk; gets thousands of progenies every year; and evaluated these progenies for 2 or 3 years. The best lines identified go to VCU (value for cultivation and use) experiments, which are conducted in as many environments as possible.

This effort of the breeding programs, so far, has been compensated by the increase in grain yield in relation to those obtained in the past. However, the differences to be detected among lines in the future will become increasingly smaller. Under such condition, the rigor in the phenotyping of progenies/lines has to be increasingly higher. Moreover, all tools that can increase the efficiency of the selection process should be used properly. It is in this context that the data analysis can contribute a lot. The use of BLUP (best linear unbiased prediction) in the different stages of the breeding program will certainly provide extra gain at no additional cost to the program.

In this chapter, we proposed to show the contributions of the data analysis using BLUP, associated with multiple selection indices in soybean breeding as a differential to increase the success of breeding. Soybean breeding is mainly supported by the following activities: evaluation and selection of parents, crosses, populations, and progenies, the latter in several generations of inbreeding. These steps are performed sequentially, and, at the end of the breeding cycle, we have a set of information that allows the accurate selection of new strains and cultivars. This chapter also addresses optimal selection procedures, via BLUP and multi-effect selection index, which simultaneously use all of this information, in order to maximize the efficiency of the selection process.

The genetic evaluation includes the estimation of variance components (genetic parameters) and the prediction of genetic values. Estimates of variance components and genetic parameters, such as heritability and genetic correlations, are essential in at least three applications: (1) knowledge of the genetic control of the traits aiming at designing efficient breeding strategies, (2) prediction of the genetic values of the candidates for the selection, and (3) determination of sample size (e.g., number of replicates) and sampling form suitable for the accurate estimation of parameters and maximization of selective accuracy.

Field experimentation, as a rule, is associated with data imbalance due to several reasons. Therefore, the optimal genotypic assessment procedure refers to the procedure of residual maximum likelihood/best linear unbiased prediction (REML/BLUP or mixed model methodology) that naturally deals with the imbalance, leading to more accurate estimations and predictions. Another advantage of the mixed model methodology is the possibility of combining multiple sources of information in a single prediction of genetic values.

The main practical advantages of REML/BLUP are as follows: they allow comparing individuals or varieties through time (generations, years) and space (locations, blocks); they allow simultaneous correction for environmental effects, estimation of variance components, and prediction of genetic values; they allow to deal with complex data structures (repeated measurements, different years, locations, and designs); they can be applied to unbalanced data and non-orthogonal designs; they allow to use simultaneously a large number of information, coming from different generations, locations, and ages, generating more precise estimates and predictions; they allow the adjustment of several alternative models, being able to choose the one that best fits the data; and, at the same time, they are parsimonious (presents a smaller number of parameters).

In autogamous plants such as soybean, countless populations derived from the cross of several pairs of lines are usually evaluated, in that several simple F_1 hybrids and their corresponding F_2 generations are generated. Plants of each F_2 population are taken for seed harvesting, and endogamous generations are advanced in order to obtain lines, mainly by means of the bulk within progenies (BDP) method (Ramalho et al. 2012, 2013).

In the selection methods, data from individual generations are used (sometimes with the aid of genealogy; *genealogic BLUP*, according to Nunes et al. (2008a, b) and Oakey et al. (2006)), and, recently, two methods have been proposed that simultaneously use data from all generations (Resende et al. 2015a, b). These methods are *BDP-BLUP* (BLUP associated with the bulk within progenies (BDP) scheme using only the progeny effects) and the *BLUP-SIPPPG* (BLUP associated with the within progenies scheme and the selection index using the effects of parents, populations, progenies, and generations—SIPPPG). In this text, these BLUP methods and their associated multi-effects indices are discussed, as well as the estimation of the necessary components of the genetic variance from the evaluation of several populations in different generations.

The BLUP/multi-effect index procedures applicable to each situation are presented sequentially following the steps of a soybean breeding program, as shown in Table 12.1.

F_0: Selection of Parents Per Se

This stage coincides with the final evaluation of lines from the previous cycle and will be discussed in the topic referring to F_6.

Table 12.1 Activities of a soybean breeding cycle

Generation	Activity	Experimental material	Design	Number of environments
F_0	Selection of parents per se	Lines in F_6 or more, of the previous cycle	Lattice	Several
F_1	Selection of F_1 populations	Hybrids of lines obtained in diallel crosses	Lattice	One
F_2	Selection of F_2 populations	F_2 populations	Lattice	One or several
F_3	Selection of individuals in progenies/F_3 populations	Individuals in progenies/F_3 populations	Lattice	One or several
F_4	Selection of progenies/F_4 populations	Progenies/F4 populations	Lattice	Several
F_5	Selection of progenies/F_5 populations	Progenies/F_5 populations	Lattice	Several
F_6	Selection of progenies/F_6 populations	Progenies/F_6 populations	Lattice	Several

F_1: Selection of Crosses or Populations

In this step, hybrids of lines obtained under diallel crosses are generally evaluated. In the case of soybeans, there is some limitation caused by the difficulty of obtaining the seeds necessary for the tests, due to the small number of seeds obtained by ways of cross. The purpose is to select populations with high mean and great genetic variability. One option is to select base on the highest value of $\mathbf{u} + \mathbf{a} + \mathbf{d}$, where $\mathbf{u}$ is the overall mean of the trait under selection, $\mathbf{a}$ is related to the additive effects of the alleles, and $\mathbf{d}$ is related to the dominance effects contributed by the loci in heterozygosis. This selection guarantees high mean, and, through higher d, it also contemplates greater genetic variability in F_∞, as a result of the greater number of loci in heterozygosity (Ramalho et al. 2012).

Since it is not possible to perform all possible crosses among the parents selected, the BLUP method for predicting unobserved hybrids can be used to obtain $\mathbf{u} + \mathbf{a} + \mathbf{d}$. This method is described below.

The procedure is based on the use of means per test, of some hybrids evaluated in different places and years. BLUP corrects the data for the fixed effects of trials (locations and years), and, later, it uses the corrected data for the prediction of untested hybrids through the use of information (parentage) associated with the genealogy of the genetic lines and the genetic and residual variances obtained by the restricted maximum likelihood (REML) procedure.

In this case, we have the following linear model (Resende 2002):

$$\begin{aligned} y &= X\beta + Zh + e \\ &= X\beta + Za_{12} + Zd_{12} \\ &= X\beta + Z_1 g_1 + Z_2 g_2 + Zd_{12} + e, \end{aligned}$$

where:
y = vector of the means per test, of line hybrids of heterotic groups 1 and 2.
β = vector of fixed test effects (locations and years).
h = vector of effects of hybrids,
where:

$$Zh = Za_{12} + Zd_{12} = Z_1 g_1 + Z_2 g_2 + Zd_{12}.$$

g_1 = vector of the genetic effects of the general combining ability (half of the interpopulational additive effects a) of each line of the heterotic group 1 in crossbreeding with lines of heterotic group 2.

g_2 = vector of the genetic effects of the general combining ability (half of the interpopulational additive effects a) of each line of the heterotic group 2 in crossbreeding with lines of heterotic group 1.

d_{12} = vector of genetic effects of specific combining ability between lines of heterotic groups 1 and 2.

e = vector of residual effects.

X, Z_1, Z_2, and Z = incidence matrices (formed by 1 and 0) for β, g_1, g_2, and d_{12}, respectively.

In this case, the BLUP predictors can be expressed in terms of Henderson's mixed model equations:

$$\begin{bmatrix} \hat{\beta} \\ \hat{g}_1 \\ \hat{g}_2 \\ d_{12} \end{bmatrix} = \begin{bmatrix} X'X & X'Z_1 & X'Z_2 & X'Z \\ Z_1'X & Z_1'Z_1 + A_1^{-1}\lambda_1 & Z_1'Z_2 & Z_1'Z \\ Z_2'X & Z_2'Z_1 & Z_2'Z_2 + A_2^{-1}\lambda_2 & Z_2'Z \\ Z'X & Z'Z_1 & Z'Z_2 & Z'Z + D^{-1}\lambda_3 \end{bmatrix}^{-1} \begin{bmatrix} X'y \\ Z_1'y \\ Z_2'y \\ Z'y \end{bmatrix}$$

where:

A_1 = parentage matrix (of Malecot) genetic additive among the lines of group 1 ($f_{AA'}$).

A_2 = parentage matrix (of Malecot) genetic additive among the lines of group 2 ($f_{BB'}$).

D = parentage matrix (of Malecot) genetic of dominance between the lines of groups 1 and 2 ($f_{AA'}, f_{BB'}$).

$$\lambda_1 = \frac{\sigma_e^2}{\sigma_{g_1}^2}; \quad \lambda_2 = \frac{\sigma_e^2}{\sigma_{g_2}^2}; \quad \lambda_3 = \frac{\sigma_e^2}{\sigma_{d_{12}}^2};$$

σ_e^2 = residual variance.

$\sigma_{g_1}^2$ = variance of the general interpopulation combination capacity that arises from the cross of the lines of group 1 with those of group 2.

$\sigma_{g_2}^2$ = variance of the general interpopulation combination capacity that arises from the cross of the lines of group 2 with those of group 1.

$\sigma_{d_{12}}^2$ = variance of specific interpopulation combination capacity.

These variance components can be estimated by the method iterative of the restricted maximum likelihood (REML), through the EM algorithm by means of the following estimators:

$$\widehat{\sigma}_{e*}^2 = \left[y'y - \widehat{\beta}'X'y - \widehat{g}_1'Z_1'y - \widehat{g}_2'Z_2'y - \widehat{d}_{12}'Z'y\right]/[N - r(X)]$$

$$\widehat{\sigma}_{g_1}^2 = \left[\widehat{g}_1'A_1^{-1}\widehat{g}_1 + \widehat{\sigma}_{e*}^2\mathrm{tr}\left(A_1^{-1}C_{11}\right)\right]/n_1$$

$$\widehat{\sigma}_{g_2}^2 = \left[\widehat{g}_2'A_2^{-1}\widehat{g}_2 + \widehat{\sigma}_{e*}^2\mathrm{tr}\left(A_2^{-1}C_{22}\right)\right]/n_2$$

$$\widehat{\sigma}_{d_{12}}^2 = \left[\widehat{d}_{12}'D^{-1}\widehat{d}_{12} + \widehat{\sigma}_{e*}^2\mathrm{tr}\left(D^{-1}C_{33}\right)\right]/n,$$

where:

tr = matrix trace operator.

$r(X) = X$ matrix rank.

C^{11}, C^{22}, and C^{33} = matrix inversion partition of the coefficients for the effects g_i, g_j, and d_{ij}, respectively.

The genetic covariance structure is presented, which derives from the fact that the genetic covariance between the hybrids $A \times B$ and $A' \times B'$, where A and A' are lines of group 1 and B and B' lines of group 2, is given by Bernardo (2014):

$$\mathrm{COV}\left[(A \times B), \left(A' \times B'\right)\right] = f_{AA'}\widehat{\sigma}_{g_1}^2 + f_{BB'}\widehat{\sigma}_{g_2}^2 + f_{AA'}f_{BB'}\sigma_{d_{12}}^2$$

The predictors presented provide the genotypic values predicted for the hybrids evaluated, by the expression $\widehat{h}_{ij} = \widehat{g}_i + \widehat{g}_j + \widehat{d}_{ij}$. They also provide the REML estimates of the additive and dominance variance components and BLUE estimates for the effects of test $\left(\widehat{\beta}\right)$. With these results, the genotypic values of untested hybrids are given by $h_{\mathrm{NT}} = C'V^{-1}yc$,

where:

$yc = (Z'Z)^{-1}Z'\left(y - X\widehat{\beta}\right)$ = vector of corrected productions (for the purposes of test) of the predictive hybrids.

C = genetic covariance matrix between the untested hybrids and the predicted hybrids (tested).

V = phenotypic variance-covariance matrix among predicted or tested hybrids.

The elements of C and the non-diagonal elements of V are given by the expression of COV[($A \times B$), ($A' \times B'$)], thus reflecting only genetic components. The diagonal elements of V are phenotypic components given by

$\sigma^2_{g_1} + \widehat{\sigma}^2_{g_2} + \sigma^2_{d_{12}} + \sigma^2_e/\ell_{ij}$, where ℓ_{ij} refers to the number of trials in which the hybrid was evaluated d_{ij}. Thus, the genotypic values of the untested hybrids are predicted from the corrected phenotypic values of the hybrids evaluated. A complete, step-by-step numerical example is presented by Resende (2002).

With the advent of genomics, parentage matrices A_1, A_2, and D should preferably be constructed using molecular marker information (Resende et al. 2014).

F_2: BLUP in the Selection of F_2 Populations

In F_2, the genomic BLUP method (G-BLUP, as discussed in Chap. 6), based on phenotypes and molecular markers, may contribute to increase selection efficiency for quantitative and oligogenic traits. Genomic selection makes it possible to identify some alleles in the F_2 generation and direct crosses among the best plants, making recurrent selection at an earlier stage. Moreover, there is an increase in the selection differential, inasmuch as it is possible to evaluate a much larger number of F_2 plants than families $F_{2:3}$.

F_3: BLUP in the Selection of Individuals in Progenies/F_3 Populations

In autogamous species, it is common to adopt recurrent selection, via some form of early selection in the F_3 generation, by exploring the great genetic variability among and within F_3 lines. Such variability contemplates 1.5 times the additive genetic variance $\left(\sigma^2_{\mathrm{a}}\right)$, 0.5 σ^2_{a} within the line and 1.0 σ^2_{a} among lines. Thus, such generation is suitable for selection, for 75% (1.5) of the total additive variance ($2\,\sigma^2_{\mathrm{a}}$) that will be available in F_∞ is already available in F_3. Thus, the selection in F_3 by means of a precise method like BLUP is relevant.

The BLUP for the selection of individual plants in F_3 lines can be derived, considering the randomized block design with several plants per plot. By using the multi-effect index, provided by Resende and Higa (1994), the optimal index or BLUP for the balanced case and considering blocks as fixed effects, for the variable Y, is given by

$$\begin{aligned} I &= b_1\delta_{ijk} + b_2 g_i + b_3 c_{ij} \\ &= b_1\left(Y_{ijk} - \bar{Y}_{ij}\right) + b_2(\bar{Y}_i - \bar{Y}\ldots) + b_3\left(\bar{Y}_{ij} - \bar{Y}_i - \bar{Y}_j + \bar{Y}\ldots\right) \end{aligned}$$

The matrix version of this index, using mixed model equations, provides the generalized BLUP for balanced and unbalanced cases. The coefficients of the index, when it comes to F_3 lines derived from the cross between two parents to generate F_1, are.

$b_1 = \frac{(1/2)\sigma_a^2}{\sigma_\delta^2}$: heritability of effect $\left(\delta_{ijk} = \left(Y_{ijk} - \bar{Y}_{ij}\right)\right)$ of individual within plot.

$b_2 = \frac{\frac{(2nb+1)}{2nb}}{\sigma_f^2 + \sigma_p^2/b + \sigma_\delta^2/nb}\sigma_a^2$: heritability of the family effect $(g_i = (\bar{Y}_i - \bar{Y}\ldots))$.

$b_3 = \frac{[(1/2)/n]\sigma_a^2}{\sigma_c^2 + \sigma_\delta^2/n}$: heritability of plot effect $\left(c_{ij} = \left(\bar{Y}_{ij} - \bar{Y}_i - \bar{Y}_j + \bar{Y}\ldots\right)\right)$.

The variance components $\sigma_f^2, \sigma_p^2, \sigma_\delta^2, \sigma_{pop}^2$ are variance among families, among plots, within plots, and among populations, respectively. The quantities n, b, and p refer to the numbers of individuals per plot, number of blocks, and number of families, respectively.

The estimation of σ_a^2, using only F_3 generation data, admits 0.25 σ_d^2 tending to zero (absence of allelic dominance) in the variation among progenies. However, even regardless of this assumption, this small fraction of the dominance variance $\left(\sigma_d^2\right)$ should not affect the ranking by BLUP, since such variance will be included (0.125 σ_d^2) also in the numerator of the weight (b_1) given to the component within line, 0.50 of the genetic variance between lines in the b_1 numerator.

In autogamous plants, they are generally simultaneously evaluated for lines that belong to several segregating (pop) populations. In this case, the BLUP used in the selection of additive genetic effects is associated with the following multi-effect index:

$$\begin{aligned} I_2 &= b_1\delta_{ijkl} + b_2 g_i + b_3 c_{ijl} + b_4 \text{pop}_l \\ &= b_1\left(Y_{ijkl} - \bar{Y}_{ij}\right) + b_2(\bar{Y}_i - \bar{Y}\ldots) + b_3\left(\bar{Y}_{ij} - \bar{Y}_i - \bar{Y}_j + \bar{Y}\ldots\right) + b_4(\bar{Y}_l - \bar{Y}\ldots), \end{aligned}$$

where:

$b_4 = \frac{\sigma_{pop}^2}{\sigma_{pop}^2 + \sigma_p^2/p + \sigma_c^2/bp + \sigma_\delta^2/nbp}$ is the heritability of the effect $(\text{pop}_l = (\bar{Y}_l - \bar{Y}\ldots))$ of the population mean.

In situations where the F_3 lineages are sown in line, without replication, the optimum index within a population includes only the effects of individual and family and is equivalent to $I_3 = b_5\left(Y_{ij} - \bar{Y}_i\right) + b_6(\bar{Y}_i - \bar{Y})$, where $b_5 = b_1$ and $b_6 = \frac{\left[2n+1\right)/(2n)\right]\sigma_a^2}{\sigma_p^2 + \sigma_\delta^2/n}$. It is important to report that in this case, the selection is not purely genetic, for the experiment did not have a replication, a fact that impairs the randomization.

F_4: BLUP in the Selection of Progenies from Several Populations in F_4

In some soybean breeding programs, the bulk within progenies (BDP) method is used and progenies are open in F_4. In this case there is no information of individual plants, but there is information of populations. Since some populations are better than others, their progenies should receive more weight in selection.

The optimum index (BLUP), combining information from populations and progenies by means of the bulk within progenies method (*BDP-BLUP method within generation*), for the selection within a generation, is

$$I = b_1(\bar{Y}_i - \bar{Y}) + b_2(Y_j - \bar{Y}_i) = \frac{\rho_p \sigma^2_{aT}}{\rho_p \sigma^2_{aT} + \rho_{gd}\sigma^2_{aT}/f + \sigma^2_e/(fb)}(\bar{Y}_i - \bar{Y})$$
$$+ \frac{\rho_{gd}\sigma^2_{aT}}{\rho_{gd}\sigma^2_{aT} + (\sigma^2_e/b) - \rho_p\sigma^2_{aT}}(Y_j - \bar{Y}_i)$$
$$= \frac{fb\rho_p h^2_{aT}}{(1 + F^* h^2_{aT}) + (b-1)\rho_{gd}h^2_{aT} + (fb-1)\rho_p h^2_{aT}}(\bar{Y}_i - \bar{Y})$$
$$+ \frac{b\rho_{gd}h^2_{aT}}{1 + F^* h^2_{aT} + (b-1)\rho_{gd}h^2_{aT} - \rho_p h^2_{aT}}(Y_j - \bar{Y}_i)$$

where σ^2_e is the residual variance (within populations) and ρ_{gd} is the intraclass genetic correlation among individuals of progenies within population in the generation g. The quantities Y_j, $\bar{Y}_i$, and $\bar{Y}$ are values of progeny means, population means, and general mean, respectively, b is the number of replicates, and f is the number of families per population. There are also the definitions:

h^2_{aT}: total heritability among and within populations.

$\rho_p = \frac{\sigma^2_{pop}}{\sigma^2_{aT}}$: proportion of total genetic variation ($\sigma^2_{aT} = \sigma^2_{pop} + \sigma^2_a$) due to the variation among populations (σ^2_{pop}), in which σ^2_a is the mean additive genetic variation within population.

$F^* = \frac{F_{IT}(1+F_{ST}) - 2F_{ST})}{(1+F_{ST})}$: F of an individual in generation g, corrected for F among populations; $\rho_{gd} = \frac{(1+F^*_{ITt})}{(1+F_{ST})}$; $F^*_{ITt} = (1 - F_{ST})F_{ISt}$; $F_{ISg} = 0.00; 0.50; 0.75; 0.875; 0.9375$, to F_2, F_3, F_4, F_5, and F_6, respectively. It is also defined the coefficient of inbreeding due to structuring in populations: $F_{ST} = \frac{\sigma^2_p}{\sigma^2_p + 2\sigma^2_a}$.

The quantity $(1 + F^*_{ITt})$ is the coefficient of the additive genetic variance among progenies and is based on $C_{tgg'}$ according to Cockerham's notation (1963), where the indices g, t, and g' refer to the generations: current, previous, and any other generation after g, respectively.

F5 and F6: Selection of Families Via BLUP Multigenerations (BDP-BLUP)

In this method, developed by Resende et al. (2015a), the complete genealogy associated with population management was used via bulk within progenies (BDP). The phenotypic means of the progenies in generations $F_{2:3}$, $F_{2:4}$, $F_{2:5}$, and

$F_{2:6}$ were used through the multigeneration index, represented by $\widehat{a} = b_3F_3 + \cdots + b_6F_6$. The weights of each generation are calculated as follows:

$$\begin{bmatrix} (K_3/h^2) & 1.00 & 1.00 & 1.00 \\ & (K_4/h^2) & 1.50 & 1.50 \\ Sim. & & (K_5/h^2) & 1.75 \\ & & & (K_6/h^2) \end{bmatrix} \begin{bmatrix} \widehat{b}_3 \\ \widehat{b}_4 \\ \widehat{b}_5 \\ \widehat{b}_6 \end{bmatrix} \begin{bmatrix} 1.00 \\ 1.50 \\ 1.75 \\ 1.875 \end{bmatrix}$$

The K_i factor for annual autogamous plants is calculated as follows:

$$K_i = \frac{(1 + F_ih^2) + (n-1)rh^2}{n}$$

The expressions also depend on the inbreeding coefficient F of the individuals of each generation and r, the intraclass additive genetic correlation among individuals of each progeny, which is equivalent to 1 for F_3 progenies, 1.5 for F_4, 1.75 for F_5, and 1.875 For F_6.

By using adjusted means (deregressing phenotypes) and weighting them by the coefficients of the index, the BLUPs are obtained from the genetic values, as shown by Resende et al. (2015a). The BLUP multigeneration of the progeny effects can be obtained from the following multivariate model (generations taken as different variables) for experiments designed in incomplete blocks: $y = X\beta + Z_1g + Z_2b + e$, where y is the data vector, β is the vector of fixed effects (general mean, experiments), g is the vector of progeny effects (random), b is the vector of incomplete block effects (random), and e is the vector of residual effects (random). X, Z_1, and Z_2 are incidence matrices for β, g, and b, respectively. If the experiments are replicated in several environments, the effects of the genotype x environment interaction can be incorporated into the model. The Selegen-Reml/Blup software computes the BDP-BLUP via model 180.

The results in Table 12.2 show the superiority of the index, the accuracy of which increases with the accumulation of generations in the index for the selection. With n equal to five replicates, the selection with the index up to F5 is adequate, providing 85% accuracy.

Table 12.3 shows the genetic correlation between each generation and the genetic values of the lines in F_∞ ($h^2 = 0.20$).

It is determined that, with the increase (by means of the increase in the number of replications) of the selection heritability toward 1 ($h^2 = 1$ in Table 12.3), the accuracy (genetic correlation) of the selection aiming at generation F_∞ also increases toward 1. In general, the results indicate that, with adequate experimentation, the selection can be performed already in F_4. This information is relevant to the recurrent selection programs in autogamous species.

Table 12.2 Accuracy of the total BLUP of accumulated generations (in the bulk within families method), for a trait with $h^2 = 0.20$ and various numbers n of replications per family

$h^2 = 0.20$						
Generation	$n = 1$	$n = 5$	$n = 10$	$n = 20$	$n = 40$	$n = 80$
F_2	0.32	–	–	–	–	–
F_3	0.51	0.58	0.62	0.66	0.68	0.69
F_4	0.65	0.76	0.80	0.83	0.84	0.85
F_5	0.75	0.85	0.88	0.90	0.92	0.93
F_6	0.81	0.90	0.93	0.94	0.95	0.96

Table 12.3 Genetic correlation between each generation and the genetic values of the lines in F_∞ (trait with $h^2 = 0.20$ or 1.00 and n replicates)

Generation	$h^2 = 0.20$; $n = 5$ (among families)	$h^2 = 1$ (among families)
F_3	0.51	0.71
F_4	0.69	0.87
F_5	0.77	0.94
F_6	0.81	0.97

F_5 and F_6: Selection of Progenies Via Total BLUP Multigenerations with Parents, Populations, and Progenies (SIPPPG)

The BDP-BLUP method was optimized for application within the population and could also be applied to the mixture of progeny of several populations. The selection method called SIPPPG has been developed to encompass the entire structure of a breeding program of autogamous plants, including all data, from parents and F_1 and F_2 generations to the effects of progenies and of inbreeding generations and populations from F_3 (Resende et al. 2015b). Traditionally, parents and F_1 and F_2 generations are analyzed separately from inbred generations from F_3.

The total index is then obtained as follows:

$$\begin{aligned}\widehat{a} = {} & \widehat{b}_{pai1}P_1 + \widehat{b}_{pai2}P_2 + \widehat{b}_{F1}F_1 + \widehat{b}_{F2}F_2 + \\ & + \widehat{b}_{F3pop}F_{3pop} + \widehat{b}_{F4pop}F_{4pop} + \widehat{b}_{F5pop}F_{5pop} + \widehat{b}_{F6pop}F_{6pop} + \\ & + \widehat{b}_{F3prog}F_{3prog} + \widehat{b}_{F4prog}F_{4prog} + \widehat{b}_{F5prog}F_{5prog} + \widehat{b}_{F6prog}F_{6prog},\end{aligned}$$

which estimates the additive genetic value of a line in F_∞ generation, considering all phenotypic information corrected from the parental generations (P_1 and P_2) to the F_6 generation.

The overall structure of the breeding program of an autogamous species, showing the distribution of total genetic variability to be harnessed by the total index among and within populations and multigenerations (SIPPPG), is shown in Fig. 12.1.

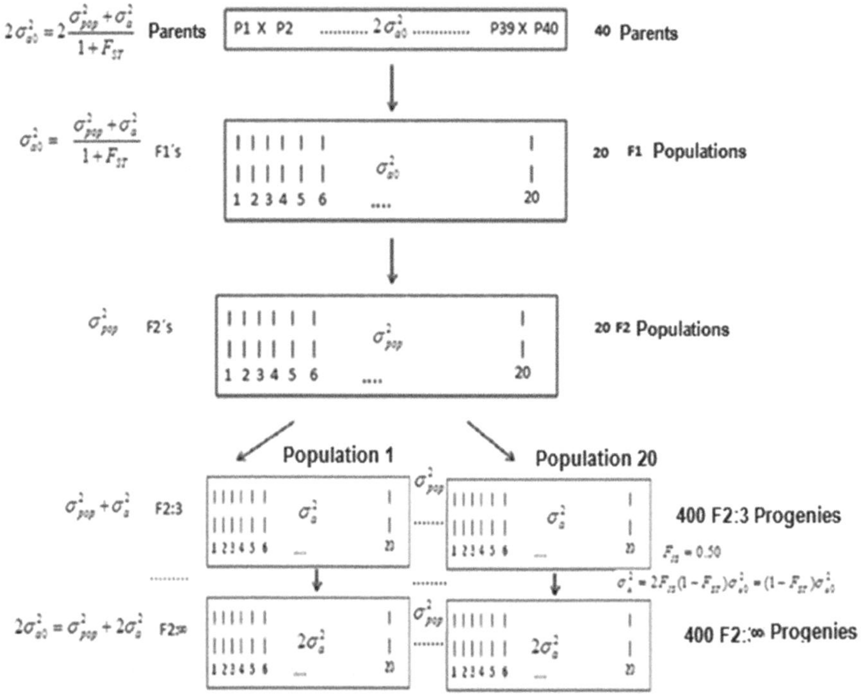

Fig. 12.1 Overall structure of the breeding program of an autogamous species, showing the distribution of total genetic variability among and within populations and multigenerations

F_1 and F_2 populations are in S_0 and present only population effects, but not progeny. Thus, the complete index, from F_1 to F_6 and more of both parents, will have 12 sources of information and weights. It may also contain F_2 plants and, in such case, it will have 13 sources of information. The method allows computing the gain in lines when the selection is made among parents and among populations F_1 and F_2. It also allows making the selection in F_3 already based on six pieces of information: three F_1 to F_3 populations, one progeny in F_3, and the two of parents. There may also be plants within F_3.

The index is represented as follows: $I = b_{\text{pop}}(\bar{Y}_i - \bar{Y})_g + b_{prog}(Y_j - \bar{Y}_i)_g$. The vector of b_{pop} weights considers the population effects in various inbreeding generations, and the vector of b_{prog} weights considers the progeny effects in several generations. As these effects are independent in the linear statistical model, the two vectors can be estimated separately, generating the 12 weights required, as shown below.

$$\widehat{b}_{\mathrm{pop}}=\begin{bmatrix}\widehat{b}_{F3\mathrm{pop}}\\ \widehat{b}_{F4\mathrm{pop}}\\ \widehat{b}_{F5\mathrm{pop}}\\ \widehat{b}_{F6\mathrm{pop}}\\ \widehat{b}_{pai1}\\ \widehat{b}_{pai2}\\ \widehat{b}_{F1}\\ \widehat{b}_{F2}\end{bmatrix}=\begin{bmatrix}\left(K_{F3\mathrm{pop}}/h_{\mathrm{aT}}^2\right) & \frac{2F_{\mathrm{ST}}}{1+F_{\mathrm{ST}}} & \frac{2F_{\mathrm{ST}}}{1+F_{\mathrm{ST}}} & \frac{2F_{\mathrm{ST}}}{1+F_{\mathrm{ST}}} & \frac{2F_{\mathrm{ST}}}{1+F_{\mathrm{ST}}} & \frac{2F_{\mathrm{ST}}}{1+F_{\mathrm{ST}}} & \frac{2F_{\mathrm{ST}}}{1+F_{\mathrm{ST}}} & \frac{2F_{\mathrm{ST}}}{1+F_{\mathrm{ST}}}\\ & \left(K_{F4\mathrm{pop}}/h_{\mathrm{aT}}^2\right) & \frac{2F_{\mathrm{ST}}}{1+F_{\mathrm{ST}}} & \frac{2F_{\mathrm{ST}}}{1+F_{\mathrm{ST}}} & \frac{2F_{\mathrm{ST}}}{1+F_{\mathrm{ST}}} & \frac{2F_{\mathrm{ST}}}{1+F_{\mathrm{ST}}} & \frac{2F_{\mathrm{ST}}}{1+F_{\mathrm{ST}}} & \frac{2F_{\mathrm{ST}}}{1+F_{\mathrm{ST}}}\\ \mathit{Sim.} & & \left(K_{F5\mathrm{pop}}/h_{\mathrm{aT}}^2\right) & \frac{2F_{\mathrm{ST}}}{1+F_{\mathrm{ST}}} & \frac{2F_{\mathrm{ST}}}{1+F_{\mathrm{ST}}} & \frac{2F_{\mathrm{ST}}}{1+F_{\mathrm{ST}}} & \frac{2F_{\mathrm{ST}}}{1+F_{\mathrm{ST}}} & \frac{2F_{\mathrm{ST}}}{1+F_{\mathrm{ST}}}\\ & & & \left(K_{F6\mathrm{pop}}/h_{\mathrm{aT}}^2\right) & \frac{2F_{\mathrm{ST}}}{1+F_{\mathrm{ST}}} & \frac{2F_{\mathrm{ST}}}{1+F_{\mathrm{ST}}} & \frac{2F_{\mathrm{ST}}}{1+F_{\mathrm{ST}}} & \frac{2F_{\mathrm{ST}}}{1+F_{\mathrm{ST}}}\\ & & & & \left(K_{pai1}/h_{\mathrm{aT}}^2\right) & 0 & \frac{1}{1+F_{\mathrm{ST}}} & \frac{2F_{\mathrm{ST}}}{1+F_{\mathrm{ST}}}\\ & & & & & \left(K_{pai2}/h_{\mathrm{aT}}^2\right) & \frac{1}{1+F_{\mathrm{ST}}} & \frac{2F_{\mathrm{ST}}}{1+F_{\mathrm{ST}}}\\ & & & & & & \left(K_{F1}/h_{\mathrm{aT}}^2\right) & \frac{2F_{\mathrm{ST}}}{1+F_{\mathrm{ST}}}\\ & & & & & & & \left(K_{F2}/h_{\mathrm{aT}}^2\right)\end{bmatrix}^{-1}\begin{bmatrix}\frac{2F_{\mathrm{ST}}}{1+F_{\mathrm{ST}}}\\ \frac{2F_{\mathrm{ST}}}{1+F_{\mathrm{ST}}}\\ \frac{2F_{\mathrm{ST}}}{1+F_{\mathrm{ST}}}\\ \frac{2F_{\mathrm{ST}}}{1+F_{\mathrm{ST}}}\\ \frac{2F_{\mathrm{ST}}}{1+F_{\mathrm{ST}}}\\ \frac{2F_{\mathrm{ST}}}{1+F_{\mathrm{ST}}}\\ \frac{2F_{\mathrm{ST}}}{1+F_{\mathrm{ST}}}\\ \frac{2F_{\mathrm{ST}}}{1+F_{\mathrm{ST}}}\end{bmatrix}$$

$$K_{\text{gpop}} = \Big(1 + \frac{F_{\text{IT}}(1 + F_{\text{ST}}) - 2F_{\text{ST}}\Big)}{(1 + F_{\text{ST}})} h^2_{\text{aT}} + (b - 1)\frac{1 + (1 - F_{\text{ST}})F_{\text{IS}t}}{(1 + F_{\text{ST}})} h^2_{\text{aT}}$$
$$+ (fb - 1)\frac{2F_{\text{ST}}}{(1 + F_{\text{ST}})} h^2_{\text{aT}}; \quad \rho_{\text{p}} = \frac{2F_{\text{ST}}}{1 + F_{\text{ST}}}.$$

In the additional inclusion of population effects in F_1 and F_2 generations and also of the parents (P_1 and P_2) of each population, it was considered that these effects are not correlated with the effects of progeny within populations (as the latter are hierarchical to populations), but are correlated with the effects of populations in various generations.

Weights of Progeny Effects in a Hierarchical Structure

$$\widehat{b}_{prog} = \begin{bmatrix} \widehat{b}_{F3\,prog} \\ \widehat{b}_{F4\,prog} \\ \widehat{b}_{F5\,prog} \\ \widehat{b}_{F6\,prog} \end{bmatrix} = P^{-1}g$$

$$= \begin{bmatrix} \left(K_{F3\,prog}/h^2_{\text{aT}}\right) & 1.00 & 1.00 & 1.00 \\ & \left(K_{F4\,prog}/h^2_{\text{aT}}\right) & \frac{1+(1-F_{\text{ST}})(0.50)}{(1+F_{\text{ST}})} & \frac{1+(1-F_{\text{ST}})(0.50)}{(1+F_{\text{ST}})} \\ Sim. & & \left(K_{F5\,prog}/h^2_{\text{aT}}\right) & \frac{1+(1-F_{\text{ST}})(0.75)}{(1+F_{\text{ST}})} \\ & & & \left(K_{F6\,prog}/h^2_{\text{aT}}\right) \end{bmatrix}^{-1}$$
$$\begin{bmatrix} 1.00 \\ \dfrac{1 + (1 - F_{\text{ST}})(0.50)}{(1 + F_{\text{ST}})} \\ \dfrac{1 + (1 - F_{\text{ST}})(0.75)}{(1 + F_{\text{ST}})} \\ \dfrac{1 + (1 - F_{\text{ST}})(0.875)}{(1 + F_{\text{ST}})} \end{bmatrix}$$

$$K_{Fg\,\text{prog}} = \frac{1 + F^* h^2_{\text{aT}} + (b - 1)\rho_{\text{gd}} h^2_{\text{aT}} - \rho_{\text{p}} h^2_{\text{aT}}}{b}$$
$$= \frac{\Big(1 + \frac{F_{\text{IT}}(1+F_{\text{ST}})-2F_{\text{ST}}\Big)}{(1+F_{\text{ST}})} h^2_{\text{aT}} + (b - 1)\frac{1+(1-F_{\text{ST}})F_{\text{IS}t}}{(1+F_{\text{ST}})} h^2_{\text{aT}} - \frac{2F_{\text{ST}}}{(1+F_{\text{ST}})} h^2_{\text{aT}}}{b}$$

By using adjusted means (deregressing phenotypes) and weighting them by the coefficients of the index, the BLUPs are obtained from the genetic values, as shown by Resende et al. (2015a). The BLUP multipopulations and multigeneration of the progeny effects can be obtained from the following multivariate model (generations taken as different variables) for experiments designed in incomplete blocks: $y = X\beta + Z_1 p + Z_2 g + Z_3 b + e$, where y is the data vector, β is the vector of fixed effects (general mean, experiments), p is the vector of population effects (random), g is the vector of progeny effects within population (random), b is the vector of incomplete block effects (random), and e is the vector of residual effects (random). X, Z_1, Z_2, and Z_3 are incidence matrices for β, p, g, and b, respectively. If the experiments are replicated in several environments, the effects of the genotype x environment interaction can be incorporated into the model. The Selegen-Reml/Blup software computes the SIPPPG-BLUP via 186 model.

Index Accuracy

The complete accuracy index estimator is given by $\hat{r}_{\widehat{aa}_\infty} = \sqrt{\hat{r}^2_{\text{prog } a\infty} + \hat{r}^2_{\text{pop } a\infty}}$,

where:

$$\begin{aligned}\hat{r}_{\text{pop } a_\infty} &= \left[\sum b_{F x\text{pop}}\rho_{\text{p}}/(1+F)\right]^{1/2} \\ &= \left[\left(b_{\text{pai1}}\rho_{\text{p}} + b_{\text{pai2}}\rho_{\text{p}} + b_{\text{F1}}\rho_{\text{p}} + b_{\text{F2}}\rho_{\text{p}} + \right.\right. \\ &\quad \left.\left. + b_{\text{F3 pop}}\rho_{\text{p}} + b_{\text{F4 pop}}\rho_{\text{p}} + b_{\text{F5 pop}}\rho_{\text{p}} + b_{\text{F6 pop}}\rho_{\text{p}}\right)/(1+F_\infty)\right]^{1/2}\end{aligned}$$

and

$$\begin{aligned}\hat{r}_{\text{prog } a_\infty} &= \left[\sum b_{Fx\,\text{prog}}\rho_{gd}/(1+F)\right]^{1/2} \\ &= \left[\left(b_{\text{F3 prog}}\rho_{\text{3d}} + b_{\text{F4 prog}}\rho_{\text{4d}} + b_{\text{F5 prog}}\rho_{\text{5d}} + b_{\text{F6 prog}}\rho_{\text{6d}}\right)/(1+F_\infty)\right]^{1/2}\end{aligned}$$

The component ρ_{gd} (in this case, equivalent to the elements of the vector g of the equation system for the coefficients b) is the correlation coefficient between the phenotypic information of the F_g generation and the additive genetic value of a line in F_∞ generation. The accuracy of the index varies only due to the individual heritability within (h^2) populations and total (h^2_{aT}) of the trait, the number of replicates or plants (b) per progeny, and the number of progenies per population (f) and ρ_{p}. The relative efficiencies of the different selection procedures can then be computed experimentally or by means of simulations. Through simulation, it is necessary to vary h^2, h^2_{aT}, b, f, and ρ_{p}.

Indirect Contribution of Sources of Information to the Value of Lines

Simulation results (numerical evaluation) regarding the indirect contribution of the sources of information for the selection of lines are presented in Table 12.4.

It has been determined that the greatest contributions are those of the effects of populations F_3 to F_6, followed by progenies of F_6 to F_3, F_2, F_1, and parents. When the variation among populations is greater than within populations, the importance of the parents F_1 and F_2 increases, which reach weights similar to those associated with the progenies. These weights are not of the index, but of each source of information if it were used alone in the selection, aiming at gain in the line to be selected. The highest weights of the population effects of generations F_3 to F_6 are due to the fact that they have been evaluated with a large number of replicates ($60 = 3 \times 20$ families), whereas the progenies had only three replications. These results are important in the early stages of the selection cycle, where information on more advanced generations is not yet available. Table 12.4 also reports accuracies, aiming at the inference about the genetic values of the future lines. In the scenario in which most of the variation is found among populations, the following accuracies

Table 12.4 Indirect contribution of sources of information to the value of lines: indirect heritabilities, accuracy (r_{gg}), and determination coefficients (r^2_{gg}) ($h^2 = 0.20$, $h^2_{aT} = 27\%$ $\rho_p = 0.33$)

Among populations	h^2 indirect/ r_{gg}/r^2_{gg}	Among progenies	h^2 indirect/ r_{gg}/r^2_{gg}	Among parents	h^2 indirect/ r_{gg}/r^2_{gg}
F_1	0.19/0.18/0.03	–	–	$Parent_1$	0.14/0.15/0.02
F_2	0.23/0.20/0.04	–	–	$Parent_2$	0.14/0.15/0.02
F_3	0.80/0.36/0.13	$F_{2:3}$	0.40/0.40/0.16	–	–
F_4	0.79/0.36/0.13	$F_{2:4}$	0.51/0.56/0.31	–	–
F_5	0.79/0.36/0.13	$F_{2:5}$	0.56/0.63/0.40	–	–
F_6	0.79/0.36/0.13	$F_{2:6}$	0.58/0.66/0.44	–	–
$h^2 = 0.20$, $h^2_{aT} = 33\%$, $\rho_p = 0.50$					
F_1	0.33/0.29/0.08	–	–	$Parent_1$	0.25/0.25/0.06
F_2	0.38/0.31/0.10	–	–	$Parent_2$	0.25/0.25/0.06
F_3	0.89/0.47/0.22	$F_{2:3}$	0.40/0.37/0.14	–	–
F_4	0.85/0.46/0.21	$F_{2:4}$	0.51/0.51/0.26	–	–
F_5	0.88/0.47/0.22	$F_{2:5}$	0.56/0.57/0.32	–	–
F_6	0.88/0.47/0.22	$F_{2:6}$	0.58/0.60/0.36	–	–
$h^2 = 0.20$, $h^2_{aT} = 43\%$, $\rho_p = 0.67$					
F_1	0.55/0.43/0.18	–	–	$Parent_1$	0.41/0.37/0.14
F_2	0.55/0.43/0.18	–	–	$Parent_2$	0.41/0.37/0.14
F_3	0.94/0.56/0.31	$F_{2:3}$	0.40/0.32/0.10	–	–
F_4	0.94/0.56/0.31	$F_{2:4}$	0.51/0.44/0.19	–	–
F_5	0.94/0.56/0.31	$F_{2:5}$	0.56/0.49/0.24	–	–
F_6	0.94/0.56/0.31	$F_{2:6}$	0.58/0.52/0.27	–	–

are provided by each source of information: 0.56 for population effects, 0.52 for the F_6 progeny effect, 0.49 for the F_5 progeny effect, 0.44 for the effect of F_4 progenies, 0.43 for the effects of F_1 and F_2 populations, 0.37 for the effects of parents, and 0.32 for the F_3 progeny effect. Thus, F_5 progenies and F_1 and F_2 populations are equally important for the selection of future lines. Same accuracies (around 0.32 to 0.37) are also contributed by the P_1 and P_2 parents and F_3 progenies. Approximately equal accuracies (0.52 to 0.56) are contributed by populations in F_3 to F_6 and F_6 progenies. The determination coefficients (r^2_{gg}) indicate the proportions of the variation among the future strains explained by each source of information. Values ranged from 2% (parents) to 44% (F_6 progenies). Thus, the parents are not very informative in this context. These results are valid for heritability (of individual plots) average within population equal to 20%.

Contribution of Information Sources in the Multigeneration and Multipopulation Index (SIPPPG)

In the comparison of the new index with that proposed by Resende et al. (2015a) involving mix of progenies of all populations as if they belonged to a single population with total variability, six scenarios were considered regarding heritability within populations and total heritability, respectively: 0.15 and 0.21, 0.15 and 0.26, 0.15 and 0.34, 0.30 and 0.39, 0.30 and 0.46, and 0.30 and 0.56. The results are shown in Table 12.5.

SIPPPG: Total Multigeneration Index with Parents, Populations (F_1 to F_6), and Progenies (F_3 to F_6)

It has been determined that when the proportion of total variation due to populations is high, the new index leads to selective superiority of up to 9% (on the multigeneration index of progenies only), which is a considerable gain. The selective accuracy reaches about 90%, a magnitude considered optimal, according to Resende and de Duarte (2007). On the traditional selection based on F6 (accuracy 0.71), the superiority reaches 28% (0.91/0.71 = 1.28). Table 12.6 shows the percentage contributions of each source of information in the index.

It has been determined, based on the last column of Table 12.6, that, given the experimentation considered (three replicates and 20 progenies per population), the most important sources of information (those with the largest percentage weights) were, in this order, progenies in F_6, progenies in F_5, populations in F_3, populations in F_4, populations in F_5, populations in F_6, progenies in F_4, progenies in F_3, populations in F_2 and parents, and populations in F_1. In this scenario, the evaluation

Table 12.5 Weights of various sources of information and accuracies (r_{gg}) of the multipopulation and multigeneration index (SIPPPG) for some scenarios

Scenarios						
h^2 within	0.15	0.15	0.15	0.30	0.30	0.30
Total h^2	0.21	0.26	0.34	0.39	0.46	0.56
h^2 between/total h^2	0.33	0.50	0.67	0.33	0.50	0.67
Weights						
P_1	0.009	0.013	0.029	0.008	0.012	0.045
P_2	0.009	0.013	0.029	0.008	0.012	0.045
F_1	0.008	0.009	0.016	0.006	0.008	0.021
F_2	0.016	0.016	0.016	0.021	0.021	0.020
F_3	0.228	0.235	0.230	0.237	0.241	0.224
F_4	0.222	0.229	0.224	0.228	0.231	0.215
F_5	0.219	0.226	0.221	0.223	0.227	0.211
F_6	0.218	0.224	0.219	0.221	0.224	0.209
$F_{2:3}$	0.109	0.120	0.139	0.095	0.109	0.130
$F_{2:4}$	0.181	0.185	0.193	0.184	0.190	0.199
$F_{2:5}$	0.239	0.239	0.238	0.277	0.272	0.264
$F_{2:6}$	0.283	0.278	0.271	0.360	0.341	0.318
r_{gg} SIPPPG	0.82	0.84	0.88	0.87	0.88	0.91
r_{gg} BDP-BLUP	0.81	0.81	0.81	0.86	0.85	0.84
Efficiency	1.01	1.04	1.09	1.01	1.04	1.08

Table 12.6 Percentage (%) contributions of information sources for the multipopulation and multigeneration index (SIPPPG)

Total h^2	0.21	0.26	0.34	0.39	0.46	0.56	
ρ_p	0.33	0.50	0.67	0.33	0.50	0.67	Mean
P_1	0.5	0.7	1.6	0.4	0.6	2.4	1.0
P_2	0.5	0.7	1.6	0.4	0.6	2.4	1.0
F_1	0.5	0.5	0.9	0.3	0.4	1.1	0.6
F_2	0.9	0.9	0.9	1.1	1.1	1.1	1.0
F_3	13.1	13.2	12.6	12.7	12.8	11.8	12.7
F_4	12.8	12.8	12.3	12.2	12.2	11.3	12.3
F_5	12.6	12.6	12.1	11.9	12.0	11.1	12.1
F_6	12.5	12.5	12.0	11.8	11.9	11.0	12.0
$F_{2:3}$	6.3	6.7	7.6	5.1	5.8	6.8	6.4
$F_{2:4}$	10.4	10.4	10.6	9.9	10.1	10.5	10.3
$F_{2:5}$	13.7	13.4	13.0	14.8	14.4	13.9	13.9
$F_{2:6}$	16.3	15.6	14.8	19.3	18.1	16.7	16.8
Total	100	100	100	100	100	100	100

of F_2 populations and parents is equally of little importance, and the evaluation of F_1 is of no use. Obviously, these conclusions are valid, given all the information available at the end of advance of the inbreeding generations to F_6.

BLUP in the Genealogical Method

In almost all the methods of study of segregating populations of autogamous species breeding programs, at a given moment during the program, the genealogy is obtained. What will be focused here is the genealogical method or pedigree, but with some adjustments, it can be extrapolated to other situations. Typically, in this method plants in the F_2 generation are chosen visually. The seeds of these plants selected are sown separately in lines, forming progenies or $F_{2:3}$ lines. The best $F_{2:3}$ progenies are selected and, also, the best individuals within these best progenies based on the visual evaluation for the characters of interest. The process is then replicated until, with successive self-fertilization generations, it results in progenies with the desired homozygous level ($F_{5:7}$ or $F_{6:7}$). The best lines will participate in regional competition trials of cultivars.

Figure 12.2 presents a basic scheme for conducting a segregating population using the genealogical method used in the simulation study carried out by Nunes et al. (2008a, b). F_2 population consisted of 64 plants, which gave rise to 64 $F_{2:3}$ progenies consisting of 40 plants each. From each progeny $F_{2:3}$, two plants were taken, generating 128 $F_{3:4}$ progenies. The process was repeated in the next generation in order to obtain 256 $F_{4:5}$ progenies containing 40 plants each.

The genealogy is recorded for each generation of population advance. Thus, it is possible to recognize the degree of genetic relationship or parentage among progenies. For example, from Fig. 12.2, it can be observed that progeny 1-1-1 is related to progenies 1-1-2, 1-2-1, and 1-2-2, but considering generation F_2 as non-inbreeding, it is not related to the other $F_{4:5}$ progenies shown in the scheme. Regardless of the work dedicated to obtain this genealogy information, in general, it has been considered to be of little utility to breeders (Ramalho et al. 2001). Since the objective of autogamous plant breeders in the evaluation of progenies in replicate experiments is to select those that associate the best additive genetic

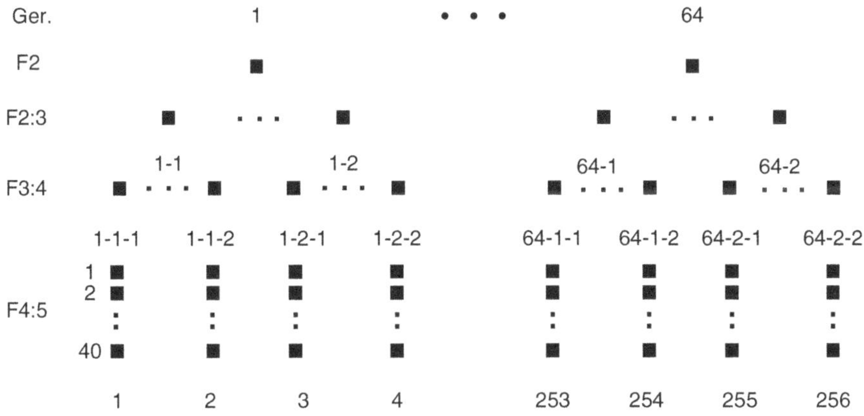

Fig. 12.2 Scheme of conduction by genealogical or pedigree method

values (VGA or a), what has been proposed is the use of genealogy to improve the efficiency of the selective process.

The inclusion of information on parentage among progenies directly in the method of analysis is enabled by the use of Henderson's mixed model approach, so that the predicted VGAs correspond to BLUP predictions. In Fig. 12.2, considering the analysis of the phenotypic data of the 256 progenies $F_{4:5}$, we have the general mixed model of the form $\boldsymbol{y} = \boldsymbol{X\beta} + \boldsymbol{Za} + \boldsymbol{e}$, where:

$\boldsymbol{y}$ = vector of phenotypic data.
$\boldsymbol{X}$ = matrix of the model relating to the fixed effects $\boldsymbol{\beta}$.
$\boldsymbol{\beta}$ = fixed effects vector.
$\boldsymbol{Z}$ = m**a**trix of the model relating to the random effects of progenies $\boldsymbol{a}$.
$\boldsymbol{a}$ = vector of random effects of progenies, with $a \sim N(0, \boldsymbol{G})$, in that $G = A\sigma_{\mathrm{a}}^2$.
$\boldsymbol{e}$ = vector of errors, with $\boldsymbol{e} \sim N(\boldsymbol{0}, R)$, in that $\boldsymbol{R} = \boldsymbol{I}\sigma_e^2$.

The matrix $\boldsymbol{G}$ refers to the matrix of genetic covariance among progenies $F_{4:5}$, denoted by $A\sigma_{\mathrm{a}}^2$, where A refers to the additive genetic parentage matrix, i.e., it contains the additive parentage coefficients among the progenies (2 × Malecot's parentage coefficient) and refers to the additive variance in F_2. In this case the matrix $\boldsymbol{A}$ is represented as follows:

$$A = I_{64} \otimes \begin{bmatrix} 1,75 & 1,50 & 1,00 & 1,00 \\ 1,50 & 1,75 & 1,00 & 1,00 \\ 1,00 & 1,00 & 1,75 & 1,50 \\ 1,00 & 1,00 & 1,50 & 1,75 \end{bmatrix},$$

where $\otimes$ is the product of Kroneker. Thus, it can be observed that matrix A is a block-diagonal matrix, which portrays the parentage among progenies $F_{4:5}$, according to the genealogy presented in Fig. 12.1, where we have four progenies $F_{4:5}$ from an ancestral F_2 plant.

The incorporation of parentage information from genealogy increases predictive accuracy, which allows progeny ranking to be closer to the fact (Nunes et al. 2008a, b). See in Fig. 12.3, where situations with different heritabilities (h^2) of the trait were simulated, in which, especially under conditions of lower heritability, correlation estimates between real VGA and those predicted via BLUP, when using parentage information (BLUPA), are superior.

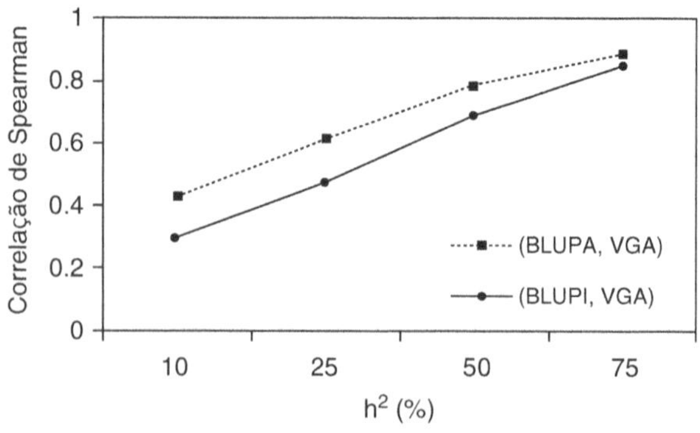

Fig. 12.3 Spearman classificatory correlation between real additive genetic values (VGA) and those predicted via BLUP, considering (BLUPA) and ignoring (BLUPI) the parentage information

Hierarchical Scheme for the Estimation of Variance Components

The genetic variance components among and within populations and the F_{ST} (inbreeding coefficient due to subdivision) required for the application of SIPPPG can be estimated as shown below. According to model $Y_{ijk} = \mu + \mathrm{pop}_i + b_j + (f/p)_{k(i)} + e$, the variance analysis for an experiment involving f progenies (f/p) $\mathrm{F}_{2:3}$ within population (pop), evaluated in b repetitions, is equivalent to:

Sources of variation	GL	QM	E(QM)
Blocks	$b-1$	Q_b	–
Populations	$g-1$	Q_p	$\sigma_{\mathrm{e}}^2 + b\sigma_{f/p}^2 + bf\sigma_{\mathrm{pop}}^2$
Progenies/populations	$(s-1)g$	$Q_{f/p}$	$\sigma_{\mathrm{e}}^2 + b\sigma_{f/p}^2$
Residual	$(b-1)(qs-1)$	Q_e	σ_{e}^2

By associating the population subdivision theory with the estimation of the variance components in the hierarchical model, we have the following summary table:

Entity	Variance composition	Variance component estimator
Within populations	$\sigma_{\mathrm{a}}^2 = (1 - F_{\mathrm{ST}})\sigma_{\mathrm{a0}}^2$	$\sigma_{\mathrm{a}}^2 = \sigma_{f/p}^2$
Among populations	$\sigma_{\mathrm{pop}}^2 = 2F_{\mathrm{ST}}\sigma_{\mathrm{a0}}^2$	σ_{pop}^2
Total = between + within	$\sigma_{\mathrm{aT}}^2 = (1 + F_{\mathrm{ST}})\sigma_{\mathrm{a0}}^2$	$\sigma_{\mathrm{aT}}^2 = \sigma_{\mathrm{pop}}^2 + \sigma_{\mathrm{a}}^2$
Between +2 within	$2\sigma_{\mathrm{a0}}^2$	$\sigma_{\mathrm{pop}}^2 + 2\sigma_{\mathrm{a}}^2$
Proportion among variances	Composition	Estimator
Among/(between +2 within)	$\frac{2F_{\mathrm{ST}}\sigma_{\mathrm{a0}}^2}{2\sigma_{\mathrm{a0}}^2} = F_{\mathrm{ST}}$	$F_{\mathrm{ST}} = \frac{\sigma_{\mathrm{pop}}^2}{\sigma_{\mathrm{pop}}^2 + 2\sigma_{\mathrm{a}}^2} = Q_{\mathrm{ST}}$
Among/total	$\rho_{\mathrm{p}} = \frac{2F_{\mathrm{ST}}}{1+F_{\mathrm{ST}}}$	$\rho_{\mathrm{p}} = \frac{\sigma_{\mathrm{p}}^2}{\sigma_{\mathrm{pop}}^2 + \sigma_{\mathrm{a}}^2}$
Inheritability within population	–	$h^2 = \frac{\sigma_{\mathrm{a}}^2}{\sigma_{\mathrm{a}}^2 + \sigma_{\mathrm{e}}^2}$
Inheritance among populations	–	$h_{\mathrm{pop}}^2 = \frac{\sigma_{\mathrm{pop}}^2}{\sigma_{\mathrm{pop}}^2 + \sigma_{\mathrm{a}}^2 + \sigma_{\mathrm{e}}^2}$
Total heritability	–	$h_{\mathrm{aT}}^2 = \frac{\sigma_{\mathrm{pop}}^2 + \sigma_{\mathrm{a}}^2}{\sigma_{\mathrm{pop}}^2 + \sigma_{\mathrm{a}}^2 + \sigma_{\mathrm{e}}^2}$

$2\sigma_{\mathrm{a0}}^2$ = original additive genetic variant of the species

Considering progenies $\mathrm{F}_{2:3}$, the following parameters can be estimated:
σ_{pop}^2: genetic variation among populations
$\sigma_{\mathrm{a}}^2 = \sigma_{f/p}^2$: mean additive genetic variation within populations
$F_{\mathrm{ST}} = \frac{\sigma_{\mathrm{pop}}^2}{\sigma_{\mathrm{pop}}^2 + 2\sigma_{\mathrm{a}}^2}$: inbreeding coefficient due to the differentiation among populations.
$\sigma_{\mathrm{a0}}^2 = \frac{\sigma_{\mathrm{pop}}^2}{2F_{\mathrm{ST}}} = \frac{\sigma_{\mathrm{pop}}^2 + \sigma_{\mathrm{a}}^2}{1+F_{\mathrm{ST}}} = \frac{\sigma_{\mathrm{aT}}^2}{1+F_{\mathrm{ST}}}$: total additive genetic variation of the original panmictic population

$\sigma^2_{aT} = \sigma^2_p + \sigma^2_a = (1 + F_{ST})\sigma^2_{a0}$: total additive genetic variation among and within populations

$\sigma^2_y = \sigma^2_{pop} + \sigma^2_a + \sigma^2_e = (1 + F_{ST})\sigma^2_{a0} + \sigma^2_e$: total phenotypic variance among and within populations

$h^2 = \frac{\sigma^2_a}{\sigma^2_a+\sigma^2_e}$: individual heritability within the population

$h^2_{aT} = \frac{\sigma^2_{pop}+\sigma^2_a}{\sigma^2_{pop}+\sigma^2_a+\sigma^2_e} = \frac{(1+F_{ST})\sigma^2_{a0}}{(1+F_{ST})\sigma^2_{a0}+\sigma^2_e}$: total heritability among and within populations

$h^2_{pop} = \frac{\sigma^2_{pop}}{\sigma^2_y} = \frac{\rho_p\sigma^2_{aT}}{\sigma^2_y} = \rho_p h^2_{aT} = \frac{2F_{ST}\sigma^2_{a0}}{(1+F_{ST})\sigma^2_{a0}+\sigma^2_e}$: heritability among populations

$h^{2*}_a = h^2_{aT} - \rho_p h^2_{aT} = \frac{\sigma^2_a}{\sigma^2_{pop}+\sigma^2_a+\sigma^2_e}$: modified heritability

$\rho_p = \frac{\sigma^2_{pop}}{\sigma^2_{pop}+\sigma^2_a} = \frac{h^2_{pop}}{h^2_{aT}} = \frac{2F_{ST}}{(1+F_{ST})}$: proportion of variances among populations and total

F_{ST} is also given by

$$F_{ST} = \frac{\sigma^2_{pop}}{\sigma^2_{pop} + 2\sigma^2_a} = \frac{\rho_p}{\rho_p + 2(1 - \rho_p)} = \frac{\rho_p}{2 - \rho_p}$$

Heritability Estimation Via Regression Among Generations

The use of intergenerational regression for the estimation of heritability is interesting because, as it involves two generations, it mitigates a little the interaction of genotypes x times of planting, avoiding the overestimation of h^2. An interesting regression is F_4 progenies on F_3 progenies, given by $b_3 = \frac{\mathrm{Cov}(F_3,F_4)}{\mathrm{Var}(F_3)} = \frac{\sigma^2_a}{\sigma^2_a+\sigma^2_{e3}} = h^2_3$. This regression provides a direct estimate of heritability. Ramalho et al. (2012) present the estimator $h^{2*} = \left(\frac{b^*}{(3-b^*)/2}\right)$, which is equivalent to $h^{2*} = \left(\frac{2b^*}{3-b^*}\right)$, in that b* is the regression of plants F_3 on progenies F_4. It is determined that the coefficient of regression is multiplied by 2 to obtain the heritability. This leads to the multiplication also of the estimation error in b^* by 2. Thus, h^2_3 is preferable in relation to h^{2*}.

Another option is to also use reverse regression of F_3 progenies on F_4 progenies. In this case, we have $b_4 = \frac{\mathrm{Cov}(F_3,F_4)}{\mathrm{Var}(F_4)} = \frac{\sigma^2_a}{1.5\,\sigma^2_a+\sigma^2_{e4}}$ and heritability is given by $h^2_4 = \left(\frac{2b_4}{2-b_4}\right)$, as shown below:

$$\begin{aligned} h^2_4 &= \frac{\mathrm{Cov}(F_3, F_4)}{\mathrm{Var}(F_4) - 0.5\,\mathrm{Cov}(F_3, F_4)} \\ &= \frac{\mathrm{Cov}(F_3, F_4)/\mathrm{Var}(F_4)}{\mathrm{Var}(F_4)/\mathrm{Var}(F_4) - 0.5\,\mathrm{Cov}(F_3, F_4)/\mathrm{Var}(F_4)} \\ &= \frac{b_4}{1 - b_4/2} = \frac{2b_4}{2 - b_4}. \end{aligned}$$

Both heritabilities can be weighted, obtaining $h^2 = \left(\frac{b_4}{2-b_4}\right) + \frac{b_3}{2}$, as shown below:

$$h^2 = \frac{h_4^2 + h_3^2}{2} = \frac{1}{2}\left(\frac{2b_4}{2-b_4}\right) + b_3 = \frac{2b_4}{4-2b_4} + \frac{b_3}{2} = \left(\frac{b_4}{2-b_4}\right) + \frac{b_3}{2}.$$

The advantage of using b_3 and b_4 simultaneously lies in the fact that it is used in two residual variance estimates (σ_{e3}^2 and σ_{e4}^2), which increases the validity of the estimate.

Another interesting method for the estimation of variance components involving progenies of several generations was presented by Ramalho et al. (2012) and is based on the weighted least squares method.

Final Considerations

In order to produce cultivars that are better than preexisting ones, the breeding programs of autogamous plants have intensified the achievement of segregating populations. Several studies have been carried out to determine the best alternative and assess more populations with a lower number of progenies or a small number of populations with a large number of progenies (Wricke and Weber 1986; Weber 1979; Baker 1984; Huhn 1996; Bernardo 2003). Although there are some aspects to be considered, obtaining a larger number of populations to the detriment of the number of progenies is the best option. For this reason, programs of large seed companies or even of programs of the smaller ones perform hundreds of biparental crosses annually. These crosses are advanced by different methods, among them, bulk and the bulk within progenies. In the latter case, progenies of the different populations are obtained from the F_2 or F_3 generation, which are evaluated until they reach a good homozygous level.

So far, the selection has been carried out with reference to the last generation in which progenies were evaluated. Recently, Resende et al. (2015a) presented a proposal for an index, which involves all the generations evaluated. However, as previously mentioned, breeders obtain numerous populations annually, and, in the progeny selection process, the merit of the population is not considered. It would be important to have a progeny selection index that includes not only the effects of progenies in different generations but also the effects of populations in all generations and the data of the parents and of the F_1 and F_2 generations simultaneously.

In this chapter, we have constructed a genetic analysis model that covers the entire breeding program structure of an autogamous species, contemplating parents, populations, progenies, and generations. Simulation and experimental results revealed that the new index favors gains in selective efficiency ranging from 5 to 28%, depending on the relative magnitude of the genetic variation among populations. The new index is especially recommended for use in recurrent selection programs in autogamous species, to which early selection is applied, leading to selective efficiency gains ranging from 8 to 28%.

References

Baker, R.J.: Quantitative genetic principles in plant breeding. In: Gustafson, J.P. (ed.) Gene Manipulation in Plant Improvement, pp. 147–176. Plenum, New York (1984)

Bernardo, R.: Parental selection, number of breeding populations, and size of each population in inbred development. Theor. Appl. Genet. **107**, 1252–1256 (2003)

Bernardo, R.: Essentials of Plant Breeding, 252 p. Stemma, Woodbury, MN (2014)

Huhn, M.: Optimum number of crosses and progeny per cross in breeding self-fertilizing crops. I. General approach and first numerical results. Euphytica. **91**, 365–374 (1996)

Nunes, J.A.R., Moreto, A.L., Ramalho, M.A.P.: Using genealogy to improve selection efficiency of pedigree method. Sci. Agric. **65**, 25–30 (2008a)

Nunes, J.A.R., Ramalho, M.A.P., Ferreira, D.F.: Inclusion of genetic relationship information in the pedigree selection method using mixed models. Genet. Mol. Biol. **31**, 73–78 (2008b)

Oakey, H., Verbyla, A.P., Pitchford, W., Cullis, B.R., Kuchel, H.: Joint modelling of additive and genetic line effects in single field trials. Theor. Appl. Genet. **113**, 809–819 (2006)

Ramalho, M.A.P., Abreu, A.F.B., dos Santos, J.B.: Melhoramento de espécies autógamas. In: Nass, L.L., Valois, A.C.C., de Melo, I.S., Valadares-Inglis, M.C. (eds.) Recursos genéticos e melhoramento—plantas, pp. 201–230. Fundação MT, Rondonópolis (2001)

Ramalho, M.A.P., Abreu, A.F., Santos, J.B., Nunes, J.A.: Genética quantitativa no melhoramento de plantas autógamas. Universidade Federal de Lavras, Goiânia (2012)

Ramalho, M.A.P., Carvalho, B.L., Nunes, J.A.R.: Perspectives for the use of quantitative genetics in breeding of autogamous plants. ISRN Genetics. **2013**, (2013)

Resende, M.D.V.: Genética Biométrica e Estatística no Melhoramento de Plantas Perenes, 975p. Embrapa Informação Tecnológica, Brasília (2002)

Resende, M.D.V., de Duarte, J.B.: Precisão e controle de qualidade em experimentos de avaliação de cultivares. Pesquisa Agropecuária Tropical. **37**(3), 182–194 (2007)

Resende, M.D.V., de Higa, A.R.: Maximização da eficiência da seleção em testes de progênies de Eucalyptus através da utilização de todos os efeitos do modelo matemático. Pesquisa Florestal Brasileira, Colombo. **28**(/29), 37–55 (1994)

Resende, M.D.V., Silva, F.F., Azevedo, C.F.: Estatística Matemática, Biométrica e Computacional, 881p. Viçosa, Suprema (2014)

Resende, M.D.V., Ramalho, M.A.P., Guilherme, S., Abreu, A.F.B.: Multigeneration index in the within progenies bulk method for breeding of self-pollinated plants. Crop Sci. (2015a). doi:10.2135/cropsci2014.08.0580

Resende, M.D.V., Ramalho, M.A.P., Carneiro, P.C.S., Carneiro, J.E.S., Batista, L.G., Gois, I.B.: Selection index with parents, populations, progenies and generations effects in autogamous plant breeding. Crop Sci. **56**(2), 530–546 (2015b)

Weber, W.E.: Number and size of cross progenies from a constant total number of plants manageable in a breeding program. Euphytica. **28**, 453–456 (1979)

Wricke, G., Weber, W.E.: Quantitative Genetics and Selection in Plant Breeding. de Gruyter, Berlin (1986)

Chapter 13
Soybean Genetic Mapping

Ivan Schuster

Abstract In over 27 years of genetic mapping of the soybean genome, a genetic map containing thousands of molecular markers, thousands of QTLs and dozens of genes was obtained. Currently, two soybean genome consensus maps are available—GmConsensus4.0 and GmComposite2003—and can be used as an anchor to map new genes or QTLs in soybean. All information on these genetic maps is available for consultation at SoyBase (http://soybase.org). Genetic map information can also be compared with the annotated genomic sequence of Williams 82 cultivar, so there are currently tools that can associate regions of genetic maps containing QTLs with the genetic sequences of the soybean genome (physical map). In addition, the new tools for high-density genotyping of SNP markers allow to obtain dozens or hundreds of thousands of markers at affordable costs and the development of linkage disequilibrium maps. Linkage disequilibrium maps allow better understanding of the structure of the soybean genome and the identification of molecular markers in linkage disequilibrium with QTLs. With the evolution of large-scale phenotyping tools (phenomics), along with high-resolution genetic maps, the genetic processes that regulate plant development can be better understood in the near future.

Keywords Molecular markers • QTLs • SoyBase • Genetic map • Physical map • Linkage disequilibrium

Introduction

The same way as all the species that were part of the first genetic mapping experiences, the first soybean genetic maps had RELP (restriction fragment length polymorphisms) as reference markers. With the advent of PCR (polymerase chain reaction) technique, a larger number of classes of markers could be used, increasing the possibility of saturating the genetic maps. RAPD (random amplified polymorphic DNA), microsatellite or SSR (single sequence repeat), and AFLP (amplified

I. Schuster, M.S., D.S. (✉)
Dow AgroSciences, Cravinhos, São Paulo 14140-000, Brazil
e-mail: ivanschuster.ivan@gmail.com

F. Lopes da Silva et al. (eds.), *Soybean Breeding*,
DOI 10.1007/978-3-319-57433-2_13

fragment length polymorphism) markers enabled to obtain a soybean genetic map with 1395 PCR-based markers, as well as 789 RFLP markers. By including SNP (single nucleotide polymorphism) markers, further 1050 markers have been included in the soybean genetic map.

Currently, all information related to soybean genetic mapping is stored in SoyBase (http://soybase.org). SoyBase is a genetic database for soybean from the USDA-ARS (United States Department of Agriculture-Agricultural Research Service). It is a global depository for genetics and genomics data of soybean. The data deposited in SoyBase is initially evaluated by professional curators, in order to check the origin of the information, before being available to be consulted. This database contains the most up-to-date information on physical and genomic genetic maps, genetic sequences, and the quantitative and qualitative traits of soybean.

The SoyBase depository contains more than 20 years of QTLs (quantitative trait loci) mapping data, for more than 90 traits. It even contains the annotated genomic sequence of "Williams 82" and the data mining tools. All the information in SoyBase is interconnected (Grant et al. 2010), and this makes comparisons and consultations easier.

Soybean Mapping History

The original genetic map of soybean was created using a population derived from crossing *Glycine max* and *Glycine soja*, using RFLP markers. While other populations were being studied, it was noticed that the variability of RFLP markers among soybean cultivars was very low, which resulted in monomorphic markers in many crosses. This means that each genetic map that was published contained only one subset (very often a small one) of the molecular markers available, making it difficult to compare the different studies.

Keim et al. (1990) published the first genetic map for soybean using molecular markers. As the frequency of RFLP markers in soybean is low, the map was obtained through an interspecific cross between a commercial soybean cultivar (A81-356022) and a soybean line from the species *Glycine soja* (PI 468916). One hundred fifty RFLP markers and three classic loci were mapped. The map contained 1200 cM, with 26 linkage groups (LG). Later, other research groups saturated this map with a further 200 RFLP markers, reaching a size of 2473 cM and 25 LGs (Diers et al. 1992; Shoemaker and Olsen 1993; Shoemaker and Specht 1995). Rafalski and Tingey (1993) also used a population obtained by crossing *Glycine max* and *Glycine soja* and could map 600 RFLP markers.

Lark et al. (1993) obtained the first genetic map among cultivars of the *Glycine max* species, using a small F2:3 population obtained from a cross between Minsoy and Noir I cultivars, in which 132 RFLP markers were mapped. Mansur et al. (1995) created a population of 240 RIL (recombinant inbreed lines), which were used intensely for the genetic mapping of soybean. Yamanaka et al. (2001) mapped 401 RFLP markers in an F2 population of 190 individuals.

Keim et al. (1997) mapped 165 RFLP markers in a population of 300 RILs obtained through a cross between PI 437654 and BSR101 (both *Glycine max*). Using this RFLP map, the authors anchored a further 25 RAPD markers and 650 AFLP markers. The map resulted in obtaining 25 LGs and a total of 3441 cM. The AFLP markers that were generated using the *Eco*RI and *Mse*I enzymes tended to get grouped together in some regions of the map, with no uniform distribution, although the AFLP markers mapped throughout all LG. Ferreira et al. (2000) included 106 RAPD markers in this map. This is the only map with a significant quantity of RAPD markers.

Akkaya et al. (1995) obtained the first soybean genetic map containing microsatellite markers, in an F2 population of 60 individuals obtained from a cross between near-isogenic lines from the Clark and Harosoy cultivars. Forty microsatellite markers were included in the genetic map of this population, together with 118 RFLP and RAPD markers, 13 classic markers controlling pigmentation and morphological traits, and seven isoenzymes.

SNP markers amplify specific loci and are randomly distributed in the soybean genome, covering the whole genome. The loci-specific detection and whole-genome sampling allowed using SNP markers to obtain a reference map, initially containing 600 SNP markers. This reference map was obtained by the joint effort of research groups from the USDA in Beltsville, MD, and Ames, IA, the University of Nebraska and the University of Utah, in the USA. This map was obtained using one, two, and, occasionally, three populations (Creagan et al. 1999). One of the populations was obtained through a cross between *G. max* and *G. soja* (developed by the USDA in Iowa). Another population is the expanded one, containing 240 RILs of the population from the cross between the Minsoy and Noir I cultivars, obtained by the University of Utah (previously mentioned). The third population, also mentioned above, is the one obtained by the cross between isolines from the Clark and Harosoy cultivars, obtained by the University of Nebraska, and contains 57 lines derived from F2.

In these reference populations, 187 markers were mapped in exclusive populations, and a much larger number of markers were mapped in more than one population. As microsatellite markers amplify specific loci in the soybean genome, it is possible to compare the position of the markers in each of the populations. Thus, a consensus map among these three populations was obtained using common markers as anchors. This consensus map contains 20 linkage groups, which one supposes to be the 20 chromosomes of the gametic cells of the soybean. In this map, 1423 markers were mapped, 606 of which were microsatellites, 689 RFLPs, and 26 classic markers (simple inherited traits).

This consensus map has been used as an anchor for various soybean genetic mapping projects since 1999, as the use of microsatellite markers allows aligning LG groups obtained in different projects to the consensus map. cDNA and genetic clones can also be mapped in the correct LG, based on this consensus map. This consensus map was updated in 2004 (Song et al. 2004) and 420 new markers were included. As a result, the consensus map had 1837 markers, 1015 of which were microsatellites, 709 RFLPs, 73 RAPDs, 24 classic markers, six AFLPs, and ten isoenzymes.

SNP markers were included in the consensus map in 2007 (Choi et al. 2007). The SNPs used in the soybean genetic map were obtained by re-sequencing STS (sequence-tagged sites), most of which were designed from sequences of single genes. Currently, 1050 SNPs derived from 1141 different genes are positioned in the 20 LGs of the soybean consensus map.

During this period, several works have been published reporting the mapping of genes and QTLs throughout the soybean genome, using the consensus map to determine the location of these genes and QTLs.

Current Genetic Map of Soybean

In 1999, and again in 2003, a group of researchers from USDA obtained a composed genetic map that combined data from several populations. Since this map contained essentially all the genetic markers, it was finally possible to position all the published QTLs in a single set of maps. Since then, recently published QTLs have been added to the composed map, together with any types of markers that are needed for describing them. This map is located at SoyBase, under the name of GmComposite2003, and it is originally based on the composed map obtained by Song et al. (2004).

The nomenclature of the linkage groups was initially proposed by the group from the USDA-ARS, in version 3.0 of the consensus map. The equivalence of the first 11 linkage groups with the soybean chromosomes was made by Zou et al. (2003), using primary trisomics. The equivalence of the other nine chromosomes was made in decreasing order of the size of the linkage groups (Table 13.1).

Recently, the number of molecular markers has grown very fast, as SNP markers are being identified and mapped. However, creating a new composed map is very time-consuming and involves a great deal of effort, because markers arise all the time. Because of that, the soybean genetic map was bifurcated into two routes: the composed genetic map (based on the 2003 version, called GmComposite2003), into which the new mapped QTLs are continuously being added, and the consensus map (currently in its fourth version), called GmConsensus4.0. This consensus map contains SNP markers and microsatellites that are also in the genomic sequence of the "Williams 82" (Wm82), which contains all the new markers based on sequencing.

Currently, the GmComposite2003 genetic map has 3245 markers (Table 13.2), 89 genes (Table 13.3), and 3027 mapped QTLs (Tables 13.4, 13.5, 13.6, 13.7, 13.8, and 13.9). The consensus map (GmConsensus4.0) does not have any QTL or mapped genes, but it is possible to align both the maps up side by side and compare the molecular markers, genes, QTLs, and sequences between them.

Since both the GmConsensus4.0 and GmComposite2003 maps have a sufficient number of markers in common, they can be aligned up side by side and compared with each other. On the main page of SoyBase (http://soybase.org) (Fig. 13.1), selecting the menu "Sequence Map/Compare Genetic Map Order & Sequence Map

Table 13.1 Equivalence among names of the linkage groups and number of chromosomes in the soybean genetic map

Number of chromosomes	Name of linkage group	Size (cM)	Criteria
1	D1a	98.41	Primary trisomy
2	D1b	140.63	Size of linkage group
3	N	99.51	Primary trisomy
4	C1	112.32	Primary t trisomy
5	A1	86.75	Primary trisomy
6	C2	136.51	Size of linkage group
7	M	135.15	Size of linkage group
8	A2	146.67	Primary trisomy
9	K	99.60	Primary trisomy
10	O	132.89	Size of linkage group
11	B1	124.24	Size of linkage group
12	H	120.50	Size of linkage group
13	F	120.03	Primary trisomy
14	B2	108.18	Size of linkage group
15	E	99.88	Size of linkage group
16	J	92.27	Size of linkage group
17	D2	119.19	Primary trisomy
18	G	105.00	Primary trisomy
19	L	101.14	Primary trisomy
20	I	112.77	Primary trisomy
TOTAL		2291.64	

Table 13.2 Number and types of molecular markers in the GmComposite2003 genetic map for soybean[a]

Type of markers	Quantity
Morphological	10
Isoenzymes	1
RFLP	789
PCR	24
RAPD	73
AFLP	3
Microsatellites	1295
SNP	1050
TOTAL	3245

[a]Data updated in November 2015

Order," it opens a page on which it is possible to get both maps lined up. In the top part of this page, there is an item called "Choose a Linkage Group." When choosing a linkage group, one can make a comparison of this linkage group between the maps and the locations of the QTLs on the GmComposite2003 map.

For example, when choosing the D2 linkage group, one gets an image like the one shown in Fig. 13.2. This figure is interactive, and when selecting markers,

Table 13.3 Genes mapped in soybean genome, in the GmComposite2003 map, following the order of linkage group and position[a]

Name of gene	Linkage group	Position (cM)	Description
PHYA_2	A1	41.00	Phytochrome A
Rhg4	A2	48.00	Reaction to infection by *Heterodera glycine*
I	A2	48.84	Color of hilum
O	A2	58.40	Color of seed integument
Y13	A2	85.00	Green, whitish, or yellowish color in seedlings and leaves
Rps8	A2	126.09	Reaction to infection by *Phytophthora sojae*
Lf1	A2	126.72	Number of leaves
Pgd2	A2	137.00	Isoenzyme—mobility of phosphogluconate dehydrogenase
Lap1	A2	143.00	Isoenzyme—mobility of leucine aminopeptidases
Ti	A2	150.00	Inhibitor of Kunitz trypsin
Ap	A2	157.27	Isoenzyme—mobility of phosphatase acid
D2	B1	0.58	Green or yellow seed embryo
Aco4	B1	29.87	Isoenzyme—mobility of aconitate hydratase
Pa1	B1	40.00	Shape of pubescence
K1	B1	64.00	Color of hilum
PHYA_1	B2	76.50	Phytochrome A
Fas	B2	85.00	Stearic acid content
Fan	B2	89.19	Linolenic acid content
Rsv3	B2	104.00	Reaction to soybean mosaic virus
LD_1	C1	22.29	Not described
Fle	C1	67.00	Presence or absence of fluorescent stearate
Dia	C1	71.08	Isoenzyme
N	C1	122.62	Color of hilum
Pgi	C2	5.50	Isoenzyme—mobility of isomerate glucose-6-phosphate
T	C2	112.44	Color of pubescence
E1	C2	113.00	Maturity and sensitivity to photoperiod
G	D1a	0.00	Color of seed integument
Me	D1a	9.40	Isoenzyme
D1	D1a	103.44	Yellow or green embryo, due to retention of chlorophyll
Pd1	D1a	120.88	Density of pubescence
Wc	D1b	27.13	Not described
ms	D1b	45.00	Male sterility
Fsp1	D1b	48.00	Fertility or sterility of the soybean
Rsv4	D1b	52.00	Reaction to soybean mosaic virus
F	D1b	75.50	Abnormal stem
wp	D1b	75.69	Color of flower
Idh1	D1b	105.41	Isoenzyme—mobility of dehydrogenase isocitrate
Rxp	D2	35.50	Reaction to infection by *Xanthomonas axonopodis* pv. *glycines*
Mdh	D2	43.18	Not described
Fr2	D2	86.19	Physiology of root—root not fluorescent in UV light

(continued)

Table 13.3 (continued)

Name of gene	Linkage group	Position (cM)	Description
Pb	E	13.60	Sharp or blunt pubescence
PHYB_1	E	18.79	Phytochrome B
Y9	E	2250	Leaves of a normal color or yellowish-green
W1	F	28.86	Color of flower
Gy5	F	46.16	11S fraction of reserve protein
Rpg1	F	66.98	Reaction to infection by *Pseudomonas syringae*
Rpv1	F	67.90	Reaction to PMV (peanut mottle virus)
Rsv1	F	69.00	Reaction to soybean mosaic virus
Rps3	F	73.00	Reaction to infection by *Phytophthora sojae*
Pa2	F	85.33	Shape of pubescence
Shr	F	103.00	Normal or atrophied seed
Cgy1	F	115.50	Presence or absence of subunit a' of B-conglycinin
rhg1	G	4.70	Reaction to infection by *Heterodera glycine*
rfs	G	8.00	Reaction to infection by *Fusarium solani* f sp. *glycines*
Rps6	G	107.50	Reaction to infection by *Phytophthora sojae*
Rps4	G	113.00	Reaction to infection by *Phytophthora sojae*
Mpi	G	113.02	Isoenzyme—mobility of isomerate mannose-6-phosphate
Pv7	H	7.83	Not described
NTS	H	19.82	Supernodulation
Ps	H	78.44	Density of pubescence
Enp	I	23.92	Isoenzyme—mobility of endopeptidase
Ln	I	50.36	Shape of leaflet (oval or narrow)
Chi	I	111.98	Not described
E107_1	J	22.87	Not described
Rcs3	J	64.90	Reaction to infection by *Cercospora sojina*
Rbs1	J	68.26	Reaction to infection by *Phialophora gregata*
Rbs3	J	68.26	Reaction to infection by *Phialophora gregata*
Rj2	J	86.00	Reaction to *rhizobium*
Rps2	J	90.80	Reaction to infection by *Phytophthora sojae*
Rmd	J	97.00	Reaction to infection by *Microsphaera diffusa*
Ep	K	10.17	Level of activity of peroxidase on seed integument
PHYB_2	K	18.29	Phytochrome B
Cda1	K	32.70	Concentration of cadmium in the seeds
Fr1	K	49.93	Physiology of root—root not fluorescent in UV light
R	K	97.12	Color of seed integument
P1	K	111.29	Density of pubescence
LbA	L	28.60	Not described
L1	L	48.27	Color of pod
gy3_1	L	65.51	11S fraction of reserve protein
Gy7	L	65.69	11S fraction of reserve protein
Dt1	L	89.12	Type of determinate or indeterminate growth
FCA_1	L	98.30	Not described
Ts	M	30.25	Not described

(continued)

Table 13.3 (continued)

Name of gene	Linkage group	Position (cM)	Description
L2	N	0.00	Color of pod
Rps7	N	16.01	Reaction to infection by *Phytophthora sojae*
Rps1	N	31.50	Reaction to infection by *Phytophthora sojae*
Rpg4	N	55.38	Reaction to infection by *Pseudomonas syringae*
Gy4	O	28.38	11S fraction of reserve protein
E2	O	136.33	Early or late flowering

Adapted from SoyBase (http://soybase.org)
[a]Data updated in November 2015

Table 13.4 Number of QTLs mapped in the soybean genome, grouped by traits[a]

Traits	Number of QTLs	QTLs nonredundant[b]
Resistance to fungi	238	90
Resistance to insects	55	42
Resistance to viruses	2	2
Resistance to nematodes	185	54
Inorganic factors	100	32
Leaves and stem	267	176
Pods	85	59
Roots	31	27
Oil	344	142
Protein	238	174
Other traits in seeds	561	333
Reproductive period	258	141
Whole plant	383	229
Productivity	232	145
Other	48	30
TOTAL	3027	1676

Adapted from SoyBase (http://soybase.org)
[a]Data updated in November 2015
[b]Number of QTLs after having removed the QTLs mapped in the same position in the linkage group, for the same trait. To remove duplicated QTLs, it was considered different when the ends of the confidence interval were at least 2 cM apart

genes, or QTLs, all the relationships between them are marked. As an example, when selecting the QTL SCN 16-1, the parts that are in yellow in Fig. 13.2 are marked.

Also taking the QTL SCN 16-1 illustrated in Fig. 13.2 as an example, when clicking on the QTL name, two options come up: "Show Detailed Information in SoyBase" and "View QTL Region in Genome Sequence." When choosing the first option, information about the origin of the work in which this QTL was mapped is shown (Fig. 13.3). When choosing the second option ("View QTL Region in Genome Sequence"), the user is redirected toward the physical map of the region

Table 13.5 QTLs relating to resistance to pathogens and inorganic factors mapped on the soybean genome

Trait	Number of QTLs	Linkage groups
Resistance to disease		
Resistance to biotic stress	3	J
Resistance to *Fusarium*	34	A2, B1, C1, C2, D1a, D1b, D2, E, G, I, L, N, O
Resistance to *Phomopsis*	3	C2, F, O
Resistance to *Sclerotínia*	36	A1, A2, B1, B2, C2, D1a, D1b, D2, E, F, G, I, K, L, M, N, O
Resistance to *Rhizoctonia*	3	A2, C2, M
Resistance to *Phytophthora*	11	A2, B1, B2, C1, C2, D1a, D1b
Resistance to insects		
Resistance to bean pyralid (insect in stored grains)	8	C1, C2, D1a, D1b, H, I, O
Resistance to *Spodoptera exigua*	2	M
Resistance to *Helicoverpa zea*	17	B2, C2, D1a, D1b, D2, E, F, G, H, J, M
Resistance to pod beetle (*Maruca vitrata*)	2	G, H
Rag1 (resistance to aphid *Aphis glycines*)	6	A2, C2, F, M
Resistance to whitefly (*Bemisia tabaci*)	7	A1, D1a, D1b, D2, J, N
Resistance to viruses		
Resistance to soybean dwarfism	1	A1
Resistance to tobacco ring stain	1	F
Resistance to nematodes		
Resistance to soybean cyst nematode	43	A1, A2, B1, B2, C1, C2, D1a, D2, E, G, H, I, J, K, M, N, O
Resistance to *Meloidogyne javanica*	5	D1a, F
Resistance to peanut root knot	3	E, F
Resistance to *Meloidogyne incognita*	3	G, O
Inorganic factors		
Tolerance to aluminum	14	A1, B2, C2, D1b, D2, F, J, K, L, O
Chlorosis of leaflet caused by Mn (manganese)	2	B2, D2
Efficient use of Fe (iron)	16	A1, A2, B1, B2, D1a, G, H, I, L, N

in which this QTL was mapped (Fig. 13.4). By following the links on these maps, one can reach the genetic sequence of the selected region in the soybean genetic map. In the map GmConsensus4.0, when selecting the closest SNP associated with the QTL of interest, one can get the list of the genes published in this region and the sequence of up to 1 Kb flanking the selected region.

Table 13.6 QTLs relating to leaf, stem, pod, and root traits mapped on soybean genome

Trait	Number of QTLs	Linkage groups
Leaves and stems		
Discrimination of carbon isotopes	4	C2, D2, F, L
Length of hypocotyl	4	B1, F, L
Area of leaflet	24	A1, A2, C2, D1b, E, F, G, H, I, J, L, M, O,
Leaflet starch	5	D2, G, J, L
Chlorophyll in leaflet	15	C2, D1a, D1b, D2, F, G, I, K, M, N, O
Length of leaflet	32	A1, A2, B1, D1a, D1b, D2, E, F, G, I, J, K, L, M, N, O
Nitrogen in leaflet	1	A1
Phosphorus in leaflet	2	F
Shape of leaflet	34	A1, A2, B1, B2, C1, D1a, D1b, D2, E, F, G, I, J, K, L, O
Content of trigonelline in leaflet	2	C2, J
Weight of leaflet	5	B1, C2, H, L
Width of leaflet	36	A1, A2, B2, C1, C2, D1a, D1b, E, F, G, H, J, K, L, M, N, O
Net rate of photosynthesis	2	J, O
Resistance of stem	8	A2, B2, C1, C2, F
Diameter of stem	2	L
Pods		
Dehiscence of pods	15	A1, 12, B2, D1b, E, J, L
Number of pods	17	A1, A2, B1, C1, C2, D1b, E, G, I, J, K, L, N, O
Relationship between weight of pod and pod wall	5	D1a, D1b, F, H, N
Weight of pod wall	3	D1a, F, G
Width of pod wall	4	A1, B1, C2, J,N
Weight of pod	3	A1, D1b, N
Start of pod formation	5	A1, A2, C2, J, L
Number of pods per node	7	A1, C1, C2, I, K, M
Roots		
Number of nodules	3	B1, E, I
Area of root	2	B1, D1a
Density of lateral roots	1	G
Length of primary root	4	B1, C2, D1a, N
Morphology of root	5	A2, C1, D1a, C, N
Weight of nodules	1	I
Content of phosphorus in root	2	B1, D2
Relationship between weight of root and weight of seedling	1	B1
Weight of root	6	A1, B1, B2, D2, N, O
Depth of primary root	2	B1, G

Table 13.7 QTLs relating to the content and composition of oil and protein, mapped in the soybean genome

Trait	Number of QTLs	Linkage groups
Oil		
Content of oil	78	All
Content of linoleic acid	25	A1, A2, B1, B2, D1b, D2, E, F, G, I, J, K, L,N, O
Content of linoleic acid	36	A1, A2, B1, B2, C1, D1b, D2, E, F, G, H, J, K, L, O
Content of oil plus protein	3	A2.C2, E
Protein		
Content of protein	83	All
Acidic fraction of the seeds	4	D2, K, L
Basic fraction of the seeds	2	I, L
Content of alanine in the seeds	6	A1, A2, B2, F, L
Content of arginine in the seeds	1	D1b
Content of asparagine in the seeds	4	A2, D1a, I, K
Content of conglycinin in the seeds	3	A2, D2, N
Content of cysteine in the seeds	2	F, G
Content of glutamine in the seeds	7	A2, D1a, D1b, J, L, M, O
Content of glycine in the seeds	5	A1, D1b, K, L
Content of histidine in the seeds	1	J
Content of isoleucine in the seeds	5	A2, B2, D1a, F, L
Content of leucine in the seeds	5	D1a, D1b, D2, G, L, M
Content of lysine in the seeds	3	D1a, E, G
Content of methionine in the seeds	3	F, G, M
Content of methionine plus cysteine in the seeds	3	D2, F, M
Content of pectin in the seeds	1	A1
Content of phenylalanine in the seeds	6	B2, D1b, J, K, L
Content of prolamin in the seeds	4	A2, B2, G, L
Content of serine in the seeds	6	A2, C1, K, L, M
Content of threonine in the seeds	5	A1, D1b, K, L
Content of tryptophan in the seeds	6	A1, C1, D1a, D1b, G, I
Content of tyrosine in the seeds	5	A2, D1b, E, K, L
Content of valine in the seeds	4	B2, D1b, F, L

Mapping Methods

The information of soybean consensus genetic map was obtained from genetic mapping of structured populations. The genetic mapping in structured populations is based on the frequency of recombination between segregant loci in segregant populations derived from biparental crosses. Various types of mapping population

Table 13.8 QTLs relating to seed traits and reproductive period, mapped in the soybean genome

Trait	Number of QTLs	Linkage groups
Other traits in seeds		
Content of phytate in the seeds	3	L, N
Content of arabinose plus galactose in the seeds	2	A1, K
Content of cadmium in the seeds	1	K
Content of calcium in the seeds	3	A2, I, M
Color of integument of the seeds	6	A2, B1, B2, C2, N
Breakableness of seed integument	13	A2, B1, C2, D1B, F, I, K, L, M, N
Hardness of seed integument	6	A2, D1B, L, N
Content of daidzein in the seeds	34	All, except G and O
Seed odor	1	A1
Content of fructose in the seeds	1	A2
Content of galactose in the seeds	1	A1
Content of genistein in the seeds	30	A1, A2, B1, B2, C1, C2, D1b, D2, E, F, G, H, I, K, M
Content of glucose in the seeds	1	O
Content of glycinin in the seeds	1	D2
Hardness of seed	2	D1a, L
Height of seed	13	A1, A2, B1, C1,C2, D2, F, L, M
Content of isoflavones	34	All, except D1a and I
Length of the seed	15	A2, C1, C2, E, F, G, L, M, N
N (nitrogen) in the seeds in R5	6	A2, B2, D1b, E, G
N in the seeds in R6	3	A2, D1b, M
N in the seeds in R7	5	B2, D1a, E, G
Oligosaccharide in the seeds	10	B2, C2, D1b, E, H, J, L
Sucrose in the seeds	19	A1, A2, B1, B2, D1b, E, F, H, I, J, L, M
Tocopherol in the seeds	17	A2, B2, C1, C2, D1a, D1b, E, F, G, I, J, N, O
Volume of the seeds	10	A1, A2, C1, C2, E, L, M
Weight of the seeds	85	Todos
Depth of the seeds	11	C1, C2, D1b, D2, E, G, L, M
Reproductive period		
Ripening of the pods	66	All
Start of flowering	35	A2, B1, B2, C1, C2, D1a, D1b, E, F, G, H, J, K, L, M, O
Insensitivity to photoperiod	7	C2, F, G, J, K, L
Start of pods ripening	4	A1, A2, C2, J
Reproductive period	20	B1, C1, C2, D1b, F, G, H, J, L, M, O
Beans filling out	5	A2, C2, F, H, M
Start of seeds forming	4	C2, I, L

can be used, ranging from F2 populations, populations derived from F2 by self-fertilization, RILs, double haploids, and backcrossing populations (Schuster and Cruz 2004).

Table 13.9 QTLs relating to whole plant, productivity, and other traits, mapped in the soybean genome

Trait	Number of QTLs	Linkage groups
Whole plant		
Branching	8	B1, C2, D2, E, G, L, O
Height of canopy	3	D1, F, M
Width of canopy	3	C2, F, I
Wilting of canopy	11	A1, B2, C1, D1b, D2, F, H, L
Content of polysaccharides in the cell wall	1	A1
Shape of flower	4	C2, D1a, I, L
Number of flowers	11	B1, B1, C1, D1b, E, G, I, K, L, M
Size of between-node	12	A1, B2, C2, D1b, D2, F, G, K
Lodging	47	All, except M
Number of knots	11	A1, B1, B2, C2, D1a, D1b, F
Height of plants	75	All
Ratio between height of plants and lodging	8	F, G, L, M
Content of phosphorus in plant	2	B1, D2
Weight of plant	5	B1, C2, D2, N, O
Density of pubescence	9	A2, D1a, D1b, E, H, K, M
Abnormal seedling	2	C2, N
Ratio between dry weight and damp weight of seedlings	3	B2, C2, F
Content of phosphorus in seedling	3	B1, D2, F
Weight of seedling	11	A2, B1, C2, D1a, D2, E, F, G, L, N, O
Productivity		
Productivity of seeds	70	All, except A2
Relationship of yield under flooding	3	B1, F
Reaction to spacing between lines	12	A2, B2, C1, C2, D1a, D1b, L, M, O
Abortion of seeds	9	A1, C1, C2, E, F, M
Number of seeds	12	A2, B1, C1, C2, F, K, L, M, O
Production of seeds	14	A1, C1, C2, D1a, F, I, J, L, M, O
Weight of seeds per plant	7	A1, A2, C1, K, N, O
Ratio between productivity and height of plants	14	B1, C2, D1b, D2, F, J, L, M
Germination of seeds	4	B1, G, K, L
Others		
Sensitivity to chlorimuron	3	E, L
Tolerance to drought	11	A1, C2, D1a, D2, H, I, J, M, N, O
Number of somatic embryos per ex-plant	9	B2, C2, D1b, D2, G, H
Efficient use of water	7	C1, G, H, J, L

Fig. 13.1 Upper menu of the main page of SoyBase (http://soybase.org)

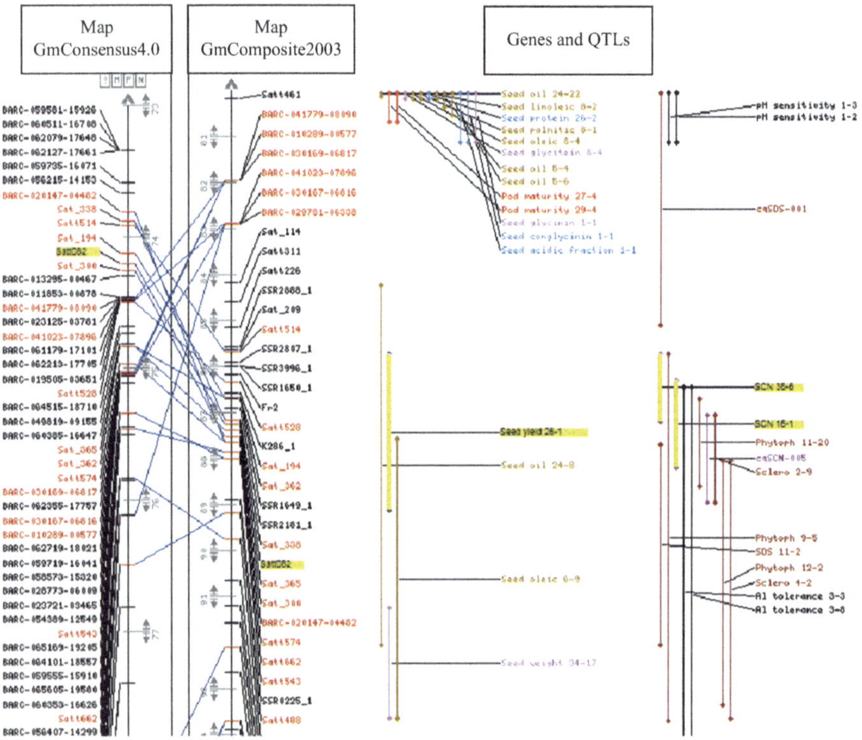

Fig. 13.2 Comparison between GmConsensus4.0 and GmComposite2003 maps. The lines between the two maps link common markers between them. On the right are shown the genes or QTLs mapped in this region of the GmComposite2003 map. The *vertical bars* on the right show the confidence intervals for locating the QTLs on the genetic map. As an example of how this tool works, when selecting SCN 16-1, all the relationships with this marker are marked in *yellow*

The QTL mapping in these structured populations was initially obtained through interval mapping (Lande and Thompson 1990), and later by composite interval mapping (Zeng 1994), which enables a relative increase in the precision of QTL mapping (Schuster and Cruz 2004). However, these strategies of linkage mapping, using populations derived from biparental crosses, consider only a small portion of the total possible alleles in certain species.

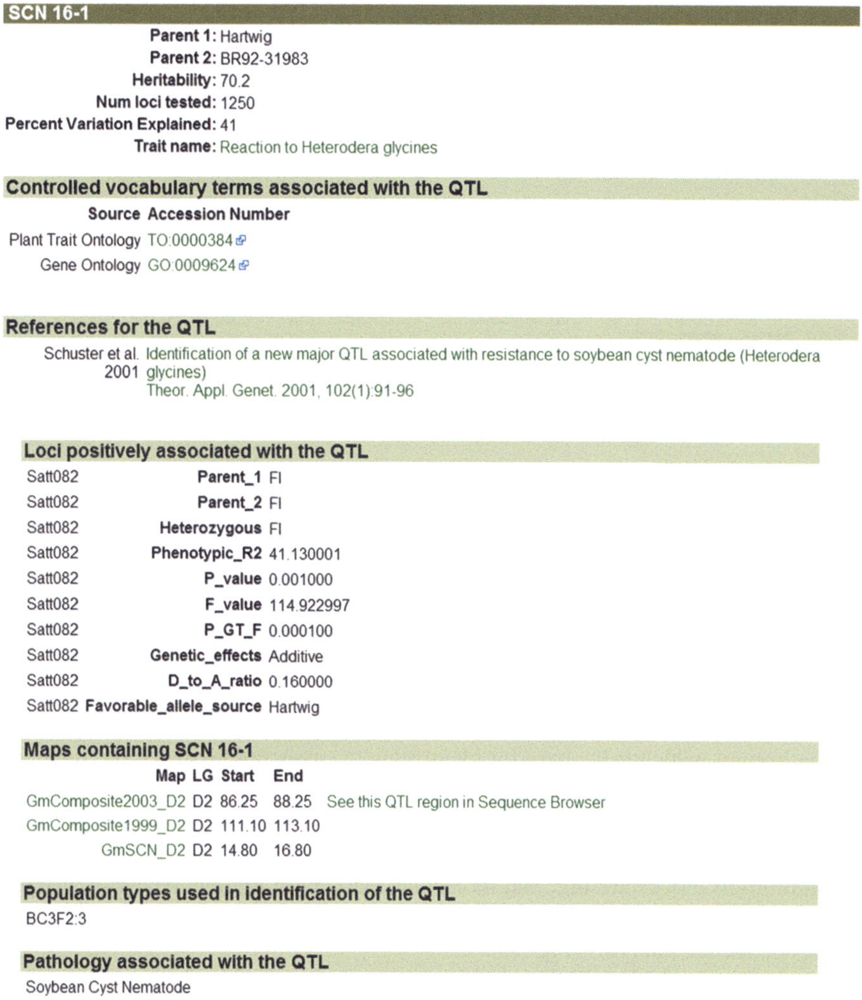

SCN 16-1

Parent 1:	Hartwig
Parent 2:	BR92-31983
Heritability:	70.2
Num loci tested:	1250
Percent Variation Explained:	41
Trait name:	Reaction to Heterodera glycines

Controlled vocabulary terms associated with the QTL

Source	Accession Number
Plant Trait Ontology	TO:0000384
Gene Ontology	GO:0009624

References for the QTL

Schuster et al. 2001	Identification of a new major QTL associated with resistance to soybean cyst nematode (Heterodera glycines) Theor. Appl. Genet. 2001, 102(1):91-96

Loci positively associated with the QTL

Satt082	**Parent_1**	FI
Satt082	**Parent_2**	FI
Satt082	**Heterozygous**	FI
Satt082	**Phenotypic_R2**	41.130001
Satt082	**P_value**	0.001000
Satt082	**F_value**	114.922997
Satt082	**P_GT_F**	0.000100
Satt082	**Genetic_effects**	Additive
Satt082	**D_to_A_ratio**	0.160000
Satt082	**Favorable_allele_source**	Hartwig

Maps containing SCN 16-1

Map	LG	Start	End	
GmComposite2003_D2	D2	86.25	88.25	See this QTL region in Sequence Browser
GmComposite1999_D2	D2	111.10	113.10	
GmSCN_D2	D2	14.80	16.80	

Population types used in identification of the QTL

BC3F2:3

Pathology associated with the QTL

Soybean Cyst Nematode

Fig. 13.3 Details obtained by clicking on QTL on the SoyBase page shown in Fig. 13.2 of the origin of mapping QTL SCN 16-1

Recently, analysis of linkage disequilibrium has been used for carrying out association mapping. Unlike linkage mapping, mapping by linkage disequilibrium infers association between genotypes (haplotypes) and phenotypic variation by evaluating the genetic polymorphisms generated in different backgrounds and through many generations of recombination (Dekkers and Hospital 2002).

Linkage disequilibrium is defined as the nonrandom association of alleles in different loci in the same chromosome. Considering a locus having the alleles A1 and A2 with frequencies of pA1 and pA2 and a second locus having alleles B1 and B2 with frequencies of pB1 and pB2, then, in equilibrium, even if the loci are

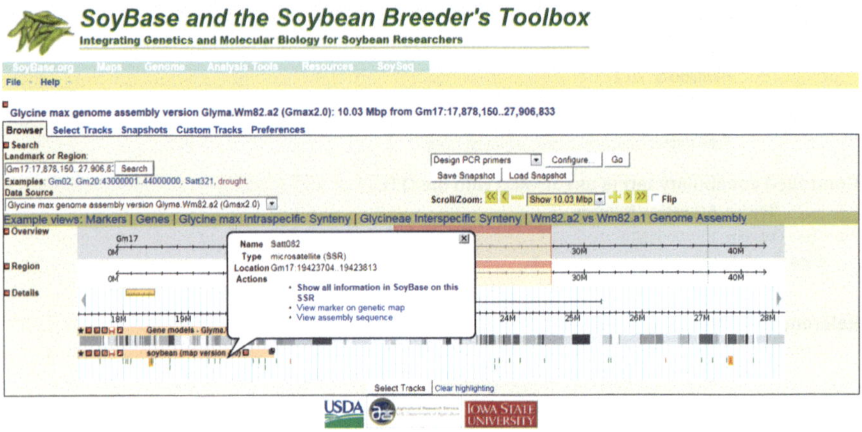

Fig. 13.4 Information on the physical map of the region containing the QTL SCN 16-1

linked, the frequency expected from a certain haplotype is the product of the frequencies of the alleles that constitute this haplotype. Thus, in equilibrium, pA1B1 = pA1pB1. Linkage disequilibrium can also be defined as the difference between the products of the haplotypes in coupling and the haplotypes in repulsion phases. In this case, in equilibrium, A1B1 × A2B2 = A1B2 × A2B1.

It is important to underline that the genetic mapping carried out with structured populations (F2, backcrossing, RILs, double haploids, etc.) is also based on linkage disequilibrium, which is the base for identifying the association, or genetic linkage, between markers and between markers and genes or QTLs. The greatest amount of linkage disequilibrium occurs in F2 populations. However, as only one generation of recombination takes place for obtaining linkage disequilibrium, the linkage disequilibrium blocks are big, and molecular markers in regions relatively distant in the genome appear linked, resulting in lower resolution of QTL mapping, with larger confidence intervals (Fig. 13.5).

For the linkage disequilibrium mapping, there is no need to prepare a mapping population. In this type of mapping, a collection of varieties or breeding lines are used for detecting blocks that are in linkage disequilibrium by using high density of markers.

Linkage disequilibrium analysis in collections of germplasm captures the recombinations that have occurred throughout the long history of recombination of the species. As the populations of the species accumulate hundreds or thousands of generations of recombination, the tendency is for all the loci in the genome to enter into equilibrium in the population. However, some loci continue in disequilibrium, even after a long history of recombinations of the species. These loci in linkage disequilibrium after many generations of recombination are closely linked and form the blocks of linkage disequilibrium (Fig. 13.5).

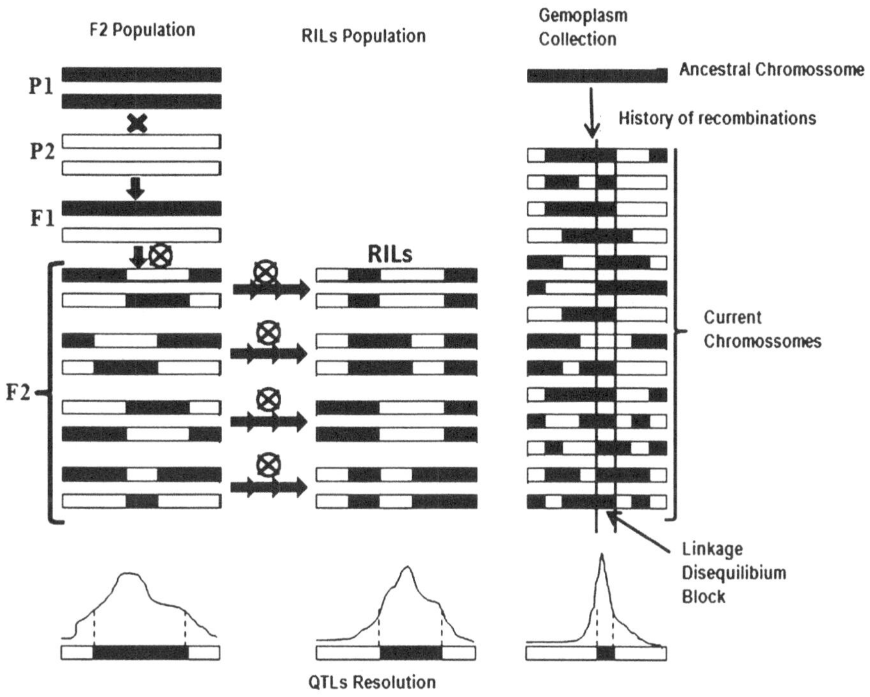

Fig. 13.5 Comparison between blocks of linkage disequilibrium and resolution of QTL mapping in F2 populations, RILs, and collection of germplasm with genetic variability. ⊗ indicates self-fertilization. The more generations there are for recombination, the greater the resolution of the mapping of genes and QTLs

That is why the blocks of linkage disequilibrium are usually small, and the molecular markers and genes or QTLs inside the same block of linkage disequilibrium must be closely linked.

The linkage disequilibrium map illustrates the blocks of linkage disequilibrium, based on the significance of the linkage disequilibrium observed between two markers (Fig. 13.6). The linkage disequilibrium between two loci is the function of time (number of generations) that has gone by since the generations of recombination started and the frequency of recombination between the loci. After several generations of recombination, in a nonstructured population, the remaining blocks of linkage disequilibrium are small. That is why the genome must be well saturated with markers so that it is possible to identify all the blocks of linkage disequilibrium. Platforms of high-density genotyping and markers that are well distributed throughout the genome, as is the case with SNP markers, need to be used. Besides that, the mapping populations must have sufficient genetic diversity and be large enough so that the blocks of linkage disequilibrium that are found represent the genetic structure of the species or of the subgroup of the species (e.g., tropical soybean).

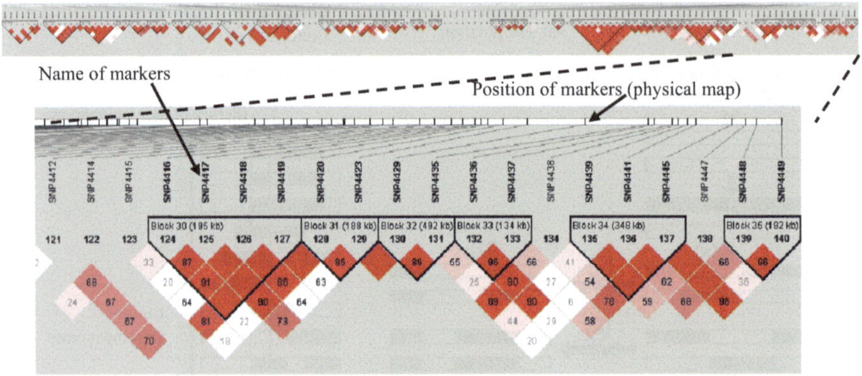

Fig. 13.6 Illustration of a linkage disequilibrium map of part of chromosome 17 (GL D2) of soybean. The upper part of the map shows a white horizontal bar with small vertical lines that indicate the position of the markers on the physical soybean map. These vertical lines are connected to the names of the markers indicated below the bar. Below the name of the markers, there is numeration in sequence showing the order of markers on the map. The lozenges (in shades of red to white) represent the linkage disequilibrium between the markers. The darker the red, the greater the significance of the linkage disequilibrium, and the paler the red, the smaller the significance, and this indicates linkage equilibrium. The triangles formed by the black lines containing one or more lozenges on the map represent the blocks of linkage disequilibrium. One block of linkage disequilibrium must be formed by a triangle that contains only red lozenges. The block furthest to the left of the figure, identified as block 30, represents a linkage disequilibrium block with four markers, and it is 195 kb in size

Linkage disequilibrium QTL mapping, also known as association mapping, identify and address QTLs based on the intensity of the correlation between mapped molecular markers and the phenotypic traits. The association of a marker with a QTL occurs when both are in the same block of linkage disequilibrium; in other words, only correlations between QTLs and markers that are closely linked should remain, thus making fine mapping easier.

The first obstacle for this association analysis is the existence, within the population, of subgroups with an unequal distribution of alleles. In these cases, false-positive associations between markers and phenotypes can be identified, even though there may not be any physical linkage between the markers and the genes (Bucker and Thornsberry 2002; Eathington et al. 2007). That is why the first stage in association mapping is to identify subgroups by doing an initial analysis of the structure of the population (Pritchard et al. 2000) (Fig. 13.7).

The association of QTLs with the molecular markers on the association mapping is generally carried out by the procedure known as GWAS (genome-wide association study). Several association methods can be used, from simple methods based on correlation analysis between the scores of the markers and the phenotype, but mainly linear models (mixed or generalized), models derived from BLUP (best linear unbiased predictor), Bayesian models, and some others. The results are generally expressed through the dispersion graph of $-\log10(P)$, by way of graphs called Manhattan Plot (Fig. 13.8).

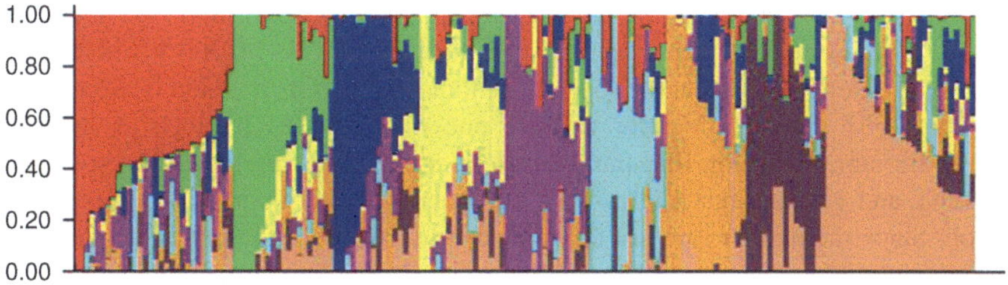

Fig. 13.7 Structure of a population of 169 Brazilian soybean cultivars. Each color represents a subgroup of cultivars, and each cultivar is shown by a vertical bar with the colors that represent the proportion of the genome that it contains from each subgroup. Some cultivars contain 100% of the genome of one subgroup (*a single color*), while most of them show a mixture of several subgroups (*several colors*). Source: Data from the author

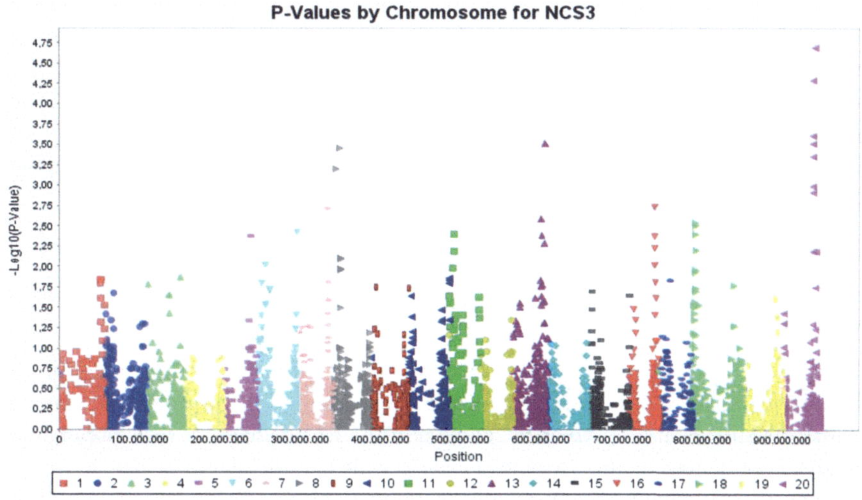

Fig. 13.8 Manhattan plot of association analysis of SNP markers and the resistance of soybean to soybean cyst nematode. Each color represents one chromosome. The higher the value of –Log10(*P*-Value), the more significant the marker. Source: Data from the author

By using a collection of varieties or breeding lines, the mapping by genomic association shows various advantages over the mapping that uses biparental crosses: (1) a wide genetic variation is shown by various haplotypes, and the markers of QTLs are not limited to just two parents; (2) the association mapping has greater resolution as it has accumulated all the meiosis throughout the generations; and (3) the phenotypic data that characterizes each cultivar, including data previously obtained, can be used without the need to test them with special mapping populations, reducing the time needed to obtain the map (Bucker and Thornsberry 2002; Sorkheh et al. 2008).

In a piece of work mapping soybean by linkage disequilibrium, we obtained blocks of linkage disequilibrium that had an average size of 300 kb (data not published). Considering that the soybean genome has 1.1 Gb, it would be necessary to have 3667 markers in order to have one marker in each block of linkage disequilibrium, assuming that these markers are well distributed in the soybean genome, and whose rarest alleles have a frequency higher than 20% in the population. Since the markers are not distributed equidistantly throughout the genome, alleles with a frequency equal or less than 20% are relatively common, and the linkage disequilibrium blocks are not uniform in size, with some small blocks (the smallest were less than 51 kb); the real number of markers must be twice or three fold of the number in theory. In other words, by using 7500–11000 markers, it is possible to have one marker in each block of linkage disequilibrium in soybean, which means having an associated marker at each QTL of the soybean genome.

Just like when mapping structured populations, the precision of the phenotypic information has a direct influence on the quality of the genetic map and on the association mapping. These two mapping models (by recombination in structured populations and by association in collections of germplasm) are not excludents. Using both the methods makes it possible to achieve greater precision in the mapping results, especially when there are rare alleles, because the association of rare alleles with a phenotypic characteristic is not precisely detected when mapping by association.

Final Considerations

Twenty-seven years have gone by since the publishing of the first soybean genetic map by Keim et al. (1990) that contained 150 RFLP markers in a population obtained by a nonspecific crossing *G. max* with *G. soja*. During that period, a consensus map containing thousands of markers, dozens of genes, and thousands of QTLs has been obtained. The information is available to community of breeders on SoyBase. The recent sequencing of the soybean genome and the availability of SNP marker kits that are able to genotype dozens of thousands of markers in a large number of soybean samples in a few weeks should generate even more information related to the soybean genome and link the information from the markers, genes, or QTLs with the soybean sequences.

Advances in the development of molecular markers, mapping methods, and analyzing capacity were not accompanied at the same pace by the increase in the capacity and quality of obtaining phenotypic data. The next challenge in QTL mapping in soybean is to obtain phenotypic data on a large scale that is extremely accurate. This science, which is revolutionizing plant breeding, is called Phenomics. In this technique, the most modern technologies are used for obtaining a greater volume of data in a shorter time and with greater precision.

References

Akkaya, M.S., Shoemaker, R.C., Specht, J.E., Bhagwat, A.A., Cregan, P.B.: Integration of simple sequence repeat DNA marker into a soybean linkage map. Crop Sci. **35**, 1439–1445 (1995)

Bucker IV, E.S., Thornsberry, J.M.: Plant molecular diversity and application to genomics. Curr. Opin. Plant Biol. **5**, 107–111 (2002)

Choi, I.-Y., Hyten, D.L., Matukumalli, L.K., Song, Q.-J., Chaky, J.M., Quigley, C.V., Chase, K., Lark, K.G., Reiter, R.S., Yoon, M.-S., Hwang, E.-Y., Yi, S.-I., Young, N.D., Shoemaker, R.C., van Tassel, C.P., Specht, J.E., Cregam, P.B.: A soybean transcript map: gene distribution, haplotype and SNP analysis. Genetics. **176**(1), 685–696 (2007)

Creagan, P.B., Jarvik, T., Bush, A.L., Shoemaker, R.C., Lark, K.G., Kahler, A.L., Kaya, N., van Toai, T.T., Lohnes, D.G., Chung, J.: An integrated genetic linkage map of the soybean genome. Crop Sci. **39**, 1464–1490 (1999)

Dekkers, J.C.M., Hospital, F.: The use of molecular genetics in the improvement of agricultural populations. Nat. Rev. Genet. **3**, 22–32 (2002)

Diers, B.W., Kein, P., Fehr, W.R., Shoemaker, R.C.: RFLP analysis of soybean seed protein and oil content. Theor. Appl. Genet. **83**, 608–612 (1992)

Eathington, S.R., Crosbie, T.M., Edwards, M.D., Reiter, R.S., Bull, J.K.: Molecular marker in a commercial breeding program. Crop Sci. **47**(S3), S154–S163 (2007)

Ferreira, A.R., Foutz, K.R., Keim, P.: Soybean genetic map of RAPD markers assigned to an existing scaffold RFLP map. J. Hered. **91**, 392–396 (2000)

Grant, D., Nelson, R.T., Cannon, S.B., Shoemaker, R.C.: SoyBase, the USDA-ARS soybean genetics and genomics database. Nucleic Acids Res. **38**(Suppl 1), D843–D846 (2010)

Keim, P., Diers, B.W., Olson, T.C., Shoemaker, R.C.: RFLP mapping in soybean: association between marker loci and variation in quantitative traits. Genetics. **126**, 735–742 (1990)

Keim, P., Schuoo, J.M., Travis, S.E., Clayton, K., Zhu, T., Shi, L., Ferreira, A., Weeb, D.M.: A high-density soybean genetic map based on AFLP markers. Crop Sci. **37**, 537–543 (1997)

Lande, R., Thompson, R.: Efficiency of marker-assisted selection in the improvement of quantitative traits. Genetics. **124**, 743–756 (1990)

Lark, K.G., Weisemann, J.M., Matthews, B.F., Palmer, R., Chase, K., Macalma, T.: A genetic map of soybean (*Glycine max* L) using an using an interspecific cross of two cultivars: 'Minsoy' and 'Noir I'. Theor. Appl. Genet. **86**, 901–906 (1993)

Mansur, L.M., Orf, J.H., Chase, K., Jarvik, K., Cregan, P.B., Lark, K.G.: Genetic mapping of agronomic traits using recombinant inbreeds lines of soybean. Crop Sci. **36**, 1327–1336 (1995)

Pritchard, J.K., Stephens, M., Donnelly, P.: Inference of population structure using multilocus genotype data. Genetics. **155**, 945–959 (2000)

Rafalski, A., Tingey, S.: RFLP map of soybean (*Glycine max*). In: O'Brian, S.J. (ed.) Genetic Maps: Locus Maps of Complex Genomes, pp. 149–156. Cold Spring Harbor Laboratory Press, New York (1993)

Schuster, I., Cruz, C.D.: Estatística genômica aplicada a populações derivadas de cruzamentos controlados, p. 568. Ed.UFV, Viçosa (2004)

Shoemaker, R.C., Olsen, T.C.: Molecular linkage map of soybean (*Glycine max* L. Merr). In: O'Brian, S.J. (ed.) Genetic Maps: Locus Maps of Complex Genomes, pp. 131–138. Cold Spring Harbor Laboratory Press, New York (1993)

Shoemaker, R.C., Specht, J.E.: Integration of the soybean molecular and classical genetic linkage group. Crop Sci. **35**, 436–446 (1995)

Song, Q.J., Marek, L.F., Shoemaker, R.C., Lark, K.G., Concibido, V.C., Dellanay, X., Specht, J.E., Cregan, J.E.: A new integrated genetic linkage map of the soybean. Theor. Appl. Genet. **109**, 122–128 (2004)

Sorkheh, K., Malysheva-Otto, L.V., Wirthensohn, M.G., Tarkesh-Esfahani, S., Martínez-Gómez, P.: Linkage disequilibrium, genetic association mapping and gene localization in crop plants. Genet. Mol. Biol. **31**(4), 805–814 (2008)

Yamanaka, N., Ninomiya, S., Hoshi, M., Tsubokura, Y., Yano, M., Nagamura, Y., Sasaki, T., Harada, K.: An informative linkage map of soybean revels QTLs for flowering time, leaflet morphology and regions of segregation distortion. DNA Res. **8**, 61–72 (2001)

Zeng, Z.B.: Precision mapping of quantitative trait loci. Genetics. **136**, 1457–1468 (1994)

Zou, Y.L., Song, Q.J., Hyten, D.L., van Tassel, C.P., Matukumalli, L.K., Grimm, D.R., Hyat, S.M., Fickus, E.W., Young, N.D., Cregan, P.B.: Single nucleotide polymorphism in soybean. Genetics. **163**, 1123–1134 (2003)

Chapter 14
Marker-Assisted Selection and Genomic Selection

Sandra Helena Unêda-Trevisoli, Fabiana Mota da Silva, and Antonio Orlando Di Mauro

Abstract Breeding techniques and molecular tools, i.e., biotechnology, have increased crop yield over the last century. Marker-assisted selection and, more recently, genome-wide association studies and genomic selection have been used as tools of breeding programs, especially in the resistant genotype selection. The genomic association study refers to significant associations between a loci and an interesting trait. Genomic selection is based on the effects of single nucleotide polymorphism (SNP) markers distributed throughout the genome, where the number of markers must be sufficiently high so all quantitative trait loci are in linkage disequilibrium with at least one marker. Thousands of markers distributed throughout the genome at reduced costs, as well as the easy access to biotechnologies, are the main way of aggregating these techniques in breeding programs. Several studies have shown that genomic selection is highly effective for improving crop yield. This tool increases the genetic gain of improved populations by increasing selection accuracy and, mainly, by reducing the generation intervals.

Keywords *Glycine max* • SNP • Genome-wide association study

Introduction

Molecular strategies were initially used as an effective and promising alternative for conducting plant breeding programs. Each year, different molecular techniques and new analysis strategies emerge to provide greater robustness to this process.

In general, a certain tendency is observed in plant improvement to integrate conventional techniques with molecular techniques and strategies. Thus, the purpose of this chapter is to provide applied knowledge on the use of the selection process assisted by molecular markers and genomic selection, as useful and

S.H. Unêda-Trevisoli, M.Sc., D.Sc., Ph.D. (✉) • F.M. da Silva, M.Sc., D.Sc.
A.O. Di Mauro, M.Sc., D.Sc.
Universidade Estadual Paulista, Jaboticabal, São Paulo 14884-900, Brazil
e-mail: strevisoli@fcav.unesp.br; motaagro@hotmail.com; orlando@fcav.unesp.br

F. Lopes da Silva et al. (eds.), *Soybean Breeding*,
DOI 10.1007/978-3-319-57433-2_14

supporting tools for programs of improvement of plant species, especially soybean crop.

DNA Markers and Marker-Assisted Selection (MAS)

Plant breeding, along with advanced agricultural practices, has shown remarkable progress in crop yields in the last century. Therefore, this area of science has currently played a key role in the search for increasing the food production. Based on the current productivity trends, predicted global population growth and environmental pressure, yield, stability, and sustainability traits should be a major focus of plant breeding, as well as the lasting resistance to disease, tolerance to abiotic stress, and efficiency in the use of water and nutrients. To achieve all of these goals, in addition to conventional breeding techniques, other tools such as biotechnology will be needed to maximize the likelihood of success. One such tool, known as DNA marker technology, is one of the most important tools in plant breeding. As the main characteristics of these markers, we can point out their stability, as they are not subject to environmental influence and because they can be used in any stage of plant development. What is more, they are generally polymorphic when compared to phenotypic markers. Some authors also mention that the potential of use of molecular markers in plant breeding is quite broad, highlighting the identification and discrimination of genotypes, quantification of genetic variability at the level of DNA sequences and their correlation with phenotypic expression, identification of parental origin and paternity tests, identification and protection of cultivars, evaluation of lines in heterotic groups, evaluation of gene flow, identification of duplicates in germplasm banks, genetic purity certification, crossbreeding monitoring, germplasm characterization, genetic diversity study, genetic mapping, and selection assistance. Moreover, they are very useful in identifying and selecting quantitative trait loci (QTLs), which may be responsible for the high proportion of phenotypic variation. Thus, the markers can provide important information regarding the number, chromosome position, magnitude of the effect and interaction among these loci.

Molecular markers in plant breeding programs can primarily provide increased efficiency and accuracy to routine activities. Thus, in several breeding programs, these molecular techniques have been used, through the process known as marker-assisted selection—MAS.

Considerations, Advantages, and Applications of MAS in Breeding

MAS has been widely used in many agricultural species. Federizzi (1998) points out that this is possibly the area with the greatest impact on breeding programs, for its use in the identification of superior genotypes, even in segregating populations. In this process, unfavorable alleles can be eliminated or reduced during the initial stages of plant development, and only those individuals that the breeder may have interest in can be kept under evaluation conditions.

In breeding programs, among the aspects to be evaluated are the cost and time for the development of cultivars and the selection of the traits of interest. In order to make this process viable, the use of the markers will be valuable when the locus of interest and the marker are strongly linked and, also, when the first generations of segregation are used to avoid erosion of the link between the locus and the mark.

One of the most important steps for the correct use of molecular markers is the establishment of the relationship between a given marker and a locus of interest, which is a laborious and careful step. Due to the recombination phenomenon, the regions that surround the locus of interest may be distinct, even among related genotypes. Therefore, a polymorphic marker between progenitors A and B may not be polymorphic between A and C. Thus, specific markers must be identified for each crossing. In many cases, however, the same marker may be useful in different crossings.

In the light of the above, one can consider that the success of MAS depends on the degree of association between the marker and the locus of interest. That is, the higher the degree of association, the lesser the chance of recombination, resulting in a more efficient selection.

In plant breeding, the search for selection assisted by molecular markers is increasing, aiming to increase the efficiency of transfer of traits of importance in the populations under study. Thus, obtaining markers linked to different superior traits in crops allows the indirect selection of desirable genotypes in early segregating generations, reducing the time, cost, and labor necessary to the development of large populations (Caixeta et al. 2009).

In the MAS, the identification of the gene of interest and the existence of markers to monitor the presence of favorable alleles should be considered. Furthermore, it is interesting to take into account aspects such as the genetic nature of the characteristic (quantitative or qualitative), gene action (additivity, dominance, or recessivity), the effect of the gene on phenotype expression, and the efficiency of the marker in the discrimination of the trait.

One of the areas in which MAS can present effective performance in breeding programs is with regard to resistance to diseases, due to its possibility of great practical application. This fact stems from the long time required and the laborious nature of conventional techniques for the selection of resistant genotypes, as well as the dependence on the need of adequate environmental conditions for the pathogen. As an example, quite often, the process of obtaining and maintaining the fungus for inoculation procedures is hampered by the characteristic of some pathogens, which require living cells for their development.

Accordingly, MAS can be considered the great tool of breeding programs, mainly in the process of characterization and selection of resistant genotypes, for it is a fast and safe process. Moreover, their use may facilitate the process known as gene pyramiding, which consists of the process of combining several genes within a single genotype. In soybean, this process was efficient for the development of a more stable resistance to the fungus that causes the Asian rust. In addition to gene pyramiding, another process that can be greatly facilitated is backcrossing, a method widely used in soybean breeding, to incorporate one or a few genes into a variety of elite or adapted genotypes. Thus, the use of MAS can greatly increase the efficiency in the selection process for recovery of the recurrent genotype, as is currently the case with modern cultivars bearing the RR gene.

According to Collard and Mackill (2008), there are five main considerations for the use of DNA markers in the MAS process:

- Reliability: The markers must be strongly linked to the target locus, preferably with genetic distance less than 5 cM. Additionally, the use of markers flanking the target gene will increase the reliability of the markers in predicting the phenotype (Fig. 14.1).

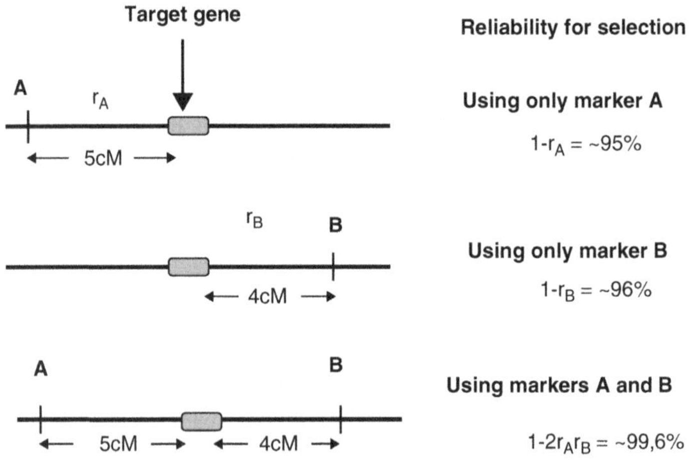

Fig. 14.1 Reliability of selection using individual and flanking markers (assuming there is no crossing-over interference). The frequency of recombination between the target gene and marker A is approximately 5% (5 cM). Therefore, recombination can occur between the target locus and marker in about 5% of the progeny. The frequency of recombination between the target gene and the B marker is about 4% (4 cM). The possibility of recombination between both markers A and B (i.e., double crossing-over) is much smaller than for individual markers (about 0.4%), and therefore the reliability of selection is much greater when using the flanking markers. Source: Adapted from Collard and Mackill 2008

- Quantity and quality of DNA: Some marking techniques require large amounts of DNA, in addition to high quality, a fact that can often hinder the practice and increase its costs in procedures.
- Feasibility of the process: The level of simplicity and the time required for the technique are important considerations. Efficient, simple, and fast methods are highly desirable.
- Level of polymorphism: The marker must be highly polymorphic regarding the group of genotypes evaluated, allowing the discrimination thereof.
- Cost: The marker must be cost-effective, in order to render its use feasible.

Main Types of Markers Used in MAS

Markers that reveal protein polymorphisms are known as biochemical markers, while DNA markers reveal polymorphisms in DNA. Biochemical markers are proteins produced as a result of gene expression, which can be separated by electrophoresis to identify the alleles. Isozymes, which are variant forms of the same enzyme, are the most commonly used, for they can reveal gene differences and work as codominant markers. However, its use is restricted mainly due to its limited number of cultivated species (Kumar 1999).

DNA markers can be classified into two categories, depending on how the polymorphism is revealed: hybridization-based polymorphisms and polymerase chain reaction (PCR) polymorphisms.

Hybridization-Based Marker Polymorphism

This class of markers includes the restriction fragment length polymorphism (RFLP) loci and the variable number tandem repeat (VNTR) loci, in which probes such as random genomic clones, cDNA clones, and those for the microsatellite and minisatellite sequences are hybridized to filters containing DNA, which are digested with restriction enzymes. The polymorphism in the case of VNTR loci is due to a difference in the number of replicates, while RFLPs are formed because of events such as point mutations, deletions, inversions, or translocations. Denaturing gradient gel electrophoresis (DGGE) is an alternative to RFLP analysis, which is able to identify polymorphisms between two DNA fragments of equal size, but differing in a single base pair (Ferreira and Grattapaglia 1998; Kumar 1999).

The polymorphism revealed by restriction fragment analyses allows the study of phenomena and biological processes: characterization of germplasm, construction of genetic maps, and evaluation of genetic diversity. These markers behave like codominants, producing quantity of relevant information, allowing the analysis of the gene action and of the allelic interactions present (Ferreira and Grattapaglia 1998).

Between these two markers, RFLP is of greater importance for the soybean crop. Keim et al. (1990) developed a genetic map for soy using RFLP and highlighted the quality of the technique. Some studies on cyst nematode resistance, including diversity analysis, were performed with good results.

The major limitation of this class of markers is basically the need to use laborious procedures and often components of the reaction that pose risk, especially radioactive probes (Ferreira and Grattapaglia 1998).

PCR-Based Marker Polymorphism

The process known as PCR (polymerase chain reaction) consists of a chain reaction of the polymerase, according to its denomination, where the amount of DNA of a certain sample is multiplied in vitro, exponentially. This technique was developed in the 1980s and expanded fast, due to its simplicity, speed, and efficiency (Mullis and Faloona 1987; Caixeta et al. 2009).

In general, the method consists in the in vitro replication of the DNA template—sequence, where each new molecule synthesized in one cycle is used as template in the next cycle, resulting in a significant increase in the number of copies. In the tube where the reaction occurs, the following components necessarily have to be present: DNA template, deoxyribonucleotides (dNTPs) for synthesis of new DNA strands, the primer oligonucleotides (primers) that will serve to guide the DNA region to be amplified, and the enzyme that polymerizes DNA (Taq DNA polymerase) in addition to the reaction buffering agents. These components are subjected to cycles with different temperatures for the occurrence of denaturation, annealing, and amplification, repeatedly, in order to guarantee the action of these reaction components and, consequently, the multiplication of template DNA. To facilitate the process, the technique was automated by the use of equipment called thermocyclers (Caixeta et al. 2009).

RAPD Markers

There are different molecular markers that can be used in the MAS process within the group of molecular markers based on PCR. In this context, RAPD (randomly amplified polymorphic DNA) is a technique that uses 10-base oligonucleotides of arbitrary sequences to amplify DNA fragments randomly distributed in the genome, without the need for prior knowledge of its sequence. The great advantages of RAPD are simple execution, fast data collection, relatively low cost and immediate applicability to any organism, and the need for small amounts of DNA (Ferreira and Grattapaglia 1998; Caixeta et al. 2009).

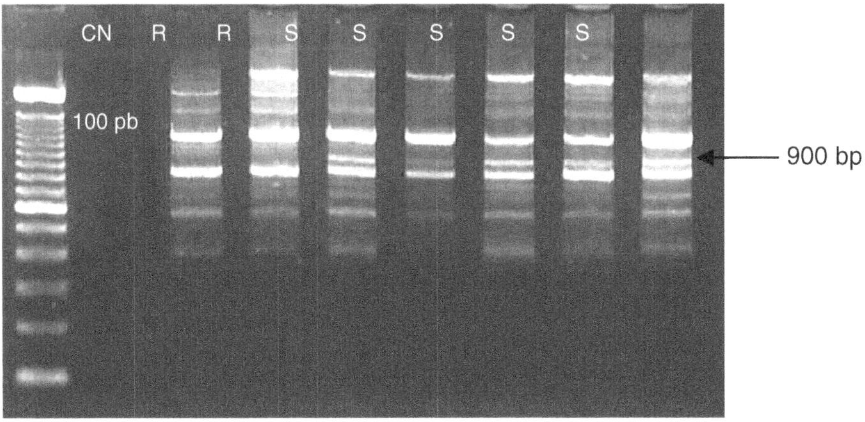

Fig. 14.2 RAPD reaction with the OPE-02 primer, testing two resistant cultivars (R) and five susceptible cultivars (S). It has been noted the polymorphism of the band of about 900 bp in the susceptible genotypes and the absence in the resistant ones. The first channel corresponds to the standard 100 bp, followed by the negative control of the reaction (CN). Source: personal archive

These characteristics make RAPD very interesting for identifying molecular markers linked to traits of interest, such as resistance genes, using the bulked segregant analysis (BSA) technique described by Michelmore et al. (1991). This technique consists in the amplification of two sets of individuals with equimolar amounts of DNA from selected genotypes based on the phenotypic expression of the trait, to detect polymorphisms linked to the target region between them. This preliminary evidence of link between the marker and the target locus is then confirmed by the segregation analysis of the entire population, estimating the frequency of recombination between both (Ferreira and Grattapaglia 1998).

In soybeans, RAPD markers associated with disease resistance were identified in relation to Asian rust, linked to the Rpp4 gene; for mildew resistance and stem cancer, several RAPD markers useful in the MAS process have been identified (Fig. 14.2).

SCAR Markers

Some characteristics related to RAPD markers such as low specificity and difficult interpretation may limit its use in certain situations, leading to the need for alternatives such as their conversion into sequence characterized amplified region (SCAR) markers, to suit the use of assisted selection. The technique consists in the isolation and cloning of the polymorphic fragment of interest, followed by the sequencing and design of the specific primers, with subsequent validation test (Ferreira and Grattapaglia 1998). Several studies on obtaining SCAR markers associated with resistance to soybean diseases can be observed in the literature, highlighting the work on resistance to the cyst nematode (Heer et al. 1998), frogeye leaf spot (Martins Filho et al. 2002), mosaic virus (Zheng et al. 2003), and stem cancer (Gavioli et al. 2007).

SSR Markers

Microsatellite molecular markers or SSR (*simple sequence repeats*) are simple sequences consisting of one to six nucleotides, which are repeated in tandem and are found in high frequency and wide distribution in eukaryotic genomes (Caixeta et al. 2009). The type of repetition (mono-, di-, tetra-, penta-, and hexanucleotides) and the frequency of motifs are variable among different taxa. Microsatellites are found in coding and noncoding regions of the genome.

The DNA sequences that flank the microsatellites are generally conserved among individuals of the same species, allowing the use of specific primers that amplify the repetitive DNA via PCR (Caixeta et al. 2009). Each microsatellite is a highly polymorphic locus, whose variations in the number of replicates determine different alleles in the profile of each individual. Such variation in the number of replicates may be caused by uneven crossing-over or slippage of the DNA polymerase replication.

SSR markers are characterized by being codominant and multi-allelic and having frequent and random distribution, allowing wide coverage of the genome (Caixeta et al. 2009). Moreover, they are based on PCR, which makes their use practical even with few amounts of DNA; they have high reproducibility and high information content per loco (Ferreira and Grattapaglia 1998).

All these characteristics make these markers efficient tools not only for studies of genetic diversity, but also for the construction of genomic maps, studies of genetic diversity, distinction and protection of varieties, evaluation of seed purity, use and conservation of germplasm, gene and quantitative trait locus (QTL) analysis, pedigree analysis, marker-assisted selection, and library analysis for gene cloning (Caixeta et al. 2009).

The high level of variation detected by SSR markers increases the resolution of the genealogy and genetic diversity study and reduces the number of markers required to distinguish genotypes. In soybean, more than 600 SSRs were developed and mapped, being distributed across all the linkage groups (Cregan et al. 1999), which allows wide coverage of the genome of the species. For this crop, several SSR markers associated with the Asian rust resistance genes were identified. In these studies, resistance genes were mapped in a population derived from FT-2 (resistant to Asian rust) and Davis (susceptible to Asian rust). Two markers (Satt079 and Satt307) flanking the dominant resistance gene for use in assisted selection were detected in the soybean C2 binding group. In the same region, the resistance gene arising from the Japanese genotype "Hyuuga" was also mapped between the Satt460 and Satt307 markers. The gene mapped on the FT2 cultivar is

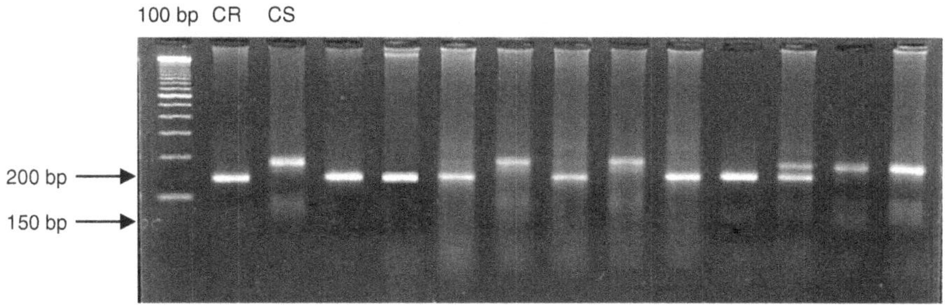

Fig. 14.3 Amplification of the Sat-162 SSR primer in DNA samples from 11 soybean segregant progenies for resistance to the cyst nematode (race 3), using as a control a resistant cultivar (CR) with a fragment of 150 bp and a susceptible cultivar (CS) with a fragment of 200 bp, aiming at the assisted selection process. Source: personal file

no longer active in Brazil due to the high diversity and virulence of the pathogen, and only the gene present in the genotype "Hyuuga" remains active. SSR molecular markers were also used to identify cyst nematode resistance genes in soybean crop (Fig. 14.3).

SNP Markers

Single-nucleotide polymorphism (SNP) markers are the most modern and most targeted class of molecular markers in the studies. These are based on the detection of the polymorphism resulting from the alteration of a single base pair in the genome sequence among individuals of the same species. Such variations occur due to mutations that may be insertions or deletions. SNPs are very frequent in the plant genome, and a SNP can be found every 100–300 bp, or less, depending on the species. There are several SNPs genotyping methods in specific regions of DNA, among them we could mention PCR-RFLP, automatic sequencing of PCR products marked with fluorescent dye, real-time PCR, and DNA microarrays, which are glass blades containing the printed target DNA, which allow the detection of numerous SNPs at the same time. A high-density map for these markers in soybean was published by Hyten et al. (2010), which served as the basis for several studies involving this marker. In recent studies, three functional SNP markers have been found, two for the Rhg1 locus and one for the Rhg4 locus, which may provide genetic information for the selection of soybean nematode-resistant genotypes (NCS). In general, one or two of these markers are already sufficient for a robust selection of plants that feature resistance to NCS (Shi et al. 2015).

In another recent work with resistance to the golden mosaic virus in India (MYMIV), the authors conducted an assay to develop large-scale SNP markers and identify potential markers linked to resistance loci for their effective use in genomic-assisted breeding focused on tolerance to the MYMIV virus (Yadav et al.

2015). Because of their robustness, currently, the SNP markers are the most used and promising class of markers, especially for resistance to diseases, but access is still somewhat limited, due to the automation required for the equipment used. However, their costs have been gradually reduced, with the option of outsourcing services, thus facilitating their access, even among different countries.

The evolution of bioinformatics associated with genomic studies was and remains pivotal for an accurate identification of the SNPs studied as well as for their associations and consequent uses within breeding programs. Currently, SNPs constitute the most recommended class of markers for mapping and genomic selection studies. Thus, many of the studies that will be mentioned below were performed from SNP markers, due to their high accuracy and efficiency.

Mapping of QTLs

Many traits of agronomic importance in plants are of a quantitative nature, whose phenotypic expression presents continuous variation, attributed to the simultaneous segregation of many genes distributed by the genome. The chromosomal regions related to their location are defined as QTLs (quantitative trait loci). With the advent of molecular markers, it was possible to map such regions (QTLs) that affect these quantitative traits. Thus, mapping a QTL means identifying its position in the genome and estimating its genetic effects, such as additive effect, dominance effect, and other effects present in the adopted model.

The use of molecular markers is, therefore, a valuable tool with great potential for genetic improvement programs, since it allows the identification of genes responsible for several characteristics of agronomic importance. Moreover, the construction of genetic linkage maps with molecular markers can help in the elaboration of strategies to be introduced in genetic breeding programs, in order to estimate how many and which are the QTLs responsible for the variation in traits as well as to locate its position in the genome, in addition to estimating their effects and their interrelationships.

One of the direct applications of genetic mapping is the localization of quantitative trait loci, including those of resistance to diseases associated with horizontal resistance. Furthermore, genetic mapping enables the selection of molecular markers (MAS) in selection and gene cloning programs, reducing the time and resources spent on the development of new cultivars.

Many are the studies with QTL mapping in soybean, for several traits, namely, oil content, protein content, resistance to soybean rust, resistance to powdery mildew, and resistance to soybean cyst nematode (races 3 and 9), and also for several agronomic traits such as plant height, weight of 100 seeds, grain yield, lodging, days for flowering, number of pods per node, and number of nodes in the main branch, among others.

Genome-Wide Association and Genomic Selection

In this topic we discussed the main concepts, methods used, and some examples of the use of this tool with soybean crop. However, this is a new topic, which is still being introduced by soybean breeding programs, both in the private and public sectors, and, therefore, there are not many published results for this crop yet. However, for some crops such as corn, wheat, barley, and forest species, there are already satisfactory results on this subject, with either simulated or real data.

Genetic-Wide Association Studies (GWAS)

GWAS (genome-wide association studies) refers to association studies between loci that govern quantitative traits (QTLs) of interest in breeding, through hypothesis testing, which detects significant effects (p-value). A very important fact in the studies of association and genomic selection is the existence of linkage disequilibrium (LD), that is, it is important that the markers are in imbalance with the loci of quantitative traits (QTLs) (Resende et al. 2013).

The purpose of the GWAS is to track the entire genome by looking for regions that affect the traits of interest in breeding programs and relate them to biological functions through genomic databases, e.g., SoyBase, Genome Blast, NCBI, and QTLdatabase. Thus, GWAS is the most comprehensive and thorough method to investigate the entire genome of a species, currently available.

The LD principle is studied to define the existing association between SNPs, also defined as a nonrandom association of alleles in different loci. When two or more specific alleles, at different loci, on the same chromosome are more frequent together than when separated, the loci are considered to be in disequilibrium. Even with the large linkage disequilibrium (LD) in soybean, large numbers of molecular markers are required for genetic analysis to cover the entire genome (Hwang et al. 2014).

In soybean crop, genome-wide association studies (GWAS) have been used to analyze several characteristics, such as disease resistance, yield, drought tolerance, and protein and oil content, among others (Wen et al. 2015).

Recently, with the development and emergence of high-throughput genotyping and sequencing techniques, the cost of SNP genotyping has been drastically reduced, making broad genomic association studies more accessible to breeding programs. Song et al. (2013) developed a SNP chip for the high-density soybean crop, called SoySNP50K, in order to characterize the soybean genome.

Genome-wide association studies (GWAS) in soybean have been conducted aiming to accelerate the improvement for the main agronomic traits. Soybean contains 20 pairs of chromosomes ($2n = 40$) in its genome, for which were found different regions of effect in different traits (Table 14.1).

Table 14.1 Broad genomic association with main characteristics in the soybean crop

Traits	Chromosome number	Authors
Oil content	02, 04, 05, 06, 08, 09, 10, 14, 15, 16, 19, and 20	Hwang et al. (2014) and Sonah et al. (2015)
Protein content	05, 06, 07, 08, 09, 10, 11, 12,13, 14, 16, 19, and 20	Hwang et al. (2014) and Sonah et al. (2015)
Flowering	02, 06, 07, 08, 09, 11, 13, 14, 18, 19, and 20	Zhang et al. (2015)
Maturity	06, 11, and 19	Zhang et al. (2015)
Height	02, 03, 04, 05, 06, 07, 08, 09, 10, 12, 13,15, 16, 17, 19, and 20	Zhang et al. (2015)
Hilum color	08 and 09	Sonah et al. (2015)
Flower color	13	Sonah et al. (2015)
Resistencia to *Sclerotinia sclerotiorum*	01, 15, 19, and 20	Bastien et al. (2014)
Resistencia to the cyst nematode	06, 08, 09,10, and 12	Vuong et al. (2015)

Genomic Selection

Plant breeding depends, to a great extent, on the phenotypic selection of quantitative traits, for example, resistance to diseases, pests, increased yield, tolerance to drought, and heat, among others. Breeding schemes that offer a high level of success for selection of traits, especially those with low heritability, have been the goal for several species.

Advances in genotyping technology, such as GBS sequencing and state-of-the-art platforms, make genomic selection more attractive due to reduced reproduction cycle as well as costs associated with phenotyping (Jarquín et al. 2014). More recently, as a result of cost reduction and greater access to SNP markers, this method has been gaining ground among genetic breeding programs.

Genomic selection (GS) consists of the selection based on the effects of single-nucleotide polymorphism (SNP) markers distributed throughout the genome (Meuwissen et al. 2001). In this method, the marker density must be sufficiently high so that all quantitative trait loci (QTLs) are in linkage disequilibrium with at least one marker.

GS can be applied to all families under evaluation within a breeding program, which presents high selection accuracy and does not require prior knowledge of the QTL positions on the map (Resende et al. 2013). GS estimates the genetic merit of an individual based on the sum of the effects of all SNP markers and not only some, such as in the marker-assisted selection. The main advantage of GS over conventional breeding is the possibility of reducing the interval of generations by means of selection based on the effects of markers even before the phenotypes are available. Once the genomic values of the individuals that still have not had their phenotypes measured have been predicted, according to the effects of the markers estimated in the training population, the candidates for selection may be selected early, thus shortening the generation interval and accelerating the genetic gain (Resende et al. 2008).

Many studies on GS have been reported in animal breeding and in plant breeding (Resende et al. 2008; Heffner et al. 2009). However, there are few reports on the use of GS in the soybean crop. Jarquín et al. (2014), when evaluating the perspectives and use of genomic selection in soybean crop for yield and other agronomic characteristics, concluded that genomic selection had good potential to improve soybean grain yield, accounting for approximately 32% of the phenotypic variation of grain production.

Populations Used in Genomic Selection: Estimation, Validation, and Selection

When applying genomic selection in a breeding program, three types of population are required, namely, training or estimation population, validation population, and selection population. In the training population, the effects of single-base polymorphic (SNP) markers are estimated based on genotyped individuals with high-density panels and phenotyped against the quantitative traits of interest in the breeding program. Subsequently, the genomic values of the candidates to selection are predicted according to the effects of the genetic markers of their genome. In the validation population, only the genomic prediction ability is determined through the correlation between the predicted—genomic and/or phenotypic—values and the estimated genetic values and/or the phenotypes, depending on the purpose. Only the individuals who are candidates for selection are used in the selection population, employing the prediction equations that were obtained in the training population.

In summary, genomic selection is basically performed in two steps: first, it is made the collection of phenotypes and the genotyping of the training population, in which the effects of markers are estimated; later, the candidates for the selection are genotyped and have their phenotypes predicted according to the effects of the markers estimated in the training population. This procedure is shown in Fig. 14.4.

The central process of the genomic selection is the calculation of genomic breeding values (GEBVs), or genomic values, which corresponds to the sum of all marker effects for a particular genotype. GEBVs are calculated using a model that was used in the training population individuals. In order to increase the accuracy of GEBVs, the training population should be representative of the candidates for the breeding program selection to which the genomic selection will be applied (Heffner et al. 2009).

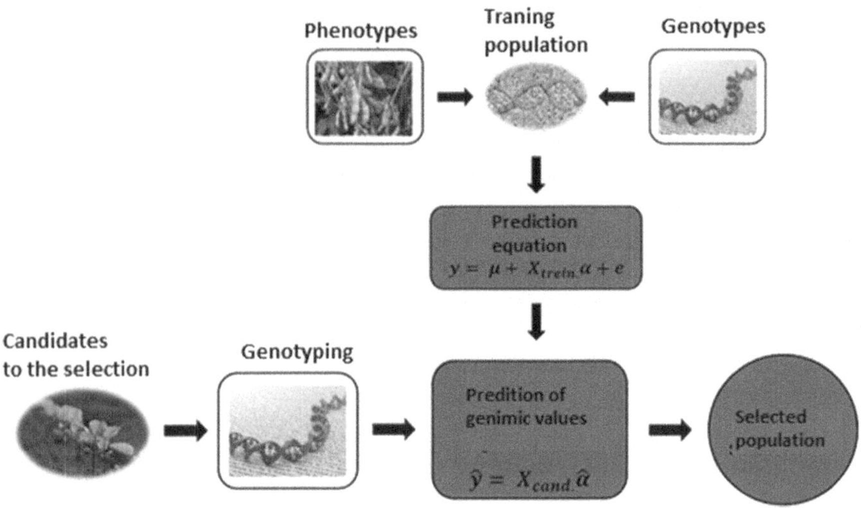

Fig. 14.4 Genomic selection process in plants, with description of training and selection populations used. Source: personal archive

Main Statistical Methods Used in Genomic Selection

For the implementation of genomic selection in breeding programs, it is necessary to apply genomic prediction methods to estimate the effects of markers and, then, GEBVs. Thus, it is necessary to establish the statistical method for genomic prediction, considering the genomic data, the complexity of the quantitative traits, and the high number of markers, which is one of the main challenges to the need of estimating several effects in a limited number of observations (N < P). In this context, several statistical methodologies have been proposed for the application of genomic selection (Meuwissen et al. 2001). They include the RR-BLUP (random ridge regression-best linear unbiased prediction); G-BLUP (genomic-BLUP); the Bayesian parametric models Bayes A, Bayes B, Bayes C, Bayes Cπ, Bayes D, Bayes Dπ, and LASSO Bayesian; and the semiparametric models, such as regression kernel Hilbert spaces (RKHS). However, there is no consensus on the most appropriate method for each population and/or characteristic (Hayes et al. 2009).

In the literature we find many studies for comparison of methods for genomic prediction used in both animal and plant breeding (Meuwissen et al. 2001; Zhao et al. 2013; Jarquín et al. 2014) by means of simulation or actual data.

According to Meuwissen et al. (2001), Bayesian methods are superior when the QTL effects are not normally distributed due to the presence of genes with large effects (Resende et al. 2013). However, Guo et al. (2012) indicate that the RR-BLUP method is equal to or greater than the others and is the most indicated for the improvement of plants, due to its easy implementation. Similar results evidencing the superiority of Bayesian methods and RR-BLUP were found by Zhao et al. (2013) in the wheat crop. In relation to the soybean

Table 14.2 Main statistical methods used in genomic selection

BLUP	It is assumed that the variance of the effects of the markers is the same for all segments of chromosomes
	Infinitesimal model with many loci of small effects
LASSO	Combines regularization with variable selection
	Estimates the effects of the most important SNPs, reducing to zero the estimates of the least important SNPs
Bayes A	Features high zero mass peak
	Supports many genes of small effect and few genes with great effects
Bayes A	A priori distribution is an inverted chi-square distribution, which results in a scaled d t-distribution for the effects of the markers
Bayes B	Similar to Bayes A
	The zero mass point is higher for the Bayes B a priori distribution, compared to Bayes A
	Many markers effects are allowed as zero
Bayes C	Characterized by the Gaussian distribution
	A common variance is specified for all loci
Bayes Cπ	A Bayes B variation
	Allows the modeling of the exponential double distribution
RKHS	Proposes to estimate the relationship between the response variable and a set of independent variables
	Allows detection of nonadditive effects

crop, Jarquín et al. (2014) concluded that G-BLUP was the best method. It is worth mentioning that, because it is a current topic, there are still few published studies published on soybean crop.

The main features for each methodology are presented in Table 14.2. Further details on the methodologies used to estimate the SNP effects can be obtained in the work of Resende et al. (2013).

Final Considerations

MAS is, undoubtedly, a process that can greatly contribute to plant breeding programs in obtaining superior cultivars. However, it ought to be considered that, in order to make the most of its potential more effectively, it is required greater integration with breeding programs, current barriers have to be well understood, and solutions have to be reached. Cost reduction and easier access to the newest techniques will be the main ways to aggregate the two processes.

As for genomic selection, it is possible to conclude that its application in plant breeding is a highly effective strategy for the selection of traits of agronomic importance. With the use of this tool, it is possible to increase the genetic gain of improved populations through better selection accuracy and, mainly, by reducing the generation interval.

References

Bastien, M., Sonah, H., Belzile, F.: Genome wide association mapping of *Sclerotinia sclerotiorum* resistance in soybean with a genotyping-by-sequencing approach. Plant Genome. **7**(1), 1–13 (2014)

Caixeta, E.T., Oliveira, A.C.B., Brito, G.G., Sakiyama, N.S.: Tipos de marcadores moleculares. In: Borém, A., Caixeta, E.T. (eds.) Marcadores Moleculares, pp. 11–94. Ed. UFV, Viçosa (2009)

Collard, B.C.Y., Mackill, D.J.: Marker-assisted selection: an approach for precision plant breeding in the twenty-first century. Philos. Trans. R. Soc. **363**, 557–572 (2008)

Cregan, P.B., Jarvik, T., Bush, A.L., Shoemaker, R.C., Lark, K.G., Kahler, A.L., Kaya, N., Vantoai, T.T., Lohnes, D.G., Chung, J., Specht, J.E.: An integrated genetic linkage map of the soybean genome. Crop Sci. **39**(5), 1464–1490 (1999)

Federizzi, L.C.: Estrutura de um programa de melhoramento de plantas e possíveis aplicações de marcadores moleculares: visão do melhorista. In: Milach, S.C.K. (ed.) Marcadores Moleculares em Plantas, pp. 3–15. UFRGS, Porto Alegre (1998)

Ferreira, M.E., Grattapaglia, D.: Introdução ao uso de marcadores moleculares em análise genética, 3rd edn, p. 220. Embrapa-Cenargen, Brasília (1998)

Gavioli, E.A., Di Mauro, A.O., Centurion, M.A.P.C., Di Mauro, S.M.Z.: Development of SCAR marker inked to stem canker resistance gene in soybean. Crop Breed. Appl. Biotechnol. **7**, 133–140 (2007)

Guo, Z., Tucker, D.M., Lu, J., Kishore, V., Gay, G.: Evaluation of genome-wide selection efficiency in maize nested association mapping populations. Theor. Appl. Genet. **124**(2), 261–275 (2012)

Hayes, B.J., Daetwyler, H.D., Bowman, P., Moser, G., Tie, R.B., Crump, R., Khatkar, M., Raadsma, H.W., Goddard, M.E.: Accuracy of genomic selection: comparing theory and results. Proc. Assoc. Advmt. Anim. Breed. **18**, 34–37 (2009)

Heer, J.A., Knap, H.T., Mahalingam, R., Shipe, E.R., Arelli, P.R., Matthews, B.F.: Molecular markers for resistance to *Heterodera glycines* in advanced soybean germplasm. Mol. Breed. **4**, 359–367 (1998)

Heffner, E.L., Sorrells, M.E., Jannink, J.L.: Genomic selection for crop improvement. Crop Sci. **49** (1), 1–12 (2009)

Hwang, E.Y., Song, Q., JIA, G., Specht, J.E., Hyten, D.L., Costa, J.M.: A genome-wide association study of seed protein and oil content in soybean. BMC Genomics. **15**(1), 1–12 (2014)

Hyten, D.L., Cannon, S.B., Song, Q., Weeks, N., Fickus, E.W., Shoemaker, R.C., Specht, J.E., Farmer, A.D., May, G.D., Cregan, P.B.: High-throughput SNP discovery through deep resequencing of a reduced representation library to anchor and orient scaffolds in the soybean whole genome sequence. BMC Genomics. **11**(38), 1–8 (2010)

Jarquín, D., Kocak, K., Posadas, L., Hyma, K., Jedlicka, J., Graef, G., Lorenz, A.: Genotyping by sequencing for genomic prediction in soybean breeding population. BMC Genomics. **15**(740), 1–10 (2014)

Keim, P., Diem, B.W., Olson, T.C., Shoemaker, R.C.: RFLP mapping in soybean: association between marker loci and variation in quantitative traits. Genetics. **126**, 735–742 (1990)

Kumar, L.S.: DNA markers in plant improvement: an overview. Biotechnol. Adv. **17**, 143–182 (1999)

Martins Filho, S., Sediyama, C.S., Moreira, M.A., Barros, E.G.: RAPD and SCAR markers linked to resistance to frogeye leaf spot in soybean. Genet. Mol. Biol. **25**(3), 317–321 (2002)

Meuwissen, T.H.E., Hayes, B.J., Goddard, M.E.: Prediction of total genetic value using genome-wide dense marker maps. Genetics. **157**, 1819–1829 (2001)

Michelmore, R.W., Paran, I., Kesseli, R.V.: Identification of markers linked to disease resistance genes by bulked segregant analysis: a rapid method to detect markers in specific genomic regions by using segregation populations. Proc. Natl. Acad. Sci. USA. **88**, 9828–9832 (1991)

Mullis, K., Faloona, F.: Specific synthesis of DNA in vitro via polymerase catalyzed chain reaction. Methods Enzymol. **55**, 335–350 (1987)

Resende, M.D., Lopes, P.S., Silva, R.L., Pires, I.E.: Seleção genômica ampla (GWS) e maximização da eficiência do melhoramento genético. Pesquisa Florestal Brasileira. **56**, 63–78 (2008)
Resende, D.V., Silva, F.F., Resende Júnior, M.F.R.: Seleção Genômica Ampla (GWS). In: Borém, A., Fritsche-Neto, R. (eds.) Biotecnologia aplicada ao melhoramento de plantas, pp. 151–188. Suprema, Visconde do Rio Branco (2013)
Shi, Z., Liu, S., Noe, J., Arelli, P., Meksem, K., Li, Z.: SNP identification and marker assay development for high-throughput selection of soybean cyst nematode resistance. BMC Genomics. **16**(314), 1–12 (2015)
Sonah, H., O'donoughue, L., Cober, E., Rajcan, I., François, B.: Identification of loci governing eight agronomic traits using a GBS-GWAS approach and validation by QTL mapping in soya bean. Plant Biotechnol. J. **13**, 211–221 (2015)
Song, Q., Hyten, D.L., Jia, G., Quigley, C.V., Fickus, E.W., Nelson, R.L., Cregan, P.B.: Development and evaluation of SoySNP50K, a high-density genotyping array for soybean. PLoS One. **8**(1), 1–12 (2013)
Vuong, T.D., Sonah, H., Meinhardt, C.G., Deshmukh, R., Kadam, S., Nelson, R.L., Shannon, J.G., Nguyen, H.T.: Genetic architecture of cyst nematode resistance revealed by genome-wide association study in soybean. BMC Genomics. **16**(593), 1–13 (2015)
Wen, Z., Boyse, J.F., Song, Q., Cregan, P.B., Wang, D.: Genomic consequences of selection and genome-wide association mapping in soybean. BMC Genomics. **15**(671), 1–14 (2015)
Yadav, C.B., Bhareti, P., Muthamilarasan, M., Mukherjee, M., Khan, Y., Rathi, P., Prasad, M.: Genome-wide SNP identification and characterization in two soybean cultivars with contrasting *Mungbean Yellow Mosaic India Virus* disease resistance traits. PLoS One. **10**(4), 1–15 (2015)
Zhang, J., Song, Q., Cregan, P.B., Nelson, R.L., Wang, X., Wu, J., Jiang, G.L.: Genome-wide association study for flowering time, maturity dates and plant height in early maturing soybean (*Glycine max*) germplasm. BMC Genomics. **16**(217), 1–11 (2015)
Zhao, Y., Zeng, J., Frenando, R., Reif, J.C.: Genomic prediction of hybrid wheat performance. Crop Sci. **53**, 1–9 (2013)
Zheng, C., Chang, R., Qiu, L., Chen, P., Wu, X., Chen, S.: Identification and characterization of a RAPD/ SCAR marker linked to a resistance gene for soybean mosaic virus in soybean. Euphytica. **132**, 199–210 (2003)

Chapter 15
Breeding GM Cultivars

Ivan Schuster

Abstract The advent of transgenic plants has brought tremendous benefits to farmers around the world but has also brought serious challenges to breeding programs. In addition to the biosafety protocols introduced in the development of transgenic varieties or hybrids, and the obligatory use of molecular tools to select the transgenic traits, new breeding processes needed to be introduced. In cross-pollinated plants, fewer transgenic lines can produce a large number of hybrids, and therefore transgenic trait introgression programs through backcrosses are the method of choice. This meant that, parallel to breeding inbreed lines and testing hybrids, a step of conversion of inbreed lines needs to be part of the allogamous plant breeding. In autogamous plants, thousands of lines need to be developed annually, and trait introgression is not feasible for new lines. Breeding of autogamous plants should be performed with transgenic traits introduced into breeding populations. Once the start of breeding for a new trait starts with a single or a small number of donors, the genetic basis of breeding populations is narrowed at each start of breeding for a new trait. In addition, in some cases, it is still necessary to eliminate old transgenic traits to replace the new ones. The elimination of plants that do not contain the desirable traits of the breeding populations makes their effective sizes smaller in the first generations of a new transgenic trait. Associated with the reduction of the genetic base due to the small number of donors, the tendency is that the speed of the genetic gain in the initial generations of breeding for a new trait will be reduced. To compensate, breeding methods that develop and recycle varieties in less time need to be used.

Keywords Backcross • Marker-assisted selection • MAS • Effective size • Genetic basis • Accelerated breeding

---

I. Schuster, M.S., D.S. (✉)
Dow AgroSciences, Cravinhos, São Paulo 14140-000, Brazil
e-mail: ivanschuster.ivan@gmail.com

F. Lopes da Silva et al. (eds.), *Soybean Breeding*,
DOI 10.1007/978-3-319-57433-2_15

Introduction

Since its launch in 1996, GM soybean crops have quickly gained farmers' preference due to the benefits offered, as the easier weed management and reduced production costs. As of 2013, the ease of insect management was also included in this benefit list for South American soybean farmers. Among all GM crop farming, soybean farmers were the fastest adopters of the new technology, and soybean occupies the largest area of GM crops in the world (Fig. 15.1). In 2014, 91 million hectares (82%) of the 111 million hectares planted with soybeans in the world were GM (James 2014).

In Brazil, the cultivation of GM soybean officially began in 2005. In the 2013–2014 crop season, 28 million hectares (93%) out of the 30.2 million hectares planted with soybeans in the country (Conab 2014) were cultivated with GM soybeans (Kleffmann Group 2014).

These numbers illustrate the importance of GM plant technology in soybean farming. Consequently, the breeding of GM soybeans is currently the standard method of soybean breeding in the main producing countries.

Biotechnology, Biosafety Regulation, and Breeding in Obtaining GM Cultivars

All activities madding with regulated GM plants (as other GMO) need to follow the biosafety laws in each country and the rules of the regulatory agencies. It applies not only in research but also in storage, transportation, and all activities with regulated GMO.

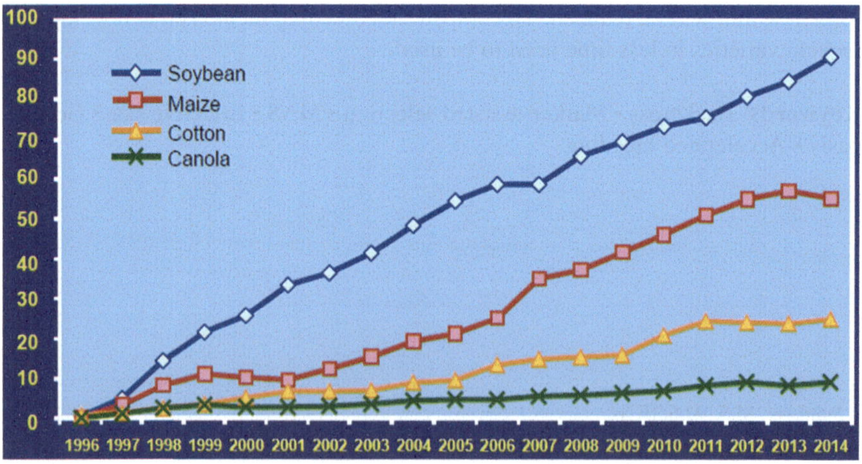

Fig. 15.1 Evolution of the main GM crops in the world, from 1996 to 2014, in millions of hectares (Source: James 2014)

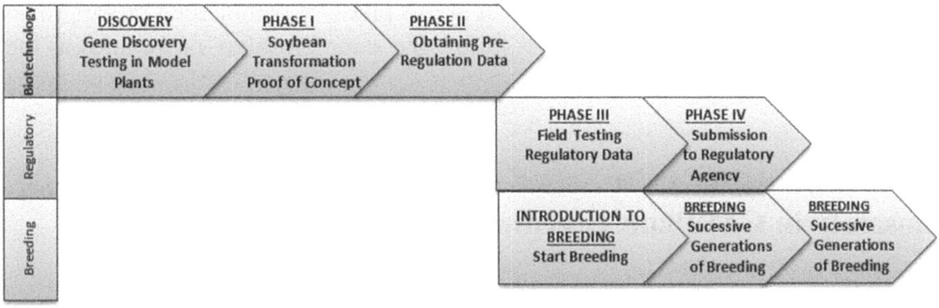

Fig. 15.2 Steps to obtain GM soybean cultivars, from the identification of the gene of interest to the breeding of plants with this new trait. This model is similar for all GM crops

Three distinct and equally important stages comprise the development GM cultivar: (a) discovery and obtaining the elite event by the biotechnology team, (b) biosafety evaluation of the new event for its deregulation by the regulatory team, and (c) obtaining GM cultivars containing the new trait by the breeding team (Fig. 15.2). The first two stages are carried out only once for any new event, while the breeding of GM plants with this new event is continuous.

The target gene is initially identified/discovered using biotechnology and genetic engineering procedures and tools. This gene is isolated, and an insert (DNA fragment) is constructed, containing the gene, one promoter, one terminator, and one selection gene, to be used in the transformation of plants. In the discovery phase, numerous candidate genes for the target trait are transferred to model plants (tobacco or *Arabidopsis*). From the evaluation of these candidate genes in the model plants, the gene to be used in the target species is selected. Next (phase I), the target species (e.g., soybeans) is transformed with the gene selected in the previous step, and the proof of concept is performed. In this proof of concept, the expression of the desired trait in the target plant (soybean), such as herbicide tolerance, insect resistance, or otherwise, is then evaluated. An evaluation is necessary on whether or not the expression of the gene in the target plant produces a sufficient amount of the protein expressed by the gene, so that the desired trait manifests itself with high efficiency. In this phase, many transformation events are obtained, to select an elite event. The elite event should contain a single copy of the insert, express itself efficiently, and not cause any changes in the transformed plant, in addition to the change desired by the transformation.

After selecting the elite event, start the activities of regulatory trials in the field, obtaining the data to be included in the dossier, that demonstrates the food and environmental safety of the new GM event, to be submitted to the regulatory agency. The regulatory agency evaluates the submitted dossier and decides whether to approve or not the new GM plant.

In parallel of the regulatory phase, breeders can start the crosses to obtain GM cultivars with this new trait. There is a risk associated with the beginning of the breeding activities prior to the commercial approval of the new transgenic event, as there is no guarantee that this trait will be approved for commercial use. For this reason, the breeding activities carried out in this phase of commercial pre-approval are also regulated activities and must follow the same biosafety procedures as the regulatory assays.

Biosafety in GM Plant Breeding

During the commercial pre-approval phase, both regulatory trials (to obtain biosafety data for trait deregulation) and breeding trials should follow biosafety standards in order to avoid that GM seeds escape into the seeds or commercial grain production chain. This is necessary because there is no guarantee that the new trait will be approved for commercial use or its food and environmental safety, before the final evaluation of the regulatory agency.

Field trials with GM soybeans containing traits not yet approved for commercialization should be isolated from other soybean crops. Normally, no more than 10 m is enough to prevent cross contamination by pollen flow. The pollen flow in soybean does not reach distances greater than 10 m (Schuster et al. 2007; Abud et al. 2003; Ray et al. 2003). In addition to being an autogamous species, soybean flowers undergo cleistogamy, i.e., fertilization occurs before the opening of the flowers, which reduces the gene flow in this species to levels close to zero. At the time of pollination, the anthers form a ring around the stigma. The pollen is then deposited directly into the stigma, resulting in a high rate of self-pollination. Pollination usually takes place the day before flowers open, and so, the stigma is only exposed after being self-pollinated (Schuster 2013).

Seeds obtained in the regulated trials that will not be reused in new assays must be destroyed and discarded in a specific area or in the area where the controlled trial was conducted (regulated area). In the case of discarding the seeds in the regulated area, the seeds should be crushed or moistened before being scattered in the area. A common method of disposal is to pack the seeds and wet them abundantly. The wetted seeds should be packaged and exposed to the sun for a few days and then scattered in the regulated area. Irrespective of whether or not the seeds are discarded in the regulated area, this area should be monitored for elimination of all volunteer plants that occur after the end of the regulated trial, until no more volunteer plants germinate under optimal soil moisture conditions.

All equipment and physical space required in activities with GM seeds and plants not yet commercially approved must be of exclusive or at least dedicated use. Equipment, rooms, and sheds of exclusive use are only used with seeds and plants of the regulated GM trait. Those of dedicated use can be used with other seeds and plants, but while they are being used with seeds and plants of the regulated trait, they cannot be used with other seeds and plants.

Anyone working in the regulated GM breeding program, including temporary and seasonal workers, should receive training on biosafety before starting activities on the regulated project.

In order to avoid unintentional noncompliance with biosafety standards, a company or institution that has an upgraded breeding program with regulated traits must have a compliance team that conducts internal audits and observes the correct application of biosafety standards.

Stewardship in GM Plant Breeding

Once the GM event has been approved by producer country, there is no need to request authorization from regulatory for the trials containing this event. However, there is no synchrony between the approval of a trait for cultivation in a producing country and the approval for consumption in an importing country. When a trait is approved for cultivation and marketing in a particular country, cultivars containing this trait can be grown and marketed. However, if an importing country has not yet approved the same trait for consumption, the product cannot be exported to this country.

Since to segregate all the soybeans containing the different GM traits is almost impossible, it is not safe to grow soybeans containing a trait that, although approved in producer country, is not approved in importing countries. This is due to the commercial risk of sending soybeans contaminated with the trait approved in producing country to an importing country that does not have this approval. At this stage, biosafety legislation no longer applies, and companies or institutions that have soybean breeding programs with this trait should adopt stewardship procedures. The stewardship program should contain procedures that prevent unintentional release of the trait under stewardship into the production chain. At this stage, in addition to breeding trials, the production of the genetic seed of the new cultivars, which occupy larger areas than the experimental ones, also starts. The same principles of isolation and exclusive or dedicated use of machines, equipment, and physical spaces, which were adopted while the product was not yet approved, should be adopted at this stage. However, at this stage, these measures are no longer aimed at biosafety of the GM trait, but at avoiding the contamination of the productive chain with seeds of this new trait and avoiding that grains containing this new trait reach an importing country still without approval.

Likewise, as for regulated traits, the company or institution with a breeding program with a trait under stewardship should have a team responsible for the compliance of the activities with this trait, both in breeding and in the production of genetic seeds.

Once approved in the importing countries, the breeding program can be conducted without biosafety and stewardship procedures. However, if in the same breeding program there is more than one trait (including non-GM plants), preventive procedures should be maintained to avoid contamination of plants of a trait with plants of the other trait.

Implication of GM Plant Breeding in the Generation of Variability

Breeding has three distinct and equally important macro steps: generation of variability, selection, and testing (Fig. 15.3). These steps are the same in conventional and GM plant breeding. However, in addition to considerations of biosafety and stewardship, and taking care to avoid cross contamination between traits, GM plant breeding also has genetic and methodological implications, especially in the generation of variability and selection stages.

In the first generations of breeding with a new GM event, a single donor is available to perform crosses, which is the very event obtained in the transformation for insertion of the new trait. Until the breeding program creates and selects its own GM germplasm to be used in the crosses, it will only use this single donor genotype in all its crosses, drastically reducing the genetic variability to be explored in the selection stage. This dependence on a single donor genotype will be maintained for some generations of crosses, and at this stage, all the segregating populations produced will be half-siblings. In addition to the reduction in variability among populations, because they are half-sibling populations, the number of populations obtained is also lower (Fig. 15.4). Besides that, this donor generally is not an elite germplasm, impacting in the probability to obtain high-yielding cultivars in the first

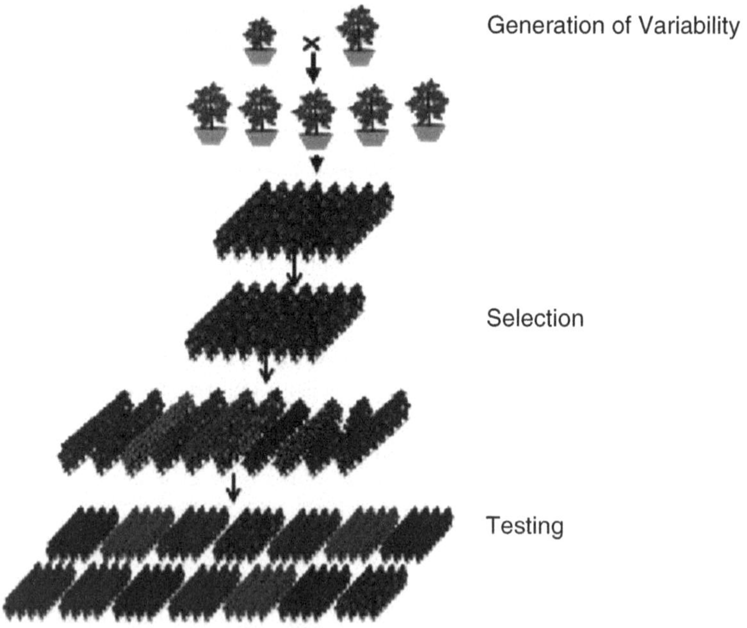

Fig. 15.3 Macro stages of a soybean breeding program, both in conventional soybean breeding and in GM soybean breeding

Breeding Program at Maturity

	V1	V2	V3	V4	V5	V6	V7	V8	V9	...	Vn
V1											
V2	1.2										
V3	1.3	2.3									
V4	1.4	2.4	3.4								
V5	1.5	2.5	3.5	4.5							
V6	1.6	2.6	3.6	4.6	5.6						
V7	1.7	2.7	3.7	4.7	5.7	6.7					
V8	1.8	2.8	3.8	4.8	5.8	6.8	7.8				
V9	1.9	2.9	3.9	4.9	5.9	6.9	7.9	8.9			
...											
Vn	1.n	2.n	3.n	4.n	5.n	6.n	7.n	8.n	9.n		n.n

$$\textit{Number of Possible Crossing Combination} = \frac{n(n-1)}{2}$$

First Generations of New Breeding Program

Variety	Donor
V1	D1.V1
V2	D1.V2
V3	D1.V3
V4	D1.V4
V5	D1.V5
V6	D1.V6
V7	D1.V7
V8	D1.V8
V9	D1.V9
...	
Vn	D1.Vn

$$\textit{Number of Possible Crossing Combination} = n$$

Fig. 15.4 Possible number of crossing combinations in a consolidated GM plant breeding program and a breeding program with a new trait

wave of new cultivars with the new trait. This limitation will disappear when the breeding program with new trait reaches their maturity and is able to use a large number of cultivars or lines containing the trait in both parents in crosses.

Implications of GM Plant Breeding in Population Size

Although a GM insert may contain more than one gene, its segregation in a segregating population is equal to the segregation of a single inherited gene. Thus, for purposes of population size estimation, it is considered that each insert segregates as a single inherited gene. However, this insert is mandatory in selected plants, that is, plants that do not contain the target should be discarded as a biosafety measure (Fig. 15.5). This implies in reduced size of the populations, after eliminating the plants that do not contain the GM trait(s), and, consequently, reduced variability obtained by the crosses. The higher the number of inserts, the larger the population size should be, in order to maintain the same variability of a conventional breeding population (Table 15.1 and Fig. 15.6).

In the first segregating generations, the selection of only homozygous plants for the GM traits that are segregating drastically reduces the size of the populations under selection (Table 15.1). This requires very large population sizes, especially when the number of inserts increases (Fig. 15.6). For example, for populations that segregate for four traits, and if only homozygous plants were selected for this four traits, an F2 population should be 256 times larger than an F2 population of conventional plant breeding.

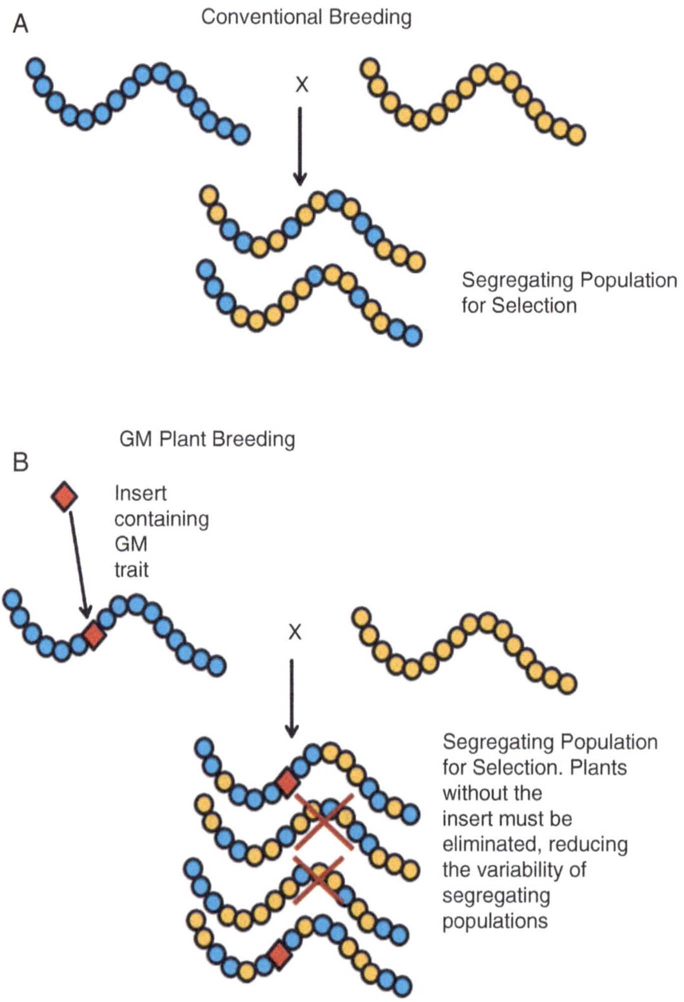

Fig. 15.5 (**a**) In conventional plant breeding, all the plants of the segregating generations can be used in the selection, and all the genetic variability obtained by the crosses can be used. (**b**) In GM plant breeding, when the trait is in only one of the parents, part of the variability generated in the crosses should be eliminated because it does not contain the trait of interest. The reduction in size of the segregating populations depends on the number of inserts that are in one of the parents and not in the other

In these early generations, it is advisable to select both homozygous and hemizygous plants for segregating GM traits, thereby reducing the size of the population required. In the same example of the F2 population segregating for four traits, if both homozygous and hemizygous plants were selected, its size should be just 3.2 times larger than an F2 population of conventional plant breeding. This

Table 15.1 Percentage of homozygous plants and of the sum of homozygous and hemizygous plants for the inserts that segregate in a population, considering that no selection for the GM traits in previous generations has been made

	% of homozygous plants for the trait (s)				% of the sum of homozygous and hemizygous plants for the trait(s)			
Selection generation	1 insert	2 inserts	3 inserts	4 inserts	1 insert	2 inserts	3 inserts	4 inserts
F2	25.0	6.3	1.6	0.4	75.0	56.3	42.2	31.6
F3	37.5	14.0	5.3	2.0	62.5	39.1	24.4	15.3
F4	43.8	19.1	8.4	3.7	56.3	31.6	17.8	10.0
F5	46.9	22.0	10.3	4.8	53.1	28.2	15.0	8.0
Fn	50.0	25.0	12.5	6.3	50.0	25.0	12.5	6.3

selection strategy requires that the selection for segregating GM traits be made again in subsequent generations.

Few changes in the breeding programs were required for the first GM soybean generation (Roundup Ready™ or RR™). In addition to aspects related to biosafety, field isolation, and segregation from conventional breeding to avoid cross contamination, the main difference between this first-generation GM plant breeding and conventional plant breeding was the inclusion of a protocol of glyphosate herbicide application in the segregating populations.

The first GM soybean cultivars were obtained by crosses between donors of the glyphosate herbicide tolerance trait with conventional soybean cultivars/lines. In this case, the application of glyphosate herbicide at the beginning of F1 plants' flowering and segregating populations eliminates the majority of male gametes that do not contain the glyphosate tolerance GM trait (Walker et al. 2006).

In spraying glyphosate at the beginning of flowering of F1 plants, approximately half of the F2 seeds obtained are homozygous for the herbicide tolerance trait, the rest being hemizygous (Fig. 15.7). There are practically no plants susceptible to glyphosate in the F2 population, and a new spraying of the herbicide on these populations eliminates the few susceptible ones.

By repeating the glyphosate application protocol at the beginning of the flowering of the F2, F3, and F4 populations, populations F3, F4, and F5 will contain 75, 87.5, and 93.75% of homozygous plants for tolerance to the herbicide (Fig. 15.7). Obviously, as in these populations the plants do not all bloom at the same time, homozygous seeds will escape due to the absence of the herbicide tolerance trait, and these will be eliminated by the spraying of glyphosate on these populations in the vegetative phase.

By selecting plants at F4 for progeny testing, 12.5% of progenies will segregate for tolerance to glyphosate and may be identified by the application of the herbicide. When progenies are derived from a single plant, if this plant is hemizygous for tolerance to glyphosate, its progeny will present approximately 25% of plants susceptible to the herbicide, which die with the application of glyphosate. Progenies that segregate for herbicide tolerance should be eliminated. Thus, the number of

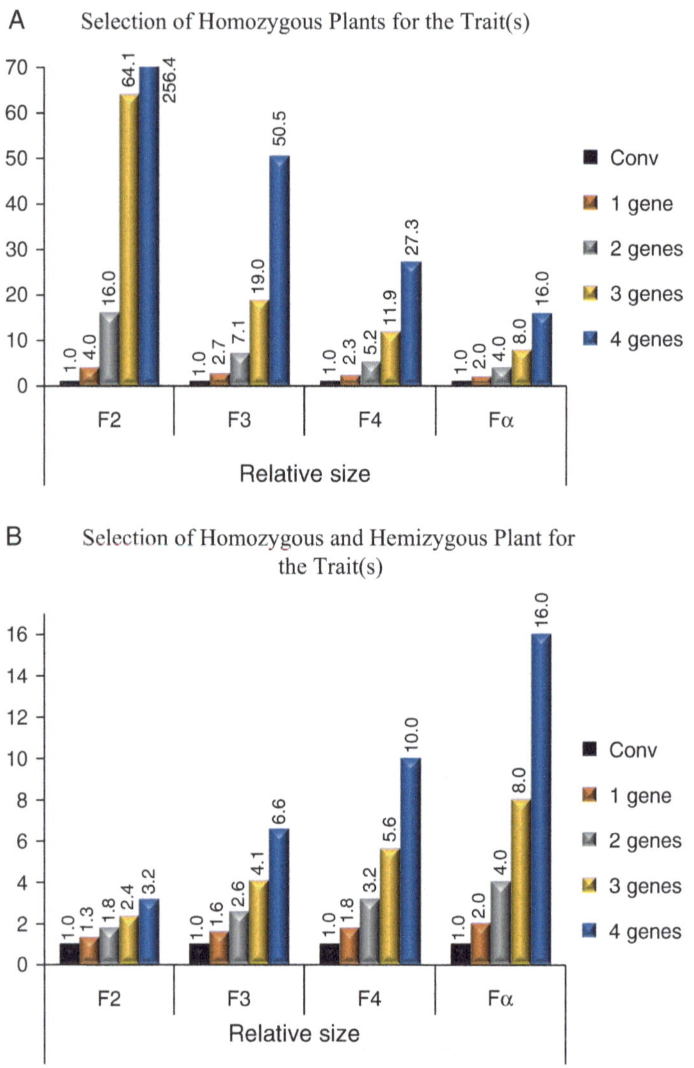

Fig. 15.6 Relative population size, considering the selection of only homozygous (**a**) or homozygous and hemizygous (**b**) for GM trait(s) that are segregating in the population, so that, after selection, the population is the same size as that of a conventional population (or with no segregating GM trait). For the estimate of population sizes in each generation, it is considered that there was no selection for the GM traits(s) in previous generations

progenies is reduced by 12.5%. To maintain the same number of progenies as used before the introduction of transgenic trait, the initial number of progenies should be increased by 12.5%. By selecting plants in F3, the total number of progenies should be increased in 25%.

Crossing RR x rr

F1 Rr

♂

♀ Gametes	R	r*
R	50%	
r	50%	

*r male gametes are eliminated after spraying Glyphosate in the beggining of the flowering.

Genotypes	Freq in F2	Freq in F3	Freq in F4	Freq in F5
RR	50%	75%	87,5%	93,75%
Rr	50%	25%	12,5%	6,25%

Fig. 15.7 RR™ trait segregation, considering the cross of an RR plant with a conventional plant, and that in each generation, beginning with F1, the plants are sprayed with glyphosate at the beginning of flowering, eliminating recessive male gametes

In South America, the second generation of GM soybeans containing a new tolerance trait to the herbicide glyphosate (second-generation RR or RR2) is associated with insect resistance (RR2 + Bt). From this point, new and more complex methodologies in breeding programs needed to be introduced.

When the second generation of GM soybeans was introduced (RR2 + Bt), the first-generation GM soybean (RR, now called RR1) already occupied more than 90% of the soybean area in Brazil, and the breeding programs already dated 15 years of RR1 soybean breeding. GM breeding programs were already mature, and the best germplasm available for crossing contained RR1 trait. Thus, most of the crosses to obtain cultivars for the second generation of GM soybean were carried out with the RR2 + Bt technology donor with RR1 plants. The number of conventional cultivars that could potentially be used in crosses was enough for only one generation of crosses, so what should be considered is the cross between RR1 plants with RR2 + Bt plants.

When RR2 + Bt donors are crossed with RR1 cultivars, the RR1 trait needs to be eliminated in the segregating populations. This is necessary because the combination of RR1 and RR2 traits is not authorized for commercial use, as this request was not made to any regulatory agency.

Crosses of RR2 + Bt plants with RR1 plants segregate to three GM traits. Considering only the segregation of traits RR1 and RR2 and considering the application of glyphosate herbicide at the beginning of flowering of F1 plants, F2 plants will contain only 8% of homozygous plants for the presence of RR2 and absence of RR1 (Fig. 15.8), instead of almost 50% when using cross between RR1 and conventional cultivars, in the earlier technology. If the insect resistance trait is still considered, only 2.0% of the plants will be homozygous for the three traits in the F2 generation (Fig. 15.9 and Table 15.2).

Crossing RR1 rr2 x rr1 RR2

F1 Rr1 Rr2

♂

F1 Gametes	1/3 R1R2	1/3 R1r2	1/3 r1R2	0 r1r2*
¼ R1R2	RR1 RR2	RR1 Rr2	Rr1 RR2	
♀ ¼ R1r2	RR1 Rr2	RR1 rr2	Rr1 Rr2	
¼ r1R2	Rr1 RR2	Rr1Rr2	rr1 RR2	
¼ r1r2	Rr1 Rr2	Rr1 rr2	rr1 Rr2	

***r1r2 male gametes are eliminated by spraying Glyphosate at the beggining of the flowering of F1 plants.**

Genotypes	Freq em F2	Freq em F3	Freq em F4	Freq em F5
RR1 RR2	8.3%	18.8%	26.6%	31.1%
RR1 Rr2	16.7%	16.7%	10.4%	5.7%
RR1 rr2	8.3%	18.8%	24.5%	27.5%
Rr1 RR2	16.7%	12.5%	7.3%	3.9%
Rr1 Rr2	25.0%	6.3%	1.6%	0.4%
Rr1 rr2	8.3%	2.1%	0.5%	0.1%
rr1 RR2	***8.3%***	***18.8%***	***25.5%***	***29.3%***
rr1 Rr1	8.3%	6.3%	3.6%	2.0%
TOTAL	100.0%	100.0%	100.0%	100.0%

In Italic Bold, the desired genotype.

Fig. 15.8 Segregation of the RR1 and RR2 events, considering that each event is in a different parent and that in each generation, beginning in F1, the plants are sprayed with glyphosate at the beginning of flowering, eliminating double-recessive male F1 gametes

By using the glyphosate application protocol at the beginning of flowering, all genotypes of the segregating populations have at least one glyphosate tolerance allele (R1_ or R2_). The selection of plants that contain the RR2 trait (in homozygosis or hemizygosis) and the absence of the RR1 trait can only be performed with molecular markers. Likewise, the selection of homozygous plants for insect resistance (BB) can also only be done with molecular markers. In the case of insect resistance, the use of immunological tests (ELISA or lateral strip test) identifies the Bt gene, but not its zygosity. For this reason, the current soybean breeding programs depend on the use of molecular markers and can no longer be performed without this tool. Thus, the use of molecular markers has become routine in soybean breeding.

The next generations of GM soybeans will continue to introduce more traits, with the same implications regarding population sizes. For each new trait, new protocols will need to be developed for the selection of specific traits. Although some of the future traits may include tolerance to other herbicides, other than

Cross RR1 rr2 bb x rr1 RR2 BB

F1 Rr1 Rr2 Bb

♂

♀ Gametes F1	1/6 R1R2B	1/6 R1R2b	1/6 R1r2B	1/6 R1r2b	1/6 r1R2B	1/6 r1R2b	0 r1r2B*	0 r1r2b*
1/8 R1R2B	RR1 RR2 BB	RR1 RR2 Bb	RR1 Rr2 BB	RR1 Rr2 Bb	Rr1 RR2 BB	Rr1 RR2 Bb		
1/8 R1R2b	RR1 RR2 Bb	RR1 RR2 bb	RR1 Rr2 Bb	RR1 Rr2 bb	Rr1 RR2 Bb	Rr1 RR2 bb		
1/8 R1r2B	RR1 Rr2 BB	RR1 Rr2 Bb	RR1 rr2 BB	RR1 rr2 Bb	Rr1 Rr2 BB	Rr1 Rr2 Bb		
1/8 R1r2b	RR1 Rr2 Bb	RR1 Rr2 bb	RR1 rr2 Bb	RR1 rr2 bb	Rr1 Rr2 Bb	Rr1 Rr2 bb		
1/8 r1R2B	Rr1 RR2 BB	Rr1 RR2 Bb	Rr1 Rr2 BB	Rr1 Rr2 Bb	rr1 RR2 BB	rr1 RR2 Bb		
1/8 r1R2b	Rr1 RR2 Bb	Rr1 RR2 bb	Rr1 Rr2 Bb	Rr1 Rr2 bb	rr1 RR2 Bb	rr1 RR2 bb		
1/8 r1r2B	Rr1 Rr2 BB	Rr1 Rr2 Bb	Rr1 rr2 BB	Rr1 rr2 Bb	rr1 Rr2 BB	rr1 Rr2 Bb		
1/8 r1r2b	Rr1 Rr2 Bb	Rr1 Rr2 bb	Rr1 rr2 Bb	Rr1 rr2 bb	rr1 Rr2 Bb	rr1 Rr2 bb		

*Male r1r2 gametes are eliminated by the spraying of glyphosate at the beginning of flowering.

Fig. 15.9 Genotypes produced in the F2 generation, taking into account the spraying of the glyphosate herbicide at the beginning of flowering of F1 plants. Double-recessive male gametes for glyphosate tolerance are not produced. The B allele confers resistance to insects. *Male r1r2 gametes are eliminated by the spraying of glyphosate at the beginning of flowering

glyphosate, and may be selected by specific herbicide application, population sizes should respect the proportion of the number of traits; otherwise, the variability after the selection for traits of interest will be reduced.

Backcrossing

One way to accelerate the development of the first cultivars of a GM trait, especially when the donor is not adapted, is to use the backcross method, especially backcrossing assisted by molecular markers. By using molecular markers to accelerate the recovery of the recurrent genome, it is possible to recover about 95% of the genome of the recurrent cultivar in three backcrossing generations, keeping the trait of interest in the recovered cultivar (Guimarães et al. 2009). By considering the generations of self-fertilization to obtain lines with the transgenic trait in homozygosis, selection of the elite line, and 1-year yield trials and seed production, a cultivar obtained by backcrossing can reach the market in 5 or 6 years.

Also in backcrossing programs, population sizes should be adjusted according to the number of inserts to be transferred and/or eliminated. Considering that one needs to obtain enough plants for the selection of recurrent genome recovery, using molecular markers, each trait under selection doubles the required size of the backcrossing populations. Taking into account that the goal of using 46 plants for

Table 15.2 Genetic frequencies obtained in the F2–F5 generations, of a cross between a RR1 rr2 bb cultivar and an rr1 RR2 BB cultivar

Genotypes	Frequency in F2	Frequency in F3	Frequency in F4	Frequency in F5
RR1 RR2 BB	2.1%	8.1%	13.4%	16.7%
RR1 RR2 Bb	4.2%	4.7%	3.2%	1.8%
RR1 RR2 bb	2.1%	6.0%	9.0%	10.7%
RR1 Rr2 BB	4.2%	4.7%	3.2%	1.8%
RR1 Rr2 Bb	8.3%	3.1%	0.9%	0.2%
RR1 Rr2 bb	4.2%	4.7%	3.2%	1.8%
RR1 rr2 BB	2.1%	7.0%	11.2%	13.7%
RR1 rr2 Bb	4.2%	4.7%	3.2%	1.8%
RR1 rr2 bb	2.1%	7.0%	11.2%	13.7%
Rr1 RR2 BB	4.2%	6.8%	5.5%	3.4%
Rr1 RR2 Bb	8.3%	3.1%	0.9%	0.2%
Rr1 RR2 bb	4.2%	2.6%	0.8%	0.2%
Rr1 Rr2 BB	6.3%	2.3%	0.7%	0.2%
Rr1 Rr2 Bb	12.5%	1.6%	0.2%	0.0%
Rr1 Rr2 bb	6.3%	2.3%	0.7%	0.2%
Rr1 rr2 BB	2.1%	2.3%	1.6%	0.9%
Rr1 rr2 Bb	4.2%	1.6%	0.5%	0.1%
Rr1 rr2 bb	2.1%	2.3%	1.6%	0.9%
rr1 RR2 BB	***2.1%***	***8.1%***	***13.4%***	***16.7%***
rr1 RR2 Bb	4.2%	4.7%	3.2%	1.8%
rr1 RR2 bb	2.1%	6.0%	9.0%	10.7%
rr1 Rr1 BB	2.1%	2.3%	1.6%	0.9%
rr1 Rr1 Bb	4.2%	1.6%	0.5%	0.1%
rr1 Rr1 bb	2.1%	2.3%	1.6%	0.9%
Total	100.0	100.0	100.0	100.0

It is considered that in all generations, including F1, the plants were sprayed with glyphosate at the beginning of flowering, eliminating the double-recessive male gametes for tolerance to glyphosate. The B allele confers resistance to insects

In italic-bold the desired genotype

selection of the recurrent genome in each population (which, by adding the two parents complete half of a PCR plate of 96 samples). For the selection of one trait (introgression or elimination), 92 RCnF1 seeds are required, as half of the RCnF1 plants will be eliminated in the selection for the presence of the gene. For two traits, 184 seeds are required and for three traits, 368 seeds. Given that, with soybeans, each cross produces on average two seeds, this means 46, 92, and 184 fertile crosses, respectively, for each recurrent parent in each backcrossing generation. That is, even if one considers the low number of plants for selection of the recurrent genome (46 plants), the number of crossed flowers in each backcrossing population should be large. For this reason, although it is a fast method, its use is only justified in the first generations of a new trait, to accelerate the obtaining of high-performance products, from recurrent cultivars with high market potential.

In the case of allogamous plants, such as corn, the conversion of lines by backcrossing is the most appropriate method, as each cross produces many seeds, and a converted elite line can generate tens or even hundreds of hybrids for testing. Thus, the conversion of a reduced set of elite lines can produce a large number of hybrids, whereas in soybean, each line only produces one cultivar (Fig. 15.10). For this reason, corn breeding programs are conducted with conventional germplasm, and as the lines advance in breeding selection and testing, it is backcrossed for introgression of the target traits.

Implication of the Increased Number of GM Traits in Soybean Breeding

The introduction of each new GM trait in a breeding program means a breeding restart, with the same implications of reduced variability due to the availability of a single donor of the technology (Fig. 15.3). As this trait donor is usually an unadapted cultivar, there is still loss of part of the genetic gain accumulated over the years.

A plant breeding program accumulates genetic gains over generations, so that after several generations of breeding, these gains are considerable. With the need to restart breeding programs with each new trait, reducing the variability of the crosses, there is an interruption of the genetic gain, until it is consolidated. Genetic gains are discontinuous, unlike a previous conventional breeding that does not need to be restarted.

A soybean breeding program having more than one GM trait will have to split its resources into two or more breeding programs (like RR1 breeding, RR2 breeding, Enlist breeding, and also conventional soybean breeding), thereby reducing the likelihood of obtaining the best genetic combinations because of the reduced size of the program. In addition, care should be elevated to avoid cross contamination of lines from one trait with another trait, at the risk of having to restart the seed purification of the lines in order to be able to advance commercially.

Breeding, by its very nature, is a multidisciplinary science. Biosafety, stewardship, and maintenance of segregation of more than one trait in the breeding program of the same institution raise their breeding program costs. In addition, they require greater specialization of research teams, expanding their multidisciplinary to include teams dedicated to the regulatory function, thus increasing the complexity of research management.

With the trend of technology replacement at shorter intervals, breeding methods should be able to develop new cultivars in ever-shorter times. To do so, the use of winter generations and selection by molecular markers (genomic selection) becomes necessary, both in the winter generations (molecular selection only) and in the summer generations (molecular selection combined with phenotypic selection), to combine genetic gains with acceleration of selection generations. If this time of obtaining new cultivars is not significantly reduced, there is a risk of having to replace a technology before the breeding program reaches its maturity.

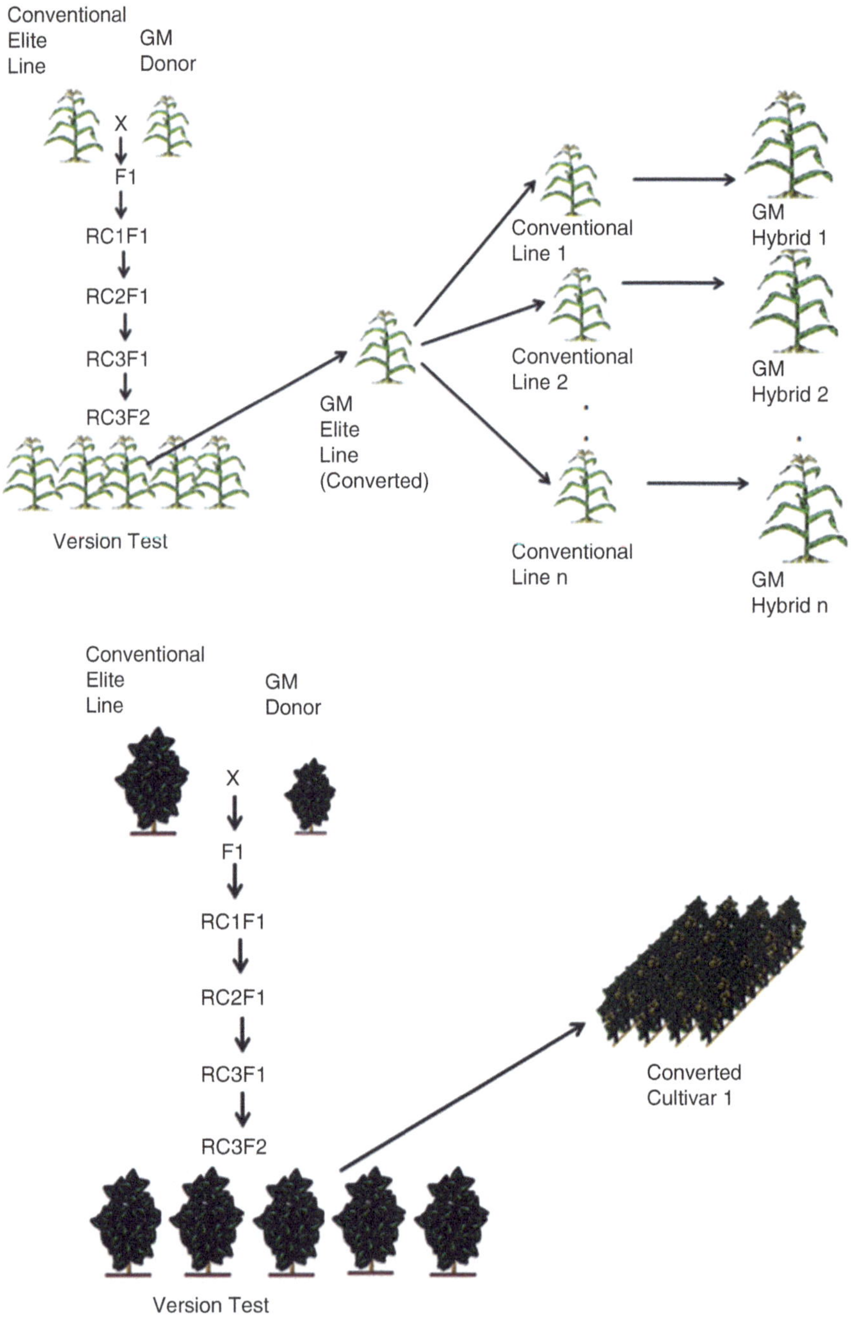

Fig. 15.10 The conversion of an elite line of corn and soybean to a GM trait by backcrossing

Final Considerations

The GM plant breeding has increased the multidisciplinary of this science. Therefore, there is a need for new specialists in the areas of compliance and regulatory, marker-assisted selection (or genomic selection), marker-assisted backcrossing, finalization of the converted lines, and data analysis. In addition, new laboratories are required, more intensive and automated mechanization of the field activities, in order to make the data available in shorter times for decision making and generation advancement in the winter. Such pieces of new high-tech equipment also demand specialized professionals in their operation and in obtaining the data in real time.

Soybean breeding in the GM plant era requires introduction of new processes into the breeding programs to increase their effectiveness, reduce the time to obtain new cultivars, and ensure compliance with the biosafety, stewardship, and genetic purity of the seeds made available by breeding for production.

Among the challenges of plant breeding in the GM plant era are the association of (a) conventional breeding for the development of the best germplasm; (b) new GM traits, with the purpose of offering new solutions for farmers and consumers; and (c) use of molecular markers to identify the zygosity of GM traits and genomic selection of native genes. This association of techniques is necessary to provide the best performance of the plants in the various growing conditions.

References

Abud, S., Souza, P.I.M., Moreira, C.T., Andrade, S.R.M., Ulbrich, A.V., Vianna, G.R., Rech, E.L., Aragão, F.J.L.: Dispersão de pólen em soja transgênica na região do Cerrado. Pesq. Agrop. Brasileira. **38**, 1229–1235 (2003)

Conab: Séries históricas: Soja. http://www.conab.gov.br/ (2014). Accessed 16 Oct 2015

Guimarães, C.T., Schuster, I., Magalhães, J.V., Souza Júnior, C.L.: Marcadores moleculares no melhoramento. In: Borém, A., Caixeta, E.T. (eds.) Marcadores Moleculares, 2ª edn, pp. 129–176. Ed. Folha de Viçosa, Viçosa (2009)

James, C.: Global Status of Commercialized Biotech/GM Crops: 2014. ISAAA Brief No. 49, Ithaca (2014)

Kleffmann Group: AMIS soybean seeds 13/14 (2014)

Ray, J.D., Kilen, T.C., Abel, C.A., Paris, R.L.: Soybean natural cross-pollination rates under field conditions. Environ. Biosaf. Res. **2**, 133–138 (2003)

Schuster, I.: Fluxo gênico e coexistência de lavouras com espécies transgênicas e convencionais. Informativo Abrates. **23**(1), 39–45 (2013)

Schuster, I., Vieira, E.S.N., Santana, H., Sinhorati, D., Silva, R.B., Oliveira, M.A.R.: Fluxo Gênico em Soja na Região Oeste do Paraná. Pesq. Agrop. Brasileira. **42**, 515–520 (2007)

Walker, D.R., Walker, A.K., Wood, E.D., Talevera, M.E.B., Fernandez, F.E., Rowan, G.B., Moots, C.K., Leitz, R.A., Owen, P.A., Baxter, W.E., Head, J.L., Boerma, H.R.: Gametic selection by glyphosate in soybean plants hemizygous for the CP4 EPSPS transgene. Crop Sci. **46**, 30–35 (2006)

Chapter 16
GM Cultivars

Pedro Patric Pinho Morais and Aluízio Borém

Abstract The term genetically modified (GM) or transgenic refers to organisms that received exotic genes using a series of laboratory techniques for cloning genes, splicing DNA, and inserting genes into them, a procedure called genetic transformation. These techniques are also known as recombinant DNA technology. The term transgenic is used in this chapter because it is the most widely used to indicate the use of recombinant DNA technology in developing the new cultivar. Soybean genetic modification technology allows the introgression of genes for specific traits from other species. GM soybean, with glyphosate herbicide tolerance, was first introduced in the USA in 1996, with the herbicide-tolerant form of 5-enolpyruvulshikimate-3-phosphate synthase (EPSPS) gen, in which function is to decrease binding affinity for glyphosate, thereby conferring increased tolerance to glyphosate herbicide. Today, most GM soybeans grown worldwide are transformed for insect resistance or herbicide tolerance, but drought tolerance, nitrogen use efficiency, trans fat-free, and reduced saturated fat have also been developed. Other potential future applications of the technology include additional nutritional enhancements, other stress tolerance, disease resistance, biofuel efficiency, etc.

Keywords Transgenic • GM varieties • Biotechnology • Genetic engineering • Genetically modified food • GMO

Introduction

A genetically modified organism, which uses the acronym GMO (genetically modified organism), means that it underwent putative changes within its genome through genetic engineering. These changes have two purposes: the first is toward

P.P.P. Morais, M.S., D.S.
Monsanto Brazil, São Paulo, SP, Brazil
e-mail: pedro.morais@monsanto.com

A. Borém, M.S., Ph.D. (✉)
Federal University of Viçosa, Viçosa, Minas Gerais 36570-900, Brazil
e-mail: borem@ufv.br

F. Lopes da Silva et al. (eds.), *Soybean Breeding*,
DOI 10.1007/978-3-319-57433-2_16

changes in the structure and/or function of the genetic material of the organism, having no introduction of genetic material from any other species in this procedure; the second purpose takes into consideration the procedure of insertion of genomic sequences of species other than the species of interest, the so-called transgenic organism. This way, all transgenic is a GMO, but not all GMOs are transgenic. Thus, the final result is the expression of a gene, which results in a protein that allows action against pests or herbicide tolerance (Rani and Usha 2013) or results in resistance to viral diseases (Aragão and Faria 2009), and, even for new generations of GMOs, the development for the production of medication and industrial products like monoclonal antibodies, vaccines, bioplastics, and biofuels (Sticklen 2005; Conrad 2005; Ma et al. 2003).

Transgenic cultivars, since their release to planting in 1994 in the USA, have grown 3–4% per year, reaching 200 million hectares in 2016 (Fig. 16.1). They are currently cultivated by 18 million farmers in 28 countries, with emphasis to the five largest planted areas in millions of hectares: USA (73.1), Brazil (42.2), Argentina (24.3), India (11.6), and Canada (11.6). Through the information in the last 19 years, there was growth 100 times greater in relation to the first year of commercialization of cultivars with this technology. They are then considered the most adopted agricultural technology in the history of modern agriculture (James 2014).

According to the International Service for the Acquisition of Agri-Biotech Applications (ISAAA), since the beginning of commercial planting of transgenic cultivars, more than ten agricultural crops were approved and marketed worldwide;

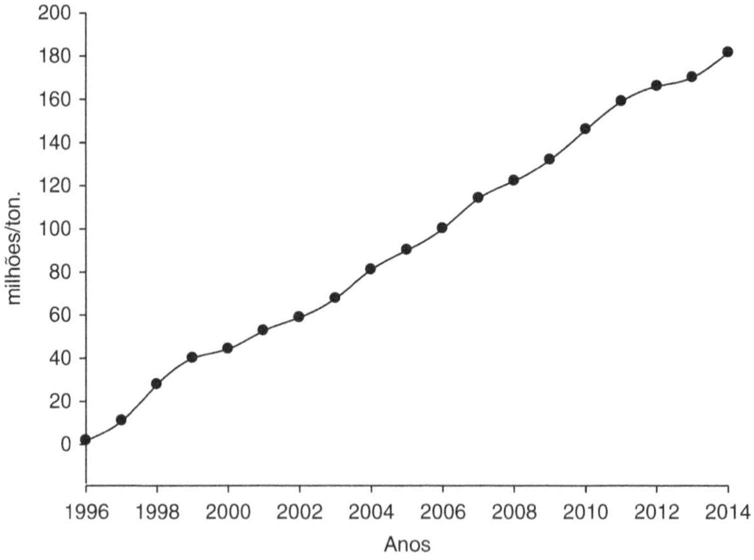

Fig. 16.1 Global area, millions of hectares planted with transgenic cultivars in the period from 1996 to 2014. Source: Adapted from ISAAA (2015)

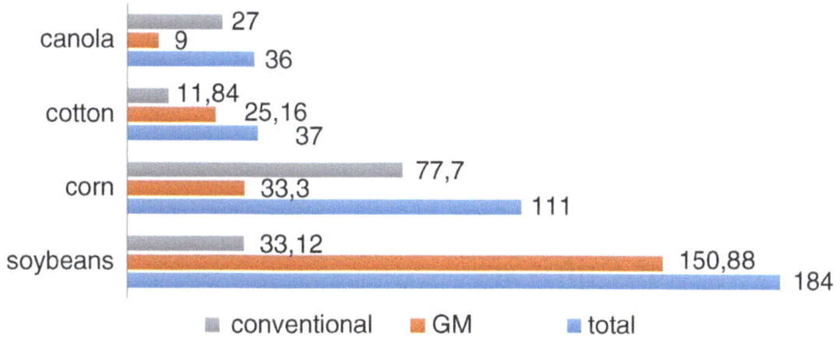

Fig. 16.2 Global area, per crop, in millions of hectares planted with transgenic and conventional cultivars. Source: Adapted from ISAAA (2015)

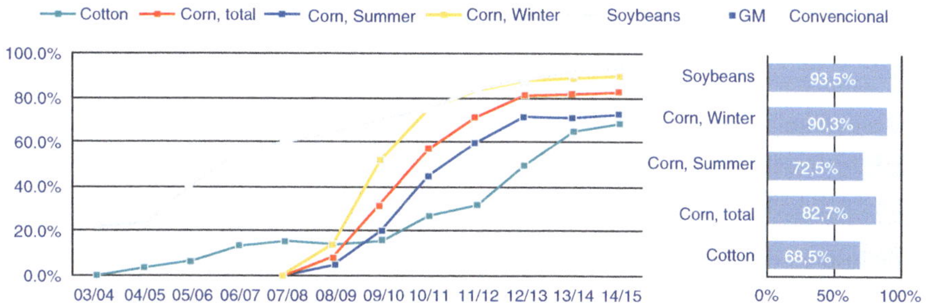

Fig. 16.3 Adoption rate of transgenic cultivars in Brazil per crop, crops from 2003/2004 to 2014/2015. Source: Céleres (2015)

among these, the most cultivated are soybeans, corn, cotton, and canola. In this context, we can see that 68% (25.16 million hectares) of the grown area with cotton are destined to transgenic cultivars. In the case of canola, the numbers turn around 25% (9.0 million hectares), 30% (33.3 million hectares) for corn, and the highest rates are reached with soybeans, that is, 82% (150.88 million hectares) (Fig. 16.2).

Currently, the rate of adherence of Americans to GM soybeans is 90%, the same levels found in Argentinian plantations. Brazil was not indifferent to the use of this technology, because during the crop of 2012/2013, the area with transgenic soya cultivars was 89.1% (24.7 million hectares), and during the crop of 2014/2015, for the first time, the rate was over 90%. According to the Céleres (2015) and CONAB (2015), of the 31.9 million hectares planted with soybeans in Brazil, 93.5% or 29.9 million hectares were cultivated with transgenic cultivars (Fig. 16.3).

This rapid adoption is based on the obvious advantages of transgenic soybeans: lower production costs (mainly herbicide costs) and less manpower per hectare (Menegatti and Barros 2007). Another great advantage is the higher yield when compared to field with high infestation of weeds, although the transgenic soybean cultivars do not aim at higher productivity. The higher yield is an indirect benefit for better weed control.

Despite all the advantages and easiness of using transgenic soybean cultivars, scientists, technicians, and producers must be cautious about the correct use of the technology and its durability and mitigation of possible resistance of weeds and insects.

Herbicide-Tolerant Cultivars

Soybean, like other species, is subject to a series of biotic and abiotic stresses that may affect its development and production. These include weeds, which can lead to severe losses in yield due to competition for light, nutrients, and water, besides making the harvest difficult. They also act as host of pests and diseases and make allelopathic pressure (Karam et al. 2006). According to Carvalho and Velini (2001), weeds are the major culprits in the decline in soybean production compared to pest and pathogen attacks. Losses in production can reach up to 37% (Oerk 2006).

Regarding the implementation and approval of transgenic events, the history in the USA began in 1994, with tomato Flavr Savr. In Brazil, it began in 1998 with the first GTS 40-3-2 event (Roundup Ready), developed by Monsanto (Table 16.1). However, after the approval of this event, there were extensive lawsuits and judicial foreclosures on the release of transgenic cultivars, with the purpose of effective permission for planting in the harvest of 2003/2004. Since then, five other events for soybean cultivars have been approved by the Brazilian National Technical Board on Biosafety (CTNBio) (Table 16.1): four tolerant to herbicides (GTS 40-3-2, BPS-CV127-9, A5547-127, A2704-12, DAS-68416-4) and a combined event (herbicide tolerant and insect resistant), MON 87701 × MON 89788 (discussed in this chapter).

The GTS 40-3-2 event features the insertion of the *cp4 epsps* gene encoding a tolerant form of the enzyme 5-enolpyruvate-shikimate-3-phosphate synthase (EPSPS) to glyphosate, allowing the plant to survive on the action of the herbicide. The gene in question was isolated from the bacterium *Agrobacterium tumefaciens* strain CP4 (bacteria commonly found in the soil, causing gall in susceptible plants). The action of glyphosate causing plant death occurs because of its ability to block the activity of the target enzyme (EPSPS) belonging to the biosynthetic pathway of the aromatic amino acids tyrosine, phenylalanine, and tryptophan. As a result, plant cells expressing the CP4-EPSPS protein continue to produce the aromatic amino acids essential to their metabolism even in the presence of glyphosate. The genetic transformation for this event occurred with microparticle bombardment (biolistic).

The second event made available in Brazil, BPS-CV127-9 (soy CV127), is derived from transformation via biolistics obtained in the laboratories of EMBRAPA Genetic Resources and Biotechnology with the introduction into the genome of soybean of gene *csr1-2* from *Arabidopsis thaliana*. The transformation was carried out in tissue from the embryonic axis taken from the apical meristem of a single soybean seed of the cultivar Conquista. The gene *csr1-2* encodes the major subunit of the enzyme acetohydroxyacid synthase (AHAS), responsible for

Table 16.1 Transgenic events deregulated for soybean in Brazil during the period 1998–2015

Commercial name	Event	Grower	Genes	Donor organisms	Characteristics	Year of approval
Roundup Ready	GTS 40-3-2	Monsanto	*cp4 epsps*	*Agrobacterium tumefaciens* strain CP4	Tolerant to herbicide glyphosate	1998
CV127	BPS-CV127-9	BASF	*crs1-2*	*Arabidopsis thaliana*	Tolerant to herbicides of the group of imidazolinones	2009
Liberty link	A5547-127	Bayer	*pat*	*Streptomyces viridochromogenes*	Tolerant to the herbicide glufosinate-ammonium	2010
Liberty link	A2704-12	Bayer	*pat*	*Streptomyces viridochromogenes*	Tolerant to the herbicide glufosinate-ammonium	2010
Intacta RR PRO	MON 87701 × MON 89788	Monsanto	*cry1Ac/ cp4 epsps*	*Bacillus thuringiensis subsp. Kurstaki strain HD73*/*Agrobacterium tumefaciens* strain CP4	Resistant to insects (Lepidoptera) and tolerant to the herbicide glyphosate	2010
Enlist	DAS-68416-4	Down AgroSciences	*aad-12/ pat*	*Delftia acidovorans*/*Streptomyces viridochromogenes*	Tolerant to herbicides 2,4-D and ammonium-glufosinate	2015

herbicide tolerance of the imidazolinone chemical group (UNESP 2009). This herbicide inhibits the action of AHAS, preventing the synthesis of the amino acids leucine, isoleucine, and valine, affecting cell division and translocation of photoassimilates to the growth points.

The A5547-127 and A2704-12 events have as their basic characteristic the tolerance to the glufosinate-ammonium herbicide obtained by introducing the gene (*pat*) via microparticle bombardment, which expresses the protein phosphinothricin N-acetyltransferase (PAT) isolated from *Streptomyces viridochromogenes* which inactivates the active component by catalyzing the acetylation of L-phosphinothricin. Biolistics was the genetic transformation procedure used in these events. Glufosinate-ammonium inhibits the enzyme glutamine synthetase, responsible for the incorporation of ammonia to glutamic acid to form glutamine. The inhibition caused by the herbicide results in glutamine deficiency and, therefore, the inhibition of photosynthesis and the accumulation of ammonia, which reaches toxic levels and leads to desiccation of the plant.

In 2015, DAS-68416-4 was approved, and soybean cultivars containing the transgenes of this event are tolerant to herbicides 2,4-dichlorophenoxyacetic acid (2,4-D) and glufosinate-ammonium. This is a stacked event, and the term pyramidal can be used as a synonym, and it is developed by transformation by *Agrobacterium tumefaciens*, which incorporated the genes *aad-12* and *pat*. The gene *aad-12* derived from a soil bacterium (*Delftia acidovorans*) encodes the protein *aryloxyalkanoate dioxygenase*-12 (AAD-12), which in turn has activity of the enzyme *α-ketoglutarate-dependent* dioxygenase, which results in the metabolic inactivation of the herbicide 2,4-D. The other gene transferred, *pat*, encoding the protein PAT, has the same characteristics previously mentioned for the events A5547-127 and A2704-12. According to the obtainer, Dow AgroSciences, the availability of DAS-68416-4 for soybean crops will enable producers to proactively manage weed populations, avoiding adverse problems or developing resistance, especially for the species resistant to the herbicide glyphosate.

Risk Management of Emergence of Resistant Weeds

Through the information collected in a website of statistical surveys on weed resistance, Weedscience (http://weedscience.org), seven weed species resistant to EPSPS synthesis inhibitors (*Amaranthus palmeri*, *Chloris elata*, *Conyza bonariensis*, *Conyza canadensis*, *Conyza sumatrensis*, *Digitaria insularis*, *and Lolium perenne ssp. multiflorum*) were identified in Brazil. For such situations to be avoided, or at least to minimize the likelihood of weed resistance, certain agricultural practices should be implemented. These practices are based on decreasing the selection pressure and on controlling resistant individuals before they can set seeds (Silva et al. 2011). Among them, we can highlight (1) rotation in the use of herbicides with different mechanisms of action, (2) sequential applications of herbicides with different mechanisms of action, (3) limitation of applications of the same herbicide, (4) crop rotation, (5) rotation of weed control methods, (6) the use of certified seeds, and (7) control of plants in adjacent areas (terraces, postharvest).

Some advantages can be obtained in the adoption of herbicide-tolerant cultivars. With the use of these cultivars, only the procedures of application of herbicide for desiccation prior to the implantation of the crop and one postemergence application are necessary (Peter et al. 2007). Concomitant to this, events that provide the use of herbicides with mechanisms of action different from that of glyphosate, as well as events that have gene staking (e.g., *aad-12* + *pat*), can provide the most extensive use of practices (1), (2), and (3) mentioned above. Thus, with different mechanisms of action, there is a greater chance that the control will be more efficient for a longer time, especially for those cultivars with more than one transgene, since the probability of a weed becoming resistant to both mechanisms simultaneously is smaller (Vargas and Roman 2006; Vargas et al. 1999).

Pest-Resistant Cultivars

The crops are subject to attacks and infestations of many diverse pests. From the beginning of the vegetative stage, several insects such as the soybean stalk weevil (*Sternechus subsignatus*), the lesser cornstalk borer (*Elasmopalpus lignosellus*), the scarab beetle (*Scarabaeoidea*), and the brown burrowing bugs (*Scaptocoris castanea* and *Atarsocoris brachiariae*) cause damage to different crops. With the vegetative growth of the plant and after flowering, other pests such as the velvet bean caterpillar (*Anticarsia gemmatalis*) and the sunflower looper (*Chrysodeixis includens*, *Rachiplusia nu*) occur in a greater proportion (Hoffmann-Campo et al. 2000). According to these authors, depending on the intensity of infestation and the phenological phase of the plants, caterpillars can cause defoliation at levels that affect production, that is, 30% of defoliation in the vegetative phase or 15% from flowering.

For pest control, the use of chemical insecticides was the most used method in the last decades. Two important aspects of chemical control stimulated scientists to seek alternative forms of pest control; these are the environmental pollution caused by the use of insecticides and their high cost (Borém 2005). Thus, biotechnology for pest control has improved over the past decades. As a result, this category allows the reduction of agricultural practices and the application of agrochemicals in the agricultural production process (Duarte 2001).

The first event to control pests in soybean via biotechnology in Brazil was MON 87701 × MON 89788 (Table 16.1), commercially called Intacta RR2 Pro. This combined event has the gene Bt *cry1Ac* (MON 87701) expressing protein Cry1Ac from *Bacillus thuringiensis* and also gene cp4 *epsps* (MON 89788) conditioning tolerance to glyphosate. This technology, according to the grower, is able to effectively control the velvet bean caterpillar (*A. gemmatalis*), the sunflower looper (*C. includens, R. nu*), the tobacco budworm (*Heliothis virescens*), and the Epinotia pod borer (*Epinotia aporema*). It is also reported that the cultivars from this event have the capacity to control the lesser cornstalk borer (*Elasmopalpus lignosellus*) and caterpillars of the genus *Helicoverpa* (*Helicoverpa zea* and *Helicoverpa armigera*). This last species (*H. armigera*), a recent phytosanitary problem in Brazil, has worried producers and scientists because of its high capacity for dissemination and destruction, causing economic losses of billions since the crop of 2012/2013 (MAPA 2015), mainly in corn, soybean, and cotton cultivations.

Risk Management of Emergence of Resistant Pests

The reasons for the emergence of insect resistance due to the crops using Bt technology are mainly due to the lack of implementation of high doses of the insecticidal protein and the nonuse of the refuge.

High Dose

A dose is considered high when the plant expresses the insecticidal protein at a concentration of at least 25 times that required to eliminate 99% of a susceptible reference population (U.S. EPA 2001). The high expression of the insecticidal protein makes any mechanism that gives the insect low or moderate resistance levels ineffective, and it is assumed that in this condition, most heterozygous insects (RS) would be eliminated, that is, the transgenic cultivar should act on all the susceptible (SS) homozygous and the heterozygous insects. This way, a small proportion of resistant homozygous (RR) (Ostlie et al. 1997) survives, which is more likely to generate heterozygous progenies which will again be targeted by high doses of protein Bt in plants (Huang et al. 2011).

Use of Refuge

The refuge comprises the area where a certain portion of the insect population is not exposed to the selection pressure, i.e., they can survive, multiply, and mate with the surviving individuals of the area with the cultivar Bt, subsequently generating susceptible heterozygotes. In the case of events having the gene Bt, the technical recommendation for refuge use is 20% of the planted area (Fig. 16.4). In addition, the following management should be adopted (Monsanto 2015): (1) The refuge area should not be over 800 m away from the transgenic cultivars (maximum distance verified by the dispersal of adult Lepidoptera); (2) The refuge area should be conducted normally with the use of insecticides or any other similar technique, and the application of Bt-based insecticides should not be allowed. This implies

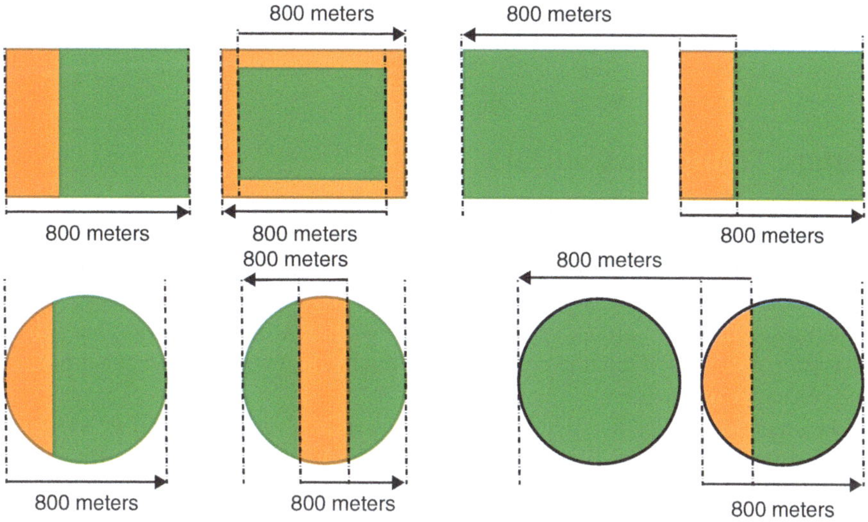

Fig. 16.4 Refuge areas in traditional crops and central pivot. Orange area: refuge area and green area: cultivar Bt. Source: The authors (2015)

selection of resistant individuals and less effectiveness of the technology over time; (3) The cultivar of soybeans planted in the refuge area must have the same vegetative cycle as the cultivar Bt, which should be at first with Roundup Ready (RR) technology; (4) The refuge area should be located in the same rural property where the cultivar Bt was planted; (5) The grower must maintain pest monitoring and adopt the level of economic damage as a parameter for intervention with chemical or biological products (not Bt based); (6) To desiccate and treat the seeds as initial methods of prevention against pests, weeds, and pathogens.

Stacking of Genes Bt

Stacking or pyramidal events are defined as events that have undergone more than one transformation or are derived from directed crossing between two transgenic lines, each transformation or each line having a specific transgene (Schrijver et al. 2007). The first generation of cultivars Bt that emerged on the market expressed a single protein, Cry1Ac. And as in the corn crop, in soybeans there are already examples of events with staking for genes Bt (*cry1Ac* + *cry1F*; *cry1A.105* + *cry2Ab2*). This staking of genes is particularly important because it allows the expression of different mechanisms of resistance against pests, resulting in more effective and lasting resistance (Aragão and Andrade 2011).

In addition to these, other procedures such as monitoring, low dose/MIP (biological control), crop rotation, seed mix, and selective, local, and temporal expression of proteins Bt and trap plants can be used for these purposes, which are widely discussed in Leite et al. (2011).

Future Transgenic Cultivars

The soybean chain around the world is of great relevance to all seed companies, since it is considered one of the largest commodities exported for grain, bran, and oil (ABIOVE 2015), and thus the adoption of transgenic cultivars is above 90%. This makes this crop attractive for the implementation of new events that bring characteristics to facilitate the crop management by the farmer and also that please the consumer market.

In order to predict possible transgenic events in a specific country, one can analyze the events already approved for the North American and in the world market. The events approved for field experimentation are an indication of which events may be available in the market in the short to medium term. Thus, regarding the herbicide tolerance events, the main trends are those for use with 2,4-D, Dicamba and their interactions with glyphosate and glufosinate-ammonium (Tables 16.2 and 16.3). These events aim to facilitate the control of weeds resistant to glyphosate due to the different mechanisms of action.

In relation to insect resistance, especially Lepidoptera, gene stacking is an alternative to maintain the long-term resistance against pests in soybean, so that the insertion of subgroups of the genes *cry1* and *cry2* are the options explored by the breeders of events (Tables 16.2 and 16.3).

Another event to be introduced on soybean cultivars is the nutritional quality, specifically for the reduction of trans fat levels. For the most part, trans fat is found in foods processed with the use of hydrogenated vegetable oils, which transform the liquid oil into solid. Thus, the oil is stable and allows the industrial use to prepare fried foods, improving the texture and extending the shelf life of the product. However, this type of fat causes serious health risks to consumers, and due to this food and medication regulatory agencies like the Food and Drug Administration—USA (FDA) are prohibiting the use of hydrogenated vegetable oil. Currently, solutions have been sought through Plenish (event DP305423) and Vistive Gold (event MON87705) (Table 16.3), in which it is possible to use its oil without going through the hydrogenation process due to the higher amount of oleic acid, providing a more stable oil.

Biosafety of Genetically Modified Plants

According to Vasconcelos and Carneiro (2013), the objective of the biosafety evaluation of genetically modified plants is to evaluate and identify the risks associated with this technology, compared to isogenic cultivars, which have a

Table 16.2 Possible transgenic events for the soybean in the next years

Event	Grower	Genes	Donor organisms	Characteristics	Countries with release
DAS68416-4 × MON89788	Dow AgroSciences	*aad-12/cp4 epsps/pat*	*Delftia acidovorans/Agrobacterium tumefaciens cepa CP4/Streptomyces viridochromogenes*	Tolerant to herbicides 2,4-D, glyphosate and glufosinate-ammonium	Canada
MON87751	Monsanto	*cry1A.105 e cry2Ab2*	*Bacillus thuringiensis* subsp. kumamotoensis	Resistant to lepidopteran insects	Canada and USA
MON87705	Monsanto	*fatb1-A e fad2-1A/cp4 epsps*	*Glycine max/Agrobacterium tumefaciens strain CP4*	Increases the levels of mono-unsaturated oleic acid and decreases the levels of saturated linoleic acid in the seed. Tolerant to herbicide glyphosate	Canada and USA
DP305423	Pioneer	*gm-fad2-1* × e *gm-hra*	*Glycine max*	It blocks the formation of linoleic acid from oleic acid and allows the accumulation of oleic acid in the seed. Tolerant to herbicides of the group sulfonylurea	Canada, Japan, and USA

Table 16.3 Soybean events granted for field experimentation (Planned Release in the Environment) in 2014 and 2015

Event	Grower	Genes	Characteristics
DAS44406-06	Dow AgroSciences	*aad-12, 2mepsps, pat*	Tolerant to herbicides 2,4-D, glyphosate and glufosinate-ammonium
DAS81419-2	Dow AgroSciences	*cry1Ac, cry1F/pat*	Resistant to lepidopteran insects and tolerant to herbicide glufosinate-ammonium
DAS44406-06 × DAS 81419 × 2	Dow AgroSciences	*cry1Ac, cry1F/aad-12,2mepsps, pat*	Resistant to lepidopteran insects and tolerant to herbicides 2,4-D, glyphosate, and glufosinate-ammonium
DAS8264 (DAS44406-6)	Dow AgroSciences	*aad-12, 2mepsps, pat*	Tolerant to herbicides 2,4-D, glyphosate and ammonium-glufosinate
DAS9582 (DAS81419-2)	Dow AgroSciences	*cry1Ac, cry1F/pat*	Resistant to lepidopteran insects and tolerant to herbicide glufosinate-ammonium
DAS8264 × AS9582 (DAS44406-6 × DAS81419-2)	Dow AgroSciences	*cry1Ac, cry1F/aad-12, 2mepsps, pat*	Resistant to lepidopteran insects and tolerant to herbicides 2,4-D, glyphosate, and glufosinate-ammonium
MON87701 × MON89788 00 × MON87708 × A841661	Monsanto	*cry1Ac, cp4 epsps, dmo*	Resistant to lepidopteran insects and tolerant to herbicides glyphosate and dicamba
MON87708 × MON89788	Monsanto	*dmo, cp4 epsps*	Tolerant to herbicides dicamba and glyphosate
SYHT0H2	Syngenta	*pat, avhppd-03*	Tolerant to ammonium glufosinate and herbicides that promote the inhibition of carotenoid biosynthesis by the inhibition of the enzyme HPPD

history of safe use. Within this context, Codex Alimentarius[1] is a guide to evaluate the risks of foods derived from genetically modified plants, which must have the following criteria:

Description of the Transgenic Plant

The type and purpose of the intended transformation must be identified in the target event. It is also necessary to describe clearly all its characteristics so as to better understand the nature of the food being subjected to the safety assessment.

Description of the Host Plant and Its Use as Food

Information about the history of the crop and its use as food, which nutrients are disposed in the line or genetic material to be modified, as well as information about possible toxic, allergic effects or other human health risks. The objective is to characterize the parental line that is not normally genetically modified and serves as a guide for the choice of parameters to be analyzed and compared to the transgenic (Costa et al. 2011), that is, to serve as a counterpart for the transgenic. This is important in the context of substantial equivalence, widely used in GM biosafety.

Description of the Donor Organism of the Gene of Interest

It is a detailed taxonomic description, including information on the history of safe use of the donor organism and possible toxic, allergic, pathogenic potential or the relation with pathogens.

Description of Genetic Modification

Information on the transformation process, genetic engineering (including the type of transformation method), source of DNA to be inserted, and information on intermediary organisms in genetic transformation; besides these, the marker

[1]Codex Alimentarius is an international forum for standardization of food trade established by the United Nations (UN), Food and Agriculture Organization (FAO), and World Health Organization (WHO) established in 1963. The forum aims to protect consumers' health and ensure equitable practices in regional and international food trade.

genes, regulatory genes, and other elements that affect the function of the transgene (size, identity, location, etc.) should be listed.

Characterization of the Genetic Modification

A process characterized by the number of insertions and copies of the inserted transgene, as well as the insertion sites being occupied; additionally, the resulting products of the transgene should be described (e.g., proteins), as well as the phenotypic characterization of the new trait, site, and level of expression of the gene product, and also infer on the integrity and stability of the transgene inserted and whether or not there are interactions with native plant genes.

Safety Assessment: Substances Expressed and Evaluation of Possible Allergenicity and Toxicity

All analyses must be done case by case and under scientific scrutiny. It is directly verified that any effect the transgenic event developed can establish on the health of those who ingest it, and for that it must be compared to the risks known and already associated to conventional foods. These risks depend on the allergic or toxic potential of the components in the modified material, nutritional quality, microbiological safety of the food, and possible side effects of gene expression (Hautea 2009; Costa et al. 2011). The content of this phase is directly linked to the concept of substantial equivalence, to which the composition of the transgenic cultivar should be compared to that of its conventional analog (Vasconcelos and Carneiro 2013). Thus, the choice of test sites should be representative, according to the environmental conditions in which future cultivars may be grown, allowing a credible evaluation of the traits and metabolic compositions of plants; as well as the same repeating assays should be performed within the sites and over a number of generations. All these treatments are effective in estimating the environmental, genotypic, and residual effects in the experiments, in order to make experimental analyses sufficiently sensitive to detect variations between conventional and transgenic material (Codex Alimentarius 2015). Additionally, at this level of evaluations, tests with animals are also important, since studies have shown that animal feed prepared with animal food or by-products from transgenic crops may be good indicators of biosafety (Calsamiglia et al. 2007; Cromwell et al. 2005; Donkin et al. 2003).

Other risk analysis procedures should be considered, in addition to those previously mentioned: (1) evaluation of the horizontal and vertical transfer potential of the genes of interest, (2) evaluation of the potential of the transgenic event to behave as an invasive plant (weedness), and (3) evaluation of possible adverse effects on nontarget organisms. All of them are widely described in Vasconcelos and Carneiro (2013).

Not unlike, many country biosafety legislations take into account the safety risk procedures for genetically modified plants under their circumstance. In general, legislation on these issues is governed by Biosafety Law and Decrees, which establishes safety standards and inspection mechanisms for construction, cultivation, production, manipulation, transportation, transfer, import, export, storage, research, commercialization, consumption, release into the environment, and disposal of genetically modified organisms (GMOs) and their derivatives, having as guidelines for the stimulation of scientific advances in biosafety and biotechnology; protection of human, animal, and plant life and health; and observance of the precautionary principle for the protection of the environment.

Coexistence

The rules of coexistence are necessary for the protection of the rights of neighboring producers from areas with transgenic cultivars to plant conventional cultivars. However, under some countries' legislation or normatives, there is provision addressing that. One example is the Decree 5950 of October 31, 2006, in Brazil, which establishes limits for the planting of transgenic cultivars in areas surrounding the conservation units, in horizontal projection from its perimeter until the damping zone is defined, and the management plan of the conservation unit is approved. In this sense, it establishes (1) 500 m for planting soybean tolerant to the herbicide glyphosate, (2) 800 m for planting insect-resistant cotton, and (3) 5000 m for planting insect-resistant cotton when there is record of occurrence of direct ancestor or wild relatives in the conservation unit.

Final Considerations

The technology involving transgenic cultivars over the years has been adopted by millions of farmers in dozens of countries. In fact all countries' major food producers are already using GM cultivars. Much is due to the benefits and easiness that these cultivars bring to mitigate biotic interference to which crops are subjected. Consequently, with increasing areas occupied by this technology, the selection pressure toward target organisms is increasing, thus increasing the likelihood of technology fail. Therefore, the management according to the agronomic technical recommendations for these cultivars is a crucial point for the durability and permanence and longevity of this technology. Nonetheless, tools and methodologies are emerging and envisioning a promising future in genetic engineering, opening prospects for increasing the range of events made available to farmers. Therefore, biosafety processes, risk evaluations, and GM rules should be able to meet the new challenges and novelties coming with new transgenic cultivars.

References

Aragão, F.J.L., Andrade, P.P.: Variedades com Eventos Piramidados. In: Borém, A. (ed.) Plantas geneticamente modificadas nos trópicos: desafios e oportunidades, 532p. Editora Suprema, São Paulo (2011)

Aragão, F.J.L., Faria, J.C.: First transgenic geminivirus resistant plant in the field. Nat. Biotechnol. **27**(12), 1086–1088 (2009)

Associação Brasileira Das Indústrias De Óleos Vegetais (ABIOVE): Cadeia produtiva de oleaginosas e biodiesel. http://www.abiove.org.br (2015). Accessed 1 Sep 2015

Borém, A.: Variedades transgênicas e meio ambiente. Biotecnolog. Ciênc. Desenvolv. **34**, 91–98 (2005)

Calsamiglia, S., Hernandes, B., Hartnell, G.F., Phipps, R.: Effects of corn silage derived from genetically modified variety containing two transgenes on feed intake, milk production, and composition, and the absence of detectable transgenic deoxyribonucleic acid in milk in Holstein cows. J. Dairy Sci. **90**, 4718–4723 (2007)

Carvalho, F.T., Velini, E.D.: Períodos de interferência de plantas daninhas na cultura da soja I—Cultivar IAC 11. Planta Daninha. **19**(3), 317–322 (2001)

Céleres: Informativo Biotecnologia. http://www.celeres.com.br (2015). Accessed 28 Aug 2015

Codex Alimentarius Commission: Guideline for the conduct of food safety assessment of foods derived from recombinant-DNA plants. http://www.who.int/foodsafety/biotech/en/codex_guidelines_plants (2003). Accessed 23 Sep 2015

Companhia Nacional De Abastecimento (CONAB): 12° Levantamento—Safra 2014/15 Grãos. http://www.conab.gov.br (2015). Accessed 10 Sep 2015

Conrad, U.: Polymers from plants to develop biodegradable plastics. Trends Plant Sci. **10**, 511–512 (2005)

Costa, T.E.M.M., Dias, A.P.M., Scheidegger, E.M.D., Marin, V.A.: Avaliação de risco dos organismos geneticamente modificados. Ciência Saúde Coletiva. **16**, 327–336 (2011)

Cromwell, G.L., Henry, B.J., Scott, A.L., Gerngross, M.F., Dusek, D.L., Fletcher, D.W.: Glufosinate herbicide-tolerant (LibertyLink) rive vs. conventional rice in diets for growing-finishing swine. J. Anim. Sci. **5**, 1068–1074 (2005)

Donkin, S.S., Velez, J.C., Totten, A.K., Stanisiewski, E.P., Hartnell, G.F.: Effects of feeding silage and grain from glyphosate-tolerant or insect-protected corn hybrids on feed intake, ruminal digestion, and milk production in dairy cattle. J. Dairy Sci. **5**, 1780–1788 (2003)

Duarte, J.O.: Effects of the Biotechnology and Intellectual Property Right Law in the Seed Industry. Tese de D.S.—University of Nebraska, Lincoln (2001)

FDA—U.S. Food and Drug Administration. http://www.fda.gov (2015). Accessed 25 Sep 2015

Hautea, D.M.: Risk analysis for GMOs : Concepts, methods and issues. In: Chowdhury, M.K.A., Hoque, M.I., Sonnino, A. (eds.) Biosafety of Genetically Modified Organisms: Basic Concepts, Methods and Issues, pp. 107–156. FAO, Rome (2009)

Hoffmann-Campo, C.B., Moscardi, F., Corrêa-Ferreira, B.S., Oliveira, L.J., Sosa-Gomez, D.R., Panizzi, A.R., Corso, I.C., Gazzoni, D.L., Oliveira, E.B.: Pragas da soja no Brasil e seu manejo integrado, 70p. Circular Técnica/Embrapa Soja, Londrina (2000)

Huang, F., Andow, D.A., Buschman, L.L.: Success of the high-dose/refuge resistance management strategy after 15 years of Bt crop use in North America. Entomol. Exp. Appl. **140**, 1–16 (2011)

ISAAA. ISAAA Brief 51-2015: Executive Summary (2015). http://isaaa.org/resources/publications/briefs/51/executivesummary/default.asp

James, C. 2014. Global Status of Commercialized Biotech/GM Crops: 2014. ISAAA brief n. 49, ISAAA, Ithaca.

Karam, D., Melhorança, A.L., Oliveira, M.F.: Plantas Daninhas na Cultura do Milho, 8p. Embrapa Milho e Sorgo, Sete Lagoas (2006)

Leite, N.A., Mendes, S.M., Waquil, J.M., Pereira, E.J.G.: O milho Bt no Brasil: a situação e evolução da resistência de insetos, 45p. Documentos/Embrapa Milho e Sorgo, Sete Lagoas (2011)

Ma, J.K.C., Drake, P.M.W., Christou, P.: The production of recombinant pharmaceutical proteins in plants. Nature. **4**, 794–805 (2003)

Menegatti, A.L.A., Barros, A.L.M.: Análise comparativa de custos de produção entre soja transgênica e convencional: um estudo de caso 19 para o Estado do Mato Grosso do Sul. Rev. Econ. Sociol. Rural. **45**(1), 163–183 (2007)

Ministério Da Agricultura, Pecuária e Abastecimento (MAPA): http://www.agricultura.gov.br (2015). Accessed 15 Sep 2015

Monsanto: Intacta RR2 PRO. http://www.intactarr2pro.com.br (2015). Accessed 9 Sep 2015

Oerk, E.C.: Crop losses to pests. J. Agric. Sci. **144**, 31–43 (2006)

Ostlie, K.R., Hutchison, W.D., Hellmich, R.L.: Bt Corn and European Corn Borer: Long-Term Success Through Resistance Management. University of Minnesota Extension Service, St. Paul (1997)

Petter, F.A., Procópio, S.O., Cargnelutti Filho, A., Barroso, A.L.L., Pacheco, L.P.: Manejo de herbicidas na cultura da soja Roundup Ready®. Planta Daninha. **25**(3), 557–566 (2007)

Rani, A.J., Usha, R.: Transgenic plants: types, benefits, public concerns and future. J. Pharm. Res. **6**, 879–883 (2013)

Schrijver, A., Devos, Y., Van den Bulcke, M., Cadot, P., De Loose, M., Reheul, D., Sneyers, M.: Risk assessment of GM stacked events obtained from crosses between GM events. Trends Food Sci. Technol. **18**, 101–109 (2007)

Silva, A.A., Ferreira, R.G., Ferreira, E.A., Vargas, L.: Variedades com Eventos Piramidados. In: Borém, A. (ed.) Plantas geneticamente modificadas nos trópicos: desafios e oportunidades, 532p. Editora Suprema, São Paulo (2011)

Sticklen, M.: Plant genetic engineering to improve biomass characteristics for biofuels. Curr. Opin. Biotechnol. **17**(3), 315–319 (2005)

UNESP: Parecer de Consultor Ad Hoc—Soja Evento BPS-CV127–9 (Soja CV127). 8p (2009)

United States Environmental Protection Agency (U.S EPA): Biopesticides registration action document-Bacillus thuringiensis plant-incorporated protectants. E: Benefit Assessment. Disponívrel em: http://www.epa.gov/oppbppd1/biopesticides/pips/Bt_brad2/5-benefits.pdf (2001). Accessed 23 Sep 2015

Vargas, L., Roman, E.S.: Identificação e manejo de plantas daninhas resistentes a herbicidas. Documentos/Embrapa Trigo, n. 60, 19p. http://www.cnpt.embrapa.br/biblio/do/p_do60.htm (2006). Accessed 12 Sep 2015

Vargas, L.; Silva, A.A.; Borém, A.; Rezende, S.T.; Ferreira, F.A.; Sediyama, T.: Resistência de plantas daninhas a herbicidas, 131p. Jard. Produções Gráficas Ltda, Viçosa (1999)

Vasconcelos, M.J.V., Carneiro, A.A.: Biossegurança de plantas geneticamente modificadas. Circular Técnica 199. Embrapa, 10p (2013)

Chapter 17
Resistance to Diseases

Éder Matsuo, Paulo Afonso Ferreira, and Tuneo Sediyama

Abstract The soybean crop is the target of many diseases, caused by fungus, bacteria, virus, and nematodes. Those diseases can cause severe loses, depending on several factors such as biotype of the pathogen, variety, environmental conditions, and the interaction of those factors. In this chapter the many diseases occurring in soybean and the strategies and methods to develop resistant cultivars are discussed. Genetic resistance has been a relevant goal in many breeding programs around the world due to the benefits in the reduction agrochemical use and costs.

Keywords *Glycine max* • Fungi • Bacteria • Virus • Nematode • Genetic resistance

Introduction

The US Department of Agriculture announced that US soybean production (*Glycine max* (L.) Merr.) is estimated at 107.1 million tons of grain. As a result, Brazil remains as the world's second largest producer of this oilseed, with an estimated production of 100.9 million tons for the 2015/2016 crop season (CONAB 2015). However, when considering the productivity (kg/ha) of soybean crop, Brazil has been leading the rank for some years. Last crop season, soybean yield in Brazil amounted to 2960 kg/ha, while the United States and Argentina produced, respectively, 2660 kg/ha and 2280 kg/ha (Bezerra et al. 2015). Among the various challenges faced by soybean farmers each season and the factors that limit

É. Matsuo, M.S., D.S. (✉)
Federal University of Viçosa - Campus Rio Paranaíba, Rio Paranaíba, Brazil
e-mail: edermatsuo@ufv.br

P.A. Ferreira, M.S., D.S.
Federal University of Viçosa - Campus Universitário do Araguáia, Barra do Garças, Brazil
e-mail: pauloafonso@ufmt.br

T. Sediyama, M.S., Ph.D.
Federal University of Viçosa, Viçosa, Minas Gerais, Brazil
e-mail: tuneo@ufv.br

F. Lopes da Silva et al. (eds.), *Soybean Breeding*,
DOI 10.1007/978-3-319-57433-2_17

productivity, affecting seed quality and increasing production costs, diseases are the most important and difficult ones to solve (Yorinori et al. 2013).

Differences in the reactions of cultivars to external factors (e.g., living organisms, disease resistance, chemicals, herbicide resistance) can be used internationally as characteristics in the DHE tests (Lovato 2011). The use of disease reaction characteristics in the cultivar description forms is required to monitor the progress of the breeding of the main species cultivated in the country, for example, soybean (Lovato 2011). Moreover, this author has reported that these descriptions of reaction are tools for decision-making by farmers when choosing the cultivar to be planted, in order to minimize the economic risks of the activity, so that this characteristic is not restricted to the need of differentiation for purposes of protection or evaluation of the cultivation value and use for commercial registration purposes. Details about the DHE test can be found in Chap. 23.

Diseases caused by bacteria, fungi, viruses, and nematodes can cause losses at different intensities, usually reaching up to 100% (Nogueira et al. 2015). In this chapter, we present, with focus on breeding, those of greater importance in the soybean crop, such as the ones caused by fungi (anthracnose, stem canker, Asian soybean rust, soybean target spot, frogeye leaf spot, brown spot, purple seed stain and leaf blight, downy mildew, white mold, powdery mildew, *Phytophthora* root rot, soybean pod, and stem blight and other root rot), bacteria (bacterial blight and bacterial pustule), virus (soybean common mosaic), and nematodes (root-knot nematode, cyst nematode, root lesion nematode, and reniform nematode). For those who aim to increase knowledge about soybean diseases, we suggest the work of Boerma and Specht (2004), Sediyama et al. (2005), Ferraz et al. (2010), Lemes et al. (2015), Sediyama et al. (2015), and Sediyama (2015).

Methods of Improvement and Conduction of the Segregant Population Aiming at Disease Resistance

The use of breeding aiming at resistance to phytopathogens, which consists on the incorporation of genetic resistance, is a feasible strategy (Castro et al. 2015), at the moment, for the majority of the diseases that attack soybean crops. However, it is worth pointing out that studies of pathogenic variability are extremely important to obtain better results with the implementation of more efficient resistance mechanisms (Castro et al. 2015).

The classic strategies used in soybean breeding programs for the development and selection of genotypes resistant to phytopathogens are methods involving hybridization (genealogical, mass, population, mixed, and SSD), backcrosses (gene pyramidation, gene rotation, and multilines), recurrent selection (Castro et al. 2015), introduction of germplasm, and selection of individual plants (Matsuo et al. 2015c). What is more, with the advent of molecular markers, it was possible to use them as a tool to assist breeders in the selection of superior genotypes (Castro et al. 2015). The main objective of using markers for assisted selection is to identify and validate markers with the traits of interest. Many marker identification studies have already been carried out, but some still require validation on other genetic backgrounds (Nogueira et al. 2015). It should be noted that even with selections made based on the genetic markers, it is important to carry out phenotypic evaluations of the promising lines with the pathogen of interest to confirm the genetic resistance (Matsuo et al. 2015b).

Although transgenic traits related to the genetic resistance to diseases have not yet launched, due to the impact caused by this technology, there is a great possibility that this tool will be incorporated into the group of important strategies for the incorporation of genes of resistance to the main diseases of soybean crops (Castro et al. 2015).

Anthracnose

Anthracnose, caused by the fungus *Colletotrichum truncatum* (Schwein.) Andrus and Moore, is considered one of the major soybeans diseases in several countries. In Brazil, it is very common in Cerrado regions, due to high rainfall and high temperatures (Dhingra et al. 2009; Henning et al. 2014).

The disease can cause loss of production of up to 100%, occurring at all stages of development of soybean plants, with the appearance of symptoms in various organs of the plant, such as leaves, branches, pods, and seeds. The leaves feature dark necrotic lesions in the limbus and ribs. This disease can occur in cotyledons, with dark-brown depressed necrotic lesions. Typically, symptoms are most commonly found in branches and pods, where irregular necrotic lesions are noted and in depression. The seeds of the pods featuring anthracnose are colonized by the pathogen, thus allowing the fungus to disperse at long distance. The disease can be controlled mainly by the use of healthy seeds and treated with fungicides, by means of crop rotation, management of crop remains, and management of plant density, avoiding over-densification and correction of soil nutrients (Dhingra et al. 2009; Henning et al. 2014).

Resistant cultivars have not been reported as an anthracnose control method (Dhingra et al. 2009; Henning et al. 2014). However, Costa et al. (2009) reported that cultivars BRS 133, BRS 153, BRS 184, BRS Macota, CD 201, CD 204, Conquista, Embrapa 48, FT-Abyara, Garantia, IAC 18, IAC 82, Iguaçú, MSOY 8001, MSOY 8400, and Sambaíba have presented a resistance reaction to *C. truncatum* inoculation. These cultivars are promising for use in integrated management programs (Matsuo et al. 2015c).

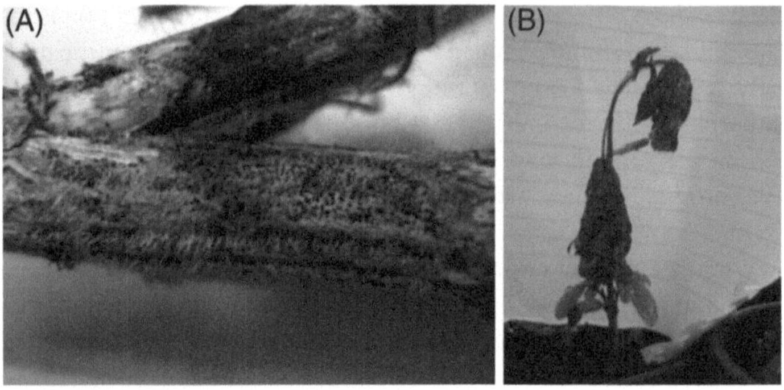

Fig. 17.1 Soybean stem canker (**a**) and (**b**) soybean plant with stem canker symptom and presence of perithecia. Photos: by Paulo Afonso Ferreira (**a**) and Éder Matsuo (**b**)

Stem Canker

Stem canker (Fig. 17.1), caused by the pathogen *Diaporthe phaseolorum* f.sp. *meridionalis* Fernández, was first reported in Brazil in the late 1980s (Dhingra et al. 2009).

The existence of sources of resistance with complete dominance and intensive work of breeding programs enabled the fast substitution of susceptible cultivars by high-yield resistant species (Yorinori and Kiihl 2001). The selection of promising lines in breeding programs can be carried out successfully, in a phenotypic manner, using the toothpick test methodology (Yorinori 1996) or with the use of molecular markers (Carvalho et al. 2002; Gavioli et al. 2007). Thus, different strategies can be used in breeding programs aiming at obtaining resistant genotypes adapted to different edafolimatic regions (Matsuo et al. 2015a). Accordingly, resistant cultivars are available and adapted to the entire producing region in Brazil (Sousa et al. 2015). Resistant soybean cultivars that are adapted to the region of interest as sources of resistance have been used in soybean breeding programs, because, in great part, hybridizations are carried out using as parents genetic material resistant to stem canker.

Asian Soybean Rust

Asian soybean rust (ASR), caused by the fungus *Phakopsora pachyrhizi* Syd. and P. Syd, was reported in Brazil for the first time in the 2000/2001 crop in the state of Paraná (Yorinori et al. 2005), and, in March 2013, the only soybean-producing state in Brazil that still had not reported the presence of ASR was Roraima (Yorinori et al. 2013). In crops, grain yield losses reached 100% in several cases (Yorinori et al. 2013). Such loss persists, even after a decade since its identification as the main disease to attack this crop, as the loss caused by rust is not only due to grain yield reduction but also due to fungicide application costs, which reduce producers' gains (Lemes and Gavassoni 2015).

Asian soybean rust is a disease that requires the adoption of different control measures, so as to maximize its cultural, chemical, and genetic control (Zambolim 2006; Lemes and Gavassoni 2015).

The development of ASR-resistant cultivars has become a goal of soybean breeding and also a challenge for breeders (Oliveira et al. 2005). Thus, the identification of genetic material that presents resistance or tolerance to rust and studies about genetic diversity are important steps in a soybean breeding program, since they guide breeders in the choice of parents that will form the breeding block (Nogueira et al. 2015).

A resistant soybean cultivar that presents a "universal" effect to ASR should be developed by pyramiding of major genes and introduction of minor genes, both with the purpose of conferring greater durability and resistance to several *P. pachyrhizi* (Yamanaka et al. 2010). For that, Vello et al. (2002) and Ribeiro et al. (2007) suggested the use of recurrent selection. Another method to conduct segregant populations for the development of tolerant soybean lines and/or resistance to ASR is the use of inoculum pressure in early generations to allow natural selection, favoring the most adapted individuals and a more efficient selection in relation to the discrimination of individuals (Ribeiro et al. 2010). Also, the use of molecular markers associated with rust resistance alleles is a potential tool to be used in assisted selection in order to identify resistance alleles in plants within segregating populations and to obtain the homozygous genotypes fixed in a few selection cycles (Sousa et al. 2015).

Currently, rust-resistant soybean cultivars are available to farmers (Fig. 17.2). However, it is recommended to apply fungicide at least once during the cycle, with the purpose of preventing resistance from being supplanted, due to the high genetic variability of the pathogen (Zambolim 2006; Henning et al. 2014).

Fig. 17.2 Soybean cultivar plants. (**a**) Rust resistant, TMG 801, and (**b**) Rust susceptible, BRS Valiosa RR Photos: Éder Matsuo

Fig. 17.3 Soybean leaf with symptoms of target spot. Photo: Paulo Afonso Ferreira

Target Spot

Caused by the fungus *Corynespora cassiicola* (Berk. and Curtis) Wei, the target spot is a disease that occurs in all soybean-planting regions of Brazil (Dhingra et al. 2009). The lesions can be observed in the aerial part of the plant, regardless of the stage of development (Fig. 17.3) (Siqueri et al. 2011). The main integrated control measures include the use of healthy seeds, resistant/tolerant cultivars, crop rotation/succession, and application of fungicides (Dhingra et al. 2009; Embrapa 2013).

Although the disease has taken important proportions for soybean farming, one does not know of soybean breeding programs that routinely evaluate lines for resistance to the target spot (Soares et al. 2009). Even so, there are reports of soybean cultivars, recommended for Brazil, that feature resistance to *C. cassiicola* (Embrapa 2013), and these, when possible, can be used to manage the disease.

Frogeye Leaf Spot

Cercospora sojina Hara is the etiological agent of *Cercospora* or frogeye leaf spot. This disease can cause average losses of 20% in soybean yield in areas with high humidity and temperature. Symptoms occur in leaves, branches, and pods, usually after flowering. At the beginning, one can notice small anasarcas that evolve into spots with a grayish center and reddish-brown edges. The main form of control is the use of resistant varieties. However, due to the presence of 25 races of the pathogen, management, with the use of healthy seeds and treated with fungicides, crop rotation and management of the crop remains, and application of fungicide in the aerial part of soybeans when necessary, is recommended to avoid supplanting resistance (Dhingra et al. 2009; Henning et al. 2014).

In Brazil, there are cultivars recommended to different producing regions and resistant to the different races of *C. sojina*. Studies conducted by Athow and Probst (1952) and Probst et al. (1965) reported that resistance is due to a gene, and the use of molecular markers is cited by Mian et al. (2008) as an alternative tool in the selection of plants of breeding programs aimed at resistance to frogeye leaf spot.

Brown Spot

Brown spot or septoriosis is caused by the fungus *Septoria glycines* Hemmi, which affects leaves, branches, pods, and seeds. The disease is more intense in conditions of long periods of rainfall, and it can cause losses of up to 20% in soybean yield. The most common symptoms are found in basal leaves, where small brown lesions appear and with irregular contours that can agglutinate, forming bigger, angular, and brownish lesions that lead to the severe early defoliation in the plants. Resistant varieties are not used in the control of the brown spot, so the most used management method is the use of healthy and treated seeds, crop rotation, and fungicide spraying in the aerial part of the plants (Dhingra et al. 2009; Henning et al. 2014).

Although the use of resistant cultivars as control is not recommended, genetic studies have been carried out. Soybean resistance to the brown spot is controlled by several minor genes, with magnitudes of heritability, in the broad sense, varying from 0.00 to 52.08% (Brogin et al. 2003). Hence, selection for resistance to brown spot should be carried out under controlled environmental conditions, such as a greenhouse, and based on the progenies from the F3 generation, from crosses involving resistance sources (Brogin et al. 2003). Martins et al. (2004) developed and validated a diagrammatic scale for measuring the severity of *S. glycines,* which is a very timely tool for the quantification of this disease due to the lack of a standard quantification system.

Purple Seed Stain and Leaf Blight

The fungus that causes the disease is *Cercospora kikuchii* (Matsumoto and Tomoy.) Gardner, which can attack leaves, petiole, stem, pods, and seeds at any stage of development, though the most characteristic symptoms are in the seed, with a coloration varying from pink to pale or dark purple, which can cover the whole seed (Fig. 17.4) (Dhingra et al. 2009). The management of the disease has been carried out through the use of healthy seeds, crop rotation, seed treatment, elimination or management of crop remains, and adequacy of plant population (Henning et al. 2014; Dhingra et al. 2009).

There is little information on genetic studies and evaluation of resistance to this pathogen. According to the authors, the PI80837 (Wilcox et al. 1975; Jackson et al. 2006) and cultivar SJ2 (Srisombun and Supapornhemin 1993) are resistant to *C. kikuchii* and controlled by a gene whose dominant allele is responsible for the resistance. In Brazil, the cultivar Emgopa 313 is indicated as a resistance pattern (Juliatti et al. 2006), and Kudo and Blum (2011), in addition to identifying cultivars

Fig. 17.4 Soybean seeds that present purple seed stain symptoms. Photo: Paulo Afonso Ferreira

resistant to the pathogen, have reported the real possibility of using genetic resistance as the main measure for control of the disease.

Downy Mildew

Downy mildew is caused by *Peronospora manshurica* (Naumov) Syd., an obligate parasitic oomycete, with a wide distribution in soybean farming areas; however, it does not have great economic importance (Dhingra et al. 2009; Henning et al. 2014). The symptoms of the disease occur in mild temperatures and high leaf moisture, beginning at the down surface of the young leaves with pale green to slightly yellowish dots, and there is sporulation of *P. manshurica* on the lower part (Fig. 17.5).

Downy mildew management is made using healthy and treated seeds, crop control, chemical control, and the use of resistant varieties (Dhingra et al. 2009). Resistant cultivars can be found, such as the cultivar UFUS-Impacta (Hamawaki et al. 2005), UFUS-Imperial (Hamawaki et al. 2007), UFUS-Mineira (Hamawaki et al. 2010), among others available to farmers.

White Mold or White Stem Rot

The white mold or white stem rot, caused by the fungus *Sclerotinia sclerotiorum* (Lib.) of Bary (Fig. 17.6), is one of the oldest diseases to affect soybeans and occurs in several producing regions of Brazil (Dhingra et al. 2009; Embrapa 2013).

Fig. 17.5 Downy mildew on soybean leaves. (**a**) Symptoms on the down surface of the leaf and (**b**) Sporulation on the lower side. Photos: Paulo Afonso Ferreira

Fig. 17.6 White mold on soybean and white and dense mycelium of the fungus on the stems. Photo: Éder Matsuo

As a control measure, it is recommended to avoid the introduction of the pathogen in the area by means of the use of seeds free from the pathogen sclerotia and to clean the equipment used in infested areas (Dhingra et al. 2009; Siqueri et al. 2011; Embrapa 2013; Henning et al. 2014), because sclerotia (primary inoculum) are difficult to eradicate after their introduction into the area (Furlan 2015). In areas where the disease already occurs, farmers must remove infected crop remains, rotate or substitute soybeans with species resistant or that do not host the fungus (e.g., corn, white oats, or wheat), increase line spacing (reducing population to the minimum recommended), adjust irrigation in order to reduce the fungus favorable moisture condition, maintain straw in the field, avoid the germination of the apothecia in the sclerodes, and use biological control with *Trichoderma* species (Görgen et al. 2009; Siqueri et al. 2011; Henning et al. 2014).

Genotypes resistant to *S. sclerotiorum* were identified by Garcia et al. (2015), while Zhao et al. (2015) reported candidate genes likely to be useful in soybean breeding for resistance to white mold. Nonetheless, there is no practical use of soybean cultivars genetically resistant to white mold (Furlan 2015). This is because the use of soybean genetic resistance to *S. sclerotiorum* is complex and should be associated with more escape mechanisms, such as resistance to lodging, canopy architecture type, plant height, and maturation and flowering periods (Furlan 2015).

Powdery Mildew

Powdery mildew, caused by *Microsphaera diffusa* Cooke and Peck, is an obligate parasite that has a wide distribution worldwide and can provide losses estimated at 30–40% of the production (Dhingra et al. 2009). The fungus is able to develop throughout the aerial part of the soybean, and the attacked plants are covered by a white powdery layer of mycelium and conidia (Fig. 17.7), which are more easily observed on both sides of the leaflets (Dhingra et al. 2009).

The most efficient method for the control of the powdery mildew is the use of resistant cultivars. Additionally, it is recommended applying fungicides when severity reaches 40–50% of the plant leaf area and avoiding late sowing (Dhingra et al. 2009; Henning et al. 2014).

Resistant genotypes in segregating breeding populations can be identified by visual selection, by subjecting the plants to tests and the resistance and susceptibility standards to the pathogen at field (Souza et al. 2015). Moreover, the molecular markers, already identified (Demore et al. 2009; Kang and Rouf Mian 2010), are of great importance as they are excellent tools in the selection of soybean segregant populations, the purpose of which is the development of resistant genotypes (Souza et al. 2015). Resistant cultivars are currently available to farmers.

Fig. 17.7 Powdery mildew on soybean plants. (**a**) Whole plant and (**b**) *Trifolium*. Photo: Éder Matsuo

Phytophthora Root and Stem Rot

Phytophthora sojae Kaufm. and Gerd. and *Phytophthora megasperma* Drechsler are the oomycetes that cause seed rot, tipping, and root rot, causing a failure in the number of soybean plants in the field. The disease is more severe under conditions of high humidity in the soil and high temperatures in compacted and poorly drained soils. The management of the disease occurs through the use of resistant varieties, treatment of seeds with fungicides, and improvement in soil drainage. However, there are no recommended control measures for adult plants (Dhingra et al. 2009; Henning et al. 2014).

According to studies, 14 *P. sojae* resistance alleles are known (*Rps1a*, *Rps1b*, *Rps1c*, *Rps1d*, *Rps1k*, *Rps2*, *Rps3a*, *Rps3b*, *Rps3c*, *Rps4*, *Rps5*, *Rps6*, *Rps7*, and *Rps8*) in eight different loci (Burnham et al. 2003; Sandhu et al. 2004). Among the several studies of identification and validation of molecular markers to resistance alleles, we mention Gordon et al. (2007) and Sugimoto et al. (2008). *Phytophthora* root rot-resistant soybean cultivars are available in Brazil.

Stem and Pod Blight

Stem and pod blight can be caused by several species of *Phomopsis*, such as *Phomopsis phaseoli* (Desm.) Sacc., *Phomopsis sojae* Lehman, and *Phomopsis longicolla* Hobbs. This disease is very common in the soybean crops all over the world, causing major grain losses when harvest is delayed due to excessive rain. Typically, the symptoms of *Phomopsis* spp. s. are found in the stems and branches, petioles, and pods, causing a necrotic lesion with black points in lines, formed by pycnidial fungi. The management of this disease is done by means of the use of resistant or tolerant varieties, the use of healthy and fungicide-treated seeds, crop rotation, and potassium fertilization (Dhingra et al. 2009; Henning et al. 2014).

The resistance of PI417479 to *Phomopsis* spp. is conferred by one or two complementary dominant genes, and this plant introduction was used for the development of the MO/PDG-0259 line (Zimmerman and Minor 1993). PI80837 is resistant, and its resistance is conferred by a dominant gene, different from that present in MO/PSD-0259 (Jackson et al. 2005). Based on several studies, Smith et al. (2008) reported that plant introductions (PI82264, Delmar, PI181550, PI227687, PI229358, PI200510, PI209908, Arksoy, PI80837, PI417479, PI360841, and 0X615) have different levels of resistance to *Phomopsis* spp.

Studies have been conducted to identify molecular markers that have also been conducted for the pod and stem blight disease, such as identification of AFLP markers (Berger and Minor 1999) and SSR (Jackson et al. 2009).

Other Root Rot

Several soil-dwelling fungi may be associated with symptoms of root and soybean stem rot, causing stand failure due to the death of young and adult plants. However, genetic improvement aimed at resistance to these phytopathogens is considered complex, for these are little specialized and attack several other hosts.

Thus, *Rhizoctonia solani* Kühn, which causes soybean root and stem rot, and *Macrophomina phaseolina* (Tassi) Goid. are responsible for charcoal rot (Fig. 17.8), and *Fusarium* species (*Fusarium solani* (Mart.) Sacc. complex species) cause red root rot or sudden death syndrome. For all of these pathogens, the ideal conditions for the diseases are soils with excess moisture, poorly drained or compacted. On the leaves, reflex symptoms, such as *Carijó* (large chlorotic and necrotic blotches between the leaf veins), also visible in plants affected by *Meloidogyne* sp. parasite (Fig. 17.9) can be observed. These diseases can be basically managed by avoiding planting in compacted or poorly drained soils and seed treatment with fungicides (Dhingra et al. 2009; Henning et al. 2014).

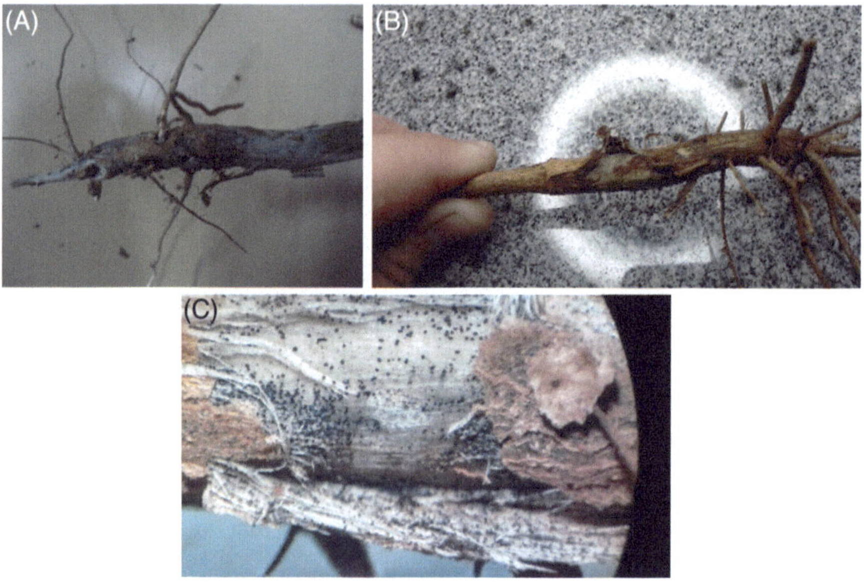

Fig. 17.8 Symptoms of soybean charcoal root and hypocotyl rot (**a** and **b**). Greasy spore of *Macrophomina phaseolina* and (**c**) *Macrophomina phaseolina* sclerodes. Photos: Paulo Afonso Ferreira

Fig. 17.9 Symptoms of *Meloidogyne* sp. in soybeans. (**a**) Roots featuring *M. javanica* knots and (**b**) Soybean *Trifolium* featuring *Carijó* symptoms. Photos: Paulo Afonso Ferreira

Bacterial Blight

Bacterial blight (*Pseudomonas savastanoi* pv. *glycinea* (Coerper) Gardan, sin: *P. syringae* pv. *glycinea*) is found in soybean crops worldwide. In Brazil it is present in all producing regions, with losses of up to 40% (Dhingra et al. 2009). The disease occurs initially on the leaves, but it can attack stems, petioles, and pods. The pathogen survives in crop residues from previous years and in infected seeds, in the field or during harvesting (Dhingra et al. 2009, Henning et al. 2014). As a control, Dhingra et al. (2009) cite the use of cultivars resistant to the prevailing race in the region, use of healthy seeds, crop rotation with nonhost species, management of cultural remains, and cleaning of tools and agricultural implements after crop treatment.

In Brazil, there are at least eight breeds of this phytobacter (Lanna filho 2015): R2, R3, R4, R6, and R7 (also previously described in the United States) and R10, R11, and R12 (new races), in that the most common is R3 race (Embrapa 2010). Thus, it is extremely difficult to predict how resistant cultivars will behave in a given area; therefore, the disease can be found in the field in cultivars considered resistant (Lanna filho 2015).

According to genetic studies, resistance to race 1 of *P. syringae* pv. *glycinea* is controlled by a gene (*Rpg1*/*rpg1*), in that the dominant allele *Rpg1* is responsible for the resistance, and resistance to race 2 apparently involves more than one gene (Mukherjee et al. 1966). Vernetti and Vernetti Junior (2009) reported that Norchief and Harosoy cultivars are carriers of the *Rpg1* allele, Merit of the *Rpg2* allele, and Flambeau of *Rpg3* and *Rpg4* alleles.

Bacterial Pustule

The bacterial pustule (*Xanthomonas axonopodis* pv. *glycines* (Nakano) Vauterin, sin: *Xanthomonas campestris* pv. *glycines*) is a common bacterial in soybean crop in several producing regions with hot and rainy climate (Dhingra et al. 2009). Its popular denomination is due to the development, in the lower part of the leaf, of small elevation resembling a small volcano, of whitish color in the center of the lesion (Henning et al. 2014). The disease can be controlled through the use of resistant cultivars, management of crop remains, use of healthy seeds, and crop rotation (Dhingra et al. 2009; Henning et al. 2014).

The genetic resistance identified in the CNS cultivar is controlled by a recessive gene (*rxp*/*rxp*) (Hartwig and Lehman 1951). With increasing efforts to expand the introgression of exotic germplasm, selection for the *rxp* allele will be required; therefore, it is desirable to identify and validate markers associated with the resistance allele, so that the selection of desirable genotypes is performed by means of the assisted selection (Narvel et al. 2001). Thus, Kim et al. (2010) presented a refinement of the genetic map in the *rxp* region with eight SSR markers and two SNPs, which may be useful in assisted selection in breeding programs. In Brazil, almost all of the recommended cultivars are resistant to bacterial pustule, and the onset of disease in the field is rare. One of the contributing factors is the use of cultivars with good resistance (Lanna Filho 2015).

Common Soybean Mosaic

The common soybean mosaic is caused by the *Potyvirus soybean mosaic virus* (SMV), found in all producing regions of Brazil. SMV can be transmitted by infected seeds and by aphids that feed on alternative hosts. In susceptible varieties, the virus may cause a dark color on seeds, near the hilum, known as brown spot, but in some susceptible varieties, it does not present such symptom. In soybean plants, SMV causes deformation and leaf blistering and plant dwarfism, due to the reduced length between knots and mosaic on the leaves, which may be dependent on the temperature of the environment. The main form of SMV control is the use of resistant soybean varieties (Dhingra et al. 2009; Henning et al. 2014).

Knowledge on the inheritance of resistance and on the genetics of the host–pathogen interaction is essential for breeding programs aimed at controlling common mosaic (Matsuo et al. 2015d). The first study of inheritance was performed by Kiihl and Hartwig (1979), and they concluded that both PI96983 and Ogden had a dominant resistance gene. Subsequently, several other scientific researches were conducted, and they identified other resistance genes. Regarding molecular markers, Matsuo et al. (2015d) cited several studies that identified and validated RAPD, RFLP, AFLP, and SSR molecular markers associated with alleles of soybean resistance to common mosaic.

In Brazil, the cultivar FT-10 was widely used in genetic breeding programs as resistant genitor (Silva et al. 2004), and soybean cultivars available in Brazil and with information on their common mosaic behavior are available in Almeida (2008) and Embrapa (2013).

Root-Knot Nematode

Soybean plants can be attacked by several species of *Meloidogyne*, but the most important are *Meloidogyne javanica* (Treub) Chitwood and *Meloidogyne incognita* (Kofoid and White) Chitwood (Dhingra et al. 2009; Henning et al. 2014). The main symptoms are knots of various numbers and sizes on the roots, necrosis among leaf veins (*Carijó* leaf) (Fig. 17.9), reduced plant size, intense pod abortion, and premature ripening of plants in dead spots (Henning et al. 2014). For accurate and accurate identification of the predominant *Meloidogyne* species in the area, samples of infested soil and roots with knots should be collected and sent as soon as possible to a nematology laboratory where they will be submitted to perineal pattern analysis and electrophoresis of isoenzymes (Dias et al. 2007). Only from the identification of the species will they define the actions for pest management, using, in combination, the various strategies, for instance, the rotation or replacement with nonhost crops and the use of resistant cultivars (Dias et al. 2007).

The development of resistant genotypes is a constant search of researchers who consider nematodes a phytosanitary problem, significantly affecting soybean production (Santos 2015). According to the genealogy analysis, almost all resistant cultivars available on the market descend from only one resistance gene source, the Bragg cultivar (Santos 2015). Thus, the work of Silva et al. (2001a) is important, because it provides details on the sources of resistance to *Meloidogyne* in addition to approaching the inheritance mechanisms of soybeans resistance to *M. javanica* facilitating the incorporation of the trait in adapted cultivars, in breeding programs (Silva et al. 2001b). Moreover, molecular tools have contributed to breeding programs in screening of lines, in order to identify those that feature resistance alleles. This can be done in the initial generations of breeding programs when it is aimed at developing resistant lines or, in advanced generations, to determine whether lines developed feature the resistance allele or not. Currently, Brazilian farmers can count on commercial cultivars tolerant and resistant to *Meloidogyne*, developed by different breeding companies.

Cyst Nematode

The soybean cyst nematode (Fig. 17.10), *Heterodera glycines* Ichinohe, can cause losses of up to 100% in soybean crops (Dhingra et al. 2009), and it is estimated that, in Brazil, the infested area has exceeded 2.0 million hectares (Embrapa 2013). As a preventive measure in an area with absence of the pathogen, it is recommended to use nematode-cyst-free seeds and to perform cleaning of agricultural machines and implements to avoid the entrance of the pathogen in the area; however, in already infected areas, the best option is to combine resistant cultivars with crop rotation and nonhost species (Henning et al. 2014).

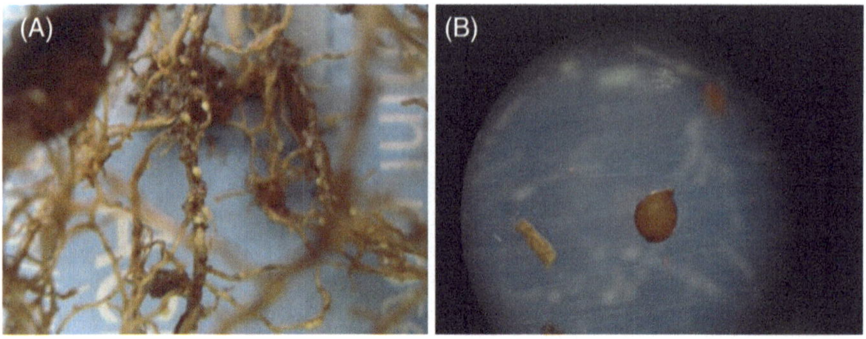

Fig. 17.10 Soybean root system with limoniform females of *Heterodera glycines* (**a**) and nematode cysts observed under a microscope (**b**). Photos: Éder Matsuo

The most used strategy for the incorporation of cyst nematode resistance has been the selection of lines from populations originating from hybridizations among adapted genotypes and resistant North American cultivars, to perform single and multiple crosses (modified backcrossing), involving different sources and adapted varieties (Dias et al. 2009). It should be pointed out that the majority of cyst nematode-resistant cultivars present the same source of resistance, PI88788 (Santos 2015). Due to the pathogenic variability of *H. glycines,* it is important to rotate the resistance genes to avoid the selection of new nematode races (Matsuo et al. 2015d).

Currently, in Brazil, there are 11 *H. glycines* races, and most of the cultivars on the market are only to race 3. However, some companies are already marketing varieties with resistance to more than one race, for example, TMG 4182 (races 1, 2, 3, 4, 5, 6, 9, and 10), TMG 4185 (races 1, 3, 4, 6, 9, and 10), FT Master (races 2, 4, 5, 9, 10, and 14), and FT Triunfo (races 1, 2, 3, 4, 5, 6, 9, and 14).

Root Lesion Nematode

Root lesions caused by the nematode *Pratylenchus brachyurus* (Godfrey) Filipjev and Stekhoven derive from the disorganization and destruction of the cells (Dias et al. 2007). In the field, the plants affected by this nematode are small in size and are located in dead spots (Fig. 17.11). The main measures for management of *P. brachyurus* involve rotation with bad or nonhost crops, soil rotation, and soil correction (Ferraz et al. 2010; Franchini et al. 2014).

The use of resistant cultivars would be a valuable measure for the management of *P. brachyurus* in soybean. However, plant breeding for resistance to *Pratylenchus* species is considered difficult, probably because they are polygraphs and relatively unspecialized, with migrating endoparasite habit, and do not settle in the host plant (Santos 2015). Nonetheless, studies (Hamawaki et al. 2010; Stetina

Fig. 17.11 Plants with undergrowth and stand failure in *Pratylenchus brachyurus* dead spots. Photo: Paulo Afonso Ferreira

et al. 2014) have shown the possibility of identifying genotypes resistant to *P. brachyurus*.

Further research should be carried out with the aim of studying the inheritance of resistance, genetic mapping, and indicating the methods of breeding and conduction of the segregant populations most recommended for the development of resistance or tolerance cultivars. However, to date, there are no varieties resistant to this nematode in the Brazilian market.

Reniform Nematode

The reniform nematode, *Rotylenchulus reniformis* Linford and Oliveira, is becoming increasingly important, especially in states where soybeans and cotton are grown in alternation. As control measures, it is recommended performing crop rotation with nonhost plant species and resistant soybean cultivars (Embrapa 2013).

The main sources of resistance to soybean cyst nematode, with the exception of PI 88788, are also resistant to the reniform nematode (Dias et al. 2007). Therefore, in order to shorten the time for the recommendation of nematode-resistant cultivars, priority should be given to studies on those genotypes that have demonstrated resistance to soybean cyst nematodes (Asmus 2008). Melo et al. (2013) identified 65 genotypes considered resistant (reproduction factor—RF—less than 1.0) and with RF mean values significantly equal to MSOY 8001. Currently, some cultivars resistant to reniform nematode are available in the Brazilian market.

Final Considerations

Soybean crops, in the Brazilian scenario, have presented production and productivities increments throughout the several years of cultivation. An important factor in this success is the integration between soybean breeders and phytopathologists, who have been working intensively in the search for resistant cultivars, in order to permit better working conditions for farmers, aiming at environmental preservation, integrated management, and reduced production cost.

References

Almeida, A.M.R.: Viroses da soja no Brasil: sintomas, etiologia e controle, 62p. Embrapa Soja, Londrina (2008)

Asmus, G.L.: Reação de genótipos de soja ao nematoide reniforme. Trop. Plant Pathol. **33**(1), 69–71 (2008)

Athow, K., Probst, A.H.: The inheritance of resistance to frog-eye leaf spot of soybean. Phytopathology. **42**, 660–663 (1952)

Berger, R.D., Minor, H.C.: An RFLP marker associated with resistance to phomopsis seed decay in soybean PI 417479. Crop. Sci. **39**(3), 800–805 (1999)

Bezerra, A.R.G., Sediyama, T., Borém, A., Soares, M.M.: Importância Econômica. In: Sediyama, T., Silva, F., Borém, A. (eds.) Soja: do plantio à colheita, pp. 9–26. UFV, Viçosa (2015)

Boerma, H.R., Specht, J.E. (eds.): Soybeans: Improvement, Production, and Uses, 3rd edn, 1144p. American Society of Agronomy, Crop Science Society of America, and Soil Science Society of America, Madison (2004)

Brogin, R.L., Arias, C.A.A., Toledo, J.F.F.: Genetic control of soybean resistance to brown spot (*Septoria glycines*): first studies. Crop. Breed. Appl. Biotechnol. **3**(1), 35–44 (2003)

Burnham, K.D., Dorrance, A.E., Francis, D.M., Fioritto, R.J., St. Martin, S.K.: *Rps8*, a new locus in soybean for resistance to *Phytophthora sojae*. Crop. Sci. **43**, 101–105 (2003)

Carvalho, G.A., Sediyama, T., Alzate-Marin, A.L., Barros, E.G., Moreira, M.A.: Identificação de marcadores RAPD ligados a um gene de resistência ao cancro da haste da soja. Fitopatol. Bras. **27**(5), 474–478 (2002)

Castro, L.H.S., Vianna, V.F., DiMauro, A.O., Unêda-Trevisoli, S.H.: Melhoramento genético da soja visando resistência a fitopatógenos. In: Lemes, E., Castro, L., Assis, R. (eds.) Doenças da Soja: Melhoramento genético e técnicas de manejo, 1st edn, pp. 179–203. Millennium, Campinas (2015)

Companhia Nacional De Abastecimento-CONAB: http://www.conab.gov.br/OlalaCMS/uploads/arquivos/15_10_16_10_52_19_safras_outu_2015.pdf (2015). Acesso em 22 Oct 2015

Costa, I.F.D., Balardin, R.S., Medeiros, L.A.M., Lenz, G., Gulart, C.A., Zemolin, C.R., SILVA, T. M.G.: Reação de germoplasma comercial de soja a *Colletotrichum truncatum*. Trop. Plant Pathol. **34**(1), 47–50 (2009)

Demore, P.S., Unêda-Trevisoli, S.H., Dimauro, A.O., Morceli, T.G.S., Muniz, F.R.S., Costa, M. M., Sarti, D.G.P., Mancini, M.C.: Validation of microsatellite markers for assisted selection of soybean genotypes resistant to powdery mildew. Crop Breed. Appl. Biotechnol. **9**, 45–51 (2009)

Dhingra, O.D., Mendonça, H.L., Macedo, D.M.: Doenças e seu controle. In: Sediyama, T. (ed.) Tecnologia de produção e usos da soja, pp. 133–155. Mecenas, Londrina (2009)

Dias, W.P., Silva, J.F.V., Garcia, A., Carneiro, G.E.S.: Nematoides de importância para a soja no Brasil. In: Fundação, M.T. (ed.) Boletim de pesquisa de soja, pp. 173–183, Rondonópolis, Fundação (2007)
Dias, W.P., Silva, J.F.V., Carneiro, G.E.S., Garcia, A., Arias, C.A.A.: Nematoide-de-cisto da soja: biologia e manejo pelo uso da resistência genética. Nematol. Bras. **33**, 1–16 (2009)
Empresa Brasileira De Pesquisa AGropecuária—EMBRAPA: Tecnologias de produção de soja—Região Central do Brasil—2011, 255p. Editora Embrapa, Londrina (2010)
Empresa Brasileira De Pesquisa Agropecuária—EMBRAPA: Tecnologias de produção de soja—Região central do Brasil 2014, 265p. Editora Embrapa Soja, Londrina (2013)
Ferraz, S., Freitas, L.G., Lopes, E.A., Dias-Arieira, C.R.: Manejo sustentável de fitonematoides, 306p. Editora UFV, Viçosa (2010)
Franchini, J.C., Debiasi, H., Dias, W.P., Ramos Junior, E.U., Balbinot Junior, A.A.: Densidade populacional do nematoide das lesões radiculares em soja e sua relação com a calagem. In: Anais da Reunião de Pesquisa de Soja, pp. 158–160. Embrapa Soja, Londrina (2014)
Furlan, S.H.: Mofo Branco. In: Lemes, E., Castro, L., Assis, R. (eds.) Doenças da Soja: Melhoramento genético e técnicas de manejo, pp. 53–72. Millennium, Campinas (2015)
Garcia, R.A., Meyer, M.C., Avila, K.A.G.B., Cunha, M.G.: Métodos de inoculação de *Sclerotinia sclerotiorum* para triagem de cultivares de soja resistentes ao mofo-branco. Pesq. Agropec. Bras. **50**(8), 726–729 (2015)
Gavioli, E.A., Dimauro, A.O., Centurion, M.A.P.C., Dimauro, S.M.Z.: Development of SCAR marker linked to stem canker resistance gene in soybean. Crop Breed. Appl. Biotechnol. **7**, 133–140 (2007)
Gordon, S.G., Kowitwanich, K., PipatpongpinyO, W., St. Martin, S.K., Dorrance, A.E.: Molecular marker analysis of soybean plant introductions with resistance to *Phytophthora sojae*. Phytopathology. **97**, 113–118 (2007)
Görgen, C.A., Silveira Neto, A.N., Carneiro, L.C., Ragagnin, V., Lobo Junior, M.: Controle do mofo-branco com palhada e *Trichoderma harzianum* 1306 em soja. Pesq. Agropec. Bras. **44** (12), 1583–1590 (2009)
Hamawaki, O.T., Juliatti, F.C., Polizel, A.C., Amorim, F.A., Shigihara, D., Santos, M.A., Hamawaki, C.L.: UFUS-Impacta: nova cultivar de soja para o Estado de Minas Gerais. Pesq. Agropec. Bras. **40**(5), 523–524 (2005)
Hamawaki, O.T., Polizel, A.C., Juliatti, F.C., Hamawaki, R.L., Brunetta, P.: UFUS-Imperial: novo cultivar de soja para o Estado de Mato Grosso. Pesq. Agrop. Brasileira. **42**(1), 137–139 (2007)
Hamawaki, O.T., Resende, D.F., Sagata, É., Sousa, L.B.: UFUS-Mineira: novo cultivar de soja para o Estado de Minas Gerais. Biosci. J. **26**(3), 424–427 (2010)
Hartwig, E.E., Lehman, S.G.: Inheritance of resistance to the bacterial pustule disease in soybeans. Agron. J. **43**, 226–229 (1951)
Henning, A.A., Almeida, A.M.R., Godoy, C.V., Seixas, C.D.S., Yorinori, J.T., Costamilan, L.M., Ferreira, L.P., Meyer, M.C., Soares, R.M., Dias, W.P.: Manual de identificação de doenças de soja, 5th edn, 76p. Embrapa Soja, Londrina (2014)
Jackson, E.W., Fenn, P., Chen, P.: Inheritance of resistance to Phomopsis seed decay in soybean PI 80837 and MO/PSD-0259 (PI 562694). Crop. Sci. **45**, 2400–2404 (2005)
Jackson, E.W., Fenn, P., Chen, P.: Inheritance of resistance to purple seed stain caused by *Cercospora kikuchii* in soybean PI 80837. Crop. Sci. **46**, 1462–1466 (2006)
Jackson, E.W., Feng, C., Fenn, P., Chen, P.: Genetic mapping of resistance to Phomopsis seed decay in the soybean breeding line MO/PSD-0259 (PI562694) and plant introduction 80837. J. Hered. **100**(6), 777–783 (2009)
Juliatti, F.C., Hamawaki, O.T., Cunha, E.P.C., Polizel, A.C., Santos, M.A., Shigihara, D.: Severidade de doenças fúngicas foliares em genótipos de soja em três localidades de plantio. Biosci. J. **22**(1), 83–89 (2006)
Kang, S.T., Rouf Mian, M.A.: Genetic map of the powdery mildew resistance gene in soybean PI242540. Genome. **53**, 400–405 (2010)

Kiihl, R.A.S., Hartwig, E.E.: Inheritance of reaction to soybean mosaic virus in soybeans. Crop. Sci. **19**, 372–375 (1979)

Kim, D.H., Kim, K.H., Van, K., Kim, M.Y., Lee, S.H.: Fine mapping of a resistance gene to bacterial leaf pustule in soybean. Theor. Appl. Genet. **120**, 1443–1450 (2010)

Kudo, A.S., Blum, L.E.B.: Reação de genótipos de soja ao crestamento foliar de Cercospora. Bragantia. **70**(2), 397–401 (2011)

Lanna Filho, R.: Fitobactérias na cultura da soja. In: Lemes, E., Castro, L., Assis, R. (eds.) Doenças da Soja: Melhoramento genético e técnicas de manejo, pp. 81–92. Millennium, Campinas (2015)

Lemes, E., Gavassoni, W.L.: Ferrugem Asiática da Soja. In: Lemes, E., Castro, L., Assis, R. (eds.) Doenças da Soja: Melhoramento genético e técnicas de manejo, pp. 35–51. Millennium, Campinas (2015)

Lemes, E., Castro, L., Assis, R. (eds.): Doenças da soja: melhoramento genético e técnicas de manejo, 363p. Editora Millennium, Campinas (2015)

Lovato, F.A.: Uso de características de resistência a doenças em testes de DHE. In: Brasil—Ministério da Agricultura Pecuária e Abastecimento. Proteção de Cultivares no Brasil, pp. 147–154. MAPA/ACS, Brasília (2011)

Martins, M.C., Guerzoni, R.A., Câmara, G.M.S., Mattiazzi, P., Lourenço, A.S., Amorim, L.: Escala diagramática para a quantificação do complexo de doenças foliares de final de ciclo em soja. Fitopatol. Bras. **29**, 179–184 (2004)

Matsuo, É., Sediyama, T., Glasenapp, S.: Resistência aos fungos. In: Sediyama, T. (ed.) Melhoramento Genético da Soja, pp. 149–180. Mecenas, Londrina (2015a)

Matsuo, É., Sediyama, T., Glasenapp, S.: Resistência aos vírus. In: Sediyama, T. (ed.) Melhoramento Genético da Soja, pp. 196–210. Mecenas, Londrina (2015b)

Matsuo, É., Sediyama, T., Glasenapp, S.: Resistência às bactérias. In: Sediyama, T. (ed.) Melhoramento Genético da Soja, pp. 181–195. Mecenas, Londrina (2015c)

Matsuo, É., Sediyama, T., Dias, W.P., Glasenapp, S.: Resistência aos nematoides. In: Sediyama, T. (ed.) Melhoramento Genético da Soja, pp. 211–238. Mecenas, Londrina (2015d)

Melo, C.L.P., Santos, W.R., Asmus, G.L.: Resistance of soybean genotypes to the reniform nematode in a controlled environment. Crop Breed. Appl. Biotechnol. **13**(1), 23–32 (2013)

Mian, M.A.E., Missaoui, A.M., Walder, D.R., Phillips, D.V., Boerma, H.R.: Frogeye leaf spot of soybean: a review and proposed race designations for isolates of *Cercospora sojina* Hara. Crop. Sci. **48**, 14–24 (2008)

Mukherjee, D., Lambert, J.W., Cooper, R.L., Kennedy, B.W.: Inheritance of resistance to bacterial blight (*Pseudomonas glycinea* Coerper) in soybean (*Glycine max* L.) Crop. Sci. **6**, 324–326 (1966)

Narvel, J.M., Jakkula, L.R., Phillips, D.V., Wang, T., Lee, S.H., Boerma, H.R.: Molecular mapping of *rxp* conditioning reaction to bacterial pustule in soybean. J. Hered. **92**, 267–270 (2001)

Nogueira, A.P.O., Sediyama, T., Gomes, J.D.: Avanços no melhoramento genético da cultura da soja nas últimas décadas. In: Lemes, E., Castro, L., Assis, R. (eds.) Doenças da Soja: Melhoramento genético e técnicas de manejo, pp. 159–178. Millennium, Campinas (2015)

Oliveira, A.C.B., Godoy, C.V., Martins, M.: Avaliação da tolerância de cultivares de soja à ferrugem asiática da soja no Oeste da Bahia. Fitopatol. Bras. **30**(6), 658–662 (2005)

Probst, A.H., Athow, K.L., Laviolette, F.A.: Inheritance of resistance to race 2 of *Cercospora sojina* in soybean. Crop. Sci. **5**, 332 (1965)

Ribeiro, A.S., Moreira, J.U.V., Pierozzi, P.H.B., Rachid, B.F., Toledo, J.F.F., Arias, C.A.A., Soares, R.M., Godoy, C.V.: Genetic control of Asian rust in soybean. Euphytica. **157**, 15–25 (2007)

Ribeiro, A.S., Moreira, J.U.V., Arias, C.A.A., Oliveira, M.F., Soares, R.M., Toledo, J.F.F.: Método de condução de populações segregantes para desenvolver linhagens de soja com tolerância e, ou, resistência à ferrugem asiática. In: XXXI Reunião De Pesquisa De Soja Da Região Central Do Brasil (ed.) Anais..., pp. 257–260. Embrapa Soja, Londrina (2010)

Sandhu, D., Hongyu, H., Cianzio, S., Battacharyya, M.: Deletion of a disease resistance nucleotide-binding-site leucine-rich repeat-like sequence is associated with the loss of the Phytophthora resistance gene *Rps4* in soybean. Genetics. **168**, 2157–2167 (2004)

Santos, M.A.: Fitonematoides. In: Lemes, E., Castro, L., Assis, R. (eds.) Doenças da Soja: Melhoramento genético e técnicas de manejo, pp. 105–117. Millennium, Campinas (2015)

Sediyama, T. (ed.): Melhoramento genético da soja, 352p. Editora Mecenas, Londrina (2015)

Sediyama, T., Teixeira, R.C., Reis, M.S.: Melhoramento da soja. In: Borém, A. (ed.) Melhoramento de espécies cultivadas, pp. 553–603. UFV, Viçosa (2005)

Sediyama, T., Silva, F., Borém, A. (eds.): Soja: do plantio à colheita, 333p. Editora UFV, Viçosa (2015)

Silva, J.F.V., Ferraz, L.C.C.B., Arias, C.A.: Herança da resistência a *Meloidogyne javanica* em soja. Nematropica. **31**, 211–219 (2001a)

Silva, J.F.V., Ferraz, L.C.C.B., Abdelnoor, R.V., Arias, C.A.A.: Identificação de marcadores moleculares de microssatélites associados à resistência de genótipos de soja a *Meloidogyne javanica*. Nematol. Bras. **25**(1), 79–83 (2001b)

Silva, M.F., Kiihl, R.A.S., Almeida, A.M.R.: Inheritance of resistance to soybean mosaic virus in FT-10 soybean. Euphytica. **135**, 339–343 (2004)

Siqueri, F.V., Yorinori, J.T., Yuyama, M.M.: Doenças da soja. In: Fundação MT (ed.) Boletim de Pesquisa de Soja 2011, pp. 301–389. Fundação MT, Rondonópolis (2011)

Smith, S., Fenn, P., Chen, P., Jackson, E.: Inheritance of resistance to phomopsis seed decay in PI 360841 soybean. J. Hered. **99**(6), 588–592 (2008)

Soares, R.M., Godoy, C.V., Oliveira, M.C.N.: Escala diagramática para avaliação da severidade da mancha alvo da soja. Trop. Plant Pathol. **34**(5), 333–338 (2009)

Sousa, L.B., Matsuo, É., Borém, A.: Vulnerabilidade genética da soja às doenças. In: Lemes, E., Castro, L., Assis, R. (eds.) Doenças da Soja: Melhoramento genético e técnicas de manejo, pp. 1–14. Millennium, Campinas (2015)

Srisombun, S., Supapornhemin, P.: Inheritance of soybean resistance to purple seed stain. Soybean Genet. Newslett. **20**, 92–93 (1993)

Stetina, S.R., Smith, J.R., Ray, J.D.: Identification of *Rotylenchulus reniformis* resistant glycine lines. J. Nematol. **43**(1), 1–7 (2014)

Sugimoto, T., Yoshida, S., Watanabe, K., Aino, M., Kanto, T., Maekawa, K., Irie, K.: Identification of SSR markers linked to the Phytophthora resistance gene Rps1-d in soybean. Plant Breed. **127**, 154–159 (2008)

Vello, N.A., Brogin, R.L., Arias, C.A.A.: Estratégias de melhoramento para o controle da ferrugem da soja. In: Congresso Brasileiro De Soja, 3; Congresso Brasileiro De Soja E Mercosoja, 2, 2002. Londrina. Anais... Embrapa Soja, Londrina (CNPSO: Documentos, 180) (2002)

Vernetti, F.J., Vernetti Junior, F.J.: Genética da soja: caracteres qualitativos e diversidade genética, 221p. Editora Embrapa, Pelotas (2009)

Wilcox, J.R., Laviolette, F.A., Martin, R.J.: Heritability of purple seed stain resistance in soybean. Crop. Sci. **15**, 525–526 (1975)

Yamanaka, N., Yamaoka, Y., Kato, M., Lemos, N.G., Passianotto, A.L.L., Santos, J.V.M., Benitez, E.R., Abdelnoor, R.V., Soares, R.M., Suenaga, K.: Development of classification criteria for resistance to soybean rust and differences in virulence among Japanese and Brazilian rust populations. Trop. Plant Pathol. **35**(3), 153–162 (2010)

Yorinori, J.T.: Cancro da haste da soja: epidemiologia e controle, 75p. Emprapa-Soja, Londrina, (Embrapa-Soja. Circular Técnica, 14) (1996)

Yorinori, J.T., Kiihl, R.A.S.: Melhoramento de plantas vidando resistência a doenças. In: Nass, L. L., Velois, A.C.C., Melo, I.S., Valadares-Inglis, M.C. (eds.) Recursos genéticos e melhoramento—Plantas, pp. 715–735. Fundação MT, Rondonópolis (2001)

Yorinori, J.T., Paiva, W.M., Frederick, R.D., Costamilan, L.M., Bertagnoli, P.F., Hartman, G.L., Godoy, C.V., Nunes, J.J.: Epidemics of soybean rust (*Phakopsora pachyrhizi*) in Brazil and Paraguay from 2001 to 2003. Plant Dis. **89**, 675–677 (2005)

Yorinori, J.T., Yuyama, M.M., Siqueri, F.V.: Doenças da soja. In: Galhardi Jr., A., Siqueri, F., Caju, J.E., Camacho, S. (eds.) Boletim de pesquisa de soja 2013/2014, pp. 174–221. Fundação MT, Rondonópolis (2013)

Zambolim, L.: Manejo integrado da ferrugem asiática da soja. In: Zambolim, L. (ed.) Ferrugem asiática da soja, pp. 73–98. Suprema, Viçosa (2006)

Zhao, X., Han, Y., Li, Y., Liu, D., Sun, M., Zhao, Y., Lv, C., Li, D., Yang, Z., Huang, L., Teng, W., Qiu, L., Zheng, H., Li, W.: Loci and candidate gene identification for resistance to *Sclerotinia sclerotiorum* in soybean (*Glycine max* L. Merr.) via association and linkage maps. Plant J. **82** (2), 245–255 (2015)

Zimmerman, M.S., Minor, H.C.: Inheritance of Phomopsis seed decay resistance in soybean PI 417479. Crop. Sci. **32**, 96–100 (1993)

Chapter 18
Resistance to Insects

José Baldin Pinheiro, José Djair Vendramim, and André Luiz Lourenção

Abstract Soybeans are attacked by several species of sucking and defoliating insects, and the sucking insects are now the main pest of the crop in tropical regions. Each year, insect damage is becoming more frequent considering the high number of species present in the crop due to the lack of adequate pest monitoring and the development of insects resistant to the insecticides. The chemical method has been widely used in insect control. However, some molecules have already been banned, others are in the process of banning, and the launch of new products is time-consuming. In this aspect, the use of resistant genotypes can be an alternative or a substitute for the chemical control of insects offering a series of advantages, both from an environmental and economic point of view. However, one of the main obstacles to obtaining resistant genotypes is related to low heritability of the traits associated with insect resistance and also with grain yield, a fact that hinders the development of resistant cultivars. Another aspect is that farmers do not accept resistant genotypes with lower yield potential than cultivars available for cultivation. Thus, in this chapter, the authors present a general discussion about the use of genetic resistance, without the use of transgenes, to obtain new soybean cultivars that highly yield and are resistant to the main pest insects in Brazil.

Keywords Soybean • Selection • Improved cultivars • Genetic resistance • Biotic stress resistance

---

J.B. Pinheiro, M.S., D.S. (✉)
Department of Genetics, College of Agriculture "Luiz de Queiroz", University of São Paulo—ESALQ, Piracicaba, São Paulo 13418-900, Brazil
e-mail: jbaldin@usp.br

J.D. Vendramim, M.S., D.S.
Department of Entomology and Acarology, College of Agriculture "Luiz de Queiroz", University of São Paulo, Piracicaba, São Paulo 13418-900, Brazil
e-mail: jdvendra@usp.br

A.L. Lourenção, M.S., D.S.
Phytosanitary Center, Agronomic Institute of Campinas, Campinas, São Paulo 13020-902, Brazil
e-mail: andre@iac.sp.gov.br

F. Lopes da Silva et al. (eds.), *Soybean Breeding*,
DOI 10.1007/978-3-319-57433-2_18

Introduction

The incorporation of soybeans in Brazilian agriculture has completely revolutionized the industry. It quickly became one of the main products in farming and for the domestic economy. Its cultivation was established as a traditional practice in the South, Southeast and, more recently, Midwest and Northeast regions. To this end, it was necessary to develop new cultivars, which have allowed productivity levels to increase in traditional areas, as well as to adapt the crop to new areas. As a result of such expansion, the damage caused by insect pests has also gradually increased.

Modern agriculture is characterized by single crop production, often over large areas, a practice that results in agricultural systems highly vulnerable to pathogens and pest insects. Despite the use of more than 2.5 million tons of pesticides per year at a cost of US$ 30 billion, pest insects, plant pathogens, and weeds continue to destroy more than 40% of the world's potential food production (Pimentel 1997). Financial losses caused by insects during the soybean crop season of 1996/1997 in Brazil are estimated at 5% (US$ 281 million)—placed third among the crops that suffer the most financial losses for this reason. To reach this calculation, Bento (2000) multiplied the annual production of each crop by the average unit prices of the products to obtain the value corresponding to the percentage lost due to insect attacks.

Soybeans are attacked by several species of insects, among which the stink bugs (*Piezodorus guildinii*, *Nezara viridula*, and *Euschistus heros*), the whitefly (*Bemisia tabaci*), and the defoliating insects (caterpillars and Chrysomelidae beetles of various species) stand out. Historically, crop pest infestations have been suppressed by chemicals. An alternative strategy, or at least complementary to the use of insecticides, arises from research aimed at the plant breeding, in order to make them more resistant to pest insects. This has been one of the factors to stabilize productivity, with significant advantages over the use of insecticides: it does not cause ecological damage, does not pose additional burden to farmers, does not imply transfer of new technology, and is compatible with other pest control methods. However, pest resistance has not been a priority for breeding programs.

Thus, in this chapter, we propose addressing the main aspects to be considered for soybean breeding aiming at insect resistance without the use of transgenes, to obtain new soybean cultivars.

Breeding Aiming at Insect Resistance

For a breeding program to be effective, breeders must have a good knowledge of the genetics of the species and of the genetic potential of the materials from which breeding will be made. Knowledge about the genetics of resistance can be of great value in these programs, as it provides a quantitative basis for recombination, selection, and stabilization of the plant against the genetic plasticity of the pest.

Two types of genetic inheritance are reported for insect resistance traits: oligogenic (major or main gene) and polygenic (minor gene).

Oligogenic inheritance is controlled by between one and three major nuclear genes, among which individual effects are relatively easy to detect. It is considerably stable to environmental influences, but, today, it is acceptable to be less stable against the development of biotypes than the polygenic heritage. Polygenic resistance is controlled by many nuclear genes, each with a small effect. It is often complex and it can be associated with quantitative traits such as yield. This type of resistance provides a considerable buffer effect against biotypes, although the environmental influence is major and able to modify its resistance expression.

A practical and immediate purpose of breeding, aiming at obtaining more resistant plants, is the determination of the degree of control provided to pest insects. In the initial stage of this process, the work should be focused on the identification of plants with the lowest rate of attack and/or damage. Initially, it is carried out in the field, with natural insect infestation, because it is necessary to evaluate large plant populations. Selection in the field is also desirable because the expression and magnitude of the resistance will be subject to changes by environmental factors.

An important limitation to the use of determination methods in the field is the occurrence of pests in low populations. This problem can be sorted out by external management, which would include the "mass" breeding of the insect or its collection, in crops in the field, and release in the experimental area, or trap crops, followed by infestations, with hosts susceptible to the attraction or use of baits; the use of nonselective insecticides, for elimination of natural enemies and competing species; and the use of cultural practices that favor the pest. These strategies are used for insects to reach suitable levels for genotype evaluation.

A final and important element in the process of selection and development of resistant genotypes is understanding the plant reaction to the damage caused by the pest. In soybeans, this damage is evaluated by the leaf area consumed, when referring to defoliation insects. In the case of sucking insects, damaged pods, leaf retention, grain filling period, the incidence of yeast spot (*Nematospora coryli*), and grain size are evaluated. These parameters must always be associated with the other agronomic traits and to the yield.

The knowledge of specific characteristics of the plants, which condition their resistance, is used to understand the genetic nature of this resistance, its precise mode of action, the improvement of selection methods, and the procedures to be used in breeding. However, several cultivars have already been released even though their genetic resistance is not known in detail. In many cases, resistance is identified in nonadapted lines with low agronomic value. This fact considerably increases the time required to incorporate resistance to cultivars with good agronomic characteristics.

Studies with soybeans aimed at resistance to insect pests are recent. They began, mainly, in the United States, in the late 1960s. As the main pest of this crop, in that country, is the Mexican bean beetle (*Epilachna varivestis*), research on resistance has focused on this pest, on which further advances were achieved. The material evaluated also proved resistant to other soybean pests. Researchers have identified PIs 171,451 (*Kosamame*), 229,358 (*Sodendaizu*), and 227,687 (*Miyako White*) as resistant to several defoliant insects: *Pseudoplusia includens* (*Chrysodeixis includens*), *E. varivestis*, *Epicauta vittata, Cerotoma trifurcata*, *Helicoverpa zea*, *Heliothis virescens* pest insects, *Helicoverpa armigera*, *Heliothis punctigera*, and *Trichoplusia ni* were, therefore, considered the main sources of resistance, based on which the breeding work began (Rossetto et al. 1981).

In this initial screening, several soybean strains and cultivars were evaluated for resistance to the Mexican beetle in North Carolina in 1967 and in South Carolina in 1968 by Van Duyn and coworkers. These three introductions are highly resistant, and present 20% of defoliation, while in commercial cultivars, this value fluctuated between 80 and 90%. Multiple insect resistance is important as the occurrence of pests varies according to production areas as well as from year to year. A program aimed at the development of resistant cultivars started with Dr. Edgar E. Hartwig, "breeder" of the United States Department of Agriculture, who had PI 229358 as a source of resistance in crossings with the Bragg cultivar.

Several breeding programs using PIs 171,451, 227,687, and 229,358 have been conducted, with considerable progress. These genotypes confer good levels of resistance against defoliating insects. Several lines and some soybean cultivars have already been released with the incorporation of this resistance. In the United States, the cultivar Crockett resulted from selection from the crossing of PI 171451 × Hampton 266A, presenting a high level of resistance to velvetbean caterpillar (*A. gemmatalis*) and to soybean looper (C. *includens*) pest insect, in addition to resistance to some diseases, such as frogeye leaf spot (*Cercospora sojina*) and stem cancer (*Diaporthe phaseolorum* var. *caulivora*) (Rossetto et al. 1982). Moreover, two other American cultivars were released, with different sources of resistance to defoliant insects: the cultivars Shore and Lamar (Pinheiro et al. 2005).

Cultivar Shore derived from the selection of the crossing of PI 80837 × Hood, developed from the V156 line, which presents a moderate level of resistance to the Mexican beetle. Cultivar Lamar is a selection of the crossing between Tracy M × F_2 progeny (Centennial × D75 10,169), which showed good resistance to the soybean looper in test cages, greenhouse, and field crops.

For the defoliating insects *A. gemmatalis*, C. *includens*, *H. zea*, *H. virescens*, and *Spodoptera exigua*, different genetic bases were determined at the resistance of PIs 171,451, 229,358, and 227,687. In crossings of PIs with Davis cultivar, it was observed that PIs 227,687 and 229,358 showed to be highly resistant to *A. gemmatalis*, *C. includens*, and *S. exigua*. PI 171451 did not show a significant level of resistance to these pests, having similar damages to that of Davis cultivar, under greenhouse conditions, and significantly lower damages under field conditions. Several researchers reported multiple resistance to pests in these three introductions.

The sucking insects to cause the most damages to soybeans are stink bugs. In Brazil, the most important species are *N. viridula*, *P. guildinii*, and *E. heros*, although other species of the genus *Dichelops*, *Acrosternum*, *Edessa*, and *Thyanta*

(Hoffmann-Campo et al. 2003) may occur. Damage caused by stink bugs can determine the total failure of the production, as they directly damage the grains and provoke the foliar retention phenomenon, which is nothing but the permanence of the plant in the vegetative phase, leading to the uneven maturity of the grains. In addition to these direct damages, by means of bites, the bugs can transmit the fungus *Nematospora coryli*, which affects not only the quality of the seeds but also the commercial value of the grains. Besides stink bugs, the whitefly *Bemisia tabaci* (Genn.) can also affect production. As soybean is an excellent host plant for this white fly, under certain conditions its populations can reach high levels and negatively affect the production. Even without reaching large populations, *B. tabaci* can multiply in soybean crops and migrate to other crops, such as bean or tomato, where even in low populations, it can convey economically impacting viruses to these crops.

It has been very difficult to achieve resistance to pod-sucking insects, although cultivars with different resistance levels, such as IAC100, IAC17, and IAC24, have already been launched. Several experimental cultivars and lines were evaluated, especially in Brazil, for the main species of stink bugs, with less susceptible materials being found. It is common to find data showing that earlier cultivars present less damage than those of medium and late cycles. This fact should be interpreted with caution, as it can result from simple host avoidance, that is, shorter-cycle cultivars undergo faster the phase most susceptible to stink bugs, which is the reproductive phase, and, as a result, suffer less damage, due to the shorter time of exposure to these insects.

The Instituto Agronômico de Campinas (IAC) was a pioneer by introducing, in its soybean breeding program, lines of research aimed at obtaining cultivars resistant to stink bugs and other insects. Studies conducted at this institution show that the search for resistance to soybean sucking bugs was successful with the introduction of PI 227687 and lines derived from PI 274454 used as sources of resistance. Moreover, the introductions of PI 229358, PI 171451, and PI 171444 have also been used as sources of resistance to stink bugs. An evaluation of the behavior of several genotypes revealed that line IAC 73-228 (derived from the crossing between Hill × PI 274454) and PI lines 171,451 and 229,358 were considered resistant to soybean sucking bugs. This material did not suffer production losses, due to the insect attack. The line IAC 73-228 also showed low leaf retention (13.3%).

It was also verified that it is possible to transfer the resistance to stink bugs and the selection against its main unfavorable characteristics. Among the selected lines, IAC 80-596-2, which descends from the UFV-1 × IAC 73-228 crossing, is noteworthy. It should also be pointed out that line IAC 78-2318 was resistant to stink bugs, in addition to presenting fewer damages to the bean shoot borer (*Epinotia aporema*), defoliating insects and lower colonization of the whitefly (*B. tabaci*), which characterizes it as a multiple-resistance line. As a result, it was used as one of the genitors in the selection cycle, which resulted in the cultivar IAC-100. Based on the results of the research aiming at the resistance to stink bugs, it could be concluded that materials with small seeds and a large number of pods presented smaller percentages of damaged seeds than materials with larger seeds when attacked by stink bugs. It is important to note that, according to Rossetto and Lara (1991), the soybean cultivar IAC-100 has a pseudo-resistance mechanism, known as damage dilution, as it produces more seeds per plant, thus, diluting the damage caused by stink bugs.

Currently, in Brazil, there are few insect resistant cultivars recommended for cultivation. Among them, we have IAC-100 and IAC-17 (Rossetto et al. 1995; Godoi et al. 2002; Veiga et al. 1999), as well as IAC-23 and IAC-24 (Miranda et al. 2003a, b), being adapted only to the Southeast region. The cultivar IAC-100 has been used in several types of research as a standard genotype of resistance to defoliating and sucking insects. As a result, its exploration as a germplasm that is a source of multiple resistance to insect pests has been common, because, in addition to this characteristic, this genotype has favorable agronomic characteristics and good grain yield (Pinheiro et al. 2005; Godoi and Pinheiro 2009). This cultivar was used in soybean breeding programs in the United States, and McPherson et al. (2007) and McPherson and Buss (2007) reported that this presents great potential as a source of resistance for sucking bug and defoliating insect.

However, these cultivars are adapted only to the Southeast region (Godoi et al. 2002) and are already obsolete in relation to the current productivity level. Thus, it is essential to develop new cultivars that add resistance to the stink bug complex, high-yield, and good agronomic characteristics and that can be recommended to all producing regions of the country.

In addition to this research institution, in 1990, USP ESALQ started a breeding program aiming at the development of insect-resistant genotypes. The elite lines have been evaluated in experiments of value for cultivation and use (VCU) for later registration, protection, and crop recommendations.

Final Considerations

The search for sources of resistance in germplasm banks is essential for the success of improvement aimed at insect resistance, leading to the development of new cultivars, thus increasing the number of resistant cultivars recommended for cultivation in Brazil and also in other soybean producing regions.

References

Bento, J.M.S.: Comedores de lucro. Des. Cult. **22**, 18–21 (2000)

Godoi, C.R.C., Pinheiro, J.B.: Genetic parameters and selection strategies of soybean genotypes resistant to stink bug complex. Genet. Mol. Biol. **32**(2), 328–336 (2009)

Godoi, C.R.C., Pereira, F.S., Umeno, F., Ázara, N.A., Lima, P.M.S., Silva, R.P., Oliveira, A.B., Araújo, I.M., Zucchi, M.I., Pinheiro, J.B.: Resistência a insetos em populações de soja com diferentes proporções gênicas de genitores resistentes. Pesq. Agropec. Trop. **1**(32), 47–55 (2002)

Hoffmann-Campo, C.B., Moscardi, J., Gazzoni, F., Corrêa-Ferreira, D.L., Lorini, B.S., Borges, I. A., Panizzi, M., Sosa-Gomez, A.R., Corso, I.C.: Integrated pest management in Brazil. In: Maredia, K.M., Dakuo, D., Mota-Sanchez, D. (Org.). IPM in the Global Arena. 1st edn, vol. 1, pp. 285–299. CABI, Trowbridge (2003)

McPherson, R.M., Buss, G.R.: Evaluating lepidopteran defoliation resistance in soybean breeding lines containing the stink bug (Hemiptera: Pentatomidae) resistance IAC-100 cultivar in their pedigrees. J. Econ. Entomol. **100**(3), 962–968 (2007)

McPherson, R.M., Buss, G.R., Roberts, P.M.: Assessing stink bug resistance in soybean breeding lines containing genes from germplasm IAC-100. J. Econ. Entomol. **100**(4), 1456–1463 (2007)

Miranda, M.A.C., Braga, N.R., Lourenção, A.L., Miranda, F.T.S., Unêda-Trevisoli, S.H., Ito, M. F.: Descrição, produtividade e estabilidade do cultivar de soja IAC-23 resistente a insetos. Bragantia. **62**(1), 19–27 (2003a)

Miranda, M.A.C., Braga, N.R., Lourenção, A.L., Miranda, F.T.S., Unêda-Trevisoli, S.H., Ito, M. F.: Descrição, produtividade e estabilidade do cultivar de soja IAC-24 resistente a insetos. Bragantia. **62**(1), 29–37 (2003b)

Pimentel, D.: Techniques for Reducing Pesticides Use—Economic and Environmental Benefits, 444 p. Wiley, New York (1997)

Pinheiro, J.B., Vello, N.A., Rossetto, C.J., Zucchi, M.I.: Potential of soybean genotypes as insect resistance sources. Crop Breed. Appl. Biotechnol. **5**, 294–301 (2005)

Rossetto, C.J., Lara, F.M.: Diluição de dano, mecanismo de pseudoresistência. In: Congresso Brasileiro de Entomologia, XIII, 1991, Recife. Resumos, p. 525. Sociedade Entomológica do Brasil, Recife (1991)

Rossetto, C.J., Miranda, M.A.C., Link, D., Rezende, J.A.M., Lourenção, A.L., Mascarenhas, H.A. A., Miyasaka, S.: Cultivares resistentes aos insetos-praga. In: Miyasaka, S, Medina, J.C., (Org.). A Soja No Brasil, pp. 327–344. Instituto de Tecnologia de Alimentos, Campinas (1981)

Rossetto, C.J., Miranda, M.A.C., Link, D., Rezende, J.A.M., Lourenção, A.L., Mascarenhas, H.A. A., Miyasaka, S.: Resistência da soja a insetos. In: Fundação Cargill. (Org.). A Soja No Brasil Central, 2ª edn, pp. 271–298. Fundação Cargill, Campinas (1982)

Rossetto, C.J., Gallo, P.B., Razera, L.F., Bortoletto, N., Igue, T., Medina, P.F., Tisseli Filho, O., Aquilera, V., Veiga, R.F.A., Pinheiro, J.B.: Mechanisms of resistance to stink bug complex in the soybean cultivar IAC-100. An. Soc. Entomol. Brasil. **24**(3), 517–522 (1995)

Veiga, R.F.A., Rossetto, C.J., Razera, L.F., Gallo, P.B., Bortoleto, N., Medina, P.F., Tisselli Filho, O., Cione, J.: Caracterização morfológica e agronômica do cultivar de soja IAC-100, 23 p. Instituto Agronômico de Campinas, São Paulo (1999) (Boletim Técnico, 177)

Chapter 19
Breeding for Tolerance to Abiotic Stress

Stênio Andrey Guedes Dantas, Felipe Lopes da Silva, Leonardo Volpato, Rosângela Maria Barbosa, Guilherme de Sousa Paula, Heloisa Rocha do Nascimento, and Marcos Deon Vilela de Resende

Abstract Soybean is the most widely grown protein and oil crop in the world, and it is, therefore, one of the crops most prone to occurrence of biotic and abiotic stresses. In recent years, with climate change, abiotic stresses have become increasingly common. The objective of this chapter is to discuss important aspects related to the soybean breeding for abiotic stresses, such as water deficit, salt, aluminum, and phosphorus efficiency use. Additionally, it addresses strategies for driving breeding programs focused on the development of tolerant cultivars, employing classic breeding with new tools, such as genomic selection. In this approach, it is essential to know the morphological and physiological changes caused by the stresses and also the mechanisms involved in plant tolerance to them. The genetic basis of these mechanisms is also discussed. In addition, the phenological stage most adequate for plant phenotyping is presented, in order to improve discrimination among genotypes, thus maximizing success in the development of new cultivars tolerant to abiotic stresses.

Keywords Climate change • Global warming • Drought • Heat • Salt • Phosphorus and aluminum tolerance

---

S.A.G. Dantas (✉) • L. Volpato • R.M. Barbosa
G. de Sousa Paula • H.R. do Nascimento
Department of Agronomy, Federal University of Viçosa, Viçosa, Brazil
e-mail: stenioandrey@gmail.com; leo.agroufv@gmail.com; rmbagronomia@yahoo.com.br; Guilherme_825@hotmail.com; helornasc@gmail.com

F.L. da Silva
Federal University of Viçosa, Viçosa, Minas Gerais 36570-900, Brazil
e-mail: felipe.silva@ufv.br

M.D.V. de Resende
Brazilian Agricultural Research Corporation/Federal University of Viçosa, Viçosa 36570-900, Minas Gerais, Brazil
e-mail: marcos.deon@gmail.com

F. Lopes da Silva et al. (eds.), *Soybean Breeding*,
DOI 10.1007/978-3-319-57433-2_19

Introduction

The constant climatic changes over the last years have been responsible for the unfavorable environmental conditions for soybean farming, implying several management difficulties. Accordingly, they emphasize the inadequate availability of essential nutrients, excess salts in the soil, high levels of radiation, water deficiency, inappropriate temperatures, compaction, and inadequate levels of organic matter in soils (Sediyama 2015), all of which considered abiotic stress.

Abiotic stress is a term that refers to unfavorable environmental conditions that affect the growth and development of plants in different ways, either by excess or by the lack of a certain factor. In order to minimize the harmful effects of abiotic stresses, plants have developed several adaptive strategies at different metabolic levels that interact during stress, such as enzymes involved in the biosynthesis of osmoprotectants (proline, sugars, sugar alcohol, glycine betaine, and polyamines), antioxidant enzymes (ROS), protective proteins (AEL and HSP), transporters, regulatory proteins, kinases, and transcription factors (Marco et al. 2015).

Knowledge of the genetic basis of these stress tolerance adaptive structures, associated with morphological, physiological, and biochemical changes, is of fundamental importance in genetic improvement, since they help the identification of target traits for the breeding process. Thus, with the use of molecular genetics tools, it is possible to further optimize the selection process and the choice of tolerant genotypes. The adequate use of this tool favors the identification of probable genomic regions where the specific genes (Quantitative Trait Loci—QTLs) that control the tolerance to stress are located.

With molecular tools being increasingly used in breeding programs, such as assisted selection, GWS, GWAS, etc., attempts have been made to develop platforms that allow phenotyping of genotypes, aiming at both the selection of the most tolerant genotype to abiotic stresses and the identification of genes to control the tolerance traits. In most cases, controlled structures have been in greenhouses and hydroponic systems to simulate the different types of stresses.

The purpose of this chapter is to briefly report important aspects related to soybean breeding, in the face of some types of abiotic stress, for example: water stress, aluminum (Al), saline, and phosphorus use efficiency, as well as discuss strategies for conducting development of tolerant cultivars.

Breeding for Water Stress Tolerance

Water stress, for soybean crop, is one of the main abiotic stresses that limit yields. Drought periods lead to dehydration or osmotic stress by reducing the availability of water to the plant's vital cellular functions, which causes loss of turgidity and modification of cell volume, protein denaturation, and impairment of cell membrane integrity (Marco et al. 2015) leading to physiological changes in the gas exchange process, directly related to photosynthesis and respiration, and morphological changes such as reduction of leaf expansion.

One of the main tools that have been used in approaches of coexistence with drought is breeding. Much progress has been made with regard to the development of more adapted and tolerant cultivars, both in the field of classical breeding and in the field of biotechnology. Classical improvement has shown significant advances; however, there are still some uncertainties as to whether the improved characteristics for a given environment may also be advantageous for others, since drought is a dynamic factor that varies according to intensity, phenological phase, soil type, and fertility, which, as a result, is a major problem for genetic improvement.

An alternative to classical breeding is genetic engineering, which is a very efficient method for traits controlled by one or a few genes. However, it is known that the drought tolerance phenomenon is a trait controlled by several genes and with strong genotypes x environments relation. There have been advances in this field, which have allowed for the simple growth of plants under stress condition (Sinclair 2011).

There is a range of studies related to breeding for water stress; however, there is still a gap to be filled between yield in favorable and unfavorable environments. Minimizing such disparity between favorable and unfavorable growing conditions, increasing production stability over different conditions, is an alternative to achieving improved materials.

In order to start the breeding program aimed at the development of tolerant water stress cultivars, it is important to know the species under study. Thus, this topic of the chapter aims at taking an approach on the physiological, morphological, biochemical, and nutritional principles that are under genetic control and that can be used as tools to achieve genetic progress in a breeding program. Moreover, we will be addressed some phenotyping methodologies for this purpose.

Every plant, when subjected to some kind of stress, no matter which, initiates a metabolic and biochemical process that triggers a number of morphophysiological adaptations to ensure its survival. In soybeans, and with water stress, this is no different. However, the level at which these adaptations are made in soybean plants varies according to the amount of stress applied, the phenological phase, the environmental condition (air temperature, radiation, wind, relative humidity, etc.) in which the plant is found, and the genotype per se.

One of the main factors affected by drought in the plants is the photosynthetic activity. This happens because, in order to save water in its tissues, it closes the stomata and, consequently, reduces the entry of CO_2 into the site of action of the rubisco enzyme (ribulose 1,5 bisphosphate), responsible for the transformation of CO_2 into energy for the plant and decreases the production of photoassimilates, causing productivity loss.

When the photosynthetic chain undergoes any sort of disturbance, as water stress, there is imbalance in the production/consumption of electrons in the cell, generating the production of reactive oxygen species (ROS). These ROS species, when in great quantity, attack the cellular structures, mainly membranes, causing damage to the cell and, consequently, to the plant. However, there are enzymes that are able to combat these ROS species, which minimizes stress to the plant and allows all of its essential structures to remain fully functional, even under stress.

Since the plant understanding the water stress, besides closing the stomata, it needs to conserve the water that is in its tissues and cells to the maximum and cannot lose it to the environment, mainly by transpiration. As a result, the plant begins to produce solutes, mainly sugars (proline, betaine, etc.), which are called osmoprotectors or osmotic adjustment (OA), which increase the cellular turgor pressure and causes water to remain inside. It was observed that soybean plants with high OA, even under stress conditions, managed to maintain high photosynthetic rate (Turner et al. 2001).

As for constitutive and morphological adaptations of the plants under drought, the features related to the root morphology, such as size, depth, angle of insertion in the soil, diameter, and number of secondary roots, are the main traits. These adaptations have already been widely studied in other crops such as wheat, maize, and rice, but they still lack study when it comes to soybeans. Evaluations of morphological adaptations, such as root structure, are more accurate than physiological adaptations, since they have a higher rate of reproducibility and stability in subsequent generations, even under optimum environmental conditions.

The joint use of all these traits aims to increase productivity itself; however, great effort and work demand are required in order to use them all at once for the genotype selection, which leads the researchers to choose to use fewer genotypes and more evaluations or do fewer evaluations on more genotypes. Sinclair (2011) suggested the determination of nitrogen (N) content in plants as an indicator of tolerance and productivity increase under water stress condition, based on the fact that the symbiotic activity is highly susceptible to water stress. Therefore, plants that can accumulate more N in stress conditions are more productive.

Researchers believe that the future of genetic improvement requires a new emphasis on phenotyping for specific and well-defined physiological characteristics (Ghanem et al. 2015). However, there is still disagreement as to which method will provide a more accurate measure when these lines are commercially tested, that is, there are several methodologies, each with its intrinsic characteristic, and which must be chosen according to the condition, availability of resources, and purposes of the program.

The stress can be applied in different forms and times of the plant development, from germination up to reproductive stages. Kosturkova et al. (2008) reported the response of genotypes capable of tolerating water stress during the germination phase with that obtained during the adult phase and observed a high correlation between the two methods. Table 19.1 shows a summary of different periods of application of stress and their implications on soybean yield components. Pardo et al. (2015) observed that the most impacting period is the grain-filling stage (R5).

In this session, we presented a summary of some of the principles for working with breeding for water stress. However, these are only a small part of a whole complex that involves the theme and deserves the inclusion of other sciences such as, mainly, the "omic" sciences, so that it is possible to obtain cultivars that are more tolerant to water stress and in less time.

Table 19.1 Effects of water deficit applied at different stages on soybean components and yield

Phenological stage	Number of seeds per plant	Yield per plant	Weight of 100 seeds
R5	45.20 A	8.56 A	15.92 A
V7 to R1	74.80 B	12.70 B	16.95 A
R3 to R5.5	80.80 B	13.65 B	17.13 A
Control without stress	90.00 B	14.29 B	19.02 A

Source: Adapted from Pardo et al. (2015)
Different letters indicate statistical difference by the Tukey test, at 5% probability

Breeding for Aluminum Stress Tolerance

Acidic soils correspond to approximately 30% of the entire world area and 50% of arable soils (von Uexküll and Mutert 1995). The toxicity of aluminum is one of the main limiting factors of the development of plants in acid soils, which affects the growth of roots, inasmuch as it interferes in the cell division and the elongation of the cells of the meristematic tissue, reduces the number of root hairs and lateral roots, causes water and nutrients absorption deficiency, and, consequently, leads to drop in grain production.

Soybean is cultivated on a large scale mainly in the Brazilian Cerrado region, mostly on latosols, which are very weathered soils, which leads them to have low levels of essential nutrients and high concentrations of H^+, which reduces the soil pH and raises Al^{3+} contents, causing toxicity to plants. Aiming at implanting the crop in the region, Hartwig et al. (2007) adopted farming practices, such as application of limestone, gypsum, and recommended fertilization based on crop requirements and the use of cultivars adapted to the conditions of the region.

The adapted cultivars are developed by plant breeding, which aiming to develop genotypes that remain stable and productive even if the soil contains aluminum. To this end, it is required a better understanding of the mechanisms developed by the plant to tolerate toxic aluminum in the soil, as well as identify and know the genetic control of this tolerance (Guimarães et al. 2011).

Al tolerance mechanisms can be grouped into two categories: (1) external tolerance mechanism, which is the one in which aluminum is prevented from crossing the plasma membrane and entering the symplast, which may include immobilization in the cell wall, selective permeability of plasma membrane, induced pH barrier in the rhizosphere by the plant itself, exudation of binder chelates, and exudation of phosphate and aluminum efflux, and (2) internal tolerance mechanism, in which aluminum enters the symplast and tolerance is found by the formation of chelates in the cytosol, binding of aluminum to proteins, and the performance of aluminum-tolerant enzymes (Rampim and Lana 2013).

Aluminum tolerance in soybean is a quantitative trait controlled by several genes and is strongly influenced by the environment. Such fact was evidenced by the discovery of several QTLs only in aluminum-tolerant soybean cultivars (Bianchi-Hall et al. 2000).

Zeng et al. (2012) identified 30 microRNAs associated with genes that manifested in the presence of toxic aluminum in the soil, demonstrating that, under stress, soybean plants trigger protection mechanisms involving microRNAs.

Ermolayev et al. (2003) identified two cDNA sequences (TCTP, clone 58 and IMPDH, clone 633) in roots of soybean cultivars that showed tolerance to aluminum. To confirm the results, the authors transferred the 633 cDNA sequence identified in tolerant soybean materials to *Arabidopsis thaliana* plants. It was observed that 86% of the genetically modified (GM) plants were tolerant to the stress condition, while only 6% of them without the GM trait survived, which confirms that the identified gene actually confers tolerance to aluminum stress.

Breeding for Tolerance to Saline Stress

In the world, today, a total of 831 million hectares have excess salts in the soil (The Food and Agriculture Organization of the United Nations, FAO, http://www.fao.org), which negatively affects productivity, mainly in arid and semiarid regions (Porcel et al. 2012).

Saline stress develops when there is excessive concentration of mineral salts in the form of ions in the soil, especially sodium (Na^+) and chlorine (Cl^-). In a situation of high salinity, the potential of water in the soil decreases, initially causing in the crop a situation of osmotic stress, which triggers consequences similar to those of drought. Then, the salt that penetrates the root is transported via xylem to the aerial part of the plant, triggering the additional toxicity effects of the specific ion, contributing to the inhibition of growth and rapid leaf senescence.

In order to minimize the effects of ionic toxicity, the absorption and compartmentalization mechanisms are activated, aiming to restore cell homeostasis (Munns and Tester 2008). In order to reestablish this cellular equilibrium, homeostasis, the plant produces osmolytes (proline, betaine, etc.), which are small biological molecules precisely selected to counteract environmental stresses such as drought and salinity. Another alternative to tolerate saline stress is to increase the efficiency of N utilization just like in water stress (Sinclair 2011).

Dong et al. (2013) used sequenced microRNAs (miRNAs) and determined that the genes involved in the nodule development process and in biological nitrogen fixation had different levels of expressivity when the plants were subjected on saline stress. These authors demonstrated that there are several genes involved in the biological process of nitrogen fixation and that, with the characterization of the miRNAs, it would be possible to identify mechanisms involved in nodule development, biological fixation and, finally, salinity tolerance.

Duanmu et al. (2015), using profile transcriptomics in wild soybean roots (*Glycine soja*) under salt stress by RNA sequencing, identified several candidate genes that act on the mechanisms of tolerance to salt stress. This allows the development of molecular markers in the future for salinity tolerance, thus providing assisted selection for this characteristic.

Both of the previously mentioned authors have used young soybean plants, which were conducted in a hydroponic system. It is a good phenotyping alternative, as it is relatively inexpensive and allows the analysis of several genotypes at a single time, besides providing different levels of stress (Liu et al. 2011).

Breeding for Phosphorus Efficiency

In tropical environments, an important factor in reducing agricultural risk, which may not be so obvious to farmers, is the genotype's capacity to adapt favorably to adverse soil conditions, to more efficiently absorb and utilize nutrients, such as phosphorus, nitrogen, potassium, etc. Hence, efficiency is defined as the ability of a given genotype to acquire the nutrient to incorporate it and use it in the production of biomass or plant material of economic yield (Blair 1993), such as grains, in the specific case of cereals.

One way to obtain efficient genotypes in the use of nutrients is genetic improvement, and the first step is to identify superior genotypes for traits of interest, be them RILs or commercial cultivars. In the case of faster genetic progress, base populations should be formed by genotypes adapted to the planting area. The knowledge about the genetic control and the type of gene action of the characteristic of interest are necessary for the selection of adequate breeding methods, to them obtain efficient genotypes for the use of nutrients.

In this topic, emphasis will be given to the efficient use of phosphorus. Phosphorus is the nutrient that most limits production in tropical regions. The use of efficiency traits of phosphorus is under polygenic genetic control, and, as a result, they are highly influenced by the environment. Thus, studies using multivariate techniques help identify more efficient genotypes to the use of phosphorus.

Several morphological, physiological, and biochemical mechanisms have already been pointed out to explain genotypic differences in the efficiency of phosphorus use. Among them are mechanisms of phosphorus absorption by plants, for example, modification of the morphological characteristics of the root system such as slope angle, depth, number of secondary roots, and absorbent hairs; increased efficiency of physiological absorption mechanisms (exudates of organic compounds and symbiotic association with mycorrhizal fungi); chemical modifications at the root-soil interface; biochemical changes under deficiency (alterations in the membrane carriers, linked to the absorption of P); and association with mycorrhizae.

Most of the characteristics associated with phosphorus use efficiency are directly related to the study of roots. Wang et al. (2010), in a review on breeding for phosphorus efficiency in soybean, argue that among all these morphological, biochemical, and physiological characteristics, root biology provides a new frontier for studies and understanding of absorption, transport, accumulation, and use of phosphorus, as well as its effects on plant growth and the development of adapted genotypes. Consequently, root phenotyping is a viable alternative to the selection of more efficient materials to the use and absorption of phosphorus.

In the identification of genetic diversity between genotypes and populations, using agronomic, morphological, and molecular traits, there are guidelines for hybridization programs resulting in populations with favorable allele accumulation for the efficiency trait in the use of phosphorus. Thus, strategies should be adopted when selecting the parents. One of them is the use of the genome-wide selection (GWS), which will be addressed in detail in the next topic.

Strategies for the Development of Abiotic Stress-Tolerant Cultivars

Breeding programs aimed at the development of cultivars tolerant to different abiotic stresses should begin with the choice of the parents to be used in the hybridizations.

Methods for selecting parents were discussed in the Chap. 8 of this book. However, phenotyping for the selection of parents tolerant to different abiotic stresses should be performed in environments that cause the stress in plants. The methodology used in the evaluation of abiotic stresses should be simple, economical, and easily adopted, allowing a large number of materials to be evaluated (Fageria et al. 2011).

In general, when working with abiotic stresses, the researcher is faced with some questions: Do the phenotyping work in the field or in a greenhouse? What are the indications that the plant is under stress? How to maintain this stress for a certain period of time? At what stage of development of the crop should stress be imposed? The responses are not simple, and, for a better illustration, the phenotyping of genotypes for the identification of water stress-tolerant genotypes in classic breeding programs is an example.

One of the proposed methods for selection and grouping of potential parents is the germination test using paper solution soaked with polyethylene glycol (PEG) (Teixeira et al. 2008a, b), which has negative osmotic potential in the solution without, however, exerting toxicity on the seeds (Villela et al. 1991). Several authors have already observed the efficacy of the low osmotic potential test for differentiation of soybean genotypes. Matthews and Powell (1986), when studying different crops, observed that the response of seeds, when placed to germinate under water deficiency, varies according to the species or cultivar. However, the same authors have also reported that the response to applied water deficit is dependent on the size and, mainly, the physiological quality of the seeds. The same was pointed out by Moraes and Menezes (2003) when reporting that regardless of the environmental condition to which the seed is exposed, germination will depend on its physiological potential.

In order to circumvent this problem of seeds with different physiological qualities, Ku et al. (2013) suggested a germination percentage scale based on results obtained from the relationship among seeds under water stress (treated with the PEG osmotic stress inducer) and seeds under normal conditions without stress. The seeds used are previously standardized with the same physiological potential, that is, from the same harvest and stored in the same conditions and for the same period of time. Such methodology cancels out the qualitative differences of the seeds, given that each treatment will be compared only with its control, to form the scale and then the scale will be used to evaluate the genotypes. Thus, the genotypes that present the lowest germination percentage reduction between optimal and stress condition will be the most tolerant.

In spite of the advantages, mainly related to the reduced time for evaluation of a genotype, this method is little used in soybean breeding programs, perhaps because a direct relation of the responses to the water deficit in the seedling phase with resistance traits in the adult phase, mainly physiological responses, still has not been determined. Moreover, there is disagreement as to the consistency of the response to this deficit in the seedling stage of cultivars with the response to water deficiency based on grain yield data (Mederski and Jeffers 1973; Oya et al. 2004; Kosturkova et al. 2008).

Phenotyping of soybean genotypes in a greenhouse has been preferred in relation to field phenotyping; in field, soil moisture control is not always efficient. Moreover, often, breeding programs do not have areas where there is an accentuated summer period and which coincides with the stages of development of water stress. In order to avoid these problems, rainout shelters can be used in the field; however, these structures increase the phenotyping costs for large numbers of genotypes.

The gravimetric control of soil water content is the most widespread and reliable method to apply water stress in pots under controlled conditions, for it allows the management of time, intensity, and duration of stress. This method has been used successfully in model species and cultivated species, such as *Arabidopsis thaliana* (Bacso et al. 2008), sunflower (Pereyra-Irujo et al. 2007), corn (Chapuis et al. 2012), and soybean (Brevedan and Egli 2003; Pardo et al. 2015).

It has been observed that soil in field capacity presents, on average, a concentration of 0.27 g water g soil^{-1} and a matric potential between zero and 0.05 MPa (Pardo et al. 2015; Pereyra-Irujo et al. 2007).

Neves-Borges et al. (2012) considered potentials of −1.5 and −3.0 MPa as moderate and severe, respectively, for gene expression studies, while Gilbert et al. (2011) when performing end-of-day measurements for the study of physiological parameters considered potentials between −0.62 and −1.05 MPa as light and moderate and greater than −1.2 MPa as severe. The level of sensitivity to water potential varies according to species and genotypes, but generally potentials less than −1.18 MPa are responsible for reducing 50% of stomatal closure in most species (Blackman et al. 2009).

According to Brevedan and Egli (2003), soybean grain yield may be more susceptible to short periods of stress, for example, 40% of soil moisture for 6 days at the grain-filling stage (R5) may compromise up to 23% of yield, and, often, this effect is relatively invisible to farmer until the time of harvest. The lack of moisture in the soil during the initial period from flowering to the beginning of seed formation (R1–R5) reduces the number of seeds, while the lack of humidity after that period (R5–R7) reduces seed size (Snyder et al. 1982; Andriani et al. 1991; De Souza et al. 1997; Brevedan and Egli 2003). The R5 stage is the one when the soybean plant is more susceptible to stress (Brevedan and Egli 2003; Pardo et al. 2015).

In studies performed by the Soybean Program of the Plant Science Department, Federal University of Viçosa, two soybean cultivars Anta 82 and MG/BR 46 Conquista were subjected to water stress at the R5 development stage, under greenhouse conditions (data not yet published). The pots were weighed every day in the early morning and late afternoon, adding the amount of water necessary to reach the level of soil moisture required by the treatments, control (no stress, soil in field capacity), and subject to stress (40% of the amount in field capacity). The duration of stress ranged from 0, 6, 9, and 12 days. In the period of application of the treatments, all plants were weighed during the 12th day, in order to guarantee the same water supply conditions for all the plants in the same period. After 12 days of stress, the plant watering was resumed.

The authors did not find a negative effect of the water stress imposed in relation to the components of soybean production. However, by means of the water fraction corresponding to the treatment under stress, the value of 0.12 g water g soil^{-1} was obtained, which could cause severe stress to the plants. As for the classification of water percentage and water potential that cause water deficit for soybean, Pereyra-Irujo et al. (2007) classified stress in levels (0.18 g water g soil^{-1} and −0.44 MPa) and severe (0.16 g water g soil^{-1} and −0.94 MPa).

In the light of the above, it can be said that it is difficult to conduct breeding programs aiming at cultivars tolerant to different abiotic stresses, since tolerance is a multigenic trait, that is, quantitative trait. What is more, plant responses to different stresses imposed are influenced by time, intensity, duration, and frequency of stress. Thus, the use of molecular genetics such as genome-wide selection (GWS) could aid in the identification of more tolerant genotypes.

The GWS technique was addressed in Chap. 14 of this book, and a different application will be discussed in this chapter, as shown in Fig. 19.1. It is the use of a population formed by cultivars and soybean elite RILs adapted to the specific regions of soybean farming, which, in this chapter, will be called the parental population. As the breeding program is focused on the Central-West Region of Brazil, the genotypes of this population should be adapted to that region. Such population should be formed by a high number of genotypes (greater than 500), and it will be used to estimate and validate the effects of the markers.

For the estimating effects of markers, an efficient phenotyping for the target trait is needed, which will infer about the tolerance of the genotypes. Thus, it is necessary to establish phenotyping strategies that allowed obtaining high accuracy in the process of estimation of the effects of the markers.

Once time validated and estimated the marker effects, tolerant and divergent genotypes will be crossed in hybridization programs. The breeding program can be

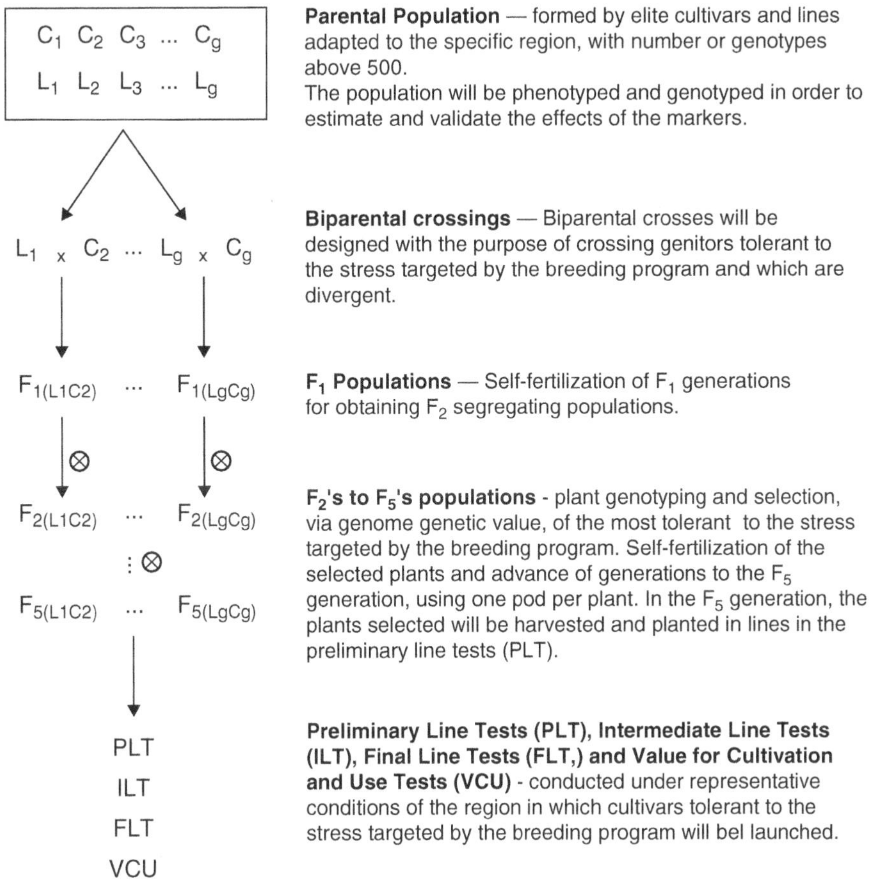

Fig. 19.1 Breeding program using genome-wide selection (GWS) for the development of soybean cultivars tolerant to abiotic stress

conducted until F5 generation in a greenhouse, using a single-pod descent (SPD) method.

In the generations F2 to F5, the plants will be genotyped, and the genome genetic value (GGV) of each plant will be determined using the effects of the estimated markers in the parental population. Therefore, there is no need to impose stress during the advancement of generations.

In the F5 generation, the more tolerant plants, using GGV estimation, will be selected and cultivated in line in the F6 generation and conducted in representative conditions of the region in which the stress-tolerant cultivars targeted by the breeding program are intended to be release. After this generation, the breeding program should be conducted as presented in Chap. 9 of this book.

Phenotyping for the estimation of the effects of marks in the parental population should be carried out based on the means of cultivars or lines in several repetitions.

To this end, data from the evaluation of lines or VCUs from the previous selection cycle can be used, or a new experiment can be carried out with the parental lines, evaluating 500–1000 lines in at least three replicates.

Another option is the genotyping of F2 plants and the phenotypic evaluation of F2:3 progenies in at least three replicates. Progenies F2:4 evaluated in at least three replicates and several sites may also be used and subjected to analysis of association with genotypes of F2 plants. This method has the advantage of mitigating the effects of the genotype x environment interaction in the estimation of genetic effects of molecular markers. Combined information of F2:3 and F2:4 in an index is even more efficient (Resende et al. 2015).

With the advent of genomics, the specific combining ability in plant breeding can be used more efficiently. The allocation of crosses is usually done by including both general (α) and specific (d) combining ability, as follows, where g is the genetic value of the cross:

$\widehat{g}_j = \sum_i^n \mathrm{pr}_j^{22}\alpha_j + \sum_i^n \mathrm{pr}_j^{12}d_j - \sum_i^n \mathrm{pr}_j^{11}\alpha_j$, e pr_j^{22}, pr_j^{12}, e pr_j^{11} are the probability of occurrence of genotypes 22, 21, and 11 in progeny j, α_j are the additive effects of allelic substitution, and d is the effects of dominance (Table 19.2).

The purpose of the selection of crosses is to select populations with high mean and great genetic variability. An option for that is the selection by the highest value of $u + a + d$, where u is the general average of the trait under selection, a relates to the additive effects of alleles (function of α), and d is related to the dominance effects contributed by the loci in heterozygosis. Such selection guarantees high mean and, through high estimation value of d, also contemplates greater genetic variability in F_∞, as a result of the significant number of loci in heterozygosity (Ramalho et al. 2012). Since it is not possible to perform all possible crosses between the selected parents, this method of genomic prediction of unobserved hybrids can be used to obtain $u + a + d$.

It is worth restating that the initial questions in this topic should be made for whatever abiotic stress focused by the breeding program: should phenotyping work

Table 19.2 Genomic prediction of genotypic values (g) of crosses

	Genotype		Probability of genotype in progenies			
Marker	Male parent	Female parent	22	12	11	Genotypic effect predicted by marker
1	12	11	0.0	0.5	0.5	$\widehat{g}_1 = 0\widehat{\alpha}_1 + 0.5\widehat{d}_1 - 0.5\widehat{\alpha}_1$
2	22	11	0.0	1.0	0.0	$\widehat{g}_2 = 0\widehat{\alpha}_2 + 1.0\widehat{d}_2 - 0\widehat{\alpha}_2$
3	21	12	0.25	0.50	0.25	$\widehat{g}_3 = 0.25\widehat{\alpha}_3 + 0.5\widehat{d}_3 - 0.25\widehat{\alpha}_3$
⋮	⋮	⋮	⋮	⋮	⋮	⋮
N	11	11	0.0	0.0	1.0	$\widehat{g}_n = 0\widehat{\alpha}_n + 0\widehat{d}_n - 1.0\widehat{\alpha}_n$
Total predicted genotypic effect						$\widehat{g}_j = \sum_i^n \mathrm{pr}_j^{22}\alpha_j + \sum_i^n \mathrm{pr}_j^{12}d_j - \sum_i^n \mathrm{pr}_j^{11}\alpha_j$

be carried out in the field or in a greenhouse? What are the indications that the plant is under stress? How to maintain this stress for a certain period of time? At what stage of crop development should the stress be imposed?

Final Considerations

Breeding programs aimed at the development of cultivars tolerant to different abiotic stresses should initially determine strategies for phenotyping of the genotypes under selection. Examples of research results have been described for the main abiotic stresses occurring in soybean: water stress, low phosphorus availability, aluminum stress, and saline stress.

The stress conditions should be imposed with the purpose of discriminating genotypes. The methodology used should be simple, economical, and easily adopted, allowing a large number of genotypes to be evaluated.

However, there is great difficulty in determining these methodologies, as there is no consensus in the literature on where to impose it, for how long, and how to maintain this stress in the plant.

Accordingly, in this chapter, we have addressed to use of genome-wide selection (GWS) as an aid to the selection of abiotic stress-tolerant genotypes. It should be pointed out that in this program, it is required for the adequate phenotyping in the population, aiming to increase the accuracy of the estimation of the effects of markers.

References

Andriani, J.M., Andrade, F.H., Suero, E.E., Dardanelli, J.L.: Water deficits during reproductive growth of soybeans. I. Their effects on dry-matter accumulation, seed yield and its components. Agronomie. **11**(9), 737–746 (1991)

Bacso, R., Janda, T., Galiba, G., Papp, I.: Restricted transpiration may not result in improved drought tolerance a competitive environment for water. Plant Sci. **174**(2), 200–204 (2008)

Bianchi-hall, C.M., Carter, T.E., Bailey, M.A., Mian, M.A., Rufty, T.W., Ashley, D.A., Boerma, H.R., Arellano, C., Hussey, R.S., Parrott, W.A.: Aluminum tolerance associated with quantitative trait loci derived from soybean PI 416937 in hydroponics. Crop. Sci. **40**(2), 538–545 (2000)

Blackman, C.J., Brodribb, T.J., Jordan, G.J.: Leaf hydraulics and drought stress: response, recovery and survivorship in four woody temperate plant species. Plant Cell Environ. **32**(11), 1584–1595 (2009)

Blair, G.: Nutrient efficiency what do we really mean? In: Randall, P.J., Delhaize, E., Richards, R. A., Munns, R. (eds.) Genetic Aspects of Plant Mineral Nutrition, pp. 205–213. Kluwer Academic, Dordrecht (1993)

Brevedan, R.E., Egli, D.B.: Short periods of water stress during seed filling, leaf senescence, and yield of soybean. Crop. Sci. **43**(6), 2083–2088 (2003)

Chapuis, R.C., Delluc, R., Debeuf, F., Tardieu, F., Welcker, C.: Resiliences to water deficit in a phenotyping platform and in the field: how related are they in maize? Eur. J. Agron. **42**, 59–67 (2012)

De Souza, P.I., Egli, D.B., Bruening, W.P.: Water stress during seed filling and leaf senescence in soybean. Agron. J. **89**(5), 807–812 (1997)

Dong, Z., Shi, L., Wang, Y., Chen, L., Cai, Z., Wang, Y., Jin, J., Li, X.: Identification and dynamic regulation of microRNAs involved in salt stress responses in functional soybean nodules by high-throughput sequencing. Int. J. Mol. Sci. **14**(2), 2717–2738 (2013)

Duanmu, H., Wang, Y., Bai, X., Cheng, S., Deyholos, M.K., Wong, G.K., Li, D., Zhu, D., Li, R., Yu, Y., Cao, L., Chen, C., Zhu, Y.: Wild soybean roots depend on specific transcription factors and oxidation reduction related genes in response to alkaline stress. Funct. Integr. Genomics. **15**(6), 1–10 (2015)

Ermolayev, V., Weschke, W., Manteuffel, R.: Comparison of Al-induced gene expression in sensitive and tolerant soybean cultivars. J. Exp. Bot. **54**(393), 2745–2756 (2003)

Fageria, N.K., Stone, L.F., Santos, A.B.: Melhoramento para tolerância à salinidade. In: Fritsche-Neto, R., Borém, A. (eds.) Melhoramento de plantas para condições de estresses abióticos, pp. 151–176. Suprema, Visconde do Rio Branco (2011)

Ghanem, M.E., Marrou, H., Sinclair, T.R.: Physiological phenotyping of plants for crop improvement. Trends Plant Sci. **20**(3), 139–144 (2015)

Gilbert, M.E., Zwieniecki, M.A., Holbrook, N.M.: Independent variation in photosynthetic capacity and stomatal conductance leads to differences in intrinsic water use efficiency in 11 soybean genotypes before and during mild drought. J. Exp. Bot. **62**(8), 2875–2887 (2011)

Guimarães, L.J.M., Guimarães, C.T., Magalhães, J.V., Parentoni, S.N.: Melhoramento para tolerância ao alumínio. In: Fritsche-neto, R., Borém, A. (eds.) Melhoramento de plantas para condições de estresses abióticos, pp. 177–197. Suprema, Visconde do Rio Branco (2011)

Hartwig, I., Oliveira, A.C., Carvalho, F.I.F., Bertan, I., Silva, J.A.G., Schmidt, D.A.M., Valério, I. P., Maia, L.C., Fonseca, D.A.R., Reis, C.E.S.: Mecanismos associados à tolerância ao alumínio em plantas. Ciências Agrárias. **28**(2), 219–228 (2007)

Kosturkova, G., Todorova, R., Sakthivelu, G., Akitha Devi, M.K., Giridhar, P., Rajasekaran, T., Ravishankar, G.A.: Response of Bulgarian and Indian soybean genotypes to drought and water deficiency in field and laboratory conditions. Gen. Appl. Plant Physiol. **34**(3–4), 239–250 (2008)

Ku, Y.S., Wen, C.Q., Lam, H.M., Li, M.W., Wen, C.Q., Liu, X., Yung, Y.L.: Drought Stress and Tolerance in Soybean. INTECH Open Access, Rijeka (2013)

Liu, G., Guan, R., Chang, R., Qiu, L.: Correlation between Na^+ contents in different organs of soybean and salt tolerance at the seedling stage. Acta Agron. Sin. **37**, 1266–1273 (2011)

Marco, F., Bitrián, M., Carrasco, P., Rajam, M.V., Alcázar, R., Tiburcio, A.F.: Genetic engineering strategies for abiotic stress tolerance in plants. In: Bahadur, B., Rajam, V.M., Sahijram, L., Krishnamurthy, K.V. (eds.) Plant Biology and Biotechnology. Volume II: Plant Genomics and Biotechnology, pp. 579–609. Springer, New Delhi (2015)

Matthews, S., Powell, A.A.: Environmental and physiological constraints on field performance of seeds. Hortscience. **21**, 1125–1128 (1986)

Mederski, H.J., Jeffers, D.L.: Yield response of soybean varieties grown at two soil moisture stress levels. Agron. J. **65**(3), 410–412 (1973)

Moraes, G.A.F., Menezes, N.L.: Desempenho de sementes de soja sob condições diferentes de potencial osmótico. Ciência Rural. **33**(2), 219–226 (2003)

Munns, R., Tester, M.: Mechanisms of salinity tolerance. Annu. Rev. Plant Biol. **59**, 651–681 (2008)

Neves-borges, A.C., Guimarães-dias, F., Cruz, F., Mesquita, R.O., Nepomuceno, A.L., Romano, E., Loureiro, M.E., Grossi-de-sá, M.F., Alves-ferreira, M.: Expression pattern of drought stress marker genes in soybean roots under two water deficit systems. Genet. Mol. Biol. **35**(1), 212–221 (2012)

Oya, T., Nepomuceno, A.L., Neumaier, N., Farias, J.R.B., Tobita, S., Ito, O.: Drought tolerance characteristics of Brazilian soybean cultivars—evaluation and characterization of drought tolerance of various Brazilian soybean cultivars in the field. Plant Prod. Sci. **7**(2), 129–137 (2004)

Pardo, E.M., Vellicce, G.R., Aguirrezabal, L., Pereyrairujo, G., Rocha, C.M.L., García, M.G., Prietoangueira, S., Welin, B., Sanchez, J., Ledesma, F., Castagnaro, A.P.: Drought tolerance screening under controlled conditions predicts ranking of water-limited yield of field-grown soybean genotypes. J. Agron. Crop Sci. **201**(2), 95–104 (2015)

Pereyra-irujo, G.A., Velazquez, L., Granier, C., Guirrezabal, L.: A method for drought tolerance screening sunflower. Plant Breed. **126**(4), 445–448 (2007)

Porcel, R., Aroca, R., Ruiz-Lozano, J.M.: Salinity stress alleviation using arbuscular mycorrhizal fungi. A review. Agron. Sustain. Dev. **32**(1), 181–200 (2012)

Ramalho, M.A.P., Abreu, A.F., Santos, J.B., Nunes, J.A.: Aplicações da genética quantitativa no melhoramento de plantas autógamas. Ed. UFLA, Lavras (2012.) 522p

Rampim, L., Lana, M.C.: Mecanismo de Tolerância Interna das Plantas ao Alumínio. Colloq. Agrariae. **9**(2), 72–89 (2013)

Resende, M.D.V., Ramalho, M.A.P., Guilherme, S., Abreu, A.F.B.: Multigeneration index in the within progenies bulk method for breeding of self-pollinated plants. Crop. Sci. **55**(3), 1202–1211 (2015)

Sediyama, T. (ed.): Melhoramento Genético da Soja. Editora Mecenas, Londrina (2015.) 352p

Sinclair, T.R.: Challenges in breeding for yield increase for drought. Trends Plant Sci. **16**(6), 289–293 (2011)

Snyder, R.L., Carlson, R.E., Shaw, R.H.: Yield of indeterminate soybeans in response to multiple periods of soil-water stress during reproduction. Agron. J. **74**(5), 855–859 (1982)

Texeira, L.R., Braccini, A., De, L., Scapim, C.A., Sperandio, D., Schuster, I., Viganó, J.: Avaliação de cultivares de soja quanto à tolerância ao estresse hídrico em substrato contendo polietileno glicol. Acta Sci. Agron. **30**(2), 217–223 (2008a)

Texeira, L.R., Braccini, A., De, L., Sperandio, D., Scapim, C.A., Schuster, I., Viganó, J., Jaremtchuk, C.C.: Avaliação de cultivares de soja quanto à tolerância ao estresse hídrico. Rev. Ceres Viçosa. **55**(3), 194–202 (2008b)

Turner, N.C., Wright, G.C., Siddique, K.H.M.: Adaptation of grain legumes (pulses) to water limited environments. Adv. Agron. **71**, 193–231 (2001)

Villela, F.A., Doni filho, L., Sequeira, E.L.: Tabela de potencial osmótico em função da concentração de polietileno glicol 6000 e da temperatura. Pesq. Agropec. Brasil. **11**(/12), 1957–1968 (1991)

Von Uexküll, H.R., Mutert, E.: Global extent, development and economic impact of acid soils. Plant Soil. **171**(1), 1–15 (1995)

Wang, X., Yan, X., Liao, H.: Genetic improvement for phosphorus efficiency in soybean: a radical approach. Ann. Bot. **106**(1), 215–222 (2010)

Zeng, Q.Y., Yang, C.Y., Ma, Q.B., Li, X.P., Dong, W.W., Nian, H.: Identification of wild soybean miRNAs and their target genes responsive to aluminum stress. Plant Biol. **12**(1), 182–190 (2012)

Chapter 20
Breeding for Nutritional Quality

Ana Cristina Pinto Juhász, Sueli Ciabotti, and Lilian Cristina Andrade de Araújo Teixeira

Abstract Besides being the major source of plant protein and oil in the world, soybean is an important functional food. Human soybean consumption is increasing in the last years, due to more availability of industrial food and cultivars with the best flavor and other seed coat colors. To further increase its consumption, breeding efforts have been targeting traits related to seed quality, cooking time, soft flavor and mild taste, and nutritional composition, such as protein and oil content, fatty acid composition, fiber content, isoflavones, reduction of antinutritional components, etc. The human population has become aware of the benefits of soybean being in their diet.

Keywords *Glycine max* • Functional properties • Chemical composition • Cooking time

Introduction

Soybean is one of the most ancient cultivated species in the world, and it was extremely important in ancient Chinese civilization, where it was a sacred bean because it was so widespread. However, what is of historic note is the role it has played in the diet in China, the country that is the biggest consumer of soybean. According to the USDA, in the 2014/2015 harvest, 30% of the world's soybean was consumed in China alone, reaching 85,900 thousand tons (SEAPA 2016). In Brazil, incorporating soybean into the diet has been more recent, its introduction having been reported in 1908, with the influx of Japanese immigrants to São Paulo.

A.C.P. Juhász, D.Sc. (✉)
Agricultural Research Company of Minas Gerais (EPAMIG), Uberaba, Brazil
e-mail: ana.juhasz@epamig.br

S. Ciabotti, D.Sc.
Federal Institute of Education, Science and Technology of Triângulo Mineiro, Uberaba, Brazil
e-mail: sueliciabotti@ifmt.edu.br

L.C.A. de Araújo Teixeira, D.Sc.
State University of São Paulo (UNESP), São Paulo, Brazil
e-mail: araujo.l.c.a@gmail.com

F. Lopes da Silva et al. (eds.), *Soybean Breeding*,
DOI 10.1007/978-3-319-57433-2_20

Soybean can be eaten in various ways. By processing soybean oil, lecithin and margarine can be obtained. When the oil is extracted, defatted flakes are left, from which we get the flour, protein isolate, and protein concentrate (Moraes et al. 2009). From the extract, yogurt and tofu (soybean cheese) are produced. Through a fermentation process of the beans, it is possible to make *tempeh, natto, miso*, and *soy sauce*. Green soybean or *edamame* (beans harvested when not yet ripe) can also be used in salads or as an aperitif, after being cooked. From the grain, it is also possible to make germinated bean sprouts, and, when cooked, the grains can be used in salads. Other ways of eating soybeans have been developed since the launch of the soybean cultivar featuring brown integument ('BRSMG 800A,' Embrapa/EPAMIG/Fundação Triângulo), by cooking them like brown beans or even preparing "Cattleman's soybean," by replacing the brown beans with soybeans when preparing the traditional dish called "Cattleman's beans." In industry, soybean is also used for processing various products, such as processed meats, bread, cakes, biscuits, cookies, dehydrated soups, chocolates, etc.

In Brazil, consumption of soybean in the human diet is still limited, despite the country being the second largest producer and the second largest exporter of soybean. In Brazil, it is used mainly in the form of oil and bran for animal feed. The main reasons for low consumption of this legume by Brazilians are that eating it is not traditional, as it was initially used only as animal feed, and its characteristic "beany" flavor.

A healthy diet must be a balanced one in order for the body to develop correctly, which means being made up of proteins, lipids, carbohydrates, minerals, and vitamins. The main source of protein in western diets comes from animals, but this is still often not accessible for many people. So, soybean is a viable alternative, as it contains an average of 40% protein in the beans. It also has a suitable balance of essential amino acids, but when compared to aminol protein, reduced amounts of sulfuric amino acids can be seen. When soybean is eaten together with cereals, however, as it is a legume, these amino acids are complemented and proteins that are good for the organism are provided (Carrão-Panizzi and Mandarino 1998).

Besides being an excellent source of nutrition, soybean is considered to be a functional food, as it has certain health benefits that prevent and/or control illnesses. ANVISA recommends a daily consumption of at least 25 g of soybean protein to reduce levels of bad cholesterol. Soybean can be used as part of a starch-reduced diet, as in cases of diabetes and preventing arteriosclerosis. It can also be used in treating obesity, by using isolates of soybean in hypocaloric diets and in arterial hypertension, as there is a relatively low level of sodium in some derivative products. Soybean also has hormonal action that is mainly the result of the isoflavones and lignins in its composition, which can show phytoestrogen activity. The isoflavones and the protease inhibitors are found in greater concentrations in soybean, and this shows to be efficient at preventing some types of cancer, like mammary and colon (Morais and Silva 1996). The phytoestrogen effect of isoflavones can also lower the risk of cardiovascular diseases and osteoporosis (Godman 2003).

Therefore, breeding soybean for human consumption, besides the agronomical traits, other aspects must also be taken into consideration, such as reducing the unpleasant taste, improving the appearance of the grains (color, size, and overall appearance), and improving the nutritional composition of the grains (increasing the amount of protein and isoflavones, reducing the amount of trypsin inhibitor, having suitable amounts of fatty acids and sugars) and physical features after cooking. These traits can be modified through breeding, the purpose of which is to obtain special cultivars of soybean for human consumption.

Technological Analyses

For the consumer to be able to enjoy the benefits of the nutritional properties of soybean, it needs to become generally accepted. This acceptability will be covered in topics regarding technological analyses: (1) what the beans look like, (2) cooking time, and (3) sensory evaluations.

Soybean Grain Appearance

The grains that are to be consumed when cooked should please the eye of the consumer. They need to be uniform, the integument and hilum must have the same coloration, the integument cannot have any imperfections in it like cracks, stains caused by disease or pests, mechanical damage or be different sizes to the other beans after being cooked, and acceptance by consumers as the parameter used.

Vello (1992) classified food-type soybean depending on their use, by the size of the seeds: (1) soybean cultivars as grains—100 seeds weighing between 10 and 19 g, used mainly by the oil and bran industries; and (2) cultivars for food—pleasant flavor, small seeds, 100 seeds weighing less than 10 g that will be eaten in the form of bean sprouts and *natto* (nuts), and large ones, 100 seeds weighing more than 20 g, to be eaten as green soybean (unripe beans) and used in the preparation of salads and other dishes together with ripe grains.

These traits are evaluated in the soybean breeding program for human food that is being developed by the EPAMIG, in which the promising lines in advanced evaluation phase go through phenotypical selection depending on the size of the seeds and the visual quality of the beans. Promising lines that have a high proportion of cracked skins, for instance, are eliminated from the program.

In response to these questions, *Embrapa/EPAMIG/Fundação Triângulo* released the cultivar 'BRSMG 790A,' in 2008, which has yellow integument and hilum and, in 2010, released 'BRSMG 800A,' which is the first Brazilian soybean cultivar with brown integument, a novelty in Brazilian soybean cultivation. The letter "A" in the naming of the cultivar indicates that it was developed especially for human consumption.

Cooking Time

This trait is very important when consuming soybeans that are for salads or when being cooked together with brown beans or for preparing "Cattleman's soybean." Cooking the brown soybean together with some brown beans is recommended, because soybean has little starch in its makeup and so does not make the traditional broth that is liked so much by Brazilians.

The cooking time for comparing the lines in the advanced phase of the breeding program is determined by the Mattson-type bean cooker, which has been adapted for soybean because its chemical composition is different from Brazilian beans. For this test, since soybeans have a firmer consistency, first the grains are cooked in a pressure cooker for 5–10 min. Then 25 grains are transferred to the Mattson-type equipment to finish off the cooking in "bain-marie" at 98 °C. This equipment is made up of 25 vertical stems that weigh 90 g each, and each one is placed on top of one soybean (Fig. 20.1), and when these grains are cooked enough, the stem pierces them. The cooking time of the genotype is ascertained when 13 out of the 25 stems have completely pierced the grains (Juhász et al. 2014).

Two advanced breeding lines of black soybean integument from *Embrapa/EPAMIG/Fundação Triângulo's* breeding program were evaluated in four places. For the MGBR09-9161 line, there was genotype × environment interaction, which meant huge variation for this trait depending on the place where it was cultivated (Fig. 20.2). Meneguce et al. (2005) observed interaction between 20 lines of food-type soybean for the cooking time trait in 2 years of cultivation. In the second year, there was a significant increase for the cooking time of some lines that were evaluated.

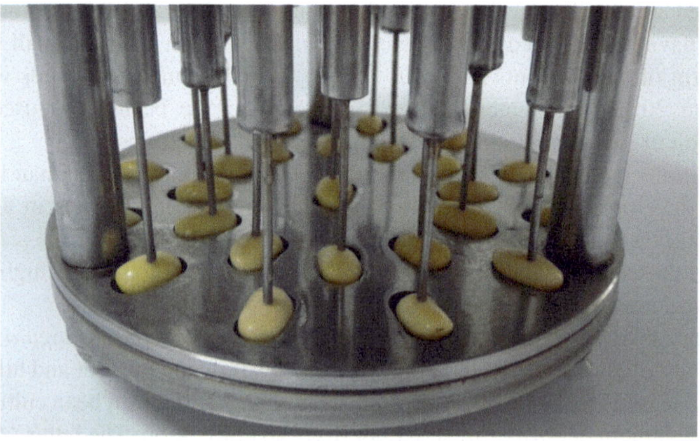

Fig. 20.1 Positioning of soybeans in the Mattson-type equipment. Uberaba, Minas Gerais. Photo: Juhász et al. 2014

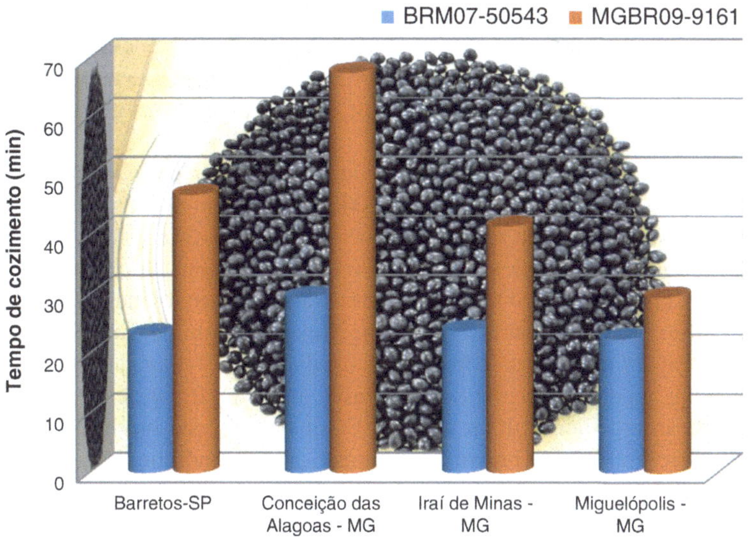

Fig. 20.2 Cooking time, in minutes, of two black soybean lines in four evaluation environments, in the 2012/2013 harvest. Uberaba, Minas Gerais

Sensory Analysis

Soybean and its derivatives have a characteristic taste called "green bean," "raw," "unripe," which is a bitter astringent taste resulting from the interaction of various chemical components in the grains (Tsutsumi 2000). One of the main causes of this taste is the action of the lipoxygenase enzyme on the oxidation of lipids. This enzyme can be made inactive by thermal processing or also by there not being any in triple-null cultivars, which occurs when there are recessive alleles. This trait is controlled by three genes (L-1, L-2, and L-3), and loco 3 is independent from loci 1 and 2.

Indication of this relationship between the taste of soybean and the action of the lipoxygenase enzyme is reported by Utumi et al. (1998), who carried out crosses between contrasting parent plants for these traits and evaluated the segregant generations. It was observed that the genetic elimination of lipoxygenase from seeds dramatically reduced the production of hexanal, one of the main compounds responsible for the characteristic taste of protein products derived from soybean, which means improved taste for human consumption in the genotypes without any lipoxygenase in the seeds.

Not all the cultivars used in human food are triple nulls for lipoxygenase, but there are differences in their composition that make them taste better, due to the higher proportion of carbohydrate, which is what occurs with the 'BRSMG 790A' cultivars (yellow coat and hilum) and the 'BRSMG 800A' (brown hilum), developed by *Embrapa/EPAMIG/Fundação Triângulo*.

At the *EPAMIG*, sensory analyses are carried out of the lines in the final phase of evaluation in the breeding program, both as extract and on the cooked grain, and, irrespective of whether there are any lipoxygenase enzymes present, there is wide variability of taste, and this makes it possible to phenotype the taste into qualitative classes (astringent, mild, sweetish, etc.).

Callegari et al. (2013) made a sensory analysis of preparations that had been compiled from the brown soybean 'BRSMG 800A' and the brown bean. The preparations were evaluated through the acceptance test (Meilgaard et al. 1991) done on 90 tasters from the ages of 14–56. The tests were based on giving the preparations, one at a time, to the tasters, who evaluated them in individual cubicles in the premises of the Federal Institute of the Minas Triangle. The mixed structured hedonistic scale of nine points—from 1 (disliked a lot) to 9 (liked a lot)—was used for evaluating appearance, color, texture, taste, and overall appearance. Preparing brown soybean together with an added 40% of brown bean broth had the same measure of acceptance as brown beans made in the traditional way, which is evidence that this way of preparing soybean reaches the proposed objective of not changing the eating habits of Brazilian consumers. It did, however, increase the level of protein in traditional beans.

Nutritional Composition

The chemical composition of ripe soybeans is normally presented in the following decreasing order of content levels: proteins (≅40%), carbohydrates (≅30%), lipids (≅23%), fibers (≅6%), ash (≅5%), and minerals and vitamins, based on dry material. These values undergo alterations in accordance with environmental and genetic effects and their interaction. The aim of plant breeding is to raise the content levels of some of these chemical compounds to the detriment of others, depending on the purpose.

This is a current inclination, seeing as the first main objectives of vegetable crop breeding were exclusively aimed toward agronomy traits of interest, such as productivity, resistance to the main pests and diseases, lines with no lodging and with appropriate insertion of pods, among other aspects.

With regard to the chemical composition for use of soybean in human food, cultivars that have higher levels and higher-quality protein, lower levels and better quality oil, and a high level of carbohydrates are normally recommended, in addition to adding traits that provide a more pleasant taste and reduced levels of anti-nutritional factors (Vello 1992).

The nutritional chemical composition can also be discerned in accordance with soy-based end products. These compositions (protein, lipids, carbohydrates, food fibers, ash, minerals (calcium, magnesium, manganese, phosphorus, iron, sodium, potassium, copper, and zinc) and vitamins (thiamin, riboflavin, pyridoxine, and C)) in soy-based foods are set out in the Brazilian Table of Food Composition (TACO),

Table 20.1 Nutrients: protein, lipids, carbohydrates, food fibers, ash, minerals, and vitamins from foods per 100 g of soy-based edible part

	Soy bread	Soy flour	Soluble soy extract[a]	Soluble soy extract (powder)	*Tofu*
Nutrients (g)					
Protein	11.3	36.0	2.40	35.7	6.60
Lipids	3.60	14.6	1.60	26.2	4.00
Carbohydrate	56.5	38.4	4.30	28.5	2.10
Food fiber	5.70	20.2	0.40	7.30	0.80
Ash	2.50	5.10	0.50	5.20	0.70
Minerals (mg)					
Calcium	90.0	206	17.0	359	81.0
Magnesium	48.0	242	15.0	216	38.0
Manganese	0.57	2.87	0.15	2.68	0.33
Phosphorus	153	539	53.0	647	130
Iron	3.30	13.1	0.40	7.00	1.40
Sodium	663	6.00	57.0	83.0	1.00
Potassium	296	1922	121	1607	182
Copper	0.16	1.29	0.08	1.19	0.18
Zinc	1.50	4.50	0.30	5.80	0.90
Vitamins (mg)					
Thiamin (B1)	0.07	0.20	*tr*[b]	*tr*	0.04
Riboflavin (B2)	0.04	0.04	*tr*	0.11	*tr*
Pyridoxine (B6)	0.79	0.03	*tr*	0.35	0.03
C	*tr*	*tr*	*tr*	9.20	*tr*

Adapted from the Brazilian Food Composition Table (2011)
[a]Natural soluble soy extract (fluid)
[b]*tr* trace. Values below quantification limits

2011, which was developed by the Nucleus of Studies and Researches of Food (*NEPA*) and the State University of Campinas (*UNICAMP*) (Table 20.1).

These values can show variation, depending on the preparation and the genotypes used. One example is the work by Pauletto and Fogaça (2012), who obtained, from a sample of *tofu*, values of fibers from 1.8 g (±1.0), of protein from 11.4 g (±2.5), and values of carbohydrates from 5.5 g (±1.0). In a sample prepared with *okara* (obtained from residue from preparing soy extract), they got values of fibers from 3.5 g (±0.3), of protein from 6.3 g (±1.5), and of carbohydrate from 7.9 g (±0.7).

It is because of its chemical composition shown, apart from other bioactive compounds, that soybean has become obvious as a functional food, as a result of the various benefits that regular consumption can bring to one's health with respect to metabolic and physiological effects.

Further details regarding the nutritional components of soybean and its relationship with plant breeding will be presented below.

Protein and Oil Content

From a nutritional standpoint, soybean contains essential components for human food, as it is an excellent source of protein in human and animal diets. The beans contain about 35–50% of protein (Krishnan 2005)—it is one of the legumes that stands out in this respect—and around 20% of oil.

The nutritional value of a protein depends, among other aspects, on the bioavailability of the essential amino acids to know the protein quality. Vegetable proteins in general are limiting in nutritional value, in sulfuric amino acids (methionine and cystine), with relation to animal protein. Sgarbieri (1996) defines it as limiting when one or more amino acids are found in the protein in quantities lower than what is required by the receptor organism or with relation to a reference standard.

The increase in amino acids has been reported involving both classic breeding and mechanisms of genetic engineering. Classic breeding aims mainly to increase the total level of protein and does not give so much importance to the amino acids involved (Krishnan 2005). An example is the work by Moraes et al. (2006a), in which the composition of amino acids in soybean lines with a high level of protein was evaluated, which promoted increase in all the amino acids, except glycine, alanine, tyrosine, and sulfates.

In order to increase the levels of sulfuric amino acids, it is possible to work with mutagenesis together with classic breeding, although using mechanisms of genetic engineering seems to be a more realistic approach. Incrementing these amino acids can occur by increasing the expression of these proteins by introducing synthetic ones (heterologous expression of protein) (Krishnan 2005).

Besides studying the amino acids that are part of the protein expression, the extent and direction of the correlation of the levels of protein and oil in the soybean seeds must be attempted (Vieira et al. 1999; Bonato et al. 2000; Moraes et al. 2006a).

There is a certain tendency to reduce the level of protein in the new cultivars that have been released, as was reported by Bonato et al. (2000), in genetic material from Rio Grande do Sul. This data proves that cultivars with higher levels of protein have to be developed to meet the demand for higher-quality cultivars for human food.

Most of the proteins in soybeans are found in special protein corpuscles contained in the cotyledon cells, which can be isolated by centrifuging in density gradient. Ultracentrifugation separates the proteins from the soybean in four fractions with sedimentation speeds equivalent to the 2S (composed of two globulins that are exempt from biological activity), 7S (globulin, polymeric protein whose structure can be broken in several ways), 11S (globulin that reveals the existence of bands in alkaline and acid conditions), and 15S (Sgarbieri 1996). The fractions 7S and 11S make up 70% of the total protein in the soybean. The rest sediments with the fractions 2S and 15S. The protein content and the relationship of globulin/albumin show variables and are influenced by many factors, among which are species, variety, type/place of soil, and use of fertilizers.

The relationship between proteins and lipoxygenase enzymes was studied by Utumi et al. (1998), who observed that there was no reduction in the total levels of protein or of reserve proteins 7S and 11S in soybean seeds with no lipoxygenase enzymes. Moraes et al. (2006a) noted that, with the increase in the level of proteins in the lines obtained, there was an increase in the levels of 11S proteins and, consequently, of the 1S/7S relationship, which improved the quality of soy flour since the 11S proteins are nutritionally better quality than the 7S proteins.

In human food, a high level of protein is desirable. Moraes et al. (2006a) obtained two isolines of a high level of protein and reduced level of oil: isoline 1 with 47.78% of protein and 16.81% of oil and isoline 2 with 46.56% of protein and 16.71% of oil, which were obtained through a soybean breeding selection program for the agro-industry, *Bioagro/UFV*.

Alves et al. (2011) evaluated eight *Embrapa* soybean cultivars and the level of protein varied between 39.41 and 44.37% and the lipids between 18.76 and 22.45%. Among these, the *Embrapa* cultivar 'BRS 258' stands out which has 44.37% of protein and 18.76% of oil.

The levels of protein and oil were also variable in six special soybean cultivars for human food, which were developed by *Embrapa Soja's* genetic breeding program. The 'BRS 284' cultivar obtained lower levels of protein (30%) and higher levels of lipids (28%), while the 'BRS 216' cultivar stood out significantly because of the higher level of (47%) and reduced level of oil (22%) (Felber et al. 2013).

Vieira et al. (1999) evaluated the levels of protein and oil in six soybean cultivars for human food and ascertained that the level of protein varied from 38.55 to 41.95% and that of oil between 22.24 and 23.80%. They concluded that the levels in special soybean cultivars for human food did not differ from those used in industry.

With regard to the level of oil, Rocha et al. (2002) evaluated 28 soybean lines in 12 environments, over four agricultural years, and observed a significant effect of the genotype × environment interaction. Places × year interaction contributed more than the isolated effects of places and years toward the environmental variation, while the genotype × year interaction was responsible for most of the genotype × environment interaction.

Levels of Fatty Acids

In Brazil, the RDC Resolution n° 360, of December 23, 2003, from the National Agency for Sanitary Inspection (ANVISA) (Brazil 2003), made it compulsory to include the total content of fats (TF) or lipids, saturated fatty acids (SFA), and trans-fatty acids (TFA) on the labels of packaged foods, and in the United States, the US Food and Drug Administration (2006) made the same requirement. This requirement has become worrying—seeing as ingesting excessive amounts of saturated fatty acids and trans in the diet have been correlated to an increased risk of cardiovascular diseases—as a strategy for preventing chronic illnesses (Aued-Pimentel Zenebo 2009).

The industrial properties and potential application of oil in foods are, to a large extent, determined in accordance with the relation and the quantity of saturated and non-saturated fatty acids, and oilseed crops cannot satisfy all the requirements of the food industry. Therefore, an interest in modifying the composition of oil in oilseed crops is increasing (Ray et al. 2008).

Soybean oil contains saturated, monounsaturated, and polyunsaturated fatty acids, saturated fatty acids: 11% palmitic acid (16: 0) and 4% stearic acid (18: 0), monounsaturated: 25% oleic acid (18: 1), and polyunsaturated: 52% linoleic acid (18: 2) and about 8% linolenic acid (18: 3) (Fehr 2007).

In order to improve the quality of soybean oil, it is necessary to raise its oxidative stability by increasing the level of oleic acid and reducing the levels of linolenic and linoleic acids. Linolenic acid negatively affects the stability, taste, and smell of soybean and makes it necessary to use hydrogenation in the refining process. Breeding can reduce the levels of linolenic acid and, therefore, the need for hydrogenation, and this improves the stability, taste, and smell, which, in turn, will reduce the levels of trans-fatty acids in soybean oil (Ray et al. 2008).

Some work on improving soybean has been carried out for this purpose. Hybrid originals between two plants with low palmitic acid levels were used, and parental cultivars with low levels of linolenic fatty acid (4%) were used as control. Two planting dates were used each year (May and June 2001, 2003, and 2004) in the Piedmont Region of South Carolina (USA). The later planting date (end of June) resulted in larger beans of superior quality, which caused an increased level of linolenic acid. Also in this study, there was a significant increase in the levels of palmitic acid, but there was no difference for stearic, oleic, and linoleic acids (Ray et al. 2008).

Identifying QTLs associated with the content of oleic acid could favor selection, aided by molecular markers, when improving plants for increasing the content of this component in soybean oil in cultivars that are destined for human food. Matta (2012) used SNP (single-nucleotide polymorphism) markers in an F_2 population arising from a cross between contrasting genotypes for the content of oleic acid. Estimates of heritability were observed that enable selecting for this trait to be done in early generations and, also, the negative correlation between levels of oleic fatty acids and other unsaturated fatty acids and in the oil content. Thus, with the increase in the content of oleic acid, there is a decrease in linoleic acid, which is favorable for the nutritional quality of the oil.

Carbohydrate, Fiber, and Ash Content

With relation to carbohydrates, it is known that soybean is made up of sugars, like glucose, fructose, and sucrose, and oligosaccharides, like raffinose and stachyose. High levels of sucrose are desirable in genotypes destined for human food as they improve the taste and smell traits in its subproducts.

Table 20.2 Centesimal composition of genotypes of soybeans (g.100 g^{-1}) regarding level of content of carbohydrates, ash, and fiber

Genotype	Carbohydrates	Ash	Fiber	Reference
'BRS 267'	–[b]	4.50	5.36	Callegari et al. (2011)
	–	6.41	–	Felber et al. (2013)
	28.08	6.45	–	Alves et al. (2011)
'BRSMG 790A'	–	5.38	6.24	Callegari et al. (2011)
	36.58	4.35	–	Ribeiro et al. (2013)
'BRSMG 800A'	35.83	4.34	–	Ribeiro et al. (2013)
BRM04–1660	–	4.81	6.08	Callegari et al. (2011)
'BRS 213'	–	5.85	–	Felber et al. (2013)
	28.39	4.90	–	Alves et al. (2011)
'BRS 216'	–	5.21	–	Felber et al. (2013)
	29.68	4.45	–	Alves et al. (2011)
'IAS-4'	32.12	5.34	5.24	Vieira et al. (1999)
'EMBRAPA-4'	29.81	5.31	5.90	Vieira et al. (1999)
'Davis'	32.08	5.57	5.47	Vieira et al. (1999)
'BR-16'	33.33	5.24	5.85	Vieira et al. (1999)
'Iguaçu'	32.54	5.56	5.66	Vieira et al. (1999)
'IAS-5'	32.16	5.43	6.38	Vieira et al. (1999)
'BRS 282'	26.83	6.35	–	Alves et al. (2011)
'BRAP-01'[a]	35.28	4.11	–	Ribeiro et al. (2013)
'MGAP-01'[a]	34.80	4.30	–	Ribeiro et al. (2013)
'MGAP-02'[a]	37.05	4.12	–	Ribeiro et al. (2013)

[a]Experimental genetic material with black skin
[b]Values not estimated

Teixeira (2008) evaluated concentrations of sucrose, raffinose, stachyose, total soluble sugars, and total protein and observed that heritability in the board sense of these traits, with the exception of stachyose, was high and positive, due perhaps to the low environmental effect. High positive correlations were found for concentrations of sucrose together with the concentrations of raffinose, stachyose, and total soluble sugars. However, negative correlations were detected between levels of sucrose and total proteins, which could hinder breeding, seeing as both components are desirable in lines being developed for human food. Moraes et al. (2006a) also report a negative correlation between carbohydrates and protein in soybean.

There is great phenotypic variability among materials destined for human food for the different content levels of carbohydrates, ash (inorganic residue that is indicative of the amount of mineral elements in a sample), and fibers (Table 20.2).

Isoflavones

Plants, as well as other organisms, produce a wide variety of compounds called secondary metabolites that arise from primary metabolites, among which are flavonoids, phenolic acids, saponins, glucosinolates, furanocoumarins, and cyanogenic compounds (Simões et al. 2004). Isoflavones are the most common form of these metabolites, and they are predominantly found in legumes, especially in soybean.

Several studies have shown the positive effects of consuming foods derived from soybean, as they contain isoflavones. The benefits include reducing the risk of cardiovascular diseases, cancer, and osteoporosis and alleviating menopausal symptoms and the like (Kim et al. 2012).

Soybean and its derivatives contain variable levels of isoflavones (daidzein, genistein, and glycine), which are bioactive compounds with various biological activities that seem to be related to their form. Isoflavones exist in four chemical forms, which totals 12 isomers: daidzein aglycones, genistein and glycyte; the dazydine β-glycosides, genistine and glycitine; the acetylated glycosylated derivatives 6″-O-acetildazydine, 6″-O-acetilgenistine, and 6″-O-acetilglycitine; and the malonylated glycosides 6″-O-malonildazydine, 6″-O-malonilgenistine, and 6″-O- malonilglicitin (Barbosa et al. 2006).

Several factors can have an influence on the chemical forms of the isoflavones, both on the grains and on the products derived from soybean. When producing grains, Avila et al. (2007) quote studies analyzing different cultivars of soybean in the same region in Brazil, where a great variation in the concentration of isoflavones was observed, which is genetically determined by and affected by environmental factors, mainly local temperature. The levels of isoflavones in soybean show high variability, since it depends on the morphological region they are extracted from (cotyledon, hypocotyl, and integument), on the variety (genetic factors), and on the environmental conditions (temperature, humidity) of cultivation.

Studies have shown that the concentration of isoflavones in soybean seeds is a quantitative trait Zhang et al. 2014; Chiari et al. 2006) with significant genotype × environment interaction. Chiari et al. (2006) report nuclear cytoplasmic genes and the interaction between them that controls the expression of the trait for content of genistine and malonylgenistine isoflavones. There is evidence of maternal effect on content of isoflavones.

According to studies, the content of isoflavones is significantly lower in seeds that are developed in high temperatures during the granulation phase. Genotypes, places, years, and the interaction between them affect the level of isoflavones in soybean. Some authors consider the effect of years to be more important than that of places. Thus, one factor that affects places and years is the sowing date, because of the difference in temperature. High to moderate heritability has been found for isoflavones, especially for daidzein (79%) and glycyte (88%), but not for genistein (22%) (Meksem et al. 2001).

Chiari et al. (2004) ascertained concentrations of six forms of isoflavone (dazydine, genistine, glicitin, malonyldazydine, malonylgenistine, and malonylglicitin) in F_2 plants derived from crossing the cultivars BARC-8ande IAC-100, which are contrasting for these contents. They ascertained that the heritability in the wide sense was higher than 90% for all the forms of isoflavone, as it is an inheritable genetic control trait, and that there is a negative correlation between content of isoflavone and content of proteins.

As it shows quantitative inheritance, it is possible to study QTLs (quantitative trait locus), which explains that about 3.5–10.5% of evaluated phenotypic variation has been reported. It has also been reported that, besides the QTLs associated with isoflavones in the linkage groups A1, B1, B2, D1a, H, K, and N, the QTLs were identified in the linkage groups M, C2, F, and G (Primoto et al. 2005). The QTLs identified in this work could be useful for developing varieties of soybean with higher levels of isoflavones, by carrying out selection assisted by molecular markers (SAM).

Besides the benefits for human health, isoflavones also have important functions for plants. They can carry out antifungal and antioxidant activities, defending the plant from attacks by fungi and insects. By inoculating them against *Phytophthora sojae*, an increase in daidzein was observed, which shows the role isoflavones play in the defense response to diseases (Connolly et al. 1999). This is a relevant factor, because the consumer often demands safe products, and the increase in the plant's natural defenses reduces the use of pesticides.

Agriculturally speaking, isoflavonoids play an important role in the process of nodulation, on account of their ability to induce the nodulation genes of *Bradyrhizobium japonicum*. The main isoflavonoid involved in this process is genistein. Special attention should be given to the '*Aoka*' cultivar in the breeding programs, as it is a good source of genes for isoflavonoids and for remaining stable in different environments (Subramanian et al. 2006).

The size of the seeds is also related to isoflavones. Kim et al. (2012) observed that small seeds (less than 13 g) show, on average, a greater concentration of total isoflavones (2520.0 $\mu g\ g^{-1}$) and concentrations of total phenolic compounds (2241.7 $\mu g\ g^{-1}$), when compared to medium-sized seeds (13–24 g) and large ones (more than 24 g).

Anti-nutritional Factors

Most legumes have their nutritional side, but they have another side that is considered to be anti-nutritional, and, when eaten raw, usually, or when inappropriately processed, they can cause adverse physiological effects in humans or reduce the bioavailability of certain nutrients (Rackis 1974).

Liener (1994) classified the anti-nutritional factors in soybean into thermolabile and thermosets. The thermolabile, which are sensitive to heat, are inhibitors of proteases, lectins, or hemagglutinins, are goitrogenic and anti-vitamin, and can be easily destroyed by heat; the thermosets are the saponins, tannins, phytoestrogens, flatulence factors (oligosaccharides raffinose, stachyose, and verbascose), lysinoalanine (formed by residue from lysine joined to the cystine or serine in alkaline conditions), allergenics, and phytate. Of these anti-nutritional factors, the most intensely studied are the proteases inhibitors, called Kunitz and de Bowman-Birk trypsin inhibitor (Wolf and Cowan 1975).

The anti-nutritional factors encountered in soybean, lectin, and Kunitz trypsin inhibitor are generally made inactive by thermal treatment. However, the residual activity of these factors can be detected in several types of products derived from soybean. Thermal treatment does not eliminate these factors and, what is more, can lower considerably the protein solubility. Genetic elimination of these anti-nutritional factors could be an alternative to thermal treatment (Moraes et al. 2006b).

In order to obtain seeds with reduced activity of the inhibitor, one needs kti-kti recessive mutant alleles that, by altering the transfer process, cause a 50% reduction in their inhibitory activity (Jofuku et al. 1989).

With the purpose of reducing costs in soybean processing and lowering the anti-nutritional factors, soybean cultivars with low trypsin inhibitor activity and that do not have any Kunitz trypsin inhibitor (KTI) inhibitor were developed. Cultivars without any Bowman-Birk inhibitor and even cultivars that are free of both these trypsin inhibitors have also been investigated (Miura et al. 2001).

Selection, assisted by specific molecular markers, to avoid having any anti-nutritional factors (Kunitz and lectin inhibitor) in soybean seeds has enabled individual heterozygotes to be identified and so has reduced one generation at each backcrossing cycle and, therefore, lessens the number of generations needed to obtain improved varieties of soybean. The segregation test has confirmed that these traits are controlled by two genes that segregate independently (Moraes et al. 2006b).

The BIOAGRO Soybean Quality Breeding Program, of the Federal University of Viçosa, has developed a line that is devoid of Kunitz Trypsin Inhibitor (KTI) and Lectin (LEC), thus improving the digestibility of soybean proteins and lowering morphological alterations of the intestinal microvilosities when compared to the genotypes that are carriers of these anti-nutritional factors. The line obtained was derived from a cross between CAC-1 HyPro without KTI and CAC-1 HyPro without lectin. Using SAM made it possible to identify the promising genotypes (Brune et al. 2010).

Galão et al. (2013) obtained lower averages of Kunitz inhibitor and phytic acid in transgenic cultivars than in conventional cultivars that had been evaluated in Londrina and Ponta Grossa. The 'BRS 244 RR' transgenic cultivar stood out, and it showed the lowest rates of phytic acid in Londrina (1.21 mg g^{-1}) as well as in Ponta Grossa (1.39 mg g^{-1}).

Another problem reported is the occurrence of flatulence caused by consuming legumes. This is due to the presence of nonreducing oligosaccharides from the raffinose family (raffinose, stachyose, and verbascosis). Although the oligosaccharides are considered to be an anti-nutritional factor, as they cause problems of flatulence in individuals, they are responsible for developing *Bacillus bifidus* in the intestinal lumen, which inhibits the development of bacteria that produce putrefactive material and which then reduces the risk of colon cancer (Morais and Silva 1996).

In the study by Jaureguy et al. (2011), the heritability and the correlations between the size of the seeds, protein, and concentrations of sucrose, raffinose, and stachyose were estimated. Ninety eight recombinant lines derived from a cross between a line with a high protein content and a cultivar with large seeds and a moderate sucrose content were evaluated, in two places in the United States, over a period of 2 years. The results indicate that, although traces of quality food were significantly affected by the year and locality, greater variability was attributed to the differences between the lines than to the locality and, by interaction, to the size of the seeds, protein, sucrose, and stachyose.

Black Soybean

Over the last few years, the demand for black hilum seed soybean has increased, as it is a functional food and has high levels of flavonoids and vitamins. Black soybean is used in traditional Chinese medicine as it is beneficial to the organism in several ways. Soybean cultivars with black or brown integument contain much higher levels of phenolic compounds and anthocyanins than those with yellow or green integument (Kim et al. 2006). Anthocyanins are associated with a vast array of biological activities, including the antioxidant function (Lee et al. 2009).

Lee et al. (2009) described for the first time nine different derivatives of anthocyanin present in the integument of the black soybean cultivar 'Cheongja 3,' which indicates the diversity of this compound in black soybean. Jeng et al. (2010) also observed differences in the total levels of phenolic compounds, anthocyanins, proanthocyanidins, and isoflavones in different cultivars of black soybean. Zhang et al. (2011) featured 60 black integument Chinese cultivars and selected a group of nine cultivars with elevated levels of photochemical and antioxidant activity.

Kim et al. (2014) analyzed ten varieties of black soybean and ascertained considerable variability among the level of complex B vitamins in them, due to the influence of the varieties, the stage of development, and the ripeness of the seeds. They also ascertained that the different colors of the cotyledon are correlated to the variation of the level of complex B vitamins.

The isoflavone content in black soybean can also be increased by chemical treatment. Jeng et al. (2013) used sodium azide (NaN_3) to obtain black soybean mutants that had an increase in the levels of isoflavone and selected two mutants (SA-M-03 e SA-M-05) with a high level of isoflavone and a satisfactory productive yield.

With the aim of meeting the demand for black-seed hilum cultivars, the *Embrapa/EPAMIG/Fundação Triângulo* partnership released, in 2014, the 'BRSMG 715A' cultivar, which has a black integument and special traits suitable for human consumption: a mild taste and less cooking time.

Final Considerations

The benefits of using soybean in human food are being evermore recognized by professionals in the areas of health and nutrition, and the soybean breeding programs for special traits have been developing differentiated cultivars that have a high level of protein; no lipoxygenase enzyme; reduced levels of linolenic acid; special cultivars for preparing *tofu, natto*, and *misso*; raised levels of isoflavones; etc., as set out in Table 20.3.

Table 20.3 Institutions responsible and cultivars with special traits for human food being developed by their breeding programs

Institution	Cultivar	Traits
Federal University of Viçosa	UFVTN 101	Triple null
	UFVTN 102	Triple null
	UFVTN 103	Triple null
	UFVTN 104	Triple null
	UFVTN 105	Triple null
	UFV TNC 105	Triple null and low linolenic acid content
	UFVTNKL 105	Triple null, without KTI and lectin
	UFVTN 105AP	Triple null, high protein content
	UFVTNK 106	Triple null and without KTI
	UFVTNKC 106	Triple null, without KTI, and low linolenic acid content
Coodetec and Federal University of Viçosa	CD 223 AP	High protein content
Naturalle[a]	NT 1475 EX	Black skin
	NT 1482 NA	Small beans (100 beans weighing <11.5 g) Suitable for producing *natto*
	NT12	Suitable for *tofu* market. Protein content >44%
	NT4	Yellow skin and hilum. Suitable for producing *miso*
	NT2	Yellow skin and hilum. Suitable for producing *miso*
	NT 1478 SP	Destined for drinks, *miso*
Embrapa/Epamig/ Fundação Triângulo	BRSMG 800A	Brown skin and hilum
	BRSMG 790A	Yellow skin and hilum
	BRSMG 715A	Black skin and hilum
Embrapa	BRS 267	Large beans, yellow hilum, and better taste
	BRS 213	Triple null
	BRS 216	Small beans
	BRS 232	High level of isoflavone and low level of trypsin
	BRS 284	High level of isoflavone
	BRS 133	High level of isoflavone
	BRS 267	Suitable for producing *edamame*. Pleasant taste Large beans and yellow hilum

[a]Data provided by *Naturalle*®

However, Brazilians still need to get used to eating soybean regularly in their diet, which could be facilitated by using cultivars with different colored integument, like the brown cultivar to be used when preparing brown beans and the black cultivar for preparing the typical dish "Brazilian black bean stew."

With the call for healthy foods, breeding programs for developing nutritionally improved cultivars must be encouraged, since the farmers are interested in cultivars that produce a large volume and that are transgenic and so fail to meet this niche in such a promising market.

References

Alves, F.P., Oliveira, M.A., Mandarino, J.M.G., Benassi, V.T., Leite, R.S., Seibel, N.F.: Composição centesimal de grãos de soja de oito diferentes cultivares. In: VI Jornada Acadêmica Da Embrapa Soja, Londrina. Anais..., pp. 7–10. Embrapa-Soja, Londrina (Documentos 328) (2011)

Aued-Pimentel, S., Zenebo, O.: Lipídios totais e ácidos graxos na informação nutricional do rótulo dos alimentos embalados: aspectos sobre legislação e quantificação. Rev. Inst. Adolfo Lutz. **2** (68), 167–181 (2009)

Ávila, M.R., Braccini, A.L., Scapim, C.A., Mandarino, J.M.G., Albrecht, L.P., Filho, P.S.V.: Componentes do rendimento, teores de isoflavonas, proteínas, óleo e qualidade de sementes de soja. Rev. Bras. Sementes. **29**(3), 111–127 (2007)

Barbosa, A.C.L., Hassimotto, N.M.A., Lajolo, F.M., Genovese, M.I.: Teores de isoflavonas e capacidade antioxidante da soja e produtos derivados. Cienc. Tecnol. Aliment. **26**(4), 921–926 (2006)

Bonato, E.R., Bertagnolli, P.F., Langr, C.E., Rubin, S.A.L.: Teor de óleo e proteína em genótipos de soja desenvolvidos após 1990. Pesq. Agropec. Bras. **35**(12), 2391–2398 (2000)

Brasil. Ministério da Saúde. Secretaria de Vigilância Sanitária. Resolução RDC no 360, de 23 de dezembro de Dispõe sobre o Regulamento Técnico sobre Rotulagem Nutricional de Alimentos Embalados. Diário Oficial da Republica Federativa do Brasil, Brasília, DF, pp. 33–34, 26 de dez, 2003, Seção 1 (2003)

Brune, M.F.S.S., Pinto, M.O., Peluzio, M.C.G., Moreira, M.A., Barros, E.G.: Avaliação bioquímico-nutricional de uma linhagem de soja livre do inibidor de tripsina Kunitz e de lectinas. Cienc. Tecnol. Aliment. **30**(3), 657–663 (2010)

Callegari, F.L., Ciabotti, S., Sá, M.E.L., Garcia, D.F., Pereira, R.E.M., Santos, A.R.R.: Avaliação da composição centesimal e fatores antinutricionais de genótipos de soja nos estádios de maturação R6 e R8. Enciclopédia Biosfera. **7**(12), 1–13 (2011)

Callegari, F.L., Ciabotti, S., Sá, M.E.L., Juhász, A.C.P., Reis, R.P., Silva, M.S.: Análise físico-química e sensorial de preparações elaboradas com soja marrom e feijão carioquinha. Rev. Sodebras. **8**(93), 14–19 (2013)

Carrão-Panizzi, M.C., Mandarino, J.M.G.: Potencial de Uso na Dieta Brasileira, 16p. Embrapa-CNPSo, Londrina (EMBRAPA-CNPSo. Documentos, 113) (1998)

Chiari, L., Piovesan, N.D., Naoe, L.K., José, I.C., Vianna, J.M.S., Moreira, M.A., Barros, E.G.: Genetic parameters relating isoflavone and protein content in soybean seeds. Euphytica. **138**, 55–60 (2004)

Chiari, L., Naoe, L.K., Piovesan, N.D., José, I.C., Cruz, C.D., Moreira, M.A., Barros, E.G.: Inheritance of isoflavone contents in soybean seeds. Euphytica. **150**(1), 141–147 (2006)

Connolly, M.S., Williams, N., Heckman, C.A., Morris, P.F.: Soybean isoflavones trigger a calcium influx in *Phytophthora sojae*. Fungal Genet. Biol. **28**(1), 6–11 (1999)

Fehr, W.R.: Breeding for modified fatty acid composition in soybean. Crop Sci. **3**(47), 72–87 (2007)

Felber, I., Pereira, J.N.; Carrão-panizzi, M.C., Oliveira, D.R., Godoy, R.L.O., Freitas, S.C., Pacheco, S., Cunha, C.P., Borges, E.: Avaliação da composição centesimal e dos teores de isoflavonas de cultivares de soja (*Glycine max* (L.) Merrill) especiais para alimentação humana. In: Americas: International Conference on Soybean Utilization, 2013, Bento Gonçalves. Proceedings.Embrapa, Brasília, CD-ROM (2013)

Galão, O.F., Teixeira, A.I., Carrão-panizzi, M.C., Visentainer, J.V.: Variation in genetic and environmental effects of beta-conglycinin (7S) and glycinin (11S) protein fractions in conventional and GM soybean cultivars grown in southern Brazil. Semin. Ciênc. Agrár. **34**(2), 683–692 (2013)

Godman, I.L.: Medicine, agriculture and functional foods. In: Borém, A., Giudice, M., Sediyama, T. (eds.) Melhoramento genômico, pp. 211–224. Universidade Federal de Viçosa, Viçosa (2003)

Jaureguy, L.M., Chen, P., Scaboo, A.M.: Heritability and correlations among food-grade traits in soybean. Plant Breed. **130**, 647–652 (2011)

Jeng, T.L., Shih, Y.J., Wu, M.T., Sung, J.M.: Comparisons of flavonoids and anti-oxidative activities in seed coat, embryonic axis and cotyledon of black soybeans. Food Chem. **123**(4), 1112–1116 (2010)

Jeng, T.L., Shih, Y.J., Wu, M.T., Wang, C.S., Sung, J.M.: Evaluations and selections for high isoflavone black soybean mutants induced by NaN_3 treatment. Am. J. Plant Sci. **4**, 35–40 (2013)

Jofuku, K.D., Schipper, R.D., Goldberg, R.B.A.: Frameshift mutation prevents Kunitz trypsin inhibitor mRNA accumulation in soybean embryos. Plant Cell. **1**, 427–435 (1989)

Juhász, A.C.P., Ciabotti, S., Pádua, G.P., Favoreto, L., Jesus, A.M.S., Fronza, V.: Melhoramento de Soja parà alimentação humana. Informe Agropecuário. **35**, 39–45 (2014)

Kim, S.L., Berhow, M.A., Kim, J.T., Chung, I.M., Chi, H.Y., Song, J., Park, N.K., Son, J.R.: Composition and content of soyasaponins and their interaction with chemical components in different seed-size soybeans. Korean J. Crop Sci. **51**, 340–347 (2006)

Kim, E., Ro, H., Kim, S., Kim, H., Chung, I.: Analysis of isoflavone, phenolic, soyasapogenol, and tocopherol compounds in soybean [*Glycine max* (L.) Merrill] germplasms of different seed weights and origins. J. Agric. Food Chem. **60**, 6045–6055 (2012)

Kim, G., Lee, J., Ahn, K., Hwang, Y., Choi, Y., Chun, J., Chang, W., Choung, M.: Differential responses of B vitamins in black soybean seeds. Food Chem. **153**, 101–108 (2014)

Krishnan, H.B.: Engineering soybean for enhanced sulfur amino acid content. Crop Sci. **45**, 454–461 (2005)

Lee, J.H., Kang, N.S., Shin, S., Shin, S., Shin, S., Lim, S., Suh, D., Baek, I., Park, K., Ha, T.: Characterisation of anthocyanins in the black soybean (*Glycine max* L.) by HPLC-DAD-ESI/MS analysis. Food Chem. **112**, 226–231 (2009)

Liener, I.E.: Implications of antinutritional components in soybean foods. Crit. Rev. Food Sci. Nutr. **34**(1), 31–67 (1994)

Matta, L.B.: Mapeamento de QTL para conteúdo de ácido oleico em soja utilizando marcadores SNP, 69f. Tese (Doutorado)—Universidade Federal de Viçosa, Viçosa, Minas Gerais (2012)

Meilgaard, M., Civille, G.V., Carr, B.T.: Sensory Evaluation Techniques, 2nd edn, 354p. CRC, FL (1991)

Meksem, K., Njiti, V.N., Iqbal, M.J., Kassem, M.M., Hyten, D.L., Yuang, J., Winters, T.A., Lightfoot, D.A.: Genomic regions that underlie soybean seed isoflavone content. J. Biomed. Biotechnol. **1**(1), 38–40 (2001)

Meneguce, B., Faria, R.T., Destro, D., Fonseca Júnior, N.S., Faria, A.P.: Interação genótipo x ano para tempo de cozimento e sua correlação com a massa e percentagem de embebição em soja tipo alimento. Ciências Agrárias. **26**(4), 463–476 (2005)

Miura, E.M.Y., Binitti, M.A.R., Camargo, D.S., Mizubito, I.Y., Ida, E.I.: Avaliação biológica de soja com baixas atividades de inibidores de tripsina e ausência do inibidor Kunitz. Arch. Latinoam. Nutr. **51**(2), 195–198 (2001)

Moraes, R.M.A., José, I.C., Ramos, F.G., Barros, E.G., Moreira, M.A.: Caracterização bioquímica de linhagens de soja com alto teor de proteína. Pesq. Agropec. Bras. **41**(5), 727–729 (2006a)

Moraes, R.M., et al.: Assisted selection by specific DNA markers for genetic elimination of the kunitz trypsin inhibitor and lectin in soybean seeds. Euphytica. **149**(1), 221–226 (2006b)
Moraes, C.S.,Pastore, G.M., Sato, H.H., Park, Y.K.: Isoflavonas de soja e suas atividades biológicas, 84p. Livraria Vilela, São Paulo (2009)
Morais, A.C.C., Silva, A.L.: Usos. In: Morais, A.C.C., Silva, A.L. (eds.) Soja, suas aplicações, pp. 125–134. Editora Médica e Científica Ltda., Rio de Janeiro (1996)
Pauletto, F.B., Fogaça, A.O.: Avaliação da composição centesimal de tofu e okara. Disc. Scientia. **13**(1), 85–95 (2012)
Primoto, V.S., Poysa, V., Gary, R.A., Jackson, C.-J., Gijzen, M., Rajcan, I.: Mapping QTL for individual and total isoflavone content in soybean seeds. Crop Sci. **45**, 2454–2464 (2005)
Rackis, J.J.: Biological and physiological factors in soybeans. J. Am. Oil Chem. Soc. **51**(1), 161A–174A (1974)
Ray, C.L., Shipe, E.R., Bridges, W.C.: Planting date influence on soybean agronomic traits and seed composition in modified fatty acid breeding lines. Crop Sci. **48**, 181–188 (2008)
Ribeiro, V.S., Vilas Boas, B.M., Maciel, M.R.P., Colpa, P.C., Figueiredo, G.C.: Composição centesimal de grãos de genótipos de soja de tegumento amarelo, marrom e preto. In: 5ª Jornada Científica e Tecnológica e 2° Simpósio De Pós-Graduação do Ifsuldeminas. Inconfidentes, Ifsuldeminas, Anais, Inconfidentes (2013)
Rocha, M.M., Vello, N.A., Maia, M.C.C., Lopes, A.C.A.: Magnitude da interação genótipos x ambientes para o caráter teor de óleo em linhagens de soja. Rev. Bras. Ol. Fibros. **6**(3), 617–625 (2002)
SEAPA. Secretaria de Estado de Agricultura, Pecuária e Abastecimento de Minas Gerais. Projeções do Agronegócio—Minas Gerais 2015 a 2025. Belo Horizonte—MG, março 2016
Sgarbieri, V.C.: Proteínas em alimentos protéicos: propriedades, degradações, modificações, 517p. Livraria Varela, São Paulo (1996)
Simões, C.M.O., Schenkel, E.P., Gosmann, G., Mello, J.C.P., Mentz, L.A., Petrovick, P.R.: Farmacognosia: da planta ao medicamento, 5th edn, 1102p. UFRGS/UFSC, Porto Alegre/Florianópolis (2004)
Subramanian, S., Stacey, G., Yu, O.E.: Endogenous isoflavones are essential for the establishment of symbiosis between soybean and *Bradyrhizobium japonicum*. Plant J. **48**(2), 261–273 (2006)
Teixeira, A.I.: Desenvolvimento de método enzimático para determinação de sacarose e correlações entre o conteúdo de açúcares solúveis e proteína em soja, 79f. Tese de D.S.—Universidade Federal de Viçosa (2008)
Tsutsumi, C.Y.: Caracterização agronômica de cruzamentos de soja tipo alimento com tipo grãos, 159f. Tese de D.S.—Escola Superior de Agricultura "Luiz de Queiroz", Universidade de São Paulo (2000)
Utumi, M.M., Barros, E.G., Oliveira, M.G.A., Sediyama, C.S., Moreira, M.A.: Efeito da eliminação genética de lipoxigenases e de polipeptídeos de reserva na qualidade da proteína de soja. Rev. Bras. Fisiol. Veg. **10**(3), 203–212 (1998)
Vello, N.A.: Ampliação da base genética do germoplasma e melhoramento da soja na ESALQ-USP. In: Simpósio Sobre a Cultura e Produtividade Da Soja: Anais, . . ., pp. 60–81. FEALQ, Piracicaba (1992)
Vieira, C.R., Cabral, L.C., Paula, A.C.O.: Composição centesimal e conteúdo de aminoácidos, ácidos graxos e minerais de seis cultivares de soja destinadas à alimentação humana. Pesq. Agropec. Brasil. **34**(7), 1277–1283 (1999)
Wolf, W.J., Cowan, J.C.: Soybean as a Food Source, p. 101p. CRC, Cleveland (1975)
Zhang, R.F., Zhang, F.X., Zhang, M.W., Wei, Z.C., Yang, C.Y., Zhang, Y., Tang, X.J., Deng, Y. Y., Chi, J.W.: Phenolic composition and antioxidant activity in seed coats of 60 Chinese black soybean (*Glycine max* L. Merr.) varieties. J. Agric. Food Chem. **59**, 5935–5944 (2011)
Zhang, J., Ge, Y., Han, F., Li, B., Yan, S., Sun, J., Wang, L.: Isoflavone content of soybean cultivars from maturity group 0 to VI grown in northern and southern China. J. Oil Fat Indus. **91**(6), 1019–1024 (2014)

Chapter 21
Agronomical Aspects of the Development of Cultivars

André Ricardo Gomes Bezerra, Tuneo Sediyama, Felipe Lopes da Silva, Aluízio Borém, Amilton Ferreira da Silva, and Francisco Charles dos Santos Silva

Abstract Soybean breeding has been carried out with the objective of increasing the frequency of favorable genes of the traits object of the program. Undoubtedly, the increase in grain yield has been the main objective of breeding programs. Other traits aiming to improve crop management or the use of the grain for industrial purposes are also targeted by breeders. Early cultivars, erect plant architecture, incorporation of biotic and abiotic stress resistance via transgenic and physical, and chemical seed quality are some examples of traits of interest in many breeding programs. The soybean develop grown in a wide variety of environments, so its ideotype should be considered dynamic and varied according to each environment and cultivation condition. Under such circumstances, breeders should address different cropping regions, cropping systems, climatic adversities, biotic stresses, and the technological level of farmers, for example.

Keywords Grain yield • Ideotype • Agronomical traits

A.R.G. Bezerra, M.S. (✉)
Fundação MS Para Pesquisa e Difusão de Tecnologias Agropecuárias, Maracaju, Brazil
e-mail: andrebezerra@fundacaoms.org.br

T. Sediyama, M.S., Ph.D. • F.L. da Silva, M.S., D.S. • A. Borém, M.S., Ph.D.
F.C. dos Santos Silva, M.S.
Federal University of Viçosa, Viçosa, Minas Gerais 36570-900, Brazil
e-mail: tuneo@ufv.br; felipe.silva@ufv.br; borem@ufv.br; fcasantossilva-ma@hotmail.com

A.F. da Silva, M.S., D.S.
Universidade Federal de São João Del-Rei (UFSJ), Campus de Sete Lagoas,
Sete Lagoas, Brazil
e-mail: amiltonferreira@ufsj.edu.br

F. Lopes da Silva et al. (eds.), *Soybean Breeding*,
DOI 10.1007/978-3-319-57433-2_21

Introduction

The goal of soybean breeding programs is to search ceaselessly for superior cultivars. Grain yield is, without a doubt, the main objective of breeding programs, followed by the search for some other traits, like earliness, resistance to biotic and abiotic stress, and the quality of seeds for industrial purposes.

The work of the breeder is based on selecting superior progenies in order to increase the frequency of favorable genes for any given trait. In other cases, the aim is to introduce new traits into the material to be improved. In this way, the breeder tries to develop a hypothetical model for each trait of interest, and this model is also known as an ideotype.

Since soybean is a plant that is sensitive to intense light, photoperiods, humidity, temperature, altitude, sowing time, and the fertility of the soil, its ideotype can be thought of as dynamic. Therefore, the ideotype can vary from one producing region to another, depending on the cultivation systems, for instance, in the case of cultivating a second crop.

Grain Yield

Grain yield is indisputably the main objective of breeding programs. In Brazil, the average productivity of soybean has been around 3000 kg/ha, with annual highs and lows caused mainly by variations in the distribution of rain during soybean growth. However, with suitably run crops and under favorable climatic conditions, bean productivity is more than 4500 kg/ha.

As this is a quantitative trait, in other words controlled by several genes, it is greatly affected by influence from the environment (Fehr 1983). The source of progenitors that is most used for increasing grain yield are the superior lines and the commercial cultivars, as they bring together a high frequency of favorable alleles that have been selected over years of breeding.

Due the genetic control of this trait, recurrent selection has been shown as promising when increasing genetic gains from the breeding programs (Guimaraes 1985; Guimaraes and Fher 1989; Rose et al. 1992).

The advantage of using recurrent selection is the fact that each selection cycle is completed in 2 years, and the improved lines will be ready to be recombined as soon as they have been identified. In conventional breeding programs, which take several years to identify and test yields from the elite lines, these lines can take from 8 to 10 years before being used as parent plants for the next selection cycle (Rose et al. 1992).

Cycle

The cycle of a certain cultivar corresponds to the number of consecutive days it takes between the seedling emerging and the pods' maturity. The average duration of the cycle can vary between 80 and over 200 days, depending on the place and the sowing time (Müller 1981; Sediyama et al. 2009). The cultivars with a long cycle grow taller and can show more lodging, and they are not harvested efficiently using automated picking machines. Most of the cultivars adapted to conditions in Brazil have a cycle of about 90–150 days. Recently, with the aim of having two crops during the rainy period in the Central Region of Brazil, preference has been given to developing cultivars with a cycle of less than 120 days, which enables good performance when sowing in the months of September and October and harvested in January and February allowing maize seeding.

Relative Maturity Group

The effect of the response to the photoperiod in the adaptation area is more accentuated in soybean than in any other crop. As soybean is classified as a short-day plant, sensitivity to the photoperiod is an aggravating factor for increasing its adaptation area (Carpentieri-Pípolo et al. 2000). For this reason, the number of days it takes to ripen is not a reliable indicator for characterizing the cycle of the plant. Besides the cycle, one needs to take into consideration the relative maturity group. For instance, two cultivars from the same cycle may belong to different relative maturity groups.

In all, 13 maturity groups are used as a method for differentiating between and identifying the cultivars that have adapted to different latitudes. These groups, beginning with the earliest cultivars, are called 000, 00, 0, I, Ii, Iii, Iv, V, Vi, Vii, Viii, Ix, and X, using the American classification. According to Sediyama et al. (2009), most of the cultivars belonging to each ripening group have generally been adapted to a latitude band of approximately 160–240 km in a north-south direction. In latitudes higher than the one being considered, the cultivars flower later than normal. However, when the latitudes are lower than the region being considered, the cultivars will ripen earlier than normal (Hartwig 1973).

According to Alliprandini et al. (2009), for the Brazilian cultivars adapted to the Southern Region (Table 21.1), the maturity groups 5.0–8.0 predominate, and their cycle can vary between 106 and 137 days. While for those adapted to the Central-West, Northeast, and Northern Regions of Brazil, the maturity groups 7.0–10.0 predominate.

Juvenile Period

The time it takes from the seedling emerging to when it is ready to receive the stimulus for inducing flowering is called the juvenile period. The short-period juvenile cultivars are those that are ready for the stimulus for flowering as soon

Table 21.1 Number of days they take to flower, period of reproductive growth, number of days to maturity, and relative maturity group of soybean cultivars adapted to the Southern Region of Brazil

Cultivar	Flowering (days)	Reproductive period (days)	Number of days to maturity	Relative Maturity Group
FT-Cometa	42	64	106	5.0
MSoy 5942	44	68	112	5.6
Fundacep 41	49	67	116	6.0
RB 501	54	65	119	6.3
CD 202	52	69	121	6.5
BRS 184	53	70	123	6.7
Carrera	57	69	126	7.0
BRS 154	53	75	128	7.2
BRS 133	60	70	130	7.4
CD 209	54	77	131	7.5
BRS 134	56	76	132	7.6
CD 205	58	78	136	8.0
MSoy 8001	58	79	137	8.1

Source: Adapted from Alliprandini et al. (2009)

as the unifoliate leaves and/or the first trifoliate appear. Short-period juvenile plants grow tall and reach high grain productivity when sown at times when the day is longer than its critical photoperiod. In this case, the sowing time is limited and cultivation during the winter is not recommended.

The long-period juvenile cultivars are ready to react to the stimulus for flowering much later, around the time of the fifth to seventh leaf or after the trifoliate leaf, even under short-day conditions. These cultivars generally become suitably tall right in mid-sowing time and can even be cultivated in the winter period, if the dampness of the soil and the temperature are not limiting. Most of the cultivars adapted to the Central-North and Northeast of Brazil show a long juvenile period.

In the literature, it is possible to find the descriptions of ten genes that influence flowering and maturity in soybean, which have been described under long-day conditions. These genes are called E1 up to E9 and *J*. Except the gene *J*, the name is formed by using the first letter of "earliness" followed by the number that shows the order in which they were discovered. Bonato and Vello (1999a) ascertained that E6 is the allele that defines earliness for flowering and maturity, which is present in the Paraná cultivar, and e6 determines late flowering and ripening in Paranagoiana and SS-1. In another work, Bonato and Vello (1999b) observed that the main genetic variation component in determining the time for flowering is additive and that the alleles that condition earlier flowering exhibit partial dominance.

Growth Types

There are very different types of growth among soybean cultivars (Sediyama et al. 2015). For example, cultivars of the determinate type show terminal racemose inflorescence and virtually stop growing in height after the initial appearance of the first flower, which generally grows on the upper third of the plant or in one of the two last nodes on the main stem. The upper leaves are similar to the others and the pods ripen from the top downward.

The cultivars of the semi-determinate growth type show terminal racemose inflorescence. However, after flowering has started, significant upward growth occurs, and the last leaves, similar to the others or smaller, can even change their shape, for instance, from oval to lanceolate. Ripening of the pods generally starts happening in the upper third, progressing upward and then down to the lower part of the plant.

Lastly, plants of the indeterminate growth type do not show terminal racemose inflorescence. After flowering has started, the number of nodes on the main stem of the plant continues increasing, and they can double in height. The upper leaves are much smaller and differ from the lower ones. The pods generally start to ripen in the lower part of the plant, progressing toward the upper end. It is common to see pods in granulation in the lower part of the plant and flowers in the upper part.

The inheritance of growth type is monogenic, with total dominance of the indeterminate over the determinate type (Vernetti and Vernetti Junior 2009). The allele, Dt1, is responsible for the indeterminate type, while the double recessive, dt1 dt1, makes the plant the determinate type. Semi-determinate growth is manifested by the dominant allele, Dt2, and the double recessive dt2 dt2 makes the plant an indeterminate type. It is also known that dt1 is partially recessive to Dt1 and is epistatic over Dt2, dt2. Therefore, for interaction among these alleles, we have the phenotypes set out on Table 21.2.

Table 21.2 Relationship between genotypes and phenotypes for growth type in soybean

Genotype	Growth type
Dt1 Dt1 dt2 dt2	Indeterminate
Dt1 Dt1 Dt2 Dt2	Semi-determinate
Dt1 Dt1 Dt2 dt2	Semi-determinate
Dt1 dt1 dt2 dt2	Semi-determinate
Dt1 dt1 Dt2 dt2	Semi-determinate
Dt1 dt1 Dt2 Dt2	Semi-determinate
dt1 dt1 Dt2 Dt2	Determinate
dt1 dt1 Dt2 dt2	Determinate
dt1 dt1 dt2 dt2	Determinate

Adapted from Vernetti and Vernetti Junior (2009)

Growth Habit

Soybean cultivars can be simple stemmed, called mono stem, or branched. In the case of branched ones, the tilt of the lateral branches indicates the growth habit. When the tilt is less than 30° with relation to the main stem, it is called semierect growth habit; when it is between 30° and 60°, it is called semierect to horizontal; and when it is greater than 60°, it is called horizontal (Sediyama et al. 2015).

According to Vernetti and Vernetti Junior (2009), two independent dominant alleles are needed to produce elevated branching, which are called Br1 Br2 (T327). The opposite condition is determined by br1 br2. Thus, a plant with Br1_ and Br2_alleles shows branching along the stalk. The double recessives of the br1 and br2 alleles bestow branching only on the lower nodes.

The breeding programs have given emphasis to developing erect and semierect growth habit cultivars, because light and pesticides penetrate the canopy better and make them more efficient. What is more, this reduces loss of beans during mechanized picking.

Height of Plant and Insertion of First Pod

Soybean plants can reach heights that vary between 20 and over 200 cm, depending on the cultivar, place of cultivation, sowing time, spacing between lines and density of plants in the line, availability of water, fertility of the soil, symbiotic activity and fixation, and other environmental conditions.

Very tall cultivars or extremely short ones cannot be picked as easily as those of medium height and, in general, are not as productive as those of medium height. When dealing with mechanized picking on relatively level ground, one can get a good crop from plants varying from 50 to 60 cm tall. However, on ground with uneven topography, the most ideal height is around 70–80 cm for picking to be efficient. Very tall plants of 90 cm or more do not show good resistance to lodging (Sediyama et al. 2009).

With regard to the height of insertion of the first pod, this can vary from near ground level to more than 30 cm up. However, the most productive cultivars and the ones that are picked most efficiently are those whose insertion varies between 10 and 15 cm.

Resistance to Lodging

The lodging of plants is an agronomical trait inherent to mechanized picking because, very often, this lodging can be an indication of having adapted badly to the place of cultivation.

Very tall plants with very thin stalks tend to lodge relatively easily, which may result in loss of beans during the picking operation. While plants with extremely thick stalks, although they are often productive, may damage the performance of the picking machines.

Some factors like plant population, soil fertility, availability of water, quality of light, and diameter of the stem can increase or decrease lodging. Generally speaking, soybean plants are more affected by lodging in heavy fertile soils, which have abundant humidity, than in light sandy soils. In sandy soils with low fertility and low humidity, lodging might occur with less intensity; therefore, indeterminate growth cultivars or later ones are preferred. In clay soil with high fertility and high humidity, the determinate growth ones or shorter ones show lower tendency toward lodging.

Resistance of plants to lodging is a trait that is inherent to cultivars and also to other ones that relate to the height of plants (Sediyama et al. 2009). Wilcox and Sediyama (1981) reported that plants of the indeterminate growth type show a variation of 0.3 points in lodging (on a scale of 1–5, in which 1 = erect plant and 5 = completely lodged plant) with an increase of 10 cm in the height of the plant. Therefore, reducing the plant height is recognized as being an effective measure for lowering or eliminating the lodging of plants (Mancuso and Caviness 1991).

Resistance to Pod Dehiscence

Pod dehiscence after the plants have ripened is an undesirable trait in commercial cultivars. This happens because in some situations, it is impossible to harvest the soybean immediately, straight after it has ripened. So, the cultivar that is to be indicated for growth must show good resistance to pod dehiscence of at least 2–3 weeks after ripening.

There is significant genetic variability in the varieties of soybean regarding resistance to pod dehiscence (Miyasaka 1958). What is more, the fertility of the soil with low K content and low relative humidity associated with high temperatures also induces early dehiscence of pods.

Bailey et al. (1997) reported that a single RFLP marker, located in linkage group J of the USD/Iowa State University genetic map, is involved in 44% of the variation in pod dehiscence. Epistasis was also noted between the main QTL and another smaller one, which are involved in controlling the trait. Recently, the *Pdh1* gene has been discovered that codifies a protein involved in the lignification of pod abscission, which increases the twisting force and causes dry pod dehiscence. Removing the *Pdh1* gene has been widely used as a source of resistance to pod dehiscence in soybean breeding (Funatsukia et al. 2014).

Resistance to Pests

The mechanisms involved in soybean cultivars resistance to insect pests are antixenosis –characterized by the insect using the host less for food, oviposition, and/or shelter; antibiosis—the deleterious effects on survival, development, and reproduction of the insect; and tolerance, the ability of the plant to keep producing when under attack from pests. Substances that are toxic for the insects, like protease inhibitors that act against defoliating insects and flavonoids act in the antibiosis mechanism, (Picanço et al. 2009).

In Table 21.3, there are some soybean genotypes that possess sources of resistance to some important crop pests.

There is also transgenic soybean that expresses the Cry1Ac gene that comes from the *Bacillus thuringiensis* bacteria, and this is a trait that bestows resistance to lepidoptera insects. This technology has been available on the market since 2010.

Resistance to Disease

Approximately 40 diseases caused by fungi, bacteria, nematodes, and viruses have been identified in Brazil. This number continues to rise as soybean is spread into new areas and as a consequence of monoculture. The economic importance of each disease varies from year to year and from region to region, depending on the climatic conditions of each harvest. Annual losses of production caused by diseases are estimated at around 15–20%. However, some diseases can bring about losses of nearly 100% (Tecnologias 2013).

Among the diseases frequently encountered are leaf blight and purple seed stain (*Cercospora kikuchii*), Asian soybean rust (*Phakopsora pachyrhizi*), anthracnose

Table 21.3 Sources of resistance of soybean genotype to insect-pests

	Insect pest			
Genotype	Bugs	Defoliator caterpillars	Whitefly[a]	Reference
IAC-100	Resistant	Resistant	Resistant Ax, Ab	Veiga et al. (1999) and Lima and Lara (2004)
BR-82 12,547	–	–	Resistant[Ax]	Lima et al. (2002) and Lima and Lara (2004)
IAC 78–2318	Resistant	Resistant	Resistant	Lourenção and Miranda (1987)
PI 229358	–	–	Resistant[Ax]	Lima and Lara (2004)
IAC 74–2832	–	–	Resistant	Lima et al. (2002)
MG-BR 46	–	Resistant[Ab]	–	Veloso (2010)

[a]Antixenosis resistance mechanisms (Ax) and/or antibiosis (Ab)

(*Colletotrichum truncatum*), stem cancer (*Diaporthe phaseolorum var. meridionalis* and *Diaporthe phaseolorum var. caulivora*), soybean pod and stem blight (*Phomopsis* spp.), pod blight (*Fusarium* spp.), brown spot (*Septoria glycines*), frog-eye leaf spot (*Cercospora sojina*), mildew (*Peronospora manshurica*), target spot (*Corynespora cassiicola*), soybean rot (*Rhizoctonia solani*), white stem rot (*Sclerotinia sclerotiorum*), sclerotic wilt (*Sclerotium rolfsii*), powdery mildew (*Erysiphe diffusa*), charcoal rot (*Macrophomina phaseolina*), red root rot (sudden death syndrome, PVR/SDS), bacterial blight (*Pseudomonas savastanoi pv. glycinea*), common soybean mosaic—VMCS (*Soybean mosaic virus*), stem necrosis—CPMMV (*cowpea mild mottle virus*), root-knot nematode (*Meloidogyne javanica, Meloidogyne incognita*), cyst nematode (*Heterodera glycines*), root-lesion nematodes (*Pratylenchus brachyurus*), and reniform nematodes (*Rotylenchulus reniformis*) (Sediyama et al. 2015).

Genetic resistance is the most economical way and the best way that is accepted by farmers, for controlling diseases. However, for most of them, there are no resistant cultivars (e.g., white stem rot, damping off, and rhizotonia root rot) or, in some cases, the number of resistant cultivars is limited (e.g., root-knot nematode and cyst nematodes) (Tecnologias 2013).

Resistance to Herbicides

Soybean cultivars in Brazil can show resistance to three herbicides: glyphosate, those from the imidazolines group, and glufosinate-ammonium. These resistances were obtained through transgenic techniques and were approved by the *CTNbio* (National Technical Commission for Biosafety).

RR (Roundup Ready) soybean, which is tolerant to the glyphosate herbicide, was developed by introducing the gene from the 5-enolpyruvylshikimate 3-phosphate (EPSP) synthase enzyme that comes from the CP4 strain of soil bacteria, *Agrobacterium tumefaciens*. Recently, the *CTNbio* approved the INTACTA RR2 PRO™ soybean. This technology brings together, all at one, protection from the main caterpillars that attack the crop, tolerance to glyphosate and advanced mapping technologies, and selection and insertion of genes into regions of the DNA that have the potential to positively impact on the productivity of beans (Sediyama et al. 2015).

Soybean, which is tolerant to herbicides from the imidazolines group, was produced by introducing the csr1–2 gene from *Arabidopsis thaliana*. This gene codifies a protein that bestows tolerance to the herbicide because of the mutation point, which results in amino acid substitution in which the serine residue in position 653 is replaced by asparagine (S653 N). LibertyLink (LL) is the name given to the soybean that is tolerant to the glufosinate of ammonium herbicide produced by introducing the PAT modified gene, which codifies the phosphinothricin N-acetyltransferase enzyme that comes from the soil bacteria, *Streptomyces viridochromogenes*.

Farmers can cultivate soybean with one more source of resistance to herbicides—the Enlist E3® technology. Recently approved, it is a genetically modified soybean that is tolerant to herbicides 2,4-D and glufosinate of ammonia. This soybean carries the aad-12 gene that codifies the protein, aryloxyalcanoate dioxygenase (AAD-12), to which it bestows tolerance to the herbicide 2,4-D (acid 2,4-Dichlorophenoxy acetic), and the PAT gene that codifies the PAT protein, to which it bestows tolerance to the herbicide glufosinate of ammonia. Furthermore, there is also an expectation of the approval of two more transgenic events that bestow resistance to herbicide; among them is Dicamba, requested by Monsanto, and Hppd, which is responsible for resistance to inhibitor herbicides from the 4-hydroxyphenylpyruvate dioxygenase enzyme.

Efficient Use of Nutrients

The Green Revolution has enabled a huge leap forward in the production of beans, mainly because of the intensified use of chemical fertilizers for overcoming the deficiencies in cultivated soils. However, availability of these fertilizers in the not-so-distant future might be compromised. In the case of nitrogen and phosphorus, for instance, in Brazil, production of these two nutrients fell from 68% of the country's total needs in 1983 to 35% in 2006, and it is likely to reach only 14% of the needs in 2025 (Lopes et al. 2000).

Scarcity of these two nutrients, therefore, is likely to be a limiting factor for increasing the productivity of Brazilian crops. That is why it will be necessary to look for new reserves, improve efficiency when applying these nutrients, reduce loss through erosion and obtain plants that use them more efficiently, or even ones that are more tolerant to the fact that availability in the soil is low (Borem and Ramalho 2011).

In the case of soybean, the supply of nitrogen is not worrying, because 72–94% of this nutrient can be obtain through biological fixation by bacteria from the bacteria of *Bradyrhizobium* kind, and the remaining 6–28% can be supplied by the soil. The case of phosphorus is more worrying, as its availability in tropical soils, as with most of the soils in Brazil, is low.

According to Borem and Ramalho (2011), most of the breeding work is aimed at improving efficient use by plants; in other words, obtaining genotypes that have great phenotypic plasticity. So, when plants are submitted to limiting cultivation conditions, they would use the scarce reserves to produce satisfactorily but when in ideal conditions, would be highly productive.

Genetic control, of both tolerance to stress and the efficient use of nutrients, is quantitative and involves several loci which are distributed throughout different regions of the genome in cultivated species (Wu et al. 2011). The quantification and understanding of the genetic relationship between these two strategies for improvement when selecting are fundamental for directing the breeding programs and increasing their effectiveness.

Seed

Physical Appearance

In soybean crops that are destined for oil extraction and bran, the seed integument must be of a yellow coloring. With relation to the thread, some countries, like the United States, prefer cultivars with a light thread, even though they are not sold at a higher price than the ones with colored threads. So far in Brazil, we have not been aware of any preference by the bran and oil industries for light-threaded soybean.

A strong element for marketing is what a batch of seeds looks like. The seed must look good and be good. Batches of seeds that have weeds, inert materials, and malformed opaque seeds mixed in with them are not well regarded by farmers (Peske et al. 2003). The size and shape of the seeds must also be taken into consideration. Seeds must preferably be rounded. Regarding the size, standardization of the batches has become a market demand. This is essential for adjusting the seed sowing machines and has also enabled adjusted and uniform stands to have emerged, and it improves the physiological potential of the seed and even contributes toward saving on seeds (Bezerra et al. 2016).

Physiological Quality and Vigor

The physiological quality of the seeds at maturity time is a very important factor, mainly because the temperatures in Brazil during the winter are not as low as in counties with a temperate climate, and so germination and vigor does not last for very long. One very efficient method for selecting the quality trait in a seed is to delay harvest until after the seeds have attained about 14% humidity, for the first time.

A delay of a 3–5-week period can discriminate the cultivars with regard to better quality seed. However, with the increase in early cultivars and harvest coinciding with rainy periods in hot weather, more specific criteria are needed for selecting the best lines.

During the maturation and harvest phases, if the humidity and air temperature is hot, the seeds tend to be of a better physiological quality. Relatively high temperatures and humidity with frequent rain produces low quality seeds that have fissures and cracks in them and a wrinkled integument. A very hot dry environment with low soil humidity, or maturation too early because of frost or applying desiccants, tends to form smaller greenish seeds (Sediyama et al. 1999).

Seeds that have been recently harvested under suitable conditions should show heightened vigor; in other words, they should be at their maximum level of activity and vigor (Carvalho and Nakagawa 2012). Consequently, the plants that are formed from highly vigorous seeds produce a greater number of seeds per plant and more seeds per plant in terms of weight, and this results in higher productivity (Cervieri Filho 2005).

Industrial Traits

Oil and Protein

In most soybean cultivars, the protein content varies between 30 and 53%, but the average content of Brazilian cultivars is 40%, while the percentage of oil in soybean seed is 20%, and it can vary from 13 to 28% (Bezerra et al. 2015).

As a rule, the productivity of soy beans is inversely correlated with the protein content. An inverse relationship is also observed between the protein content and oil content in the seed, and this has limited the development of cultivars for high levels of some of these traits (Wilcox 1998). As for the oil content and productivity of beans, depending on the genotypes, they can be high and positive or vary from positive to negative or be absent.

However, it is possible to obtain genotypes with high protein content and maintain the productivity averages of beans and the physiological potential of the seeds and their recurrent parents (Mello Filho et al. 2004).

In a way that is analogous to productivity, for protein content in seeds, it is possible to increase the genetic gains by the recurrent selection method. According to Wilcox's (1998) results, eight selection cycles resulted in an increase of 5.8 g/kg of protein in the seed and a significant decrease of 2.3 g/kg of oil per cycle. The strong inverse relationship between protein and oil in seeds was more pronounced in late plants than in early ones. An average of 53% of the plants in the sixth selection cycle showed between 8 and 480 g/kg, or rather, the highest protein content in the seeds.

Animal Feed

Products derived from soybean make up, on average, 20% of poultry and pig food, and it is very important for animal performance and production costs. The great advantage that comes from soybean products is the quality of its protein and oil. Excepted the sulfurated amino acids, soybean protein is rich in the main essential amino acids that are needed by poultry and pigs. What is more, the high energy value, due to the richness of its oil, makes soybean attractive for animal feed (Abreu et al. 2009).

However, together with the protein and oil, some undesirable components are also found in the beans. Anti-nutritional factors, like trypsin inhibitors, phytic acid, and raffinose oligosaccharides are present in soybean. Thus, it is important to determine the composition of soybean genotypes with regard to the biochemical components relating to the quality as well as to the anti-nutritional factors (Paula 2007).

According to Paula (2007), the genotype CS 02564 shows lower levels of raffinose, stachyose, and phytic acid and is considered to be promising for the breeding programs for soybean destined for human and animal consumption.

Human Consumption

All soybean that is cultivated for producing oil and bran can be destined for human consumption. However, there are some types that show better traits, both visual ones and in terms of bean quality, which are preferred for human consumption (Sediyama et al. 2009).

In breeding programs that aim to develop special cultivars for human consumption, the breeder evaluates the agronomic characteristics, as well as other ones relating to taste, smell, appearance, color, and size of bean (Juhász et al. 2014).

Soybean for food must have a pleasant taste when eaten directly in human food, and there are two classifications: small (100 seeds weigh less than 10 g), which are eaten in the form of bean shoots and *natto* (nuts) and large (100 seeds weigh more than 20 g), which are eaten as a vegetable (unripe grain) and in salads (ripe grain) Yokomizo et al. 2000).

The cultivars destined for *tofu* or soy cheese must have a light thread and be large and rounded, and the seed integument should come off easily when immersed in water, and they generally have a higher protein content.

Nowadays, there are various bred cultivars that do not have an undesirable taste. The UFVTN 101, UFVTN 102, UFVTN 103, BRS 213, and BRS 257 cultivars do not have any lipoxygenase enzymes which are responsible for the unpleasant taste in soybean products. These cultivars are excellent-quality raw material for the industrial processing of food (Carrão-Panizzi and Pípolo 2009; Sediyama et al. 2009; Silva et al. 2012).

Adaptability and Stability of Bean Production

Adaptability is the genotypes' ability to take advantage of stimuli from the environment; and stability is the genotypes' ability to manifest highly predictable behavior, even when there are variations in the environment (Cruz et al. 2012).

The genotype × environment interaction is one of the main problems in breeding programs of any species, whether at the selection phase or when recommending cultivars (Allard and Bradshaw 1964). The G × A interaction makes it difficult to select genotypes that are widely enough adapted. The effects of this occur because of the different responses from the same set of genes in different environments (Duarte and Vencovsky 1999).

Soybean cultivars show different grain yield mainly when planted in different regions. One can generally obtain maximum grain yield in environments that have been developed or selected. However, it is quite frequent to observe that some cultivars, although they do not show high grain yield in all environments, have such a satisfactory average performance that they surpass the others by demonstrating good stability.

Therefore, when indicating cultivars, preference should be given to those that show maximum grain production and, also, those that, based on the average of different environments, such as sowing time, soil fertility, distribution of rain, and others, show high grain production. (Sediyama et al. 2009).

Naming Cultivars

In order for a cultivar to be protected, it has to have a name that is different from any others that already exist for the same species or similar species, even in the international context (Article 15, subsection Ii, of Law n° 9.456/97; article 7°, §1°, item "b", of Decree n° 2366/97; and article 13, 1, of Decree n° 3.109/99) MAPA (2015). It is possible to check the existence of pre-existing names on the international cultivars database (http://www.upov.int/pluto/en/), which is maintained and updated by The International Union for the Protection of New Varieties of Plants (*UPOV*).

According to instruction from MAPA (2015), the name given to a new cultivar is very important in the process of protecting and marketing the species. In Brazil, clarification on the matter can be obtained from the National Service for the Protection of Cultivars (*SNPC*). The existence of two cultivars with the same name is prohibited. The legislation states that the cultivar must have its own name, does not add value, has its own quality, and is not be linked to any geographical spots. The cultivar cannot be called any name that is different from the one under which it has been protected.

Some examples of names used in Brazil are A (*Aventis*), BCR (*Bacuri Pesquisa and Melhoramento*), BR and BRS (*Empresa Brasileira de Pesquisa Agropecuária—EMBRAPA*), BRSGO (*EMBRAPA* Goiás), BRSMG (*EMBRAPA* Minas Gerais), BRSMA (*EMBRAPA* Maranhão), BRSMT (*EMBRAPA* Mato Grosso), BRMS (*EMBRAPA* Mato Grosso do Sul), CAC (*Cooperativa Agrícola de Cotia*), CD (*Coodetec*), CS (*Coopadap* Sementes), DM (*Dois Marcos*), EMGOPA (*Empresa Goiana de Pesquisa Agropecuária*), FMT (*Fundação Mato Grosso*), FT (*FT Pesquisa e Sementes*), GT (*Genética Tropical*), IAC (*Instituto Agronômico de Campinas*), IAS (*Instituto Agronômico do Sul*, ICA (*ICA Melhoramento Genético Ltda.*), MSOY (Monsanto), OC (*Organização das Coperativas do Estado do Paraná*—OCEPAR), P (*Pioneer*), TMG (T*ropical Melhoramento e Genética*), BMX (*Brasmax*), RSF (*Rendimento Sem Fronteira*), UFU and UFUS (*Universidade Federal de Uberlândia*), Ufv, Ufvs, Ufvtn, and UFVTNK (*Federal University of Viçosa*).

Nowadays, one can also see a tendency, after abbreviations, to add numbers that represent the maturity group and the order in which they were commercially launched. They might also contain initials, like RR (Roundup Ready), IPRO (*Intacta Produtiva*), LL (LibertyLink), which indicate that transgenic events are present in the cultivar.

Final Considerations

Success in developing new cultivars goes beyond a basic knowledge of genetics and plant breeding. The breeder has to plan how to conduct the program in a way that meets the needs of different seeding regions, cultivation systems, climatic problems, and technological level of the farmers, for instance. What is more, one needs to make an exception for the time elapsed between the beginning of the program, which is when the hybrids are made, and the launch of the new cultivars. This is because, very often, a new cultivar might not meet the demand of the farmers, which sometimes changes depending on the market, available technology, phytosanitary problems, etc.

References

Abreu, M.L.T., Juarez, L.D., Oliveira, R.F.M.: Soja e seus derivados na alimentação de aves e suínos. In: Sediyama, T. (ed.) Tecnologias de produção e usos da soja, pp. 273–290. Ed. Mecenas, Londrina (2009)

Allard, R.W., Bradshaw, A.D.: Implications of genotype—environmental interactions in applied plant breeding. Crop Sci. **4**(5), 503–508 (1964)

Alliprandini, L.F., Abatti, C., Bertagnolli, P.F., Cavassim, J.E., Gabe, H.L., Kurek, A., Matsumoto, M.N., Oliveira, M.A.R., Pitol, C., Prado, L.C., Stecking, G.: Understanding soybean maturity groups in Brazil. Enviroment, cultivar classification and stability. Crop Sci. **48**, 801–808 (2009)

Bailey, M.A., Mian, M.A.R., Carter Jr., T.E., Ashley, D.A., Boerma, H.R.: Pod dehiscence of soybean: Identification of quantitative trait loci. J. Hered. **88**(2), 152–154 (1997)

Bezerra, A.R.G., Soares, M.M., Sediyama, T.: Tamanho e vigor da semente. In: Sediyama, T. (ed.) Produtividade da soja, pp. 142–155. Editora Mecenas, Londrina (2016)

Bezerra, A.R.G., Sediyama, T., Borem, A., Soares, M.M.: Importância econômica. In: Sediyama, T., Silva, F.L., Borem, A. (eds.) Soja: Do Plantio à Colheita, pp. 9–26. Ed. Ufv, Viçosa (2015)

Bonato, E.R., Vello, N.A.: E6, a dominant gene conditioning early flowering and maturity in soybeans. Genet. Mol. Biol. **22**(2), 229–232 (1999a)

Bonato, E.R., Vello, N.A.: Aspectos genéticos do tempo para o florescimento em variantes naturais de soja. Pesq. Agropec. Bras. **34**(6), 989–993 (1999b)

Borem, A., Ramalho, M.A.P.: Estresses abióticos: desafios do melhoramento de plantas nas próximas décadas. In: Fritsche-Neto, R., Borem, A. (eds.) Melhoramento de plantas para condições de estresses abióticos, pp. 9–28. Ed. Ufv, Viçosa (2011)

Carpentieri-pípolo, V., Almeida, L.A., KIIHL, R.A.S., Rosolem, C.A.: Inheritance of long juvenile period under short day conditions for the BR80–6778 soybean (*Glycine max* (L.) Merrill) line. Euphytica. **112**(2), 203–209 (2000)

Carvalho, N.M., Nakagawa, J.: Sementes: ciência, tecnologia e produção, 5ª edn, 590p. Ed. Funep, Jaboticabal (2012)

Carrão-Panizzi, M.C., Pípolo, A.E.: Cultivares de soja especiais para alimentação humana. Londrina: Embrapa Soja, 4p. (Folder n° 03/2009), (2009)

Cervieri Filho, E.: Desempenho de plantas oriundas de sementes de alto e baixo vigor dentro de uma população de soja. 42f. Tese de D.S.—Universidade Federal de Pelotas (2005)

Cruz, C.D., Regazzi, A.J., Carneiro, P.C.S.: Modelos biométricos aplicados ao melhoramento genético, vol. 1, 514p. Ed. Ufv, Viçosa (2012)

Duarte, J.B., Vencovsky, R.: Interação genótipos x ambientes: uma introdução à análise AMMI, 60p (Série Monografias, 9). Sociedade Brasileira de Genética, Ribeirão Preto (1999)

Fehr, W.R. Mejoramento. In: Norman, A.G. (ed.) Fisiologia, mejoramento, cultivo y utilizacion dela soja. Rio de Janeiro: Editorial Hemisferio Sur, pp. 119–155. (Traducción: Resnik, F.C.Z.) (1983)

Funatsukia, H., Suzukic, M., Hirosea, A., Inabac, H., Yamadad, T., Hajikad, M., Komatsua, K., Katayamae, T., Sayama, T., Ishimotoa, M., Fujino, K.: Molecular basis of a shattering resistance boosting global dissemination of soybean. PNAS. **111**(50), 17797–17802 (2014)

Guimaraes, E.P., Fher, W.R.: Alternative strategies of recurrent selection for seed yield of soybean. Euphytica. **40**(1), 11–120 (1989)

Guimaraes, E.P.: Genetic improvement of soybean from populations developed by alternative recurrent selection strategies. 107f. Theses Ph.D., Iowa State University (1985)

Hartwig, E.E.: Varietal development. In: Caldwell, B.E. (ed.) Soybeans: Improvement, Producion, and Uses, pp. 187–210, Madson, American Society of Agronomy (1973)

Juhász, A.C.P., Ciabotti, S., Pádua, G.P., Favoreto, L., Jesus, A.M.S., Franza, V.: Melhoramento de soja para alimentação humana. In: Inovações, tecnologias e sociedade: 40 anos EPAMIG. EPAMIG (Informe Agropecuário, edição especial), vol. 35, pp. 39–45 (2014)

Lima, A.C.S., Lara, F.M.: Resistência de genótipos de soja à mosca branca *Bemisia tabaci* (Genn.) biótipo B (*Hemiptera: Aleyrodidae*). Neutrop. Entomol. **33**, 71–75 (2004)

Lima, A.C.S., Lara, F.M., Barbosa, J.C.: Oviposition preference of *Bemisia tabaci* (Genn.) B biotype (*Hemiptera: Aleyrodidae*) on soybean genotypes. Neutrop. Entomol. **31**, 297–303 (2002)

Lourenção, A.L., Miranda, M.A.C.: Resistência de soja a insetos: VIII. IAC 78-2318, linhagem com resistência múltipla. Bragantia. **46**, 65–72 (1987)

Lopes, A.S., Daher, E., Bastos, A.R.R., Guilherme, L.R.G.: Suprimento e extensão das reservas de nutrientes no Brasil. In: Prochnow, L.L., Casarin, V., Stipp, S.R. (Orgs.) Boas práticas para uso eficiente de fertilizantes, vol. 2, 1st edn, pp. 238–307. Ipni, Piracicaba (2000)

Mancuso, N., Caviness, C.E.: Association of selected plant traits with lodging of four determinate soybean cultivars. Crop Sci. **31**, 911–914 (1991)

Mello Filho, O.L., Sediyama, T., Moreira, M.A., Reis, M.S., Masoni, G.A., Piovesan, N.D.: Grain yield and seed quality of soybean selected for high protein contente. Pesq. Agropec. Brasil. **39** (5), 455–450 (2004)

Miyasaka, S.: Contribuição para o melhoramento da soja no estado de São Paulo. 47f. Tese D.S.—Escola Superior de Agricultura "Luiz de Queiroz", Universidade de São Paulo (1958)

Ministétio Da Agricultura Pecuária e Abastecimento—MAPA. Denominação de cultivares. Disponível em: http://www.agricultura.gov.br/vegetal/registros-autorizacoes/protecao-cultivares/denominacao-cultivares (2015). Acesso em 10 Nov 2015

Müller, L.: Taxonomia e morfologia. In: Miyasaka, S., Medina, J.C. (eds.) A Soja No Brasil, 1st edn, pp. 65–104. Instituto de Tecnologia de Alimentos, Campinas (1981)

Paula, S.A.: Composição bioquímica e fatores antinutricionais de genótipos de soja. 74f. Dissertação M.S., Universidade Federal de Viçosa (2007)

Peske, S.T., Rosenthal, M.D., Rota, G.R.M.: Sementes: fundamentos científicos e tecnológicos, 1ª edn, 415p. Ed. Ufpel, Pelotas (2003)

Picanço, M.C., Fernandes, F.L., Morais, E.G.F., Campos, M.R., Xavier, V.M.: Manejo integrado das pragas. In: Sediyama, T. (ed.) Tecnologias de produção e usos da soja, pp. 119–132. Ed. Mecenas, Londrina (2009)

Rose, J.L., Butler, D.G., Ryley, M.J.: Yield improvement in soybeans using recurrent selection. Aust. J. Agric. Res. **43**, 135–144 (1992)
Sediyama, T., Matsuo, E., Oliveira, R.C.T., Glasenapp, J.S.: Características agronômicas de cultivares. In: Sediyama, T. (ed.) Melhoramento genético da soja, pp. 73–82. Ed. Mecenas, Londrina (2015)
Sediyama, T., Teixeira, R.C., Barros, H.B.: Cultivares. In: Sediyama, T. (ed.) Tecnologias de produção e usos da soja, pp. 77–99. Ed. Mecenas, Londrina (2009)
Sediyama, T., Teixeira, R.C., Reis, M.S.: Melhoramento da soja. In: Borem, A. (Org.) Melhoramento de espécies cultivadas, vol. 1, 1st edn, pp. 487–533. Ed. Ufv, Viçosa (1999)
Silva, J., Prudencio, S., Carrão-Panizzi, M., Gregorut, C., Fonseca, F., Mattoso, L.: Study on the flavour of soybean cultivars by sensory analysis and electronic tongue. Int. J. Food Sci. Technol. **47**, 1630–1638 (2012)
Tecnologias De Produção De Soja—Região Central Do Brasil 2012 e 2013: Embrapa Soja: Embrapa Cerrados: Embrapa Agropecuária Oeste, 262p. Embrapa Soja, Londrina (2011)
Veiga, R.F.A., Rosseto, C.J., Razera, L.F., Gallo, P.B., Bortoleto, N., Medina, P.F., Tisselli Filho, O., Cione, J.: Caracterização morfológica e agronômica do cultivar de soja IAC-100, 23p. Iac, Campinas (1999)
Veloso, E.S.: Resistência de cultivares de soja a *Spodoptera frugiperda* (J. E. Smith) (*Lepidoptera: Noctuidae*). 56f. Dissertação M.S., Faculdade de Engenharia de Ilha Solteira, Universidade Estadual Paulista, Ilha Solteira (2010)
Vernetti, F.J., Vernetti Junior, F.J.: Genética da soja: caracteres qualitativos e diversidade genética. Embrapa Informação Tecnológica, Brasilia (2009)
Yokomizo, G.K., Duarte, J.B., Vello, N.A.: Correlações fenotípicas entre tamanho de grãos e outros caracteres em topocruzamentos de soja tipo alimento com tipo grão. Pesq. Agropec. Brasil. **35**(11), 2235–2241 (2000)
Wilcox, J.R.: Increasing seed protein in soybean with eight cycles of recurrent selection. Crop Sci. **38**, 1536–1540 (1998)
Wilcox, J.R., Sediyama, T.: Interrelationships among height, lodging and yield in determinate and indeterminate soybeans. Euphytica. **30**, 323–326 (1981)
Wu, Y., Liu, W., Li, X., Li, M., Zhang, D., Hao, Z., Weng, J., Xu, Y., Bai, L., Zhang, S., Xie, C.: Low-nitrogen stress tolerance and nitrogen agronomic efficiency among maize inbreds: Comparison of multiple indices and evaluation of genetic variation. Euphytica. **180**, 281 (2011)

Chapter 22
Seed Multiplication and Maintenance

Laércio Junio da Silva, Reginaldo Castro de Souza Júnior, and Hamilton Carvalho dos Santos Júnior

Abstract The seeds are the instruments that lead to the farmer all the technology invested during the breeding of a certain cultivar. Thus, seed multiplication and maintenance stages are critical to the success of any breeding program. The breeding companies must maintain a minimum stock of genetic seeds, which corresponds to the first generation of a cultivar, and these should be maintained with high purity. These seeds are multiplied to meet the demand for seeds by grain producers. Therefore, special cares are essentials during multiplication of these seeds, aiming the maintenance of their identity and genetic purity. In this sense, some procedures are necessaries during all stages of seed production to reduce to the minimum the risks of contamination, as isolation of the fields destined for seed production, cleaning of machinery, identification and removal of plants that do not correspond to the cultivar planted, and certification of the genetic purity of seed lots. A general discussion of these topics is addressed in this chapter.

Keywords Genetic purity • Commercial seeds • Seed production • Seed certification

Introduction

The seed multiplication step is fundamental to the success of any breeding program, since it is the means that takes to the farmer all the technology invested during this step, for example, resistance to pests and diseases, tolerance to abiotic stresses, and, also, the GM traits inserted in the new cultivars. Thus, it is necessary to take major care in the multiplication of these seeds, in order to maintain their genetic identity. Breeding enterprises must maintain a minimum stock of genetic seeds, which corresponds to the first generation of a cultivar, and these must be kept with the highest purity level. These seeds are multiplied to meet the demand of grain producers. Therefore, during this process, there is a risk of loss of the genetic identity of the seeds, due to several factors, such as the presence of plants of other

L.J. da Silva, M.S., D.S. (✉) • R.C. de Souza Júnior • H.C. dos Santos Júnior
Federal University of Viçosa, Viçosa, Minas Gerais, Brazil
e-mail: laercio.silva@ufv.br; reginaldo.souza@ufv.br; hamilton.junior@ufv.br

F. Lopes da Silva et al. (eds.), *Soybean Breeding*,
DOI 10.1007/978-3-319-57433-2_22

cultivars in the fields, mutations, and segregation. Thus, special care should be taken at all stages of seed production to minimize these risks.

Genetic Seeds

Genetic seeds are the first generation of a cultivar, shortly after the breeding program, and are the basis for the seed production system of a given cultivar, and a minimum stock of genetic seeds with the highest genetic purity must be maintained. These seeds are multiplied by a limited number of generations in order to reach sufficient quantity and meet market demand. At the end of this process, the seeds that are produced and marketed have to be genetically pure, in other words, they should not contain seeds of other cultivars or plants with any descriptor that fail to meet the standards of the cultivar, known as atypical plants.

Genetic Seed Production

The production of genetic seeds is fundamental to the maintenance of the genetic identity of the cultivar, and there must be a stock of seeds that enable the multiplication thereof, that is, if there are not enough genetic seeds to meet the demand, the entire production system will be affected.

The production and maintenance of genetic seeds, within the company responsible for the creation of the cultivar and under the direct supervision of the breeder, involves the cultivation of individual plants with evaluation of genetic purity at all stages.

The production of genetic soybean seeds is divided into two phases (Krzyzanowski and Toledo 2009). The first is divided into two stages, as subsequently described:

The first step consists of the installation of a plant selection block. To this end, eight rows of 12 m in length are sown. Within the block, approximately 400–500 plants are selected based on the phenotypic characteristics of the cultivar. The seeds are individually harvested per plant and evaluated. All seeds and plants that fail to meet the standard for the cultivar descriptors should be discarded.

For the second stage, 40 seeds are separated from each selected plant, which must be sown in individual 3 m lines, consisting of progeny lines. During cultivation of the lines, the descriptors of the cultivar must be evaluated in the field, and any variation observed in the plants implies the disposal of the whole individual line. The homogeneous lines are collected individually, and the seeds are again compared to the descriptors, and all seeds of the lines which present any variation, when compared to the cultivar standard, are discarded.

The second phase of the genetic seed production consists of the installation of the selection blocks, formed from the seeds harvested in the individual lines of the

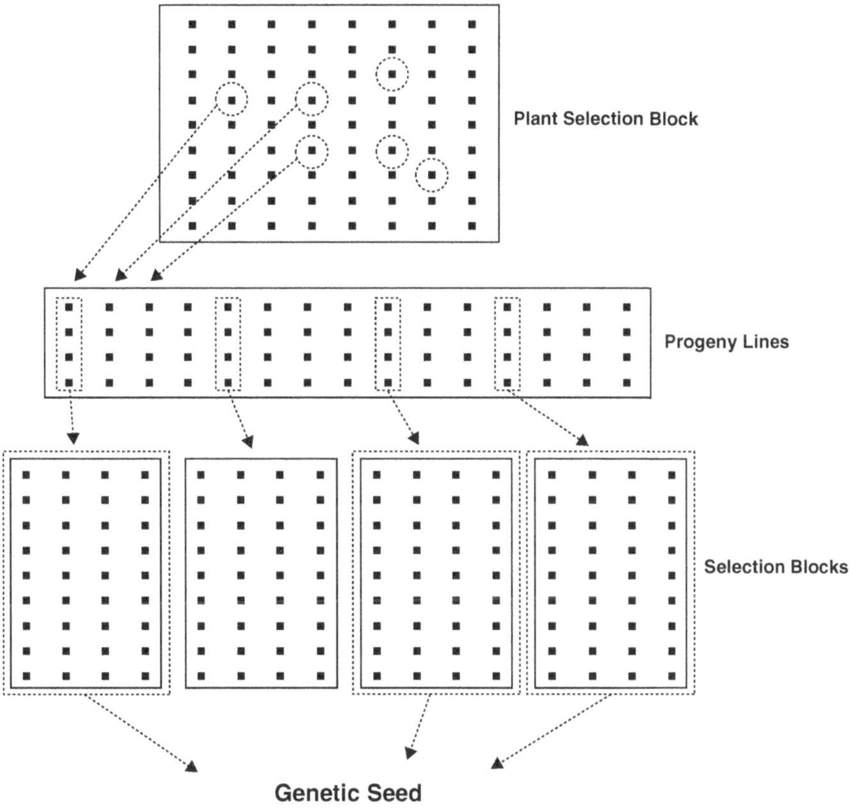

Fig. 22.1 Scheme for production and maintenance of genetic soybean seeds

second stage of the first phase. Each block is formed by four rows of 25 m each. During plant cultivation, the descriptors should be evaluated and compared to the cultivar standard, and, should any variation be determined within the selection block, the entire block should be discarded. Finally, the genetic seeds are constituted by the reunion of the seeds harvested in each homogeneous block in the second phase of plant selection. The phases and steps for the production of genetic soybean seeds are illustrated in Fig. 22.1.

Seeds Multiplication

Seed multiplication is the link between the breeding program aimed at the development of a new crop and the farmer. Seed is, therefore, the means that brings to farmers all the genetic potential of a cultivar with superior traits (Peske et al. 2012). In the multiplication stage, small amounts of seed must be multiplied in sufficient volume to meet market demand. In this process, several factors can affect the genetic quality of the seeds, such as crosses between plants of different cultivars or, even, late segregation, that can occur in soybean.

Brazil has a fairly organized seed production system ruled by specific legislation. Thus, the Lei de Sementes e Mudas (Law on Seeds and Seedlings), Law 10.711/2003, together with its regulation (IN 5153/2003 (Brazil 2007)), establishes the norms and procedures to be adopted.

According to this legislation, for a new cultivar to be marketed, it must be listed in the National Register of Cultivars (RNC). To this end, it should receive its own name, which cannot be expressed in numerical form only; the denomination given must be different from that of pre-existing cultivar and must not lead to error on the intrinsic traits or origin of the cultivar (Brazil 2007). For registration purposes, the technical report of the value for cultivation and use (VCU) tests is also required.

In addition to being registered, which allows the commercialization thereof, the cultivar can be protected, according to Law No. 9456, of April 25, 1997, which guarantees to the breeder the economic return for the work dedicated for several years for its development. Accordingly, the breeder receives royalties by authorizing the multiplication of the seeds of the cultivar developed under his responsibility. Moreover, there is the case of patents, given that the GM traits in cultivars are patented by the companies that develop the technology and, therefore, the seed producers that want to multiply seeds of GM cultivars, in addition to the payment of royalties to the breeder, must pay the technology rate to the company that holds the patent.

In Brazil, seed multiplication must be performed in compliance with rigorous generation control, whereby seeds are produced under two classes: certified and not certified. Figure 22.2 shows the control of generations in the multiplication of seeds of autogamous plants, such as soybean, in Brazil.

Production of Basic and Commercial Seeds

As previously reported, genetic seeds consist of the first generation of a cultivar. Thus, basic seeds are produced from genetic seeds and are usually produced by the company holding the cultivar, or the seed producer, who must be registered within the National Register of Seeds and Seedlings (RENASEM), may be allowed to produce it. The basic seeds, as well as the other categories, must be produced with major care, aiming at maintaining the genetic purity of the cultivar. These seeds are the basis to produce the seeds of other categories, which is done aiming at increasing the quantity of seeds to meet market demand (Table 22.1).

The certified of first-generation seeds are produced from the basic seeds, which can be multiplied once more, giving rise to certified of second-generation seeds. Certified seeds must be produced under the supervision of the technical head of a company accredited by the Ministry of Agriculture as a certifier or, yet, by the technical head of the very seed producer, when this latter, besides being enrolled

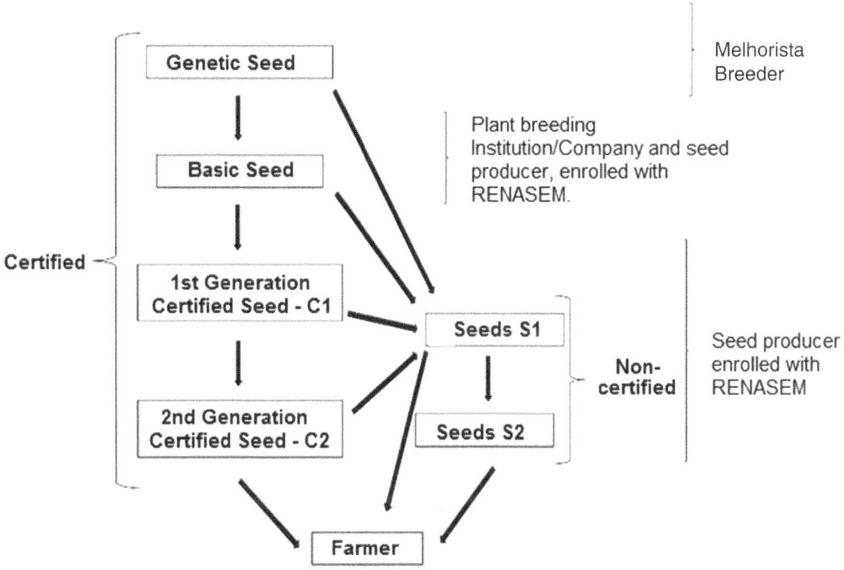

Fig. 22.2 Generation control in the soybean seed production system (Reis et al. 2005)

Table 22.1 Demand for soybean seeds and quantity to be produced in each seed category

Calculation of necessary area (ha)	Area (ha)	Calculation of the seed quantity (t)	Necessary quantity (t)	Seed class
	300,000	300,000 × 0.06	18,000	C2
18,000 t/1.2 t/ha	15,000	15,000 × 0.06	900	C1
900 t/1.2 t/ha	750	750 × 0.06	45	Basic
45 t/1.2 t/ha	37.5	37.5 × 0.06	2.25	Genetics
2.25 t/1.2 t/ha	1.9			

Source: Adapted from Peske et al. (2012)

with RENASEM as a seed producer, is also accredited as a seed certifier for private production (Brazil 2007).

In addition to the certified class, soybeans from any category of certified class may be multiplied by up to two generations, giving rise to seeds S1 and seeds S2, which are produced outside the certification process and under the supervision of the technical responsible of producer registered in RENASEM.

The purpose of this generation control in multiplication of seeds is to allow seeds of high physical and genetic purity to get to the farmer, which contributes to the maintenance of the genetic identity of the cultivar in sufficient amount to meet the market demand. As an example, we present the calculation of the demand for soybean seeds by the market and the quantity of seeds of each category to be produced.

In order to illustrate, we will demonstrate the calculation of the quantity of soybean seeds required in each seed class in a 300,000-ha crop and a 64% commercial seed use rate, that is, 64% of the farmers will buy seeds to install in their grain production fields. Hence, there will be 300,000 × 0.64 = 192,000 ha planted with certified seeds. For the calculation, the sowing density of 60 kg/ha and the average yield of 1.2 t/ha of seeds were considered (Table 22.1).

Thus, to meet a demand of 18,000 t of C2 seeds, to be sown in 30,000 ha, it is necessary to start the seed multiplication process with 2.25 t of genetic seeds, in an area of only 37.5 ha. Therefore, it is evident the importance of maintaining the genetic purity of these seeds, which would be compromised if their production was not properly carried out.

Genetic Purity

New soybean cultivars are continually being launched every year. Currently, 1474 soybean cultivars registered with the RNC and 675 protected under the National Service for the Protection of Cultivars (SNPC) are set for commercialization. Thus, the number of commercially available cultivars is large, and the genetic basis is narrow among those commercially available. Accordingly, tests of genetic purity and discrimination of cultivars have become more difficult and complex.

Seeds of other cultivars, contamination of a lot with seeds from the cross between different cultivars, or even late segregation affect the genetic purity of a lot of soybean seeds. Thus, specific procedures are required at the production stage so that genetic contamination is minimal.

Maintenance of Genetic Purity

The success of the soybean breeding program, as well as that of seed propagated crops, is related to the maintenance of genetic purity in subsequent seed multiplication stages. This ensures that the characteristics introduced by the breeder in the new cultivar are maintained throughout the generations. Moreover, it is important for the cultivar to fully express all its attributes of agronomic quality, such as cycle, productivity, disease resistance, grain type, and organoleptic and seed quality (Kryzanowski et al. 2008).

Some procedures must be rigorously adopted during seed production in order to guarantee the genetic quality in their multiplication, in addition to high purity.

Isolation

Soybean plant presents the cleistogamy phenomenon, that is, the pollen grains are released on the stigma of the flower before its opening, so the reproduction of the crop occurs by self-fertilization. There are reports that the occurrence of cross-fertilization is in the range of 1% (Matsuo et al. 2013). This factor is considered in the determination of isolation, that is, physical distance between different seed production fields.

According to the Normative Instruction 45 of MAPA (Brazil 2013), soybean seed production fields, of different cultivars or categories of seeds, must have a minimum distance of 3 m between them. Observing these rules is essential to minimize the contamination by pollen *foreign* to the seed production field, that is, coming from neighboring fields, so that the seed produced has high genetic purity.

Inspection of Seed Production Fields

Inspection of seed production fields is the most important operation at this stage and has the purpose of comparing the field conditions with the standard required by the legislation regarding the number of contaminations (Brazil 2013). The contaminants in seed production fields are undesirable plants, that is, the atypical plants, those attacked by pests and diseases, and those of other crops and invasive plants. MAPA prepared the Manual for Inspection of Seed Production Fields, which includes all the procedures to be adopted in this operation (Brazil 2011) and which should be followed carefully.

In soybean production fields, at least two inspections must be carried out (Brazil 2013). The first in the flowering phase and the second in the preharvest stage. The flowering stage is characterized for allowing the evaluation of most descriptors of soybean plants, such as cycle, size, growth habit, flower coloring, among others. Thus, at this stage the inspection must be carried out thoughtfully, in order to correctly identify and quantify atypical plants, that is, which have any trait different from those of the cultivar being produced. If it fails to meet the standards, the field may be condemned and not allocated for seed production.

Ideally inspections in seed production fields should be carried out during all phenological stages of the crop, but they are more efficient when done at stages where morphological traits are more evident and the distinction of atypical plants becomes more efficient. In the case of soybeans, the initial stages (V1 and V2), flowering (R1 and R2), and the final maturation stages (R6, R7, and R8) are the phenological stages that allow greater efficiency in distinguishing atypical plants within seed production fields.

Removing Unwanted Plants

Roguing is the operation of elimination of undesirable or contaminating plants in a seed production field, that is, those from other cultivars or those that fail to meet cultivar standard or, also, plants contaminated by pests and diseases (Carvalho and Nakagawa 2012).

This operation must be performed manually, and the identification of the contaminants must carried out in the entire production field, especially in the pre-flowering, flowering, and preharvest phases (Matsuo et al. 2013). In general, it is recommended that this operation be done by small groups, accompanied by a responsible person and by people with training in evaluation soybean descriptors. It cannot be done by sampling, so, the entire seed production field must be traversed and the contaminating plants removed from the area.

Cleaning of Machinery and Equipment

In addition to respecting the minimum isolation required between production fields, both sowing and seed harvesting operations should be carried out with great care to avoid mechanical mixing. Thus, the strict cleaning of all the components of the machinery used in these operations is essential. This same procedure must be adopted when changing cultivars. Also, the cleaning of the equipment used in the post-harvest operations, as well as in processing, besides the use of new packaging for the storage of seeds, among other precaution, is crucial in order to minimize the risks of contamination of the seed batches.

All these procedures are adopted to meet the production and commercialization standards for soybean seeds (Table 22.2). The presence of contaminants above that allowed by the legislation can lead to condemnation of the field for seed production, in case corrective measures cannot be taken or, also, condemnation of the seed lot.

Certification of Genetic Purity of Seed Lots

The increase in the number of registered cultivars and the problems that involve the national and international commercialization of seeds make the ability to distinguish and identify cultivars fundamental for the monitoring and control of this trade (Vieira et al. 2006).

According to Schuster et al. (2004), so that all the efforts exerted to obtain an improved variety are translated into benefits for agricultural production, the seeds used by producers must feature high quality, and, therefore, their certification is fundamental as a controlling agent.

Table 22.2 Standard for production and commercialization of soybean seeds in Brazil (Brazil 2013)

1	Maximum batch weight (kg)				30,000
2	Minimum weight of samples (g)				
	Sample submitted or average				1000
	Work sample for purity analysis				500
	Work sample for determination of other seeds by number				1000
3	Deadline for requesting the registration of fields (days after planting)				45
		Categories/indices			
		Basic	C1	C2	S1 and S2
4.1	Survey				
	Maximum area of land (ha)	50	100	100	150
	Minimum number	2	2	2	2
	Minimum number of subsamples	6	6	6	6
	Number of plants per subsample	1000	500	375	250
	Sample population	6000	3000	2250	1500
4.2	Rotation (agricultural cycle)	–	–	–	–
4.3	Insulation or surround (minimum in meters)	3	3	3	3
4.4	Atypical plants (out of type) (at most)	3/6000	3/3000	3/2250	3/1500
4.5	Plants of other species				
	Cultivated/wild/harmful tolerated	–	–	–	–
	Harmful forbidden	–	–	–	–
5	Seed parameters				
5.1	Purity				
	Pure seed (% minimum)	99.0	99.0	99.0	99.0
	Inert material (%)	–	–	–	–
	Other seeds (% maximum)	0.0	0.1	0.1	0.1
5.2	Determination of other seeds by number (maximum)				
	Seeds of other cultivated species				
	Others	0	0	1	2
	Vigna unguiculata	0	0	0	0
	Wild seed	0	1	1	1
	Harmful seed tolerated	0	1	1	2
	Forbidden seed	0	0	0	0
5.3	Germination (% minimum)	75	80	80	80
	Validity of the germination test (maximum in months)	6	6	6	6
	Validity of the germination reanalysis test (maximum in months)	3	3	3	3

The genetic purity of soybean seeds can be affected by adventitious seeds in the lot, which are the result of the natural cross between plants of different cultivars or the mixture of seeds of different cultivars.

The certification of the genetic purity of seed lots is done with the purpose of, in addition to evaluating the lots for contamination by seeds from other cultivars, detecting GM seeds in conventional seed lots. Thus, there are several tools available, and the certification of the genetic purity of soybean seeds can be performed either by means of the analysis of morphological descriptors or through the use of molecular markers, especially the analysis of proteins and nucleic acids (DNA and RNA) (Vieira et al. 2006).

Morphological Markers

Morphological markers are the most frequently used to certify the genetic purity of seed lots and evaluate the phenotypic characteristics of seeds, seedlings, or plants. In soybeans, one of the main characteristics used in the discrimination of cultivars is the seed hilum color. This feature is also very important in seed analysis laboratories, to differentiate cultivars and detect varietal mixtures during the certification process. Apart from the color of the hilum, other characteristics are associated to the seeds, such as size, shape, color, and appearance of the tegument and hilum shape (Moreira et al. 1999).

Morphological descriptors are the most recommended for seed purity analyzes, but they present some disadvantages, such as the influence of the environment, besides requiring time and space to be evaluated. Apart from that, soybean is a narrow genetic base crop, which further hinders the separation of cultivars by means of such descriptors (Vieira et al. 2009a, b). Although recommended, the use of these descriptors brings some limitations, such as suffering from the effect of the environment, not being stable, and many evaluated in the adult phase of plants, which requires time and physical space. Another problem is the facts that the number of soybean descriptors recommended, which add up to a total of 38, may not be enough.

Molecular Markers

Molecular markers are extremely important and efficient tools for evaluating the genetic purity of soybean seeds. The evaluation based on molecular markers of DNA is conclusive, as these do not undergo environmental influence, and data are reproducible and stable (Schuster et al. 2004).

Tests of genetic purity of soybean seeds based on molecular markers can be performed with the formation of groups of seeds called bulk, as suggested by Schuster et al. (2004). According to these authors, the analysis in bulks provides results with greater speed and the same precision of the individual analysis of seeds, besides reducing time and costs.

The intension here is not going into details on the subject but, rather, only to report what is currently available in relation to the use of molecular markers in the genetic purity analysis of soybean seed lots. More details on the use of these techniques are provided by Vieira et al. (2006).

Protein Electrophoresis

Biochemical markers of proteins and enzymes for the identification of soybean cultivars are recommended by UPOV, but not yet by ISTA, probably due to the lack of standardized and tested protocols (ISTA 2007). The use of biochemical markers offers advantages such as obtaining results within 24–48 h and simplicity of operation.

Storage proteins, which are a direct gene expression product allowing some expression regardless of the influence of the environment, are considered stable markers, taking into account the stability criterion required for the minimum descriptors predicted in the LPC (Vieira et al. 2009a).

The use of protein electrophoresis in the identification of cultivars is based on the fact that they are products of the genes and therefore can be considered markers for the genes that encode them. In most species, seed storage proteins exhibit considerable polymorphism with respect to charge, size, or both parameters. Moreover, they are encoded at various locus, are comparatively present in large quantities, and are readily extractable. In the case of autogamous species such as soybean, a narrow genetic base can occur, making it difficult to certify genetic purity by the use of this technique.

The isoenzymes, on the other hand, should be used with great discretion for this purpose, since certain enzymatic systems may present variations due to the deterioration stage in which the seeds are or when there are microorganisms in association with the seeds.

According to Vieira et al. (2009a), biochemical markers of storage proteins and isoenzymes as well as molecular markers are efficient in the characterization of soybean cultivars. According to these authors, enzymatic systems such as esterase, alcohol dehydrogenase, urease, malate dehydrogenase, alkaline phosphatase, leucine aminopeptidase, superoxide dismutase, phosphoglucoisomerase, and 6-phosphogluconate dehydrogenase can be used to distinguish different soybean cultivars.

DNA Polymorphism

There are several technologies based on the DNA polymorphism used for identification of cultivars and determination of genetic purity in seed lots. The main advantage of using DNA is that it is not affected by the environment, which allows a more objective and accurate analysis. Among the techniques used, Vieira et al. (2006) highlight (a) restriction fragment length polymorphism (RFLP), (b) random amplified polymorphic DNA (RAPD), (c) microsatellites or simple sequence repeat (SSRs), (d) amplified fragment length polymorphism (AFLP), and (e) single-nucleotide polymorphism (SNP).

Molecular markers for the evaluation of the genetic purity of seed lots are not routinely used in laboratories for seed analysis, mainly because of the high costs. However, there has been great interest in the research in the study of these markers, aiming to evaluate their applicability and efficiency in the purity analysis of seed lots.

Among the indicated molecular markers, microsatellites (SSR) have been the most recommended ones, and their application depends on the development of specific loci for each species. SSR technology is burdensome in its implementation; however, in the case of soybeans, different pairs of primers are available in the market, which makes it possible to use them in the identification and analysis of genetic purity. In addition, SSR markers can also be used to detect genetic contamination in seed production fields (Vieira et al. 2009b).

Microsatellite molecular markers are efficient for the evaluation of soybean seeds with variation in hilum coloring and detection of varietal mixtures (Rabel et al. 2010). They are indicated for this type of analysis, because they are highly polymorphic in soybean, well reproducible, and codominant and present low cost (Cregan et al. 1999). Schuster et al. (2004), in a study using the microsatellite technique in the evaluation of the genetic purity of soybean seed batches, verified that the analysis of DNA extracted from seeds, considered atypical by the visual analysis method, with microsatellite molecular markers is efficient in determining the genetic purity of soybean seed batches. Furthermore, according to the author, the method of bulk analysis of atypical seed DNA samples allows results to be obtained with the same precision of the individual analysis and at lower cost.

GM Presence Evaluation

The number of transgenic cultivars has increased significantly in recent years. However, some markets or farmers opt not to adopt this technology. Therefore, in the production of conventional cultivar seeds, controlling the genetic purity of the seeds in relation to the presence of GM traits is essential (Vieira et al. 2006).

Normative Instruction No. 42 of December 1, 2006, establishes a maximum 0.1% level of GMO presence in conventional seed batches with a 95% confidence level.

Thus, several techniques are indicated for the evaluation of GMOs in seed lots. Among them, bioassay is the simplest one, and it is used in the detection of glyphosate-tolerant soybeans, in which the germination test is conducted with herbicide-moistened substrate (Torres et al. 2003), or the seeds are soaked in solution containing the product (Matsuo et al. 2013). Among the techniques for precise identification of GM traits in conventional seed batches, the most common one is immunochromatography (Valle 2010), but other methods can be cited, such as enzyme-linked immunosorbent assay (ELISA) and those based on PCR reaction. Among the methods that have been developed and that have not yet been made available for routine use, we highlight DNA chips or microarray (Vieira et al. 2006).

Immunochromatography is a method for the detection of specific proteins using antibodies. The immunochromatographic kits, commonly called "lateral flow strips," have great application in the detection of proteins expressed in genetically modified (GM) plants which are not found in plants of conventional cultivars (Valle 2010).

Final Considerations

The seeds should be carefully multiplied to avoid losing the genetic identity of the cultivar. Such care involves procedures in fields and laboratories, which guarantee the multiplication of both GM seeds, under the responsibility of the breeder, and of commercial ones, with high genetic purity. Thus, knowledge of these techniques, such as isolation of fields, *roguing*, and laboratory-based tests, used to evaluate seed batch contamination is critical.

References

Brazil: Ministério da Agricultura, Pecuária e Abastecimento. Legislação brasileira sobre sementes e mudas, p. 318. Secretaria de Defesa Agropecuária, Brasília (2007)

Brazil: Ministério da Agricultura Pecuária e Abastecimento. Guia de inspeção de campos para produção de sementes, 3rd edn, p. 41. Mapa/ACS, Brasília (2011)

Brazil: Instrução Normativa 45 de 17 de setembro de 2013. Padrões de identidade e qualidade para a produção e a comercialização de sementes, p. 6. Diário Oficial da República Federativa do Brasil, Brasília (2013)

Carvalho, N.M., Nakagawa, J.: Sementes: ciência, tecnologia e produção, 5th edn, p. 590. Funep, Jaboticabal (2012)

Cregan, P.B., Jarvik, T., Bush, A.L., Shoemaker, R.C., Lark, K.G., Kahler, A.L., Kaya, N., Vantoai, T.T., Lohnes, D.G., Chung, J., Specht, J.E.: An integrated genetic linkage map of the soybean genome. Crop Sci. **39**, 1464–1490 (1999)

International Seed Testing Association (ISTA): Species and variety testing. In: International Rules for Seed Testing, pp. 1–32. ISTA, Bassersdorf (2007)

Kryzanowski, F.C., França Neto, J.B., Henning, A.A., Costg, N.P.: O controle de qualidade agregando valor à semente de soja, p. 12. Embrapa Soja, Londrina (2008). (Circular técnica, 54)

Krzyzanowski, F.C., Toledo, J. F.F.: Semente genética de soja—origem e procedimentos técnicos de produção - série sementes, p. 4. Embrapa Soja, Londrina (2009), (Embrapa Soja. Circular Técnica, 66)

Matsuo, E., Oda, M.C., Marchiori, R., Sediyama, T., Nogueira, A.P.O.: Manutenção da pureza genética e multiplicação de sementes. In: Sediyama, T. (ed.) Tecnologias de Produção de Sementes de Soja, 1st edn, pp. 205–220. Mecenas, Londrina (2013)

Moreira, C.T., Souza, P.I.M., Farias Neto, A.L., Almeida, L.A.: Ocorrência de variações na coloração do hilo de sementes de cultivares de soja [*Glycine max* (L.) Merrill], p. 4. Embrapa Cerrados, Planaltina (1999), (Comunicado Técnico, 5)

Peske, S.T., Villela, F.A., Meneghello, G.E.: Sementes: fundamentos científicos e tecnológicos, 3rd edn, p. 573. Universitária/UFPel, Pelotas (2012)

Rabel, M., Vieira, E.S.N., Lana, U.G.P., Paiva, E., Sehnem, M.A.S., Schuster, I.: Marcadores moleculares microssatélites na avaliação de sementes de soja com variação na coloração do hilo. Revista Brasileira de Sementes. **32**(2), 19–25 (2010)

Reis, M.S., Campos, S.R.F., Borém, A., del Giúdice, M.P.: Produção e Comercialização de Sementes. In: Borém, A. (ed.) Melhoramento de Espécies Cultivadas, 2nd edn, pp. 897–930. Editora UFV, Viçosa (2005)

Schuster, I., Queiroz, V.T., Teixeira, A.I., Barros, E.G., Moreira, M.A.: Determinação da pureza varietal de sementes de soja com o auxílio de marcadores moleculares microssatélites. Pesq. Agrop. Brasileira. **39**(3), 247–253 (2004)

Torres, A.C., Nascimento, W.M., Paiva, S.A.V., Aragão, F.A.S.: Bioassay for detection of transgenic soybean seeds tolerant to glyphosate. Pesq. Agrop. Brasileira. **38**(9), 1053–1057 (2003)

Valle, L.A.: Imunocromatografia para avaliação de presença adventícia de OGM em sementes. Informativo Abrates. **20**(3), 84–85 (2010)

Vieira, M.G.G.C., von Pinho, E.V.R., Salgado, K.C.P.C.: Técnicas moleculares em sementes. Informe Agropecuário. **27**(232), 88–96 (2006)

Vieira, E.S.N., Pinho, É.V.R.V., Carvalho, M.G.G., Silva, P.A.: Caracterização de cultivares de soja por descritores morfológicos e marcadores bioquímicos de proteínas e isoenzimas. Rev. Bras. Sementes. **31**(1), 86–94 (2009a)

Vieira, E.S.N., Schuster, I., Silva, R.B., Oliveira, M.A.R.: Variabilidade genética em cultivares de soja obtida com marcadores microssatélites genotipados em géis de agarose. Pesq. Agrop. Brasileira. **44**(11), 1460–1466 (2009b)

Chapter 23
Registration and Protection of Cultivars

Amilton Ferreira da Silva, Tuneo Sediyama, Aluízio Borém, Felipe Lopes da Silva, Francisco Charles dos Santos Silva, and André Ricardo Gomes Bezerra

Abstract In Brazil, 1657 soybean cultivars are registered in the Ministry of Agriculture, Livestock and Food Supply (MAPA) in 2017. Every year a large number of new cultivars are submitted to registration and protection. Thus, from the moment that this new cultivar is obtained, the registration allows the production and commercialization of the seeds in the country. This process ensures the genetic identity and the varietal quality of the cultivars. In addition, the protection of cultivars guarantees the intellectual rights to the breeders and enables public and private research companies to benefit, as it guarantees the continuity of breeding programs, with the partial or total use of these materials. For the registration of a particular cultivar in the National Register of Cultivars (RNC), it must first be submitted to tests to determine the Value of Cultivation and Use (VCU), that is, the intrinsic value of the combination of the agronomic traits of the cultivar with its properties of use in agricultural, industrial, and commercial activities and for unprocessed product consumption. In order to be able to be protected, the cultivar must undergo field tests in which it must have distinctiveness (differences of any other coexistent at the date of the application for protection), uniformity (uniformity among plants within the same generation), and stability (maintenance of traits in successive generations). Thirty seven descriptors are used for the DUS test, 30 of which were compulsory and seven additional. Thus, knowledge of the plant morphology is essential to accurately describe the traits of the cultivar and always ensure the launch of new varieties, which are more productive with desirable traits.

A.F. da Silva, M.S., D.S. (✉)
Federal University of São João del-Rei, Campus de Sete Lagoas, Sete Lagoas, Minas Gerais, Brazil
e-mail: amiltonferreira@ufsj.edu.br

T. Sediyama, M.S., Ph.D. • A. Borém, M.S., Ph.D. • F.L. da Silva, M.S., D.S.
F.C. dos Santos Silva, M.S. • A.R.G. Bezerra, M.S.
Federal University of Viçosa, Viçosa, Minas Gerais 36570-900, Brazil
e-mail: tuneo@ufv.br; borem@ufv.br; felipe.silva@ufv.br; fcasantossilva-ma@hotmail.com; bezerra.agro@yahoo.com.br

F. Lopes da Silva et al. (eds.), *Soybean Breeding*,
DOI 10.1007/978-3-319-57433-2_23

Keywords Soybean cultivars • Brazilian legislation • Value of cultivation and use • Descriptors

Introduction

The law in force, Law No. 9456/97, called the Plant Variety Protection Law (LPC, the acronym in Portuguese), establishes that cultivar is the "variety of any genus or superior plant species that is clearly distinguishable from other known cultivars by a minimum margin of descriptors, by its denomination, which is homogeneous and stable as regards the descriptors through successive generations and is of species that can be used by the agroforestry complex, described in a specialized publication available and accessible to the public, as well as the hybrid component line."

From the moment this new cultivar was obtained, the registration allows the production and commercialization of the seeds in the country. It is an important process for breeding programs, as it ensures the genetic identity and the varietal quality of the cultivars. Furthermore, the protection of cultivars, which guarantees the intellectual rights of breeders, enables public and private research companies to benefit from the access of resources derived from rights on the cultivars they develop, thus, giving partial or total sustainability to the continuity of breeding programs and the subsequent launching of new cultivars of interest to Brazilian agribusiness (Carvalho et al. 2009).

Registration of Cultivars

The National Register of Cultivars (RNC) was established by means of Administrative Rule No. 527, of December 30, 1997, given the need to make the latest advances in plant genetic research available to farmers, in addition to adjusting the Brazilian Systems for Assessment and Recommendation of Cultivars and Cultivar Registration to the normative rules under MERCOSUL interregional agreements. Currently, RNC is governed by the Brazilian Seed Law No. 10.711/03, which enables cultivars for the production, processing, and marketing of seeds and seedlings across the national territory.

The purposes of the RNC were to replace the old systems for evaluation and recommendation and registration of cultivars by means of the establishment of a registration information system provided by the breeder or holder of the rights of exploitation of the cultivar; promote the prior registration of national and foreign cultivars, being thus qualified for the production and marketing of certified seed and seedlings and other classes, in Brazil; establish the preparation of the updated list of species and cultivars available on the market, the registration of information on the value for cultivation and use (VCU) of cultivars, and the periodical publication of the Brazilian plant variety list; and regulate the market by preventing the marketing of certified and supervised seed and seedlings of unregistered cultivars (Borém and Miranda 2005).

For the registration of a certain cultivar with the RNC, it must first be submitted to tests to determine the value for cultivation and use (VCU), that is, the intrinsic value of the combination of the agronomic traits of the cultivar with its properties of use in agricultural, industrial, and commercial activities and for unprocessed product consumption.

The following are the minimum requirements for determining the VCU of a soybean cultivar established by MAPA (Ministry of Agriculture, Livestock and Food Supply).

Tests

1. Number of sites: one place in each edaphoclimatic region of importance for the crop
2. Minimum period of accomplishment:
   (a) Conventional cultivars: 2 years
   (b) Essentially derived cultivars: 1 year, provided the recurrent parent is enrolled with the National Register of Cultivars (RNC)

Experimental Design

Statistical design: randomized blocks with at least three replicates or another design with equal or greater experimental accuracy.

Plot size: each plot must have a minimum of 4.0 m^2 (characterize the usable area).

Ripening groups: to be tested, cultivars and lines will be classified into up to six ripening groups: super-early ripening, early ripening, mid-early ripening, mid-late ripening, and late ripening.

Control cultivars: two cultivars registered in the RNC should be used as comparative standards and by ripening group, and they should be kept during the test period of each of them.

In the case of essentially derived cultivars, also include the recurrent parent (initial cultivar) as a control.

Characteristics To Be Evaluated

Descriptors (see Table 23.1): fill in if the cultivar is not protected in Brazil.

Table 23.1 Soybean descriptors (*Glycine max* (L.) Merrill)

Features		Classification rank
Compulsory features		
1.	Seedling: hypocotyl anthocyanin pigmentation	Absent
		Present
2.	Only for cultivars which present anthocyanin in the hypocotyl Seedling: intensity of hypocotyl anthocyanin pigmentation	Weak
		Medium
		Strong
3.	Flower: predominant color	White
		Purple
4.	Leaf: green color intensity	Weak
		Medium
		Strong
5.	Leaf: lateral leaflet shape	Narrow lanceolate
		Lanceolate
		Triangular
		Pointed oval
		Round oval
6.	Leaf: lateral leaflet size	Small
		Medium
		Large
7.	Plant: type of growth	Determined
		Semi-determined
		Undetermined
8.	Plant: height	Low
		Medium
		High
9.	Plant: growth habit (branches inclination)	Erect
		Semi-erect
		Horizontal
10.	Plant: color of pubescence on the main stem	Gray
		Light brown
		Medium brown
11.	Plant: pubescence density on the main stem	Low
		Medium
		High
12.	Pod (with pubescence): color	Light gray
		Dark gray
		Light brown
		Medium brown
		Dark brown
13.	Vegetative cycle: from emergence to flowering	Early
		Medium
		Late

(continued)

Table 23.1 (continued)

Features		Classification rank
14.	Total cycle: from emergence to ripening	Early
		Semi-early
		Medium
		Semi-late
		Late
15.	Relative ripening group	4.0–10.0
16.	Transgenesis	Absent
		Present
17.	Seed: predominant size	Small
		Medium
		Large
18.	Seed: shape	Spherical
		Flat spherical
		Elongated
		Flat elongated
19.	Seed: integument brightness	Low
		Medium
		High
20.	See: integument color (excluding the hilum)	Yellow
		Greenish yellow
		Green
		Dark brown
		Medium brown
		Dark brown
		Black
21.	Seed: genetic hilum color	Gray
		Yellow
		Light brown
		Medium brown
		Imperfect black
		Black
22.	Seed: Peroxidase reaction	Positive
		Negative
23.	Reaction to bacterial pustule in greenhouse (*Xanthomonas axonopodis* pv. *glycines*)	Resistant
		Moderately resistant
		Susceptible
24.	Reaction to frogeye leaf spot (*Cercospora sojina*) (mix of races: Cs-2, Cs- 4, Cs-7 and Cs-15)	Resistant
		Moderately resistant
		Susceptible
25.	Reaction to frogeye leaf spot (*Cercospora sojina*) (mix of races: Cs-23, Cs-24 and Cs-25)	Resistant
		Moderately resistant
		Susceptible

(continued)

Table 23.1 (continued)

Features		Classification rank
26.	Reaction to stem cancer (*Phomopsis phaseoli* var. *meridionalis*/*Diaporthe phaseolorum* var. *meridionalis*)	Resistant
		Moderately resistant
		Susceptible
27.	Reaction to stem necrosis (*Cowpea mild mottle virus*)	Resistant
		Moderately resistant
		Susceptible
28.	Reaction to root-knot nematode (*Meloidogyne javanica*)	Resistant
		Moderately resistant
		Susceptible
29.	Reaction to root-knot nematode (*Meloidogyne incognita*)	Resistant
		Moderately resistant
		Susceptible
30.	Reaction to soybean cyst nematode - race 3 (*Heterodera glycines*)	Resistant
		Moderately resistant
		Susceptible
Additional features[a]		
31.	Reaction to soybean common mosaic virus (VMCS, soybean mosaic virus)	Resistant
		Moderately resistant
		Susceptible
32.	Reaction to soybean cyst nematode—race 1 (*Heterodera glycines*)	Resistant
		Moderately resistant
		Susceptible
33.	Reaction to soybean cyst nematode—race 14 (*Heterodera glycines*)	Resistant
		Moderately resistant
		Susceptible
34.	Reaction to bacterial pustule: field resistance test (*Xanthomonas axonopodis* pv. *glycines*)	Resistant
		Moderately resistant
		Susceptible
35.	Reaction to brown stem rot (*Cadophora gregata, sin. Phialophora gregata*)	Resistant
		Moderately resistant
		Susceptible
36.	Reaction to soybean target spot (*Corynespora cassiicola*)	Resistant
		Moderately resistant
		Susceptible
37.	Reaction to soybean root rot (*Phytopthora sojae*)	Resistant
		Moderately resistant
		Susceptible

[a]The presentation of the information contained in the item "Additional Information" is not mandatory. However, these characteristics may be considered for differentiation, if the evaluation of the other characteristics of the Table of Descriptors is not enough. Therefore, it is suggested that such information be presented whenever the breeder has the possibility to evaluate it. Other differentiating characteristics, including disease resistance, may be used for differentiation, provided that the candidate cultivar and the most similar cultivars are planted under the same conditions

3. Agronomic traits:

   Vegetative cycle: number of days from emergence to flowering (50% of flowering plants).

   Total cycle: number of days from emergence to ripening (Note: ripening, 95% of dry pods).

   Height of plants (cm).

   Height of insertion of the lower pods (cm) (Note: evaluated in the useful area).

   Degree of lodging:

   - All or almost all plants upright
   - All or almost all plants slightly inclined or up to 25% of the plants lodged
   - All plants moderately inclined or 25–50% of the plants lodged
   - All plants strongly inclined or 50–80% of the plants lodged
   - More than 80% of the plants lodged

   Dehiscence degree of pods (evaluated at 15 days post ripening):

   - 0% threshing
   - 1–3% threshing
   - 4–10% threshing
   - 11–20% threshing
   - More than 20% of threshing

   Weight of 100 seeds (in grams, base 13% of seed moisture).
   Reaction to diseases and nematodes: greenhouse and/or in the field.

4. Yield assessment:

   The yield of the cultivars and lines will be calculated based on the yield of the useful area of the plots, standardized to 13% of humidity and transformed into kilograms per hectare.

   Experiments whose variation coefficients are greater than 20% should not be computed in the overall analysis of sites and, consequently, in the yield calculation for the region.

   The cultivar will be registered in the National Register of Cultivars (RNC), which, in the tests for determination of the VCU, achieved yield equal to or greater than the mean of the controls. Otherwise, the applicant for registration must indicate the trait(s) of relevance that justifies its registration with the RNC.

5. Evaluation of technological/industrial quality

   The industrial quality of the soybean cultivars will be expressed by the oil and protein contents in grains, in percentage, and on the weight of the dry matter of the grain including the contents of control cultivars.

   Samples for these analyses can be collected from only one replicate (block) of each site (experiment), each year.

Information Update

New information on the cultivar, such as changes in the adaptation region, reaction to pests, diseases, nematodes, limitations etc., should be sent in the same VCU forms, to be attached to the registration document.

Crop Protection

On April 25, 1997, the Brazilian government enacted the first legislation that guaranteed the rights of breeders of new plant varieties, Law No. 9456, regulated by Decree No. 2366, as of November 5, 1997. The Law also created, in the Ministry of Agriculture, Livestock and Food Supply (MAPA), the National Service for Protection of Plant Varieties (SNPC), to which it attributed the competence for the protection of cultivars in the country. The SNPC's mission is to secure the free exercise of the intellectual property rights of breeders of new phylogenetic combinations in the form of distinct, homogeneous, and stable vegetable cultivars, watching over the national interest in the field of protection of cultivars (MAPA 2010).

This law is based on the model approved by the International Union for the Protection of New Varieties of Plants (UPOV), an international organization headquartered in Geneva, Switzerland, responsible for implementing the International Convention for the Protection of New Varieties of Plants, the first version of which dates from 1961 and underwent three amendments: 1972, 1978, and 1991. With the LPC, Brazil adhered to the UPOV/78 Convention (Assis et al. 2010). Accordingly, UPOV members are obliged to protect Brazilian cultivars, and, on the other hand, Brazil undertakes to protect cultivars from these countries, facilitating the exchange of new materials generated by Brazilian and foreign research (Piloro et al. 2007).

The granting of a certificate of protection of cultivars guarantees the authorization for reproduction or vegetative multiplication in the country of plants or their parts (MAPA 2010).

In Brazil, it is subject to protection: (1) the new cultivar, as defined in article 3, subsection 5, of Law No. 9456/97; (2) the essentially derived cultivar; and (3) cultivars not covered by these two groups, but whose applications for protection are submitted no later than 12 months after the disclosure of the descriptors of the species, when the maximum time of commercialization, in the date of submission of the application, has been in up to 10 years. This latter form of protection will only produce effects for essentially derivative cultivars, that is, it is a protection that will have its effects, as for exercising the resultant rights, only in a relationship between breeding companies. Such protection does not encompass seed producers and, naturally, would not include the farmer who tries to sell his material as seed. It should also be noted that this protection will be granted only for the remaining period, i.e., if the cultivar has already been marketed in the country for 8 years, and the total term of protection for this species is 15 years, in which case, the term granted is of additional 7 years (MAPA 2010).

By law, the term of protection is 15 years for most species, mainly grains (oilseeds, cereals, and others) (MAPA 2010).

The protection of cultivars, which guarantees the intellectual rights of breeders, allows public and private research companies to benefit from the resources derived from the rights to the cultivars they develop, giving partial or total sustainability to the continuity of the breeding programs and the subsequent launching of new cultivars of interest to Brazilian agribusiness (Carvalho et al. 2009). Before the LPC, the cultivar used to be in the public domain as soon as it was released and could be multiplied by anyone (Borém and Miranda 2005).

How to Apply for Crop Protection

The request for the protection of a vegetable cultivar in Brazil is made through the physical submission of a number of documents to the National Service for Protection of Plant Varieties (SNPC), of the Ministry of Agriculture, Livestock and Supply, in Brasília. The application for protection may be made by the very breeder, his legal representative or by the transferee of the right to the cultivar (MAPA 2010).

In Brazil, the protection is based on a sworn declaration, that is to say, the person responsible for the information provided to the SNPC is the breeder or the transferee of the right on the cultivar, who can be held criminally liable in case information fails to correspond to the object of application for protection (candidate cultivar) (MAPA 2010).

For the application for protection, it is necessary to fill out forms available on the website of the Ministry of Agriculture, on the Cultivar Protection page *(www.agricultura.gov.br > Serviços > Proteção de Cultivares > Formulários para a Proteção de Cultivares)*. There, it is possible to access the CultivarWeb system, which allows users to obtain the cultivar protection application form, store data for later application, and follow the progress of the requirements sent to the SNPC (MAPA 2010).

The protection application record will be provided by the SNPC, once the complete documentation (forms + supplementary documents) has been received, provided the electronic protection application form has already been submitted (MAPA 2010).

Necessary Requirements for the Cultivar Subject to Protection

- To be a product of genetic improvement
- To be of a species that can be protected in Brazil
- Not to have been marketed abroad for more than 4 years, or for more than 6 years, in the case of vines or trees
- Not to have been commercialized in Brazil for more than 12 months
- To be different
- To be homogeneous
- To be stable

The last three requirements are proven through specific experiments gathered in what we call DUS (distinctiveness, uniformity, and stability) tests. In Brazil breeders are responsible for carrying out the tests, but abroad, the tests are carried out by government authorities who send the results upon request of the SNPC and payment of a fee by the applicant for protection (MAPA 2010).

DUS (Distinctiveness, Uniformity, and Stability) Tests

In Brazil, these tests are performed by the breeders at experimental stations. These are field trials in which the traits of distinctiveness (differences of any other coexistent at the date of the application for protection), uniformity (uniformity among plants within the same generation), and stability (maintenance of traits in successive generations) of the cultivar are tested. They follow their own methodology for each species and require an in-depth knowledge of the species, its behavior, groups, and varieties, and, in some cases, the use of reference cultivars for the characterization of the new cultivar (MAPA 2010) is indispensable.

Performance of Distinctiveness, Uniformity, and Stability (DHE)

According to MAPA, for the tests, the following rules must be followed:

1. Each trial shall include at least 300 plants, at the recommended standard seed density for the region of adaptation of the cultivar and will be conducted under conditions that ensure the normal development of the plants. The size of the plots should be such that the plants or parts of plants can be removed for measurements and counts, without prejudice to the observations, which should be made at the end of the development period. At least two replicates, with at least four rows each, shall be used to avoid edge effects in the evaluation of the traits.

Separate plots for observations and measurements may only be used if they have been subjected to similar environmental conditions.
2. The tests will be conducted for at least two similar periods of cultivation in the region of adaptation of the cultivar.
3. The tests will, typically, be conducted in the same experimental area and in the same sowing times. If an important characteristic of the cultivar cannot be observed at that location, it can be evaluated at an additional location.
4. Additional assessments for special purpose may be established.
5. All observations for determination of distinctiveness and stability will be made in at least 20 plants or in groups of 20 plants.
6. For determination of the uniformity of the characteristics observed, the plots should have a normal sowing density and should be applied in a standard population of 0.5% with probability of acceptance of at least 95%. In the case of plots with 300 plants, the maximum number of atypical plants allowed will be four. The atypical plant methodology should be applied only to the qualitative and pseudoqualitative traits. In the case of quantitative traits, appropriate statistical methods should be applied, or example cultivars should be used for purposes of comparison.

Grouping Traits

1. In the selection of similar cultivars to be planted in the DUS test, grouping traits are recommended (DOC 2232/2009).
2. Grouping traits are those in which the expression levels observed, even when obtained at different locations, can be used for the organization of the DUS test, individually or in conjunction with other traits, so that similar cultivars are planted in groups.
3. The following traits are considered useful as grouping:

   (a) Transgenesis
   (b) Relative ripening group
   (c) Plant: type of growth
   (d) Flower: predominant color
   (e) Plant: color of pubescence on the main stem
   (f) Seed: genetic color of the hilum

In the case of soybean, the descriptors mentioned in Table 23.1 are mandatory and additional.

Due to the large number of cultivars registered each year, these descriptors may not be sufficient for differentiation of a new cultivar. Consequently, some studies have been carried out to identify new descriptors for soybeans.

Nogueira (2007) found that the hypocotyl length, epicotyl length, unifoliolate leaf petiole length, trifoliolate leaf petiole length, length of the terminal leaflet rachis of the first trifoliolate leaf, and the form of the unifoliolate leaf may be useful in the differentiation of soybean cultivars. Silva (2013) also observed some traits as possible differentiators of cultivars, being:shape of the unifoliolate leaf base; phylotaxis presented by the first pair of trifoliolate leaves; pod depression; hypocotyl length; plant height and epicotyl length; length of the first internode, of the petiole of the first trifoliolate leaf, of the petiole of the unifoliolate leaf, of the rachis of the first trifoliolate leaf, and of the angle formed by the insertion of the unifoliolate leaf petioles; pod length; pod diameter; pod thickness; pod waist thickness; pod curvature; and hilum width and length.

Foreign Cultivars

Cultivars protected in other countries or with ongoing protection, with DUS tests carried out by foreign institutions, recognized before the competent national authority, are protected upon the submission of the results of the tests carried out by these institutions. The reports are requested by the National Service for Protection of Plant Varieties (SNPC) directly to the foreign institution. The service is charged by foreign institutions, which send invoices—referring to the issuance of the reports and referral to the SNPC—directly to the breeder or responsible person appointed by them (MAPA 2010).

Stages and Processing Deadlines

The processing stages at MAPA can be accompanied by the applicant or their legal representative through CultivarWeb, by accessing the personal page, using the password, as described below.

After granting the protection, public information on the cultivar, such as Certificates of Protection, Excerpts and Publications in the Official Gazette, are available in the protected cultivars register on MAPA's portal (www.agricultura.gov.br) (MAPA 2010).

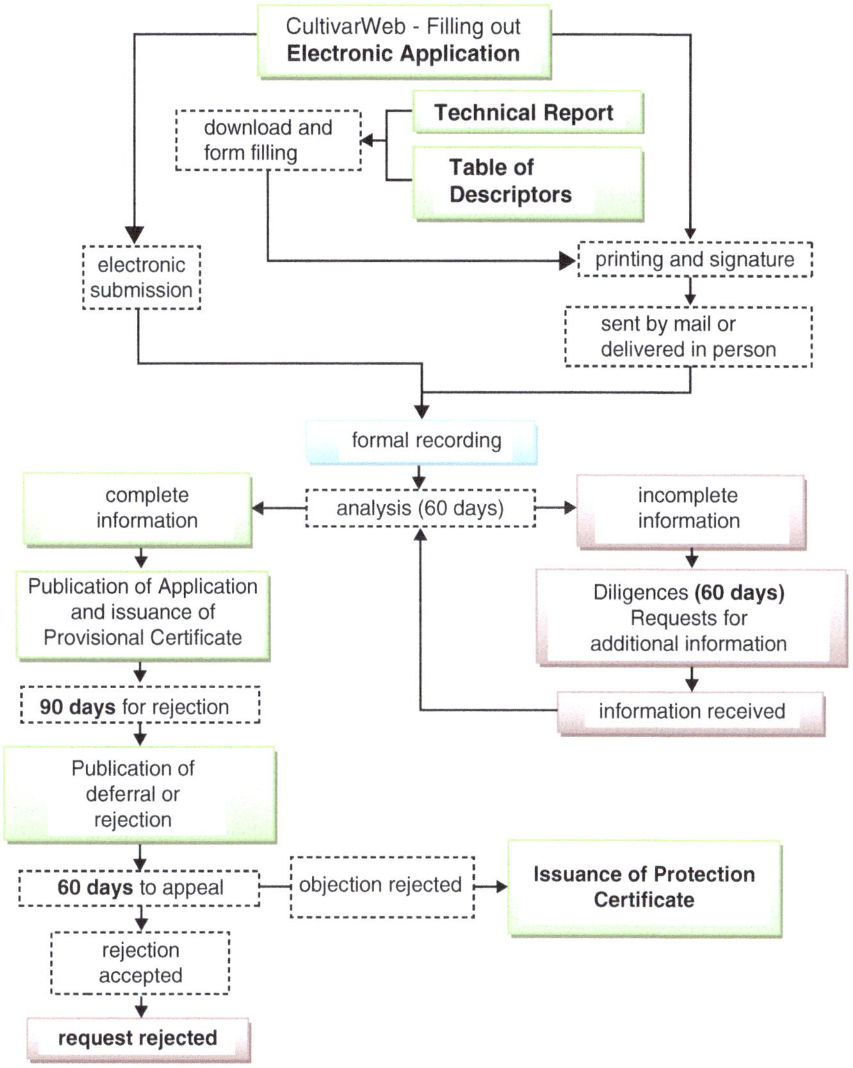

Source: MAPA, 2010.

Final Considerations

Due to the great economic importance of soybean crops, there are a large number of cultivars launched each year by public and private institutions in Brazil. Thus, knowledge of plant morphology is essential to accurately describe its characteristics, since, due to the large number of cultivars released on the market, it is increasingly more difficult to differentiate some of the available descriptors, for it must be clearly distinguishable from other known cultivars of the same species. Therefore, the production and marketing of the seeds of a cultivar will only be allowed after registration. The protection of the cultivar is an important step because it will guarantee the intellectual property of the breeder.

References

Assis, G. M. L., Valls, J. F. M., Carvalho, M. A., Valentim, J. F., Andrade, C. M. S.: Descritores morfológicos para a condução de ensaios de distinguibilidade, homogeneidade e estabilidade de Arachis pintoi Krapov. & W.C Greg, 28p. (2010), (Embrapa Acre, documento 117)

Borém, A., Miranda, G.V.: Melhoramento de Plantas, p. 525. UFV, Viçosa (2005)

Carvalho, S.I.C., Bianchetti, L.B., Reifschneider, F.J.B.: Registro e proteção de cultivares pelo setor público: a experiência do programa de melhoramento de Capsicum da Embrapa Hortaliças. Hortic. Bras. **27**(2), 135–138 (2009)

Ministério da Agricultura, Pecuária e Abastecimento (MAPA)—Informações aos Usuários de Proteção de Cultivares: http://www.agricultura.gov.br/arq_editor/file/INFORMACOES_AOS_USUARIOS_SNPC_nov2010.pdf (2010). Accessed 11 Dec 2015

Nogueira, A.P.O.: Análise discriminante na caracterização de novos descritores em soja (*Glycine max* (L.) Merrill), 96f. Dissertação de M.S., Universidade Federal de Viçosa, Viçosa (2007)

Piloro, A.E., Arias, C.A.A., Carneiro, G.E.S., Toledo, J.F.F.: Desenvolvimento de germoplasma e cultivares de soja, p. 10 (2007), (Embrapa Soja. Circular Técnica, 52)

Silva, F.C.S.: Influência do tamanho de sementes e de características agronômicas em descritores adicionais de soja, 46f. Dissertação de M.S., Universidade Federal de Viçosa, Viçosa (2013)

The manufacturer's authorised representative in the EU is Springer
Nature Customer Service Centre GmbH, Europaplatz 3, 69115 Heidelberg,
Germany. If you have any concerns regarding our products, please
contact ProductSafety@springernature.com

Printed and bound by CPI Group (UK) Ltd, Croydon, CR0 4YY
30/09/2026
02225657-0013